Developments in Primatology: Progress and Prospects

Series Editor: Louise Barrett

More information about this series at http://www.springer.com/series/5852

Urs Kalbitzer • Katharine M. Jack
Editors

Primate Life Histories, Sex Roles, and Adaptability

Essays in Honour of Linda M. Fedigan

Springer

Editors
Urs Kalbitzer
Department of Anthropology and McGill School of Environment
McGill University
Montreal, QC, Canada

Department of Anthropology and Archaeology
University of Calgary
Calgary, AB, Canada

Katharine M. Jack
Department of Anthropology
Tulane University
New Orleans, LA, USA

ISSN 1574-3489 ISSN 1574-3497 (electronic)
Developments in Primatology: Progress and Prospects
ISBN 978-3-319-98284-7 ISBN 978-3-319-98285-4 (eBook)
https://doi.org/10.1007/978-3-319-98285-4

Library of Congress Control Number: 2018958642

This Springer imprint is published by the registered company Springer Nature Switzerland AG
The registered company address is: Gewerbestrasse 11, 6330 Cham, Switzerland

This book is dedicated to Linda M. Fedigan: collaborator, mentor, and friend to us all.

Preface

In December 2016, a group of more than 50 friends, students, colleagues, and scholars gathered in the picturesque Rocky Mountain town of Banff, Alberta, Canada, to celebrate the career of Dr. Linda Marie Fedigan (see Fig. 1). The bulk of the celebration involved a 2-day symposium that included 18 research presentations by a variety of Linda's former graduate students, postdoctoral fellows, collaborators, and colleagues (see full write up in Jack and Kalbitzer 2017). We are honored to bring together many of those papers here in this *Festschrift* or "Book of Friends" (see Szathmáry, Chap. 1 this volume). A quick glance at this book's table of contents should make readers wonder how a single person could impact research on such a wide variety of topics and taxa given that most scholars are fortunate to have influenced research in a single area. Like the symposium itself, this volume is focused around four main themes, all of which have been inspired and invigorated by the research of Linda Fedigan: (1) Primate Life Histories; (2) Sex Roles, Gender, and Science; (3) Primate-Environment Interactions; and (4) Primate Adaptations to Changing Environments. Each of these thematic sections, and Linda's association with and impact on these topics, is introduced by the individuals who served as discussants at the symposium (Pavelka, Barrett, Huffman, and Henzi).

In addition to the scientific content of this book, we also wanted to add a personal note to this Festschrift. Therefore, in organizing and editing this volume, we asked the first authors of each manuscript, as well as the authors of section introductions, to respond to the following prompt:

> *As we mentioned to you in our earlier communications about the edited book, we would like to include short overviews of all first authors' connections to Linda. The format for these is completely at your discretion (but keep in mind that they will be published). They will likely be included in the book preface. These should be around 250 words or less. You can comment on when/where/how you met Linda, how she inspired your own research or career, or whatever you feel is appropriate as a way of honoring her impact on our field of study.*

The collection of 19 paragraphs that we received (see below) is as variable as the authors themselves. Contributors to this volume reside throughout the world (Canada, Japan, the United Kingdom, and the United States) and include established academics holding (or retired from) various professorships, postdoctoral

Fig. 1 Symposium participants posed for a photo at the Banff Center, December 2, 2016. Linda Marie Fedigan is in the center of the second row from the front. (Photo by John Addicott)

fellows, laboratory technicians, and members of various conservation units at highly regarded zoos around the world. These contributors include former students, collaborators, colleagues, and friends whose associations with Linda began as early as 1977 and continue through to the present. Despite this diversity in contributors, a common sentiment resonates throughout these "reflections on Linda" – collegiality. Indeed, Linda's collegial, collaborative, supportive nature of was remarked upon almost 20 years ago by Donna Haraway in her 1989 book *Primate Visions*:

> *It would be difficult to find a primatologist more conscious than Linda Marie Fedigan of the delicate determinations of what may count as primate science and as nature by the niceties of words and publishing practices.* (Haraway 1989, p 316)

Our review of Linda's Festschrift conference featured in *Evolutionary Anthropology* was entitled "How to cultivate a tree" (Jack and Kalbitzer 2017). We initially used this title as a means to highlight the enormous impact that Linda has had on the field of primatology, as illustrated by the sheer number of students she has mentored and that they in turn are now mentoring, and so on. This impact, we envisioned, was like the branches and roots of a tree (see Fig. 2). However, in preparing this book, and reading the reflections written by contributors, we now more fully realize that Linda's *cultivation* goes well beyond a single tree (even with its many branches and deep roots). Linda's collegial approach to science has profoundly affected the direction and practice of primatology as clearly illustrated by the sentiments expressed below and the impressive collection of original scientific manuscripts presented in this volume.

Fig. 2 "Tree of Fedigan." The scholars listed in large, green leaves at the end of a branch, completed their doctoral studies under Linda Fedigan's supervision (Fedigan's academic children, 11 in total). The smaller leaves on each of those branches, in turn, give the names of graduate students supervised by these PhDs (Fedigan's academic grandchildren). The multitude of other leaves on the tree list the names of Linda's master's advisees (29 in total) and students who completed honors theses under her supervision, as well as postdoctoral fellows with whom she has worked. (The image was created by Tracy M. Wyman, Department of Anthropology and Archaeology, University of Calgary)

Contributor Responses: "Your Connection to Linda Fedigan?"

Pamela J. Asquith
Adjunct Professor, Environmental Studies, University of Victoria, Canada; Senior Associate, Linacre College, Oxford, UK

"I first met Linda and Larry Fedigan in Nairobi in 1984 at the Xth IPS congress, organized by James Else and Richard Leakey. I was concluding three years postdoctoral fieldwork on Japanese primatologists in their labs and field sites, both within and without Japan. I had just been offered a sessional teaching position at the University of Calgary to begin later that year, and well remember Linda and Larry's warm welcome to the same province. Linda at that time taught at the University of Alberta. Toward the end of my sessional appointment, I applied to work with Linda as an Izaak Walton Killam postdoctoral scholar. The Killam postdoctoral awards are open to applicants in either science or social science and of any nationality. Together, Linda's and my work addressed both science (Linda's field work on the Arashiyama West Japanese macaques) and social science (my study of institutional and cultural influence on Japanese primatology). One result of our collaboration was the organization of a conference, held at the Banff Centre in 1986, that brought together for the first time Japanese and Western primatologists who had observed the Japanese macaque troops in Arashiyama, Kyoto, and in Dilley, Texas. The latter group was known as the Arashiyama West troop that had fissioned from the original Kyoto troop and on which Linda had conducted her doctoral research. The meeting was funded by the Wenner-Gren Foundation and our co-edited book, *The Monkeys of Arashiyama: Thirty-five Years of Research in Japan and the West* (SUNY Press 1991), is based on this meeting. My two-year (1985-87) postdoc with Linda was followed by a SSHRC Canada Research Fellowship, also held at the University of Alberta. Throughout these and many subsequent years, Linda's collegiality, quiet mentorship, insightful discussions, and unfailing interest in our respective research areas have been marvelous supports for my, and many others', contributions to the academy."

Louise Barrett and Peter Henzi

Professors, Department of Psychology, University of Lethbridge, Alberta, Canada. Louise is also a Canada Research Chair in Cognition, Evolution and Behaviour

"Although her work had been known to us for some time, we first met Linda Fedigan in 2007 when, very soon after our arrival at the University of Lethbridge, she invited us up to meet the Calgary primatologists. While we were surprised that news of our arrival had made it up the Deerfoot Trail, we were also touched by Linda's generosity and pleased for the entrée to a larger community. This initial contact had an immediate impact: we established personal friendships and academic connections that have lasted and strengthened over the years. These have provided us with a welcome sense of place, negating the possibility of any misguided institutional rivalries. As a consequence, Lethbridge and Calgary students now move smoothly between the two institutions and have collaborated in the field, as have we and their supervisors. Clearly, these connections require individually congenial colleagues but they depend equally on an underlying ethos that places high value on the academy as a venue for fruitful cooperation and not as an arena for a zero-sum game. Linda's leadership here has been unambiguous and, like all powerful maternal effects, should leave its mark for some time to come."

Mackenzie L. Bergstrom

Adjunct Assistant Professor, Department of Biology/Project *Dragonfly*, Miami University and Conservation Program Manager - Graduate Programs Advisor Community Engagement, Institute for Conservation Research, San Diego Zoo Global, US

"Linda was my supervisor at the University of Calgary during my MA (2005 – 2009) and PhD in Biological Anthropology (2009 – 2015). It was an incredible experience working with her to study the behavioral ecology of the white-faced capuchins at Santa Rosa. I feel honored to have had the opportunity to work on such a well-established field project with access to long-term data on demographics, behavior and characteristics of the capuchins' habitat. Linda focused on the development of her field project as well as her students as professionals, and she laid the foundation for our success through the support and resources she provided. She gave me space to explore my interests and data, but she still provided direction regarding larger concepts and through her unique perspective of the long-term patterns at the site."

Fernando A. Campos

Assistant Professor, Department of Anthropology, University of Texas at San Antonio, US

> "My connection to Linda began indirectly, when I first got my feet wet with primate field work. I spent most of 2002 to 2005 volunteering on studies of wild capuchin monkeys, first in Costa Rica and later in Ecuador. The two directors of those studies had previously branched off of Linda's academic tree: Susan Perry, Linda's former postdoc, and Kathy Jack, her former PhD student. These connections set my career path in motion—Linda was the pinnacle of graduate advisors for an aspiring capuchin researcher—and I was extremely excited to receive the news, while in an internet café in Guayaquil, Ecuador, that I would be admitted as her graduate student at the University of Calgary. Linda was my advisor and primary mentor throughout my Masters and PhD from 2005 to 2014. It's hard to imagine a better advisor, and I'm especially grateful for two instances of support and generosity. First, after my wife completed her PhD and was offered a postdoc in Japan, Linda was fully supportive when I proposed spending the last two years of my PhD writing my dissertation while living 8,000 km away from Calgary. Second, in 2015, Linda brought me in to a remarkable working group, which includes some of the founders of Primatology, that resulted in a great postdoc and that surely played an important role in procuring a tenure-track job to begin later this year. It has been a privilege working with Linda, and I look forward to many more years of productive collaboration."

Colin A. Chapman

Professor, Department of Anthropology, McGill University, Canada.

Colin is also a Canada Research Chair in Primate Ecology and Conservation and a honourary lecturer in the Department of Zoology at Makerere University, Uganda

> "Being one of Linda's first students, I feel that I have been lucky in so many ways. Not only did I have Linda as a mentor to get my own career off the ground, but I had the pleasure to watch those that followed me. I helped set up the program in Santa Rosa when it first started, and it has been a pleasure throughout the years to think of Linda's students using the trails we first cut and benefiting from the animals we first habituated. Maybe somewhat like a distant uncle, I thoroughly enjoyed watching Linda's many students succeed and forge their own academic careers and personal lives. I have many fond memories from my early years, such as inheriting the motorbike from Linda on St. Kitts or conducting the very first censuses in Santa Rosa. Looking back, one memory that is still poignant was a conversation I had with Linda during the first field course with the Japanese macaques in Texas when I was an undergrad. Linda took me aside one day and said she was worried that I was not serious enough or dedicated enough to go on in primatology – I think now Linda can look back on that conversation and think that she should not have worried."

Jeremy Hogan

Research Technician of the PACE Lab and Santa Rosa Capuchin Project, Department of Anthropology and Archaeology, University of Calgary, Canada

> "Much to my surprise, while contacting potential supervisors for graduate school, my runaway first choice replied right away to set up a meeting. This supervisor, of course, was Linda, who proved to be welcoming and helpful from the beginning, and thus began my path into "capuchinology". Following my master's program, fate intervened yet again and the Santa Rosa research technician position opened up, a role I could not be happier in. Beyond remaining with an influential and interesting project (and staying up to date on the monthly monkey drama in the park), I get to continue working with a true role model, both

professionally and personally. Linda is always looking to provide the greatest opportunities for success for anyone that she works with, spending considerable time and effort to further their goals. Thank you Linda, for being the best boss imaginable!"

Michael A. Huffman
Associate Professor, Section of Social Systems Evolution, Primate Research Institute, Kyoto University, Japan

"It has been a great pleasure knowing Linda. Our first meeting was at the 1981 American Society of Primatologists annual congress held in San Antonio in June that year. It was my first conference, and I was awe struck, having the opportunity to present a paper on my observations of the Arashiyama-Kyoto troop at a symposium on Arashiyama West studies organized by Claud Bramblett. I had just returned from 2.5 years in Japan as an undergraduate exchange student, having spent 12 months of that time up on Arashiyama observing the Arashiyama B troop while auditing Prof. Junichiro Itani's graduate seminars in Anthropology at Kyoto University. I was thrilled to have the chance to attend such a scientific meeting and to be in the company of so many awesome people whose names I had read about in the literature on Japanese macaques. Claud and his wife hosted me at their home in Austin for a few days, and I finally got to see the Arashiyama West troop in Dilley. This tradition of hospitality and solidarity has not changed, in no small part to the efforts of Linda. It is always a great opportunity working with Linda and her students, former or current, on a project, whether it's a book, a field trip, being an associate editor under her for AJP, or a conference-seminar visit; she projects a strong sense of fairness, equality and compassion. She has been a great colleague and an amazing role model for so many of us."

Katharine M. Jack
Professor, Department of Anthropology, Tulane University, New Orleans, LA, US

"I vividly remember my first meeting with Linda (it's one of those events engrained in my mind, likely from spiked cortisol!). It was 1992, and I was with my now husband, Craig Lamarsh (a former undergraduate student of Linda's), and we were interviewing with her for entrance into the MA program at the University of Alberta. We were terrified! Although we weren't accepted into the program, we ended up working closely with Linda over the following 3 years while we conducted research at the Arashiyama West site under Mary Pavelka's guidance (one of Linda's first doctoral students). Fortunately, Linda did accept me as a doctoral student in 1996 (my husband went on to veterinary school), but even today, after working with Linda for more than 25 years, I still get nervous giving talks when I know she is in the room. What is it about this "tiny but mighty" woman that makes us shake at the knees? She is, frankly, one of the kindest, gentlest, most giving people that I know, so this reaction is rather curious. There is just something about Linda that makes those of us around her strive to work hard, to do our best, to try not to disappoint. This response is not due to a conscious push on her part but, rather, I think it is our reaction to her intellect and her influence on the field of primatology. I continue to be in awe of this amazing scientist and am so thankful that she convinced me to study the capuchins of Santa Rosa (rather than all-male bands of patas monkeys in Africa…yes, you can imagine the look on her face when I told her that plan!). The foundation that Linda created with the Santa Rosa Primate Project and her collaborative spirit are truly remarkable. I am (we all are) so very fortunate to be the recipients of her intellect, guidance, patience, and compassion."

Steig E. Johnson

Associate Professor, Department of Anthropology and Archaeology, University of Calgary, Canada

"I first encountered Linda's work in an undergraduate course in primate behaviour through her textbook, *Primate Paradigms*. The book had a huge impact and really brought me into the field. It is remarkable still in its ability to present nuanced arguments and evidence weighing on the major research themes in primate behavioural ecology and evolution. Linda's importance for my development as a scholar of course did not end there, and it has been a true privilege to have worked with her for over a decade in the Department of Anthropology and Archaeology at the University of Calgary. She has been the model for how to run a high-impact research program and successfully mentor graduate students, while always demonstrating the highest standards for ethics, professionalism, and respect for others."

Urs Kalbitzer

Postdoctoral Fellow, Department of Anthropology, McGill University, Canada

"In 2014, a few months after finishing my PhD, I was hiking with a friend in middle of nowhere in the Norwegian mountains when I switched on my phone for the first time in days. I had postponed this trip for a Skype interview for a postdoc position with Linda, but the interview had only lasted 10 minutes, which I interpreted as a clear indicator that I would not be offered the job (I didn't know about Linda's efficiency of doing such things!). Now, however, with only one bar of reception, I received a text message from my wife, Annie, who told me that I was offered the position. This was probably the best location to receive such an offer, and I asked Annie to send Linda the email I had prepared in case they'd offer me the job (of course, I had also prepared one in the case I was rejected). When we arrived in Canada, Linda and John welcomed us at the airport, and it immediately felt like meeting people we already knew. Since then, I've enjoyed working with Linda: she is encouraging, but honest if she thinks something is unrealistic, and she always stays calm when things get stressful, which can be contagious. Of the same importance, however, was that Linda and John always did a great job helping us to feel at home in Calgary. They welcomed us with a bag full of food for our first days in Calgary, organized parties especially for us on several occasions, took us out for skiing and biking, and supported us after a truck ran over me. Both Annie and I are very grateful for this experience, and that Linda offered me to work with her in 2014."

Shoji Kawamura

Professor, Department of Integrated Biosciences, University of Tokyo, Japan

"I had my lab in 1999 as an associate professor in a new graduate school (Graduate School of Frontier Sciences) in University of Tokyo. I was struggling to extend my research area to a unique direction based on my expertise in molecular evolutionary studies on color vision and in anthropology. New World monkeys were one of my major study targets to explore because of their unique polymorphism of color vision as reviewed in this book chapter. I was aiming to connect genetic and experimental analyses in the lab and behavioral observation of wild monkeys in the field. I first contacted a Japanese primatologist who had a study field in Colombia. However, I had to abandon this plan due to the political upheaval around that time. After months of many email contacts to several primatologists, I had a fortune to encounter Dr. Linda Fedigan in 2003 April through Dr. Mikael Huffman and Dr. Yukiko

Shimooka of the Primate Research Institute, Kyoto University. Linda kindly accepted my proposal to conduct a color vision study at her field site in Santa Rosa, Costa Rica. She also mediated my collaboration with Dr. Filippo Aureli, who started his field study of spider monkeys at the same site. This was a start of an incredibly successful collaboration."

Gráinne M. McCabe

Head of Conservation Science at the Bristol Zoological Society, UK

"Linda has had a significant impact on my career as a primatologist. From my first job as a field assistant in Santa Rosa National Park, Costa Rica, to my current position as Head of Field Conservation and Science at Bristol Zoological Society, Linda has had a role to play; and for that I will always be grateful. But beyond opening doors, Linda's path, focus and areas of interest have influenced my research direction. In the 70s and 80s, most primate studies focused on chimpanzees and baboons. At this time, however, Linda began studies on species that were less well studied: Japanese macaques, not much of a focus outside of Japan, and soon after, white-faced capuchins in the Neotropics, where very few primatologists were undertaking research. Linda's work establishing long-term monitoring has led to an enormous wealth of information about the sociality, ecology and life history of these species. I now find myself in a very similar position with my study species – the Sanje mangabey. When I began my PhD research in 2007, we knew very little about this monkey endemic to Tanzania. It is one of seven species of Cercocebus mangabeys found across Africa, all of which are threatened with extinction. Linda's example of creating detailed long-term datasets through systematic research examining life history traits helped me to see the value of such information, which is critical when species are facing serious conservation threats. With this in mind, we started the Sanje Mangabey Project to collect these data to determine the population viability of this endangered species."

Amanda D. Melin

Assistant Professor, Department of Anthropology and Archaeology, University of Calgary, Canada

"I can say without any hesitation that Dr. Linda Fedigan has been the most influential and valued mentor I have had the honour to interact with in my professional life. She has inspired me, guided me, and helped me over the past 15 years, and I could not feel more fortunate to be her colleague. Linda has also become a very good friend, for whom I am incredibly grateful. We also both love cats and frequently bond over cat anecdotes and images; I appreciate that to no end. I first met Linda Fedigan in 2003, when our mutual acquaintance, Mary Pavelka, suggested I speak with Linda about my interest in pursuing graduate school. (I'll leave out the part about being hurt that Mary didn't want me as a student…I had originally approached Mary about being my graduate school advisor after taking her field school in Monkey River, Belize! Just kidding, Mary). After meeting Linda in 2003, the rest was history. I completed a master's degree and Ph.D. under Linda's supervision. Through Linda, I met two other very important mentors, colleagues, and friends in my life - Dr. Shoji Kawamura in 2004, and shortly after, Dr. Kathy Jack. I was also introduced to a place in the world that now feels like a second home, Guanacaste, Costa Rica. In 2011, I joined Linda and Kathy as co-directors of primate research in Sector Santa Rosa, Area de Conservacion, Guanacaste, and this has been one of the most important and enjoyable aspects of my career. Thank you, Linda, for believing in me and for being such an outstanding role model. I hope I can pass on to the trainees I work with some of the wisdom, understanding, and academic passion I gained through working with you."

Mary S. M. Pavelka

Professor, Department of Anthropology and Archaeology, University of Calgary, Canada

"I met Linda in May of 1981 when I was 22 years old and lucky enough to be a student in the only field school she ever taught. Without qualification, I can say that that experience determined much of the trajectory of the rest of my life. I abandoned my plans to attend law school, turning instead to graduate school and research on the Arashiyama West Japanese monkey colony that I had come to know in that field school. Under Linda's supervision, I studied female courtship behavior at the MA level and the social manifestations of aging at the PhD level, graduating in 1988. Miraculously, the following year a job was advertised for an Assistant Professor of Primatology in the Anthropology Department at the University of Calgary, just a 3 hour drive from the University of Alberta where I studied with Linda. Even more miraculously, I got it. My collaborations with Linda continued and our relationship matured into a strong lifelong friendship. We continued to travel to South Texas to work with the AW monkeys and to use the rich genealogical database to investigate reproductive termination (it wasn't menopause so we didn't call it that) in the females. We produced a series of papers that we privately referred to as The Terminator Series. In 1991, since Linda was not interested in teaching the undergraduate field school again, I took up the reins, and in that first year of teaching the field school (which I continued for over 20 years), Katharine Jack, one of the editors of this volume, was a student. Such has been the cascading effect of Linda on my life and others. In 1997, I moved from studying the provisioned AW Japanese macaques to wild howler and spider monkeys in Belize, again inspired by Linda's move to establish the Santa Rosa study site in Costa Rica. In 2000, the Canadian government introduced a new Canada Research Chairs program, which I leaped on as an opportunity to recruit Linda from the University of Alberta to the University of Calgary and allow us to establish ourselves as a unique concentration of primatology in Canada. Eventually, I served a six-year term as Department Head, and we came full circle, in an administrative sense, with Linda reporting to me. But make no mistake about it, I am the one that has continued for my whole adult and professional life to benefit from the mentoring and friendship that only someone like Linda Fedigan can provide."

Karen B. Strier

Vilas Research Professor and Irven DeVore Professor of Anthropology at the University of Wisconsin-Madison, US

"My reflections about Linda coalesce around three of her many unique contributions. First, Linda has been an influential spokesperson for the centrality of primatology within anthropology and for the importance of neotropical primate studies within primatology. These positions may seem obvious and noncontroversial today, but that is in large part because people like Linda were so effective in securing our niche at a time when it was being challenged. I first read some of her essays on these topics as a graduate student and continue to find them instructive today. In 1996, I had the opportunity to see Linda's quiet mode of leadership at a Wenner Gren Foundation conference she co-organized on theory, method, and gender in primatology (Strum and Fedigan 2001). The conference was held in Brazil, which provided an opportunity to bring Linda and a few other participants to my field site. After her visit, my students named a muriqui in her honor. A second facet of Linda's influence has been her decency and high ethical standards. She is one of the most careful people I know when it comes to citing ideas and work. There are few others who take the time or trouble she does to confirm a quote or insure that a grant proposal of hers will not compromise someone else's ongoing project. A similar level of thoughtfulness runs through her classic 2010 review of ethical issues in the field and is familiar to all of us who have had the opportunity to collaborate with her. Finally, there is Linda's long-term research on primate life histories and demography, which has been represented in the Primate Life History Database since our Primate Life History Working Group was founded more than a decade ago (Strier et al. 2010). Her pioneering comparisons of sympatric primate species led to the longitudinal monitoring of multiple groups of capuchin monkeys within the population at Santa Rosa. Documenting the demographic trends of these monkeys in the context of both ecological and social transitions and fluctuations has inspired not only her students and colleagues in Costa Rica, but also many others, including myself, who work with diverse species around the world."

Emőke J. E. Szathmáry

President Emeritus, Professor Emeritus, and Senior Scholar (Anthropology) at the University of Manitoba, Canada

"In 1977 Linda Fedigan's presentation, "Reproductive strategies and consort partner selection in Arashiyama West troop," was one of three non-human primate behavior papers at the 3rd meeting of the Canadian Association for Physical Anthropology (CAPA). Back then, the CAPA was a tiny organization. Most meeting attendees heard all presentations and, during coffee breaks and meals, spoke with the contributors. I was as curious about non-human primate research methodologies as in the results obtained, for I was skeptical about the validity of textbook claims. Some months later Linda wrote me, and requested my comments on what she had written about sexual selection, especially the concept of female choice. My response lacked insight regarding the dissonance in evolutionary theory that posited mate choice, competition for mates, and male dominance, but which were presumed to be operative in Anthropoidea. Theories and assumptions without validation just irked me, especially when they extended to Homo sapiens. On the other hand, I was intrigued by Linda's core issue – was "female choice" a Hobson's choice – and what would research show? I read more about female primates, which proved useful when I became Editor-in-Chief of the Yearbook of Physical Anthropology, and later, the American Journal of Physical Anthropology. I also maintained contact with Linda and was witness to her increasing research stature. Her findings have shown that females have essential roles in ensuring their species' survival and success. That this was missed decades ago reflects, in part, a lack of awareness of the influence of gender and cultural perspectives on science – an impact that Linda was among the first to demonstrate. Linda's commitment to rigor in research, clarity in explanation, and her dispassionate examination of theories as "dynamic exercises in human reasoning" have held my attention since I read her first book. She has my admiration and respect."

Sarah E. Turner

Assistant Professor, Department of Geography, Planning and Environment, Concordia University, Canada

"Linda's research and writing have been an inspiration since I first encountered her book, *Primate Paradigms*, as an undergrad student taking an elective primatology course. Linda's work was a beacon: she challenged assumptions about sex and gender in primate research, while always keeping the science and the monkeys themselves central. She showed me that it was possible to be a rigorous scientist in animal behavioural ecology and a feminist. As I pursued my undergraduate studies and masters research on Japanese macaques, I came back again and again to Linda's research and writing for guidance and inspiration. I felt so lucky when Linda agreed to supervise my PhD research a few years later. Her expertise and scholarship as a scientist and her kindness as a mentor and advisor throughout my PhD and beyond make me feel very fortunate to be part of her academic lineage and to continue to work with her on monkey-related projects for years to come."

Kim Valenta
Assistant Research Professor, Department of Evolutionary Anthropology, Duke University, US

"Every few weeks in Santa Rosa, I'd prepare to call Linda from the field. Melanie Meeking (then Luinstra) and I would compile a list of things that had happened, questions we had, and ideas we needed Linda's feedback on. I'd pick up a phone card in town the week before and make my way to the landline on the porch of the comedor, nervously clutching my list and phone card. And when Linda picked up, which she always did, I'd say "Hello Linda, this is Kim Valenta, calling from Santa Rosa National Park, Costa Rica." And every time, Linda would say "You know you can just say 'Hi, it's Kim'. I know who you are." Linda never gave me a reason to be nervous about talking to her – she was as kind and gracious to me as she is to everyone. Nonetheless, her status impressed and terrified me (and continues to!). Of all of the things Linda taught me - which are not to be underestimated, given that she singlehandedly developed me into a field biologist who could write grants, publish papers, and give a scientific talk without fainting – the one I hold dearest is the generosity, humility, and patience it takes to support developing scientists. I can only hope that one day I am able to be as helpful, kind and assuring to the students who phone me with their trembling hands and bullet-point lists as Linda was to me. It truly meant the world."

Eva C. Wikberg
Assistant Professor, Department of Anthropology, University of Texas at San Antonio, US

"I first met Linda when I started graduate school at the University of Calgary, and she has been a member of my PhD committee, postdoctoral advisor, and mentor. Some of my first interactions with Linda involved skiing and discussing cloudberry liqueur. She has over the years provided me with insightful advice on small and large matters, welcomed me into her collaborative research team, and generously shared her exceptional long-term data. Linda continues to be a great source of inspiration for me, in particular her research model and pioneering studies showing the importance of cooperation in the lives of monkeys and researchers alike."

New Orleans, LA, USA — Katharine M. Jack
Montreal, QC, Canada — Urs Kalbitzer

References

Haraway DJ (1989) Primate visions: gender, race, and nature in the world of modern science. Routledge

Jack KM, Kalbitzer U (2017) How to cultivate a tree: Celebrating the career of Linda Marie Fedigan. Evol Anthropol Issues, News, Rev 26:139–142. doi: https://doi.org/10.1002/evan.21524

Strier KB, Altmann J, Brockman DK, et al (2010) The Primate Life History Database: A unique shared ecological data resource. Methods Ecol Evol 1:199–211. doi: https://doi.org/10.1111/j.2041-210X.2010.00023.x

Strum SC, Fedigan LM (2000). Primate encounters: models of science, gender and society. The University of Chicago Press, Chicago

Szathmáry EJE (2018) In admiration of Linda Marie Fedigan. In: Kalbitzer U, Jack KM (eds) Primate life histories, sex roles, and adaptability - essays in Honour of Linda M. Fedigan. Springer, New York, pp 1–13

The original version of the book was revised. Perface page text has been updated.

Acknowledgments

We thank all the authors for their dedicated work on their manuscripts and their patience throughout the preparation of this book. It was a genuine pleasure to participate in the symposium with all of you and to work with you on your chapters. In addition to our own reviews, each manuscript in this book was anonymously peer-reviewed by two or more external reviewers. We are very grateful to all of our colleagues who provided these reviews, which greatly enhanced the quality of the chapters.

We also would like to thank all those involved in the planning and preparation of the Festschrift Symposium for organizing such a wonderful event: John Addicott, Louise Barrett, Peter Henzi, Jeremy Hogan, Steig Johnson, Amanda Melin, Mary Pavelka, Pascale Sicotte, and Tracy Wyman. We would like to give special thanks to Jeremy Hogan, who made sure that the symposium happened without any glitches, and Mary Pavelka, who played an integral role in the development of this book. We know that all of the book contributors were highly motivated to work very hard on their manuscripts because of this symposium (and their friendship with Linda!).

We are also grateful to Janet Slobodien and Jeffrey Taub from Springer for working with us on this book volume and for their patience. Louise Barrett was also extremely helpful throughout the publication process, answering our many questions. She and Peter Henzi encouraged us to submit the original book proposal to the *Developments in Primatology* series. Of course, we owe our greatest thanks to Linda Fedigan – it is only through her encouragement, unwavering collaborative spirit, and friendship that any of this has been possible.

Contents

Contributors

Pamela J. Asquith School of Environmental Studies, University of Victoria, Victoria, BC, Canada

Linacre College, Oxford, UK

Louise Barrett Department of Psychology, University of Lethbridge, Lethbridge, AB, Canada

Mackenzie L. Bergstrom Department of Anthropology and Archaeology, University of Calgary, Calgary, AB, Canada

Sarah Bortolamiol Department of Geography, McGill University, Montreal, QC, Canada

UMR 7533 Laboratoire Dynamiques Sociales et Recomposition des Espaces, Paris Diderot University, Paris, France

UMR 7206 Eco-Anthropologie et Ethnobiologie (MNHN/CNRS/Paris Diderot), Paris, France

Lauren J. N. Brent Department of Psychology, Washington Singer Laboratories, University of Exeter, Exeter, UK

Kerry A. Brown Department of Geography and Geology, Kingston University, Kingston Upon Thames, UK

Fernando A. Campos Department of Anthropology, The University of Texas at San Antonio, San Antonio, TX, USA

Department of Biology, Duke University, Durham, NC, USA

Colin A. Chapman Department of Anthropology and McGill School of Environment, McGill University, Montreal, QC, Canada

Makerere University Biological Field Station, Fort Portal, Uganda

Wildlife Conservation Society, Bronx, NY, USA

Section of Social Systems Evolution, Primate Research Institute, Kyoto University, Kyoto, Japan

Kenneth L. Chiou Department of Psychology, University of Washington, Seattle, WA, USA

D. P. Croft Department of Psychology, Washington Singer Laboratories, University of Exeter, Exeter, UK

Linda M. Fedigan Department of Anthropology and Archaeology, University of Calgary, Calgary, AB, Canada

David Fernández Department of Applied Sciences, University of the West of England, Bristol, UK

Peter Henzi Department of Psychology, University of Lethbridge, Lethbridge, AB, Canada

Jeremy Hogan Department of Anthropology and Archaeology, University of Calgary, Calgary, AB, Canada

Michael A. Huffman Primate Research Institute, Kyoto University, Kyoto, Japan

Katharine M. Jack Department of Anthropology, Tulane University, New Orleans, LA, USA

Steig E. Johnson Department of Anthropology and Archaeology, University of Calgary, Calgary, AB, Canada

Urs Kalbitzer Department of Anthropology and Archaeology, University of Calgary, Calgary, AB, Canada

Department of Anthropology and McGill School of Environment, McGill University, Montreal, QC, Canada

Shoji Kawamura Department of Integrated Biosciences, Graduate School of Frontier Sciences, The University of Tokyo, Chiba, Japan

Gráinne M. McCabe Department of Field Conservation and Science, Bristol Zoological Society, Bristol, UK

Amanda D. Melin Department of Anthropology and Archaeology, University of Calgary, Calgary, AB, Canada

Department of Medical Genetics, University of Calgary, Calgary, AB, Canada

Alberta Children's Hospital Research Institute, University of Calgary, Calgary, AB, Canada

Monica S. Myers Department of Anthropology and Archaeology, University of Calgary, Calgary, AB, Canada

Masayuki Nakamichi Osaka University, Osaka, Japan

Hisami Nobuhara Awajishima Monkey Center, Hyogo Prefecture, Japan

Toshikazu Nobuhara Awajishima Monkey Center, Hyogo Prefecture, Japan

Mary S. M. Pavelka Department of Anthropology and Archaeology, University of Calgary, Calgary, AB, Canada

Simon M. Reader Department of Biology, McGill University, Montreal, QC, Canada

Karen B. Strier Department of Anthropology, University of Wisconsin-Madison, Madison, WI, USA

Emőke J. E. Szathmáry St. Paul's College, University of Manitoba, Winnipeg, MB, Canada

Sarah E. Turner Department of Biology, McGill University, Montreal, QC, Canada

Kim Valenta Department of Anthropology and McGill School of Environment, McGill University, Montreal, QC , Canada

Makerere University Biological Field Station, Fort Portal, Uganda

Duke University, Department of Evolutionary Anthropology, Durham, NC, USA

Shasta E. Webb Department of Anthropology and Archaeology, University of Calgary, Calgary, AB, Canada

Eva C. Wikberg Department of Anthropology, University of Texas at San Antonio, San Antonio, TX, USA

Rachel E. Williamson Department of Anthropology and Archaeology, University of Calgary, Calgary, AB, Canada

Chapter 1
In Admiration of Linda Marie Fedigan

Emőke J. E. Szathmáry

It is a great pleasure to be included in a Festschrift that has Linda Fedigan as its focus – the person who influenced the thought and action of students through their formative years of studies and whose intellect and research acumen were magnets that attracted scholars and scientists to join her in collaborative work over many years. The chapters in this volume, each contributed by the people Linda touched directly, illustrate advances in fields of expertise within primatology, and each chapter stands as an individual testament to Linda's influence in the myriad ways that an exceptional mentor and thought leader influences others.

My own contribution to this volume differs in scope from the other chapters because I am not a primatologist. Questions about sex differences and evolutionary theory brought Linda and me together over 40 years ago, but my focus on the genetics of the indigenous peoples of North America, as Linda expanded her work in non-human primate behaviour, precluded the possibility of collaborative research. To my regret we have never worked at the same institution, and though we have each changed universities during the time we have known each other, we have remained located thousands of kilometres apart. Nevertheless, our conversation continued over the years, as it ranged from navigating the shoals of the academy and the research enterprise to the influences we had on our lives and the choices we made in our careers. My path ultimately led to academic administration, and my tribute to Linda reflects how other anthropologists and the broader community regard her contributions to her discipline and to her society. That perspective also contributes to the hermeneutics of this Festschrift.

In November 2016, just before the Festschrift conference, I attended a lecture given by a visiting astrophysicist, who also happens to be a Jesuit priest. He spoke about the universe and the perceived cleavage in the secular world between science

E. J. E. Szathmáry (✉)
St. Paul's College, University of Manitoba, Winnipeg, MB, Canada
e-mail: emoke.szathmary@umanitoba.ca

U. Kalbitzer, K. M. Jack (eds.), *Primate Life Histories, Sex Roles, and Adaptability*, Developments in Primatology: Progress and Prospects,
https://doi.org/10.1007/978-3-319-98285-4_1

and faith. He argued that science addresses theories and facts to provide explanations for phenomena, while faith transcends this process to provide meaning about things that are not scientifically testable. Those comments, especially on those matters that are scientifically testable, struck a chord with me because the natural world, the *living* world, also proffers questions that are scientifically testable. The explanations that are provided today, for example, on the evolution of non-human primate behaviour, do rest on evidence procured from observations that tested particular hypotheses. Those that were sustained form aggregates of meaning – for example, that, among primates, "female behavior and relations with their kin, their young, and their mates were increasingly seen as highly variable combinations of cooperation and competition" (Fedigan 2008:362). But it was not always so.

The very question that brought Linda Fedigan and me together in the mid-1970s was Linda's questioning of Darwin's theory of sexual selection: Is the concept of "female choice" of the male that is victorious in male-male competition, a Hobson's choice – that is, no choice – for the female has to take or not to take the victorious male. That primate females have passive roles in reproduction and are just the gene transmission vehicles that dominant males approach for mating was very much the prevailing view about primate sexual behaviour in those days. Even the explanation of the concept of dominance – that it meant "priority of access" for food and sex – assumed that males determined mating patterns, and non-primatologists like me, who discussed the concept of dominance in our undergraduate classes in physical anthropology, were completely unaware that the evidence then available from field research on male rank and reproductive success yielded contradictory results (Fedigan 1982). The textbooks we used simply took the concepts as true.

Linda Fedigan, however, was a person who did not take explanations of primate social behaviour at face value and probed all things that have bearing on the facts. I think she began to do that long before I discovered that not all facts about non-human primate behaviour were true. It had not occurred to me, for example, that facts could arise from improper research designs, which would invalidate using such facts as evidence for or against a particular hypothesis. In my undergraduate days, when the role of big male baboons was considered relevant to understanding the behaviour of our early hominin ancestors, and a re-articulation of Darwin's hunting hypothesis underpinning the evolution of human behaviour loomed large (Washburn and Lancaster 1968), I was dismayed to learn that flaws in research designs were especially frequent in early studies of non-human primate behaviour. A reading today of Washburn, Jay, and Lancaster's 1965 paper on findings arising from field studies of Old World monkeys and apes is astounding because its advice illustrates the shortcomings of the manner in which field research had been conducted. For example, the authors observed that it is desirable to make repeated visits to groups of free-ranging primates, that recognition of individual animals is essential for learning everything from group membership to troop durability and flexibility, that individual recognition is required for assessing the establishment and change of dominance systems, and that long-term observations of a species over different seasons and habitats are necessary to understand the interplay between biology and social learning. Given what needed to be known, field studies proliferated, and by 1972, Phyllis Dolhinow (formerly, Jay) could write that "What at one time was conceived as *the* primate pattern was transformed into variations on themes" (Dolhinow 1972:352). The observed variability within and between species' behaviours made "what it means to behave as a primate…even more difficult than to include every primate in one simple definition of the mammalian order Primates" (Dolhinow 1972:352). And yet, in the same volume, Washburn and Hamburg (1972:296) not only concluded that "successful aggression has been a major factor in primate evolution" but that "Man…" also inherited that "…biological base, modified by the great development of the social brain and language."

The excitement that arose from findings of field studies on non-human primate behaviour and the increasing questioning of the role of female primates within the evolution of their species let alone the relevance of their behaviours to the evolution of the human species was part of the ethos of the 1970s, the decade in which Linda Fedigan earned her doctorate. That occurred in 1974 at the University of Texas (Austin).

1.1 As the Twig Is Bent…

One would not have predicted that the child of a German-Italian war bride and her US Air Force husband, born in 1949 in Enid, Oklahoma, would become a renowned primatologist. Linda's toddler years were spent near her father's family in rural Georgia, while he was away serving in Korea. It could not have been an easy time for her mother, but after her father returned, he was periodically based in Germany, so that as a child Linda lived close to her German and Italian grandparents and other European relatives. She learned German – one of the several languages she speaks – and with language and experience came multiple ways of perceiving the world.

When Linda was in junior high, another socially and intellectually expanding experience presented itself during her father's term of duty in San Antonio. Those were the years of Sputnik, the first artificial space satellite that was launched by the Soviet Union in 1956. To many Americans that achievement showed that, "Ivan was smarter than Johnny" (Fedigan, Pers. Comm. December 2016), and enormous emphasis was placed on renewing excellence in learning. It should be no surprise to anyone who knows Linda that she was selected for a special scholarship at a private school that sought to train young scientists. She thrived in the small class setting, primed with motivated teachers, academically high-achieving classmates, and drills in science, languages, and math. So much so that, when her father was again assigned for a tour of duty in Germany, she was too advanced for 11th grade at the American high school on base. Nevertheless, Linda's mother was unwilling to let her daughter go off to college.

Only after a year of "one rather boring high school on base" did 17-year-old Linda escape to the American College in Paris, where she spent 2 years. Linda regards her time at the American College as a formative experience, but it is worth noting that her mother did require her to live in a dormitory run by Polish nuns! In Paris Linda learned French, took her first course in anthropology, and "fell in love with the idea of cultural variation and human sociality" (Fedigan, Pers. Comm. December 2016). She also fell in love with Larry Fedigan (1938–1990), an Irish citizen teaching English as a second language in Paris. They married when Linda was 19, and they travelled together to the University of Texas, Larry to earn a doctorate in teacher education, and Linda, earning three degrees at Austin, culminating in her PhD at the age of 25.

Canada was the lucky recipient of this exceptional young couple in 1974 when Linda accepted a tenure-track appointment at the University of Alberta. She and Larry were already familiar with the province, for they had worked as sessional instructors at the University of Calgary between the years Linda earned her MA and her PhD. Her doctoral thesis focused on Japanese macaques – a group that had been transported from their home base near Arashiyama, Japan, to the arid plains around Laredo, Texas, where she and Larry were also managers of the Arashiyama transfer project. Linda believes that her two Laredo years were foundational, giving her an experience beyond what one typically acquires when earning a doctorate. While there she was "initiated into the cooperative ways that Japanese primatologists conduct their research", and she received a "treasure trove of 20 years of longitudinal life history data" on these monkeys. Most importantly these experiences became the rock on which her "belief in the values of long-term data and collaboration among scientists" is built (Fedigan, Pers. Comm. December 2016). In that view, and its practice, she has never wavered.

1.2 Questioning Theory, Practice, and Meaning

Linda's approach to her studies has transformed understanding of the nature of non-human primate social dynamics and the interpretations arising from them regarding the behaviour of our own hominin ancestors. She did so by critical appraisal of how science was done – by questioning, analysing, and changing the methods and process involved in behavioural research. She also challenged the meanings attributed to primate behaviour, realizing that the assumed scientific objectivity that grounded extant interpretations of behaviour was gender based.

In 2000, Linda stated in a short essay that she was drawn to observational studies on non-human primates because of her introverted personality. She was interested in sociality and relations between the sexes, and it was easier to watch monkeys than to ask women questions about private matters, such as "sex, men, and morality" (Fedigan 2000:452). The range of appraisal undertaken by that self-described introverted personality over the past 45 years is astonishing, and for convenience I order them under the common questions of "what", "where", "who", and "why":

- *What* do we really know about male and female non-human primate behaviour? What could be reasonably ascertained from field studies that were a season's duration, or conducted over a year, or extended over a lifetime?
- *Where* was primate research undertaken – on New World or Old World monkeys, animals living in tropical (e.g. 10° 50′ N, 85° 38′ W [Costa Rica]) or northern

regions (e.g. 35° 00′ N, 135° 66′ W [Arashiyama, Japan]), on animals affected by human habitation or were free-ranging, or at sites before and/or after a major catastrophe affecting the animals' habitat? Did location yield differences in behaviour between males and females in the same species and in different species?

- *Who* conducted research – male or female – and what are their cultural origins? Does the identity of the investigator have an impact on the questions they pose, what they observe, and how they interpret it?
- *How* does evolution produce social behaviour? Why would females want to live in social groups with dominant males? Are evolutionary theories about sexual selection biased? Is the female choice embedded in them really a Hobson's choice? How can we measure fitness in males versus females?

These are deliciously tantalizing questions, spiced as they are with more than a bit of rising awareness that something was wrong with the theoretical propositions of the 1960s and 1970s, for example, that there was a "universal primate pattern" of behaviour that could still be identified in non-human primate species and which characterized the social organization and behaviour of humanity's ancestors.

Looking back from today, over the past 45 years of research on non-human primate behaviour, it is clear that Linda Fedigan's appraisal and questioning of the theories concerning the evolution of primate sociality and the process whereby these had been tested showed that both the theory and the findings were flawed. To properly test any theory, the process of research, among other matters its duration, its focus primarily on one sex, and even the language used to formulate research questions lest primatologists project "Western gender role stereotypes" (Fedigan 2009:260), had to change. And they changed.

Linda would be the first to protest that she was one of several women primatologists challenging dogma, and credit does not belong to her alone. True, but she began alone, writing the preface to her first book, *Primate Paradigms*, in 1980, and then waiting another 2 years before Eden Press, out of Montréal, published it. That the book was important was indubitable because in January 1983 a review of it was published in *Science*. The reviewer contrasted Linda's text with Sarah Hrdy's 1981 work, *The Woman That Never Evolved*, and noted that, "unlike Hrdy, Fedigan does not attempt to present provocative new syntheses and speculations, but rather concentrates on dissecting the processes by which cultural attitudes and empirical data interact" (Berman 1983:281). Why would *Science* review a book written by a young woman just 9 years past her PhD, working at a Canadian university on the northern prairie, a book published by an obscure Canadian press? Perhaps because, as Carol Berman observed, Linda had achieved her aim: she had indeed published a book that is "accessible and interesting to non-specialists, yet acceptable and even helpful to my colleagues in primatology" (Fedigan 1982:i); moreover she had written a "masterly, useful contribution to the primate literature" (Berman 1983:281). The University of Chicago Press agreed with that assessment, and it published the second edition of *Primate Paradigms* in 1992.

Today we can rightly say that thoughtful, "introverted" Linda, who never confused her opinion for evidence, has altered the process of research on non-human primates. From her studies, conducted with her current and past students, as well as colleagues from afar, conclusions have been drawn that have changed our understanding of male and female roles in primate societies, as we have also come to understand the behavioural diversity manifest in non-human primate species. Many of the areas Linda began exploring have led to new avenues of inquiry, not the least of which has been Linda's exploration of the role of women in primatology specifically and in anthropology more generally.

1.3 Primate Studies, Feminism, and Scientific Objectivity

In this volume, Pamela Asquith has described the importance of Linda's contributions to studies of gender and primatology and noted the broad themes that intersect in Linda's writings: sex roles of primates, the influence of feminist thought on primatological thinking, and the impact of primatology on feminist thought. Asquith also observed that Linda was the first scholar "to examine both images of women in theories of human evolution and the role of female scientists in constructing those images" (Asquith 2016).

There are few accounts of the history of primate studies, but it is worth noting that no mention was made of androcentrism in the constructions of theory in Gilmore's (1981) review of the rise of primatology in physical anthropology. Though Thelma Rowell's (1974) paper on the concept of social dominance was extremely influential, the closest she came to suggesting that a theory may be due to "unconscious anthropocentrism" – note, not "androcentrism" – is to ask in a sidebar (within brackets) "is our own species more than usually bound by hierarchical relationships, at least among the males, who have written the most about this subject?" (Rowel 1974:132). However, in the early 1980s, some women primatologists explicitly stated that they had concerns about the emphasis on the baboon model for understanding primate behaviour, given that the male dominance and aggression seen among baboons were not observed in several other primate species (Fedigan and Strum 1999). Sarah Hrdy (1981), for example, noted that theories about human evolution based on non-human primate models were male-centred. Her focus, however, was not on questioning the validity of such theories but on redefining behaviours that differ between adult males and females, some of which previously had been considered pathological. Among langurs, the species she studied, an incoming adult male that takes over a troop not only kills or drives off the incumbent alpha male, but he also kills the infants of lactating females, while females form alliances to protect them. Hrdy argued that infanticide by males and the formation of protective alliances by females indicated different *reproductive strategies* between the sexes. This interpretation was certainly novel, and it caught significant attention. In contrast, Linda's focus on the influence of cultural attitudes towards the interpretation of non-human primate behavioural data in general, and her explicit cautioning about the use of language that reflects an unconscious bias in the study of sex roles (Fedigan 1982:22–25), was less noted in the popular press. It was, nevertheless, noted by women primatologists and anthropologists. Fact is, Linda's thoughts articulated in 1982 marked a turning point in the construction of primatological theory and led to a more comprehensive understanding of primate society.

We no longer question that androcentrism in human evolutionary theory was rampant in the 1980s, and the idea that scientific theory and intellectual inquiry were not necessarily objective nor free of unconscious bias had not gained much traction until the twenty-first century among "serious" scientists. Women who argued that science was socially constructed faced dismissal of their ideas as biased by feminist thought. At the same time, the rise of sociobiology and the explanatory power this approach provided (e.g. females as reproductive strategists) had made sociocultural colleagues increasingly hostile towards biological anthropologists, including non-human primate behaviourists (Pavelka 2002). The conflicts led Linda and Shirley Strum to hold an international workshop in 1998 "to coax participants from all sides of the 'Science Wars' to exchange ideas rather than insults" (Fedigan 2000:452). They created the conditions whereby primatologists and scholars of science could consider the roles played by theory, method, and identity (gender and ethnicity) in the history of primate studies. The resultant book was not just about ideas in primatology but also provided considerable insight into the nature of science itself (Strum and Fedigan 2000). By 2009, Linda could argue that not only are there more female than male primatologists but that in this discipline there is also more awareness of gender than in other disciplines. However, though women primatologists have incorporated "tools of gender analysis" (Fedigan 2009:256) into their research, most dismiss the notion that their way of thinking is influenced by feminism. Linda regards this as a paradox that can be explained by three factors: their denial may well be the result of concerns that what is feminized is discounted; that the "science wars" are political conflicts, with feminist thought being "political" while sociobiological thought being "nonpolitical"; and that though field work raises complex questions, its problems can be ignored because science is "free from sociocultural influences" (Fedigan 2009:267). Clearly, language continues to matter.

The themes explored in this conference – life histories, sex roles, gender and science, primate-environment interactions, and changing environments – represent the research areas that arise from Linda's focus on basic questions over the past 45 years: who, what, where, and how research on non-human primate sociality is undertaken and what have we learned from the findings based on her approaches. They are a testament to the influence she has had on researchers' thought and field work practice, as well as the collective sociality that drives their approach to their science. There are other recognitions, too, of the contributions Linda has made to the study of primate sociality and practice.

1.4 Recognition and Its Reasons

On May 12, 2016, Linda Fedigan was named a member of the Order of Canada, the highest civilian honour in the Canadian honours system. She received this recognition:

> For her contributions to advancing our understanding of the behaviour and society of several primate species, and for her dedication as a mentor to the next generation of primatologists.

> Linda Marie Fedigan has made enduring contributions to the study of primates, some of our closest evolutionary relatives. Canada Research Chair in Primatology and Bioanthropology at the University of Calgary, she established a world-class research station in Costa Rica's Àrea de Conservación Guanacaste. Known for her field work, she has conducted groundbreaking long-term studies of the life history and reproductive patterns of female monkeys, which have increased our understanding of how primates adapt to their environments. An exemplary mentor, she is also known for her academic study of the role of women in science. (The Governor General of Canada 2016).

What detail lies behind this honour? It focuses on two facets of the many-faceted jewel of Linda's achievements that I have outlined above. Do I dare expound to a group of primatologists the nature of her scientific achievements beyond what I have already said? Well, yes, I dare. In layman's language let me sum up what any layman can understand, and then I will turn my attention briefly to the second of the reasons the Order of Canada was conferred on Linda – her mentorship of students.

I first met Linda in Banff in 1977, when the Canadian Association for Physical Anthropology held its annual meeting there. She was already interested in sex roles, in the life history of non-human primates, in reproductive senescence (menopause) in primate females, and in the behavioural ecology of primates. These themes are involved in understanding the evolution of sociality. For a trait to evolve, those with the trait have to be more successful in reproducing than those without it. Thus for any primate species, understanding sociality requires determining how female and male primates manage to live together and what affects their reproduction within social groups, and, most importantly, it demands an accurate count of the number of offspring produced by each animal.

Today it is hard to believe that 45 years ago most studies of the behaviour of Old World monkeys (African and Asian species) were seasonal or otherwise short in duration rather than undertaken over many years. With observational time limited, most field primatologists focused on the behaviour of more easily seen animals. These were ground-dwelling species (e.g. baboons) and were often larger animals (e.g. males), those that were higher ranked (all males relative to females and higher-ranked females relative to other females). However, focusing on animals that were more easily seen because they lived on the ground and were large and dominant led to much information about males and little about females. Males were also poor candidates for the assessment of reproductive success, simply because paternity in the days before genomics was difficult to establish with any confidence. On the other hand, females that gave birth were observable with their babies for months and, dependent on the species, were often observable for years. Linda knew that natural selection manifests itself by differential reproduction and survival. To obtain information that would test the evolution of sociality required knowledge of the number of offspring produced by each female, not just the dominant ones. Further, because monkeys reproduce until they die, data collection had to be obtained over their lifetimes. Finally, attention to individual life histories was necessary to permit understanding of behavioural patterns that influenced offspring survival to reproductive age.

For these reasons, after she earned her doctorate in 1974, Linda began to focus increasingly on the behaviour and reproductive performance of females. She had done her thesis on social roles in a group of 150 Japanese macaques (*Macaca fuscata*) that had been transported to a ranch in South Texas in 1972 (Arashiyama West Primate Research Station) – a flat, hot, arid region significantly different from the environment in which *Macaca fuscata* had evolved. The group afforded further opportunity for detailed investigations, including access to a database ("Arashiyama East") kept on the group's predecessors in Japan. Linda was one of the first western-trained primatologists to emphasize the importance of life histories to the study of sociality. She was also one of the first to recognize that Japanese primatologists approached their field subjects – the snow monkeys – differently than did western-trained primatologists.

In 1991, Linda and her colleague at the University of Alberta, Pamela Asquith, co-edited a book entitled *The Monkeys of Arashiyama: 35 Years of Research in Japan and the West.* That text was the first to provide a detailed overview of the research findings that could be obtained from longitudinal, life histories of macaques. It was instrumental in turning the attention of western primatologists to the value of such data. Today Linda continues working on an even longer database – collected over 50 years through the Arashiyama East-West project.

Linda Fedigan's second major contribution to the study of primate life histories, reproduction, and behavioural ecology began in the early 1980s when she became concerned that some of her findings may not be generalizable because the Arashiyama West group was provisioned, not just translocated from where the species had evolved. Accordingly in 1983, she established, in Costa Rica, the Santa Rosa Primate Field Project, where she and her students could begin studying three species of natural (wild) populations of New World primates – capuchin monkeys (*Cebus* sp.), howler monkeys (*Alouatta* sp.), and spider monkeys (*Ateles* sp.). Because these species are arboreal, research on their behaviour would provide critical information on sociality in an arboreal ecological setting in comparison to what is known about the terrestrial macaques. Since its inception, the Santa Rosa Primate Field Project has provided important information on the social structure, behavioural ecology, life histories, and conservation of these notoriously difficult-to-observe animals.

Is anyone who is not a primatologist interested in such studies? Given the public interest in studies of non-human primate behaviour generated by Jane Goodall's research on chimpanzees, it should surprise no one that Omni Film Productions made a film about Linda Fedigan's studies on New World monkeys 15 years after the Costa Rican project began. It was broadcast on the Discovery Channel in 1998 in the series "Champions of the Wild" (National Film Board 1998). A long stream of scientific journal publications – more than 100 of Linda's 158 published journal articles as of 2016 – have also emanated from this project, including a book, *The Complete Capuchin Monkey*, published in 2004 by Linda and her two co-authors, Dorothy Fragaszy and Elisabetta Visalberghi. By 2017, 19 MA students and 7 PhD students had also completed their thesis research at Santa Rosa.

Can I give you an example of a definitive finding that has emerged from Linda's studies on female reproduction and life histories? When Linda began her studies with the Arashiyama West group of Japanese macaques, the first papers that suggested differences in male and female patterns of reproductive behaviour were just emerging. Today it is well established that the reproductive strategies of male and female primates differ, but not in the ways short-term studies done 45 years ago suggested. Then it was thought that higher female rank meant greater reproductive success, and males were always aggressors to babies not their own. Linda Fedigan's research has shown considerably more complexity exists than this. In capuchin monkeys, as one example, closely related females help to rear infants to reproductive maturity; they protect the young against attacks from males outside the group, as do large numbers of males when they are resident within the group. The best predictor of reproductive success for males remains dominance, but for females, Linda's studies showed that the best predictor is the presence of supportive kin and protective males (Jack and Fedigan 2006).

Research on primate life histories has also led to other questions. Establishing that non-human primates reproduce to the end of their lives required long-term research in many different species, and it required assessment of the factors that lead to an absence of reproductive senescence compared to human females. Some asked whether there is any evolutionary advantage to female human primates living past menopause. Mary Pavelka and Linda first tackled such issues in a 1991 paper for the *Yearbook of Physical Anthropology*, and in 2013, Linda was still writing on the subject with nine other authors in a paper published in the *Proceedings of the National Academy of Sciences (USA)* (Alberts et al. 2013). The ten authors of this paper used "long-term, individual-based data" from natural populations of seven non-human primate species as well as data from humans that were living in pre-industrialized social groups (Dobe! Kung Bushmen). It is worth noting that Linda's Costa Rican database that was used in this study is one of only seven such longitudinal databases around the world!

What of the second facet of Linda's contributions recognized in her Order of Canada citation – mentorship of students? Social scientists are sometimes parodied for their tendency to count things, but counting matters, as does quantity – just ask anyone who has done chi-square tests on the same set of results obtained in samples of 100 versus 1000. Over her career, Linda Fedigan has supervised to completion the theses of 10 Bachelor of Arts (honours) students, 29 Master of Arts students, and 12 doctoral students, as well as 5 post-doctoral fellows. The number of students Linda has supervised is very large for any discipline and is remarkable for a biological anthropologist because each of the students, but especially the graduate ones, required funding for their research projects. Field work and laboratory work are expensive – and there is no doubt that Linda Fedigan's research grants helped to make her students' graduate research possible. In addition to students, one must note that Linda has also trained a large number of laboratory technicians – more than any other biological anthropologist in Canada.

Many of Linda's students have obtained positions of responsibility outside the academic sector in Canada and abroad, for example, in wildlife biology, conservation biology, and zoo administration, and even as policy advisor for the Canadian federal government. Many have risen to prominence at universities in Canada, the USA, and elsewhere in the world. Perhaps the most extraordinary example of Linda's success in mentoring students to earn advanced degrees is that of her Costa Rican field assistant, who is an instructor at the University of Costa Rica in Heredia. Individually and collectively, Linda's former students are a testament to her mentorship, and the beneficiaries of her more than $2 M of research success.

Yes, I must mention dollars, because between 1981 and 2015 alone, Linda received $2,160,000 in research grants, counting monetary awards from all her sources of funding. Those of you who have come here from other countries may not appreciate the difficulty faced by Canadian biological anthropologists in obtaining federal funding to support their research. Though the National Science and Engineering Research Council of Canada (NSERC) now has a numbered category for "anthropology", it still has no panel to evaluate research in biological (aka physical anthropology). Nevertheless, Linda has received NSERC support unabated since 1983. There is no other biological anthropologist in Canada who can match this record. The fact is, Linda has demonstrated over and over again to the animal behaviourists and experimental psychologists on the review panels who examine her project proposals that her projects are important scientifically, and she has achieved the goals of the project for which she received earlier funding. The culmination of her research achievements to date was Linda's receiving more than $900,000 from the Canada Foundation for Innovation when she received a Tier I Canada Research Chair – an award that brought her to the University of Calgary from the University of Alberta. These Tier I Chairs are conferred for 7-year terms and are renewable. They are given, as described on the CRC website, to "outstanding researchers acknowledged by their peers as world leaders in their fields" (Canada Research Chairs 2016). And she is.

The Order of Canada citation recognized major elements of Linda Fedigan's contributions to her science and to her students; however, my comments would not be complete without noting her encouragement of young women to study science and the inspiration she has provided the general public. Linda was involved with WISEST (Women in Science, Engineering and Technology) when she was a professor at the University of Alberta. The film about her as a "Champion of the Wild" on the Discovery Channel, and her role as well as the roles her female graduate students have played in the Cybermentor programme at the University of Calgary, has drawn the attention of many, including young girls who became aware that they could also "study monkeys" and thereby contribute to the world's knowledge.

Has Linda's research eminence been recognized in other ways nationally and internationally? I have already mentioned the Tier I Canada Research Chair she was awarded in 2002, which she held until her retirement at the end of 2015. She received the American Society of Primatologists' Distinguished Primatologist Award in 2013. She was the first – and to date only – Canadian executive editor of the *American Journal of Primatology*, the official publication of the American Society of Primatologists (2004–2007). In Canada, her research eminence was recognized in 2005 when she was elected a fellow of the Royal Society of Canada.

1.5 The Tree Inclined: At Retirement

One cannot do justice to Linda Fedigan's eminence in a chapter such as this. It is ironic that my comments have provided only a minimalist sketch of the elements of personal history that shaped Linda and the woman she became, in comparison to the exquisite detail Linda has provided about the life history of female monkeys. There is no question that Linda's contributions to her science and society have been recognized, but it would be false to conclude that her road to success has been free of tribulations, both personal and professional. Linda began her career as a sessional instructor, and she knows well the insecurity that time-limited contracts bring to young scholars. Securing continuing support from Canadian federal granting agencies for research on monkeys was no mean feat and was particularly difficult at the beginning, when Linda had no track record of research success. It is often assumed that successful scholars enjoy continuing equilibrium in their personal lives, which enable their focus on their studies, but Linda has known separation and death as well as the joys of deep and lasting friendships. The intellectual bond formed between her and Dr. John Addicott when she sought his statistical advice at the University of Alberta eventually blossomed into a marriage that has lasted for 26 happy years. John, an ecologist and professor emeritus of Biological Science at the University of Calgary, continues with statistical advice not only to Linda but also to her colleagues and students (e.g. DeGamma-Blanchet and Fedigan 2006). He developed the collaborative primate census and life history "PACE" database (Fedigan and Jack 2013) that is still maintained at the University of Calgary, and he has co-authored papers with Linda and other collaborators (e.g. Fedigan et al. 2014), along with publishing his own ecological research. Their partnership is special, as is the relationship that Linda and John, alone and together, have forged with her students and colleagues over their academic lives.

1.6 Conclusions

Though much remains to be said about Linda, each participant in the Festschrift conference, many of whom have contributed chapters to this volume, knew that her dedication to research is matched by her concern for her students. She has not claimed their work as her own, she has partnered with them in the publication of their research with the student's name going first more often than not, and she has supported them in their search for a career path that will allow them to fulfil their potential. Linda's relationships with her colleagues have been, and are, meaningful and deep, and each participant in that conference could cite his or her relationship with Linda as the reason for their presence there.

To contribute to a book that is dedicated to an honoured person and will be presented to that person during her lifetime is a special undertaking. The meaning of "Festschrift", indicates this in the original German, but I especially liked the Latin translation provided by Wikipedia: "*liber amicorum* (literally: 'book of friends')". The papers that were given over the 2 days of the Festschrift conference will indeed comprise a book written by friends, united as much by respect for Linda Fedigan's knowledge and contributions to the field of primate behaviour, as by their feelings of affection for her. I am one of them.

References

Alberts S, Altmann J, Brockman D, Cords M, Fedigan L et al (2013) Reproductive aging patterns in primates reveal that humans are distinct. Proc Natl Acad Sci U S A 110(33):13440–13445

Asquith PJ (2016) A woman of science: sorting fact and illusion in gender and primatology. Linda Fedigan Festschrift Symposium 2016, Banff, Alberta.

Berman CM (1983) Behavioral patterns. Science 219(4582):281

Canada Research Chairs (2016) Types of Chairs. http://www.chairs-chaires.gc.ca/program-programme/index-eng.aspx. Accessed 18 Nov 2016

DeGama H, Fedigan LM (2006) Effects of forest fragment area, isolation, age, habitat type and water availability on monkey density in a tropical dry forest. In: Estrada A, Garber P, Pavelka M, Luecke L (eds) New perspectives in the study of Mesoamerican primates: distribution, ecology, behavior and conservation. Springer, New York, pp 165–188

Dolhino P (1972) Primate patterns. In: Dolhinow P (ed) Primate patterns. Holt, Rinehart & Winston, Toronto, pp 352–392

Fedigan L (1982) Primate Paradigms. Eden Press, Montréal

Fedigan LM (2000) A view of the science: physical anthropology at the millennium. Am J Phys Anthropol 113:451–454

Fedigan LM (2008) Primatologists who focus on females/gender. In: Rosser SV (ed) Women, science and myth. ABC, CLIO, Inc., Santa Barbara, pp 357–364

Fedigan LM (2009) The paradox of feminist primatology. In: Wyer M, Barbercheck M, Giesman D, Öztürk H, Wayne M (eds) Women, science and technology, 2nd edn. Routledge, New York, pp 256–270

Fedigan L (2016) Personal communication, Nov 2016

Fedigan LM, Asquith PJ (eds) (1991) The monkeys of Arashiyama: 35 years of research in Japan and the west. SUNY Press, Albany

Fedigan LM, Jack KM (2013) Sexual conflict in white-faced capuchins: it's not whether you win or lose. In: Fisher ML, Garcia JR, Chang RS (eds) Evolution's empress: Darwinian perspectives on women. Oxford University Press, New York, pp 281–303

Fedigan LM, Strum SC (1999) A brief history of primate studies: national traditions, disciplinary origins, and stages in North American field research. In: Dolhinow P, Fuentes A (eds) The nonhuman primates. Mayfield Press, Mountain View, pp 258–269

Fedigan LM, Melin AD, Addicott JF, Kawamura S (2014) The heterozygote superiority hypothesis for polymorphic color vision is not supported by long-term fitness data from wild neotropical monkeys. PLoS One 9(1):e84872. https://doi.org/10.1371/journal.pone.0084872

Fragaszy D, Visalberghi E, Fedigan LM (2004) The complete capuchin monkey: the biology of the genus Cebus. Cambridge University Press, Cambridge

Gilmore HA (1981) From Radcliffe-Brown to sociobiology: some aspects the rise of primatology within physical anthropology. Am J Phys Anthropol 56:387–392

Hrdy S (1981) The woman that never evolved. Harvard University Press, Cambridge

Jack K, Fedigan LM (2006) Why be alpha male? Dominance and reproductive success in wild white-faced capuchins (*Cebus capucinus*). In: Estrada A, Garber P, Pavelka M, Luecke L (eds) New perspectives in the study of Mesoamerican primates: distribution, ecology, behavior and conservation. Springer, New York, pp 367–386

National Film Board (NFB) (1998) Champions of the wild: Costa Rican monkeys. http://www3nfbca/sg/25700pdf. Accessed 15 Nov 2016
Pavelka MM (2002) Resistance to the cross-species perspective in anthropology. In: Fuentes A (ed) Primates face to face. Cambridge University Press, New York, pp 25–44
Pavelka MSM, Fedigan LM (1991) Menopause: a comparative life history perspective. Yearb Phys Anthropol 34:13–38
Rowell TE (1974) The concept of social dominance. Behav Biol 11:131–154
Strum SC, Fedigan LM (eds) (2000) Primate encounters: models of science, gender, and society. The University of Chicago Press, Chicago
The Governor General of Canada (2016) Order of Canada. Honours Find a Recipient. https://www.gg.ca/honour.aspx?id=15918&t=12&In=Fedigan. Accessed 12 Nov 2016
Washburn SL, Hamburg DA (1972) Aggressive behavior in old world monkeys and apes. In: Dolhinow P (ed) Primate patterns. Holt, Rinehart & Winston, Toronto, pp 276–296
Washburn SL, Lancaster CS (1968) The evolution of hunting. In: Washburn SL, Jay PC (eds) Perspectives on human evolution 1. Holt, Rinehart and Winston, Toronto, pp 213–229
Washburn SL, Jay PV, Lancaster JB (1965) Field studies of old world monkeys and apes. Science 150:1541–1547

Part I
Primate Life Histories

Mary S. M. Pavelka

In Anthropology, life history research has always meant two quite distinct things, one shared with the social sciences and one with the biological sciences. As explained by Linda Fedigan herself (1997), to social and cultural anthropologists, a life history is a qualitative research tool exploring the biography or autobiography of individual lives as they were lived. For the biological anthropologists interested in the cross-species perspective provided by primatology however, life history research shares more with the biological approach. The life history perspective in the biological sciences is more concerned with the quantitative analysis of species level characteristics investigating the trade-offs that individuals of the species make between survival, growth, and reproduction. Key life history research typically includes variables such as infant survivorship, age at first and last reproduction, reproductive lifespan, inter-birth interval, life expectancy, and maximum lifespan. While these are often thought of as species level characteristics, with species typical values (e.g., wild Japanese macaques typically give birth every 2 years, but the orangutan inter-birth interval is closer to 8 years), long-term studies of known individuals are helping us to understand variation among individuals in the expression of some traits and relative absence of variation in others. In the end, as evolutionary anthropologists, we want to know what characteristics or combination of characteristics lead to differential reproductive success as we attempt to understand the evolution of primate life histories.

> "I have always been drawn to questions about how and why some females, some groups, and some populations reproduce better than others over time." (Fedigan 2014:188)

Linda's interest in primate life history, as the quote above indicates, has always been solidly rooted in understanding the lives of individual monkeys. When she first

M. S. M. Pavelka (✉)
Department of Anthropology and Archaeology, University of Calgary, Calgary, AB, Canada
e-mail: pavelka@ucalgary.ca

began to study primates, she was gifted access to the Arashiyama West group of Japanese monkeys, including the 18 years of detailed genealogical and life history data that the Japanese researchers had meticulously collected and generously shared. This introduction to the field led to Linda's lifetime respect for and appreciation of not only long-term research on known individuals throughout their lives but for cooperation, collaboration, and the sharing of data. In terms of life history research, it began with the Japanese researchers patiently teaching Linda to recognize every individual in the group, which quickly revealed how these individuals were differentiated by matriline, dominance, and personality. And for another 25 years she and others continued to build the detailed records of the lives of every monkey born into the group, including, of course, date of birth, survivorship, age at first and last reproduction, reproductive lifespan, inter-birth intervals, and maximum lifespan. Linda and collaborators published a series of papers that analysed the completed lives of 95 individuals, known and monitored from birth to death. While much of this involved complex survival analyses of life table data, just about every adult in the sample was known to Linda as a distinct individual, and sorting through printouts of results was often punctuated by fond memories of individual monkeys who formed part of the data set. So the social vs biological science perspectives on life history research somewhat converge as we understand population level life history traits from the perspective of individual lives lived in the context of a social group and its ecological and social history. And as all readers know, Linda, inspired by the richness of the Japanese database and its potential to address issues of primate life history established, in 1983, what was to become one of few multi-decade studies of known individuals in an arboreal new world primate, resulting in a white-faced capuchin monkey database comparable to those constructed by baboon and macaque researchers in the Old World.

One of the other pioneers in the establishment of the long-term study of new world monkeys was Karen B. Strier, who has been studying the northern muriqui of Brazil's Atlantic forest, since 1982. Strier is the author of the first of the three chapters in this part on Life History (Chap. 2, this volume). In her chapter, titled "The Temporal Scale of Behavioral and Demographic Flexibility: Implications for Comparative Analyses and Conservation", Strier emphasizes the difference between behavioural variance/flexibility and demographic variance/flexibility *and* the importance of distinguishing between variance based on spatial comparisons and flexibility that requires temporal comparisons. Further, she argues the need to consider the extent to which components of individual life histories are free to vary in response to external stimuli, and thus to be indicative of demographic flexibility. For example, the length of inter-birth intervals is critically important to questions of individual variation in reproductive success and to the potential reproductive rate of species threatened with extinction or trying to recover from population declines. Strier points out that there are four phases of the inter-birth interval: gestation, lactation, anovulatory, and cycling, not all of which are free to vary. Gestation length in particular appears to be fixed, and it is the three non-gestational components that may be more plastic and indicative of demographic flexibility. Additionally, intraspecific comparisons show that the only variable found to really influence the length of the inter-birth interval for individuals is the survival of the previous offspring.

Whether or not a life history trait is constrained or free to vary in response to local conditions comes up in the second chapter in this part, "Post-fertile Lifespan in Female Primates and Cetaceans" by Pavelka, Brent, Croft, and Fedigan (Chap. 3, this volume). Based on analyses of the 95 completed lives of the Japanese monkeys described above, along with examination of published literature on post-reproductive lifespan in primates, Fedigan and Pavelka have long argued for the uniqueness of the menopause in women, based on its universality and timing in the life course. This conclusion has been supported by recent analyses of available long-term data sets on several wild primate species, as discussed in the chapter in this volume. We have also suggested that 50 years of age might be a fixed endpoint in mammalian female reproduction, and that menopause in humans may be a by-product of increases in longevity and a phylogenetically constrained reproductive system that ceased operating at 50 years of age, likely as a pleiotropic effect. However, our collaboration with Lauren Brent and Darren Croft to produce the chapter for this volume has revealed some important findings. For example, it is clear from the available data on baleen fin and Bowhead whales that reproduction well past the age of 50 is possible in some mammals and that there must be a more variable oocyte supply in different mammalian taxa than previously known. Additionally, although a post-fertile stage of life is very rare in mammals, it does occur in some exceptionally long lived taxa, such as humans and two species of toothed whales, and to have evolved independently at least three times. Furthermore, mathematical models taking both costs and benefits of continued reproduction in old age into account are showing that direct selection to cease reproduction is a strong possibility in species such as the resident killer whales.

Strier, as well as McCabe and Fernandez (Chap. 4, this volume, "Seasonal Patterns of Infant Mortality in Wild Sanje Mangabeys, ***Cercocebus sanjei***"), points out that infant survivorship is a life history trait that is also a fundamental factor influencing population growth, and the population viability of primate species. What are the main factors affecting the survivorship of an infant? In many species it may be infanticide (such as is relatively common during group take overs in the capuchins of Santa Rosa National Park that have been studied by Linda and her students for over 35 years (see Jack and Fedigan, Chap. 6 this volume). However, the effects of food on female (and infant) condition are also critically important, which are themselves tied to seasonality, habitat quality, group size, and population densities (see Bergstrom et al., Chap. 11 this volume) all of which are all being affected by anthropogenic pressures and climate change (see Kalbitzer and Chapman, Chap. 14 of this volume). McCabe and Fernandez investigate the timing of infant births and infant mortality relative to seasonal changes in the food supply and find that the timing of conception with peak food abundance can influence infant survival. But in their study, something else seems to be going on. Thirty-four percent of infants died in the first year of life, and these deaths were highly clustered within the early dry season, which coincides with an initial drop in fruit production. However, it appears that a seasonal infestation of early dry season ticks is probably the more important cause of infant mortality, or possibly an interacting effect of a tick-borne disease and a decrease in nutritional intake during the early dry season. Habitat and climate change may be affecting the prevalence of the responsible ticks in the Sanje mangabey habitats, reminding us that there will likely be a host of cascading effects of climate change including effects on wildlife disease (Behie et al. 2014). Studies such as these demonstrate that climate and habitat change are introducing variation in the expression of some life history traits and subsequently on demography and ultimately species survival.

References

Behie AM, Kutz S, Pavelka MS (2014) Cascading effects of climate change: do hurricane-damaged forests increase risk of exposure to parasites. Biotropica 46(1):25–31

Fedigan LM (2014) Questions my mother asked me: an inside view of a thirty-year primate project in Costa Rica. In: Strier KB (ed) Primate ethnographies. Pearson/Prentice Hall, Upper Saddle River, pp 186–196

Fedigan LM (1997) Changing views of female life histories. In: Morbeck ME, Galloway A, Zihlman AL (eds) The evolving female: a life history perspective. Princeton University Press, Princeton, NJ, pp 15–26

Chapter 2
The Temporal Scale of Behavioural and Demographic Flexibility: Implications for Comparative Analyses and Conservation

Karen B. Strier

Abstract Many primate species exhibit high levels of behavioural variation that correlate with demographic variance across populations and within populations over time. Analyses of these relationships have typically treated behavioural variance as the product of individual responses to local ecological and demographic conditions and therefore indicative of behavioural flexibility. However, in contrast to the extrinsic nature of the variance in ecological conditions such as climate, food availability, or predation pressure, demographic variance can reflect both extrinsic processes and intrinsic properties of individual life histories, such as the timing and rate of fertility and dispersal decisions. If we follow the rationale for interpreting behavioural flexibility, then demographic flexibility would correspondingly reflect the responses of components of individual life histories that are free to vary to external stimuli. The dynamics between behavioural and demographic flexibility can shift over the course of individual lifespans and are expected to affect the size and composition of groups and populations and therefore levels of intrasexual competition and many other features of sociality. These dynamics also affect group and population growth rates, with implications for the recovery and management of small populations and the resilience of large populations subject to rapidly or unpredictably changing conditions. Longitudinal and cross-sectional data from the critically endangered northern muriqui (*Brachyteles hypoxanthus*) and other primate species provide comparative insights into patterns of behavioural and demographic flexibility that can advance our understanding of primate adaptive potentials and constraints over ecological and evolutionary time.

Keywords Behavioural variation · Demographic variation · Dispersal · Interbirth interval · Fertility

---

K. B. Strier (✉)
Department of Anthropology, University of Wisconsin-Madison, Madison, WI, USA
e-mail: kbstrier@wisc.edu

U. Kalbitzer, K. M. Jack (eds.), *Primate Life Histories, Sex Roles, and Adaptability*, Developments in Primatology: Progress and Prospects,
https://doi.org/10.1007/978-3-319-98285-4_2

2.1 Introduction

Many primate species exhibit high levels of behavioural variation that correlate with demographic variance or variation in the size and composition of groups across populations and within populations over time (e.g. Kamilar and Baden 2014; Strier et al. 2014; Strier 2017). Analyses of these relationships have typically treated behavioural variance as the product of individual responses to local ecological and demographic conditions and therefore as being indicative of the behavioural flexibility of a species (Chapman and Rothman 2009; Strier 2009; Kappeler and Kraus 2010; Wong and Candolin 2015). However, this perspective obscures an important distinction between the causes of ecological and demographic variance. Specifically, in contrast to the causes of variance in ecological conditions such as climate, food availability, disease, and predation, which are strictly extrinsic in nature, the causes of demographic variance can reflect both the effects of extrinsic processes and the intrinsic properties of individual life histories. These intrinsic individual properties may include the timing and rate of fertility and dispersal decisions, which can change in response to ecological and demographic conditions over time. Thus, although it may be possible to detect what seem like cause-effect relationships between stimuli and responses, the reality is that life history responses to local conditions are based on dynamic feedback loops between demography and the effects of flexibility in life histories on demography.

Not surprisingly, the dynamics of these interacting stimuli and responses are nonlinear and therefore difficult to model (e.g. Strier et al. 2014; Lee and Strier 2015). In this sense, they are similar to other nonlinear phenomena such as the relationships that have been documented between male numbers, population density, and infanticide in langurs (Moore 1999). However, if we use the rationale that a species' behavioural flexibility is the product of individual responses to varying external stimuli, then demographic flexibility should be the product of individual life history responses (Saether et al. 2013).

The dynamics between behavioural flexibility and demographic flexibility can shift over the course of individual lifespans and are expected to affect the size and composition of groups and populations and therefore levels of intrasexual competition and many other features of sociality (e.g. Moore 1999; Strier 2003, 2011; Gogarten et al. 2014; Lee and Strier 2015). These dynamics also affect group and population growth rates, with implications for the recovery and management of small populations and for the resilience of large populations that may be subject to rapid or unpredictable fluctuations (Strier 2000; Snell-Rood 2013).

In this paper, I apply the underlying rationale for interpreting evidence of behavioural flexibility (e.g. Strier 2017) to interpretations of demographic flexibility and its potential to reflect how components of individual life histories may vary in response to variation in local demographic conditions. I use longitudinal data from the critically endangered northern muriqui (*Brachyteles hypoxanthus*), supplemented with examples from other primate species, to offer comparative insights into patterns of behavioural and demographic flexibility that can advance our understanding of primate adaptive potentials and constraints over ecological and evolutionary time. This approach follows other recent investigations into the effects of phenotypic plasticity on population persistence and resilience (e.g. Dingemanse and Wolf 2013; Forsman 2015; Maldonado-Chaparro et al. 2017). By identifying the subset of demographic variance that is the product of individual responses, or the components of life histories that are free to vary, this approach aims to establish a theoretical framework for distinguishing demographic flexibility from demographic variation per se.

2.2 History of Behavioural and Demographic Changes in Northern Muriquis

The northern muriqui population at the Reserva Particular Patrimônio Natural-Feliciano Miguel Abdala (RPPN-FMA) in Caratinga, Minas Gerais, Brazil (19°50′S, 41°50′W), has undergone tremendous changes in its group and population demography over the past 35 years. Our record of these changes provides a basis for evaluating flexibility in individual life histories during the shift from past conditions of population growth to current conditions of habitat saturation. The northern muriqui is endemic to the Atlantic Forest of southeastern Brazil, which ranks among the top biodiversity hotspots and among the most threatened ecosystems on the planet (Myers et al. 2000). Fewer than 1000 northern muriquis are distributed among 11 populations or metapopulations in the states of Minas Gerais and Espírito Santo (updated from Mendes et al. 2005). The study population is among the largest and best protected of these, in addition to having been the subject of the most intensive, long-term research (Strier and Mendes 2012). Its isolation from other muriqui populations also makes it extremely vulnerable to the genetic, ecological, and demographic risks of small populations (Franklin 1980; Shaffer 1981; Giplin and Soule 1986; Dobson and Lyles 1989; Saccheri and Hanski 2006) and extremely valuable as a natural laboratory because we have been able to study this closed population in its entirety. Non-invasive genetic studies on this population are presently underway (Chaves et al. 2011; Strier et al. 2011), and analyses of the forest's productivity and regeneration potential have been conducted (Boubli et al. 2011).

Systematic studies were initiated on one of the two original muriqui groups (Matão group) present in the forest in 1982, yielding detailed individual life history data from July 1983 through the present (Strier and Mendes 2012). The size of the Matão group has increased steadily from 22 individuals in July 1982 to more than 120 individuals as of April 2017, and some of its original members are still alive, including adult females >40 years old at present. Systematic studies on the other groups in this population were initiated in 2002, yielding demographic and life history data on the entire population from 2003 through the present. One of these groups (Jaó) was present in 1982; the other two groups were established when the Jaó group fissioned in 1988 (M2 group) and 2002 (Nadir group; Strier et al. 2006).

The increase in the number of muriqui groups from two to four can be attributed to increases in group sizes, which caused one of the groups to fission on two occasions (Strier et al. 2006; Tokuda et al. 2014), and ultimately led to a sevenfold increase in the population's size (from an estimated 50 individuals in 1982 to a peak of some 350 individuals in 2016). The population growth likely reflects a combination of factors including its recovery from past disturbances, such as the forest's initial fragmentation, possible hunting pressures in the distant past, and selective logging, which declined over the years and ceased entirely with the forest's designation as a federally protected private reserve, or RPPN. These declines in disturbances coincided with highly favourable demographic conditions including female-biased birth sex ratios and extremely low mortality, which may have been a consequence of low population density and low predation pressures (Strier and Mendes 2012; Strier 2017).

Northern muriquis live in unusually peaceful, egalitarian societies in which males are philopatric and females typically disperse from their natal groups (Strier 1999a). They are the largest New World primates, and in addition to their female-biased dispersal, other aspects of their life histories (e.g. age at maturity and first reproduction, birth intervals, and estimated lifespans) are more similar to those of apes than to monkeys of similar body size (Strier 2003; Bronikowski et al. 2016).

Despite the documented group and population expansion, rates of overt aggression and contest competition have remained low (Strier et al. 2001a). Nonetheless, other behavioural changes can be directly attributed to changing demographic conditions. For example, by the time the Matão group had doubled in size, the size of its home range had nearly doubled in size but without a corresponding increase in day ranges. Instead, the group's previously cohesive pattern of association had shifted to more fluid associations involving variably sized parties (Dias and Strier 2003). Such facultative adjustments in party size resemble the fission-fusion dynamics found in other primates and non-primates (Aureli et al. 2008) and may reflect similar behavioural mechanisms for reducing intragroup feeding competition and travel costs (Isbell 1991; Gillespie and Chapman 2001). A subsequent increase in the use of terrestrial substrates has also been documented, indicative of an expansion of the group's vertical niche and consistent with an intensification of the muriquis' range use necessitated by habitat saturation (Tabacow et al. 2009).

Previous analyses of muriqui demographic variables from 1987 to 2010 revealed significant changes beginning around the millennium (Strier and Ives 2012). These include a shift from female-biased to male-biased birth sex ratios and increased mortality across all age-sex classes, both of which were consistent with predicted responses to increasingly high population density and habitat saturation. However, we also detected a significant increase in fertility that was unexpected considering the usual effects of high densities on resource competition and corresponding reproductive rates. The continued growth of the population could be explained by this unexpected increase in fertility, which we hypothesized might instead reflect at least the short-term benefits of the expansion of the muriquis' vertical niche as they increased their terrestrial activities.

2.3 Flexibility and Intra- and Interspecific Variance in Fertility

Interbirth intervals (IBIs), together with age at first reproduction (AFR) and reproductive lifespan, are key life history traits that constrain female primate fertility (Fedigan and Rose 1995; Alberts et al. 2013). They are also, ultimately, among the key determinants of the rate of population growth and are therefore important variables for assessing the ability of small populations of endangered species to increase in size (Dobson and Lyles 1989) and to respond to environmental changes (Morris et al. 2011; Saether et al. 2013).

Both IBI and AFR are expected to be sensitive to local demographic conditions. These conditions can have direct, density-dependent effects such as those that impact survival and reproductive rates (Strier and Ives 2012). Alternatively, they may have more indirect effects, such as those when shifts in the sources of infant mortality affect maternal investment patterns, and thus IBI, or when shifts in the relative strength of intra- versus intergroup competition affect the timing and success of female dispersal, and thus AFR (e.g. Barrett et al. 2006a, b; Fedigan and Jack 2012).

For IBIs, the relative contributions of intrinsic or "care-dependent" mortality versus extrinsic or "care-independent" sources of infant mortality underlie intraspecific variation in maternal investment patterns (Barrett et al. 2006a, b). When high maternal investment increases infant survivorship, reproductive rates tend to be slower than when infant survivorship is care-independent. However, the relationship between infant survival and birth intervals can be confounded by conditions that infants experience during weaning, which vary with ecological conditions such as habitat quality and resource predictability, as well as with demographic conditions that affect levels of competition (Fedigan and Rose 1995; Lee 1996, 1999; van Noordwijk and van Schaik 2005; Barrett et al. 2006a, b).

Shifts in AFR might be expected in response to changes in the relative (and absolute) levels of intra- and intergroup competition as group size and group densities increase, with alternative predictions for earlier or later AFR. For example, rising levels of intragroup competition might stimulate females to disperse from their natal groups at younger ages, resulting in earlier AFR if early dispersal is successful. Higher group densities (resulting from the increase from two to four groups with overlapping ranges within the forest) might also lead to more frequent intergroup encounters, facilitating female dispersal and thus earlier AFR. Alternatively, rising levels of intergroup competition might make it more difficult for dispersing females to join and become established their new groups. Hostility from residents in these groups could lead females to remain longer in their natal groups as anovulatory, prepubescent nonbreeders, resulting in later AFR due to later dispersal and maturation.

Trade-offs between growth and reproduction could also influence these patterns in a variety of ways. For example, higher population density and associated higher levels of competition could result in slower growth rates, thereby delaying dispersal age and age at sexual maturity, either or both of which would delay AFR. However, if dispersal and sexual maturity ages are tied to a more constrained period of growth, then the variance in AFR might be less flexible.

IBIs have received more detailed comparative attention than life history traits such as AFR or reproductive lifespan, in part because IBIs can be measured over comparatively much shorter time scales. In the northern muriqui, for example, AFR occurs at a median of about 9 years ($n = 30$, range = 6.96–11.98 years; updated from Strier et al. 2006, for first births between 2012 and 2016). AFR is roughly 2–3 years after the age at which most females disperse from their natal groups (Strier et al. 2015). Discovering the muriqui AFR took decades because of the length of time required to accumulate a reasonable sample size of females whose ages were known from birth in their natal groups until their subsequent first reproductions in the groups into which they dispersed (Strier and Mendes 2012). Indeed, in the 9 years it takes to document one AFR, multiple females with roughly 3-year IBIs could have given birth to up to three surviving offspring.

Overall, the mean (± sd) of "successful" IBI in muriquis is 3.01 ± 0.57 years ($n = 235$ IBIs, with successful IBI defined as encompassing the period between the birth of the first infant in an interval that survived until at least the birth of the second infant in the interval; median = 2.98; range = 1.30–5.06 years; based on biography data as described in Strier et al. 2010). However, the average IBI declined by more than 6 months (from 1205 days to 1015 days) during the 1987–2010 period when the population was expanding rapidly (Strier and Ives 2012). In contrast to the flexibility in average IBI (and its apparent role in regulating fertility), the average AFR during this same period of population expansion remained remarkably inflexible, despite the consistent pattern of high interindividual variation (mean = 8.88 years, range = 7.00–12.36 year; Strier and Ives 2012). Indeed, despite interindividual variation (which differs from flexibility; Strier 2017), there have been no detectable temporal shifts in AFR associated with the increases in population size that undoubtedly affect intra- and intergroup dynamics (Fig. 2.1).

The temporal inflexibility in AFR suggests it is a more conservative life history trait than IBI, at least in muriquis and possibly in other female-dispersing species (Strier 2008), with potentially interesting implications for understanding the ecological constraints that influence the initial growth-reproduction trade-off. This may explain the apparently conservative nature of AFR in female muriquis, which typically disperse prior to the onset of puberty and sexual activity, and may not reproduce until nearly 3 years after joining their new group (Strier and Ziegler 2000). This delay between dispersal and AFR does not appear to be related to prolonged stress because social assimilation of immigrant females rarely extends beyond their initial dispersal season (Printes and Strier 1999), and faecal cortisol levels were either not elevated or followed the seasonal pattern of elevation during the mating season exhibited by philopatric males (Strier and Ziegler 2000). Clearly, the relationship between dispersal regimes, which exhibit a strong phylogenetic signal in primates, and AFR merits closer scrutiny with long-term data sets from wild populations of primates in which females typically disperse (Di Fiore and Rendall 1994; Lee and Kappeler 2003; Lee and Strier 2015).

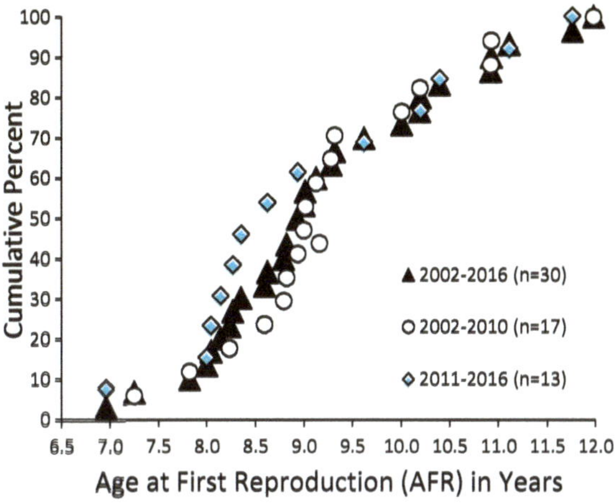

Fig. 2.1 Age at first reproduction. Solid triangles show the cumulative percentage of AFR for 30 females for which female birthdates were known to within 58 days (median = 2 days; range = 0–58 days) and first offspring birthdates were known to within 71 days (median = 8 days; range = 0–71 day) and whose first reproductions occurred between 2002 and 2016. Open circles show the cumulative percentage of 17 females with AFRs between 2002 and 2010. Grey diamonds show the cumulative percentage of 13 females with AFRs between 2011 and 2016. Overall ($n = 30$), mean ± sd AFR = 9.22 ± 1.26 years (median = 8.97 years; range = 6.96–11.98 years). Although the median AFR was about half a year earlier from 2011 to 2016 (median = 8.61 years) compared to 2002–2010 (median = 9.01 year), the differences were not significant (Mann Whitney $U = 94.5$, $p > 0.05$; $Z = 0.6487$, $p = 0.5157$)

2.4 Constraints on IBI Flexibility

Interspecific variation in IBI has generally been attributed to species-specific life histories, but intraspecific variation in IBI has received considerably more attention, in part because of its hypothesized links with individual attributes such as maternal rank, age, or parity, which also mediate the consequences of local, social, and ecological conditions (e.g. group size and resource availability, respectively) on female fertility (reviewed in Fedigan and Rose 1995; see also Fedigan et al. 2008). Nonetheless, in analyses of the sources of variation in the IBIs of sympatric capuchin monkeys, howler monkeys and spider monkeys at Santa Rosa National Park, Costa Rica, Fedigan and Rose (1995) found that the only variable that significantly influenced IBI in any of these species was the survival of the previous offspring.

The effect of infant mortality on IBI is nearly ubiquitous across the primate order, except in the case of strong reproductive seasonality, where the resumption of postpartum cycling in females is dependent on environmental cues or energetics instead of on the release from lactational amenorrhea. In strongly seasonal breeders, the timing of an infant's death relative to the timing of the mating season may dictate whether or not an offspring's death affects the mother's next IBI (Strier 2004; Brockman and van Schaik 2005; Gogarten et al. 2012).

Long-term studies of the Santa Rosa capuchin monkeys have revealed that infant survival is most affected by social factors such as turnover in male group membership and female social bonds (Fedigan et al. 2008; Brasington et al. 2017; Kalbitzer et al. 2017). The sensitivity of IBIs to these measures of social instability or stability is mediated by whether they decrease or increase infant survivorship, respectively, and is therefore likely to exhibit high variance among individuals and as social conditions change over time.

Interspecific comparisons of IBIs typically focus on the relatively invariant components of a species' life history (e.g. Lee and Kappeler 2003), whereas intraspecific comparisons are usually made by examining the variation in those components of IBIs that are free to vary among populations, groups, or individuals over the course of their lives. As depicted by Fedigan and Rose (1995), a primate's IBI can be partitioned into its gestational, lactational/anovulatory, and cycling phases, with gestation being the most conservative and invariant phase of the IBI. In Fedigan and Rose's comparisons of three sympatric primates at Santa Rosa National Park, Costa Rica, gestation accounted for 33% of the mantled howler monkey's 19.0-month IBI, 20% of the white-fronted capuchin monkey's 26.4-month IBI, and 22% of Geoffroy's spider monkey's 34.7-month IBI, leaving 67–80% of the IBI for the more variable, lactation/anovulatory, and cycling phases. Northern muriquis fall within this range, with their 7.2-month gestation (Strier and Ziegler 1997) accounting for 19.91% of the median 35.76-month IBI.

The initial phase of lactation could also be considered an invariant component of the IBI because most primate infants are entirely dependent on their mothers for nutrition during at least the first weeks of life (Lee 1996, 1999; Barrett et al. 2006a). Conversely, in some species, extended postweaning maternal investment may be necessary for offspring survival, with direct consequences for the corresponding IBIs. The exceptionally long IBIs of orangutans, for example, have been attributed to the extra years of maternal attention required before their offspring reach full ecological competence (van Noordwijk and van Schaik 2005). We have similarly proposed that muriquis may experience a secondary "weaning" period, corresponding to an offspring's achievement of full locomotor independence (Guedes et al. 2008). The timing of locomotor independence, which is constrained by body size, limb length, and neuromotor development, may be especially challenging to achieve in complex, three-dimensional arboreal habitats with uneven canopies. Variation in the timing of locomotor independence, as well as in the degree to which mothers adjust their own travel and locomotor patterns to accommodate their offspring's developing locomotor skills, may contribute to intraspecific differences in the timing of postpartum cycling and conception (Guedes et al. 2008).

The unexpected increase in female muriqui fertility in our study population, coincidental with the muriquis' expansion of the vertical niche to include a significant increase in time spent on the ground, raises intriguing possibilities about the flexibility of their IBIs (Strier and Ives 2012). In addition to expanding their nutritional base, terrestrial infant care (including infant carrying) may be less energetically costly, without risks of arboreal falls, resulting in faster resumption of postpartum cycling and/or shorter cycling-to-conception delays. Indeed, flexibility in maternal investment strategies, resulting in changes in IBIs, is expected to shift

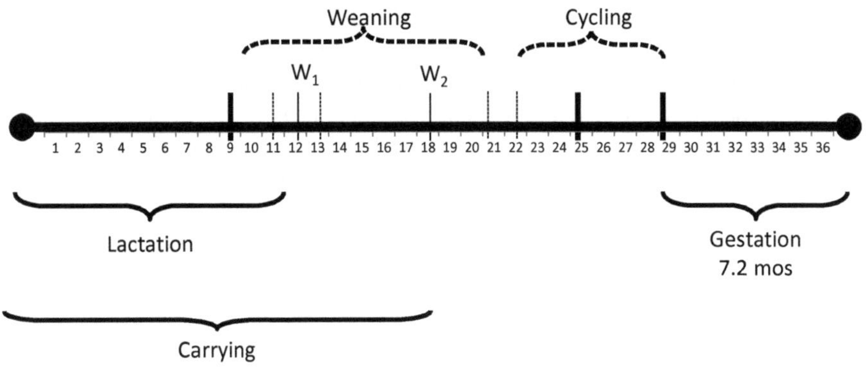

Fig. 2.2 Components of the muriqui birth interval. The first 9–12 months of lactation and gestation (7.2 months; Strier and Ziegler 1997) are considered to be relatively inflexible, while the lag between the resumption of ovarian cycling and conception (estimated by counting backward from birthdate) is subject to extrinsic, environmental stimuli. By contrast, the interval between weaning to solid foods (W_1) and weaning to independent locomotion (W_2) is more flexible and, when shortened, may result in shorter IBI. Female condition and seasonality may also affect the delay from weaning to cycling and the number of cycles to conception. Solid vertical lines indicate the timing of relatively invariant events; dotted lines indicate greater potential variance

with changes in the sources of infant mortality and the abilities of mothers to mediate these (Barrett et al. 2006b). Both the resumption of postpartum cycling and of cycling-to-conception delays involve components of the muriqui IBI that are potentially free to vary, at least within potential constraints imposed by their variable reproductive seasonality (Fig. 2.2).

2.5 Variance in Responses to Seasonality

Seasonal variation in the availability of preferred foods is known to contribute to both inter- and intraspecific differences in postpartum cycling and the number of cycles to conception. In extreme cases, seasonal food scarcities and negative energy balance can constrain reproduction altogether (Knott 1998; Emery Thompson et al. 2009, 2012). Interspecific comparisons often show variation in the degree of reproductive seasonality among sympatric species (e.g. sympatric capuchin monkeys, howler monkeys, and spider monkeys at Santa Rosa National Park, Costa Rica; Fedigan and Rose 1995). Nonetheless, the ability to respond to interannual variation in food availability is itself highly variable and likely to be species-specific.

Flexibility in the timing of cycling or conception, and ultimately, birth seasonality, should be predicated on the type of stimulus that triggers ovulation. Invariant cues, such as day length, should result in less flexibility, regardless of interannual variation in more variant extrinsic cues, such as climate and rainfall, or intrinsic cues, such as levels of fat storage, which vary in response to female energetic conditions at any particular time (e.g. Xiang et al. 2017).

At the RPPN-FMA in southeastern Brazil, muriqui births have been documented in every month of the year, but they are usually concentrated during the peak dry season months, from June to September. By contrast, sympatric brown howler monkey births occur year-round and without a seasonal peak (Strier et al. 2001b). With a 7.2 mo gestation (Strier and Ziegler 1997), this means that most muriqui conceptions occur during the peak rainy season months, from November to February (Fig. 2.3). Nonetheless, there is a significant interannual variation in rainfall patterns, and years with heavy precipitation over an extended rainfall season may contribute to extended or concentrated conception seasons and corresponding birth seasons (Fig. 2.4; Strier 1999b; Strier et al. 1999).

Such flexibility in reproductive seasonality is easy to explain in adaptive terms if extended conception seasons increase the chances that a female unable to conceive at the optimal time will nonetheless reproduce later that year instead of having to wait for the next breeding season. Flexibility in responsiveness to seasonality also implies that muriqui reproductive seasonality is not regulated by invariant cues such

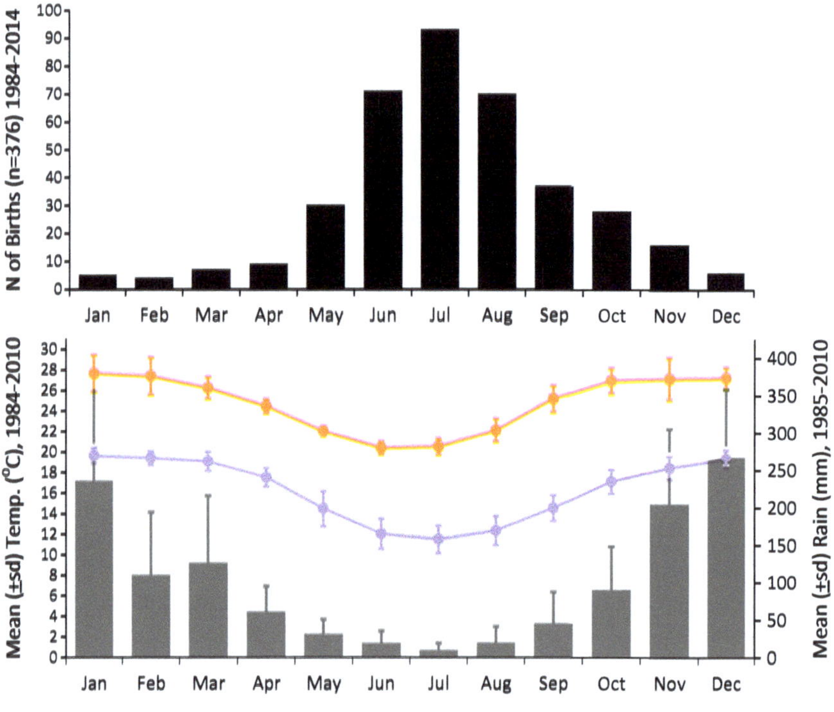

Fig. 2.3 Seasonality in rainfall and reproduction. Muriqui births (top figure, black bars) are concentrated during the cold, dry months (lower figure: monthly mean ± sd maximum temperature [upper line], minimum temperature [lower line], and rainfall [grey bars]). Birth data are based on all live births in each of the four study groups beginning with the first known birthdate in that group through September 2014, as follows: Matão group, from June 1984 (N = 184), Jaó group from September 2002 (N = 75), Nadir group from July 2003 (N = 66), and M2 group from September 2003 (N = 51)

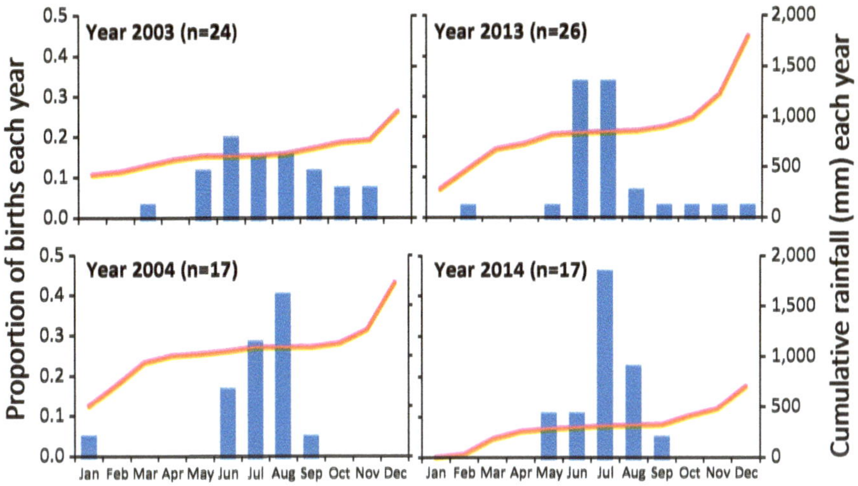

Fig. 2.4 Interannual variation in birth seasonality. Monthly distribution of births (bars, plotted as a proportion of the N of live births each year) relative to cumulative annual rainfall (lines). The concentration of births is influenced by both rainfall and the size of the birth cohort. (Updated from Strier 2000; Strier et al. 2001, b)

as day length. However, the distribution of conceptions in any particular year does not always align with rainfall during the preceding months. This suggests that muriqui lactation energetics may be more affected by environmental conditions than ovulation, as has been described in other primates (Lee 1996). Longer temporal patterns, perhaps encompassing the previous year, or other factors, such as habitat quality, anthropogenic pressures, or group size and population densities, may also influence interannual variation in the timing of conceptions and births (Campos et al. 2017).

The apparent flexibility of seasonal timing in muriqui conceptions and birth does not extend to the seasonal timing of female dispersal. Remarkably, 100% of the 43 females, whose natal group dispersal dates could be reliably determined, dispersed during the rainy season months from October through May (Strier et al. 2015). Their avoidance of the peak dry months, from June through September, is difficult to explain, although prime variables include seasonally limited food resources and more restricted ranging patterns, which would also reduce the frequency of inter-group encounters. Nonetheless, the apparent lack of flexibility in their dispersal seasonality is consistent with the phylogenetically conservative nature of primate dispersal regimes in general, and with the invariance of AFR, which is linked to female dispersal in muriquis. In other words, seasonal constraints on the timing of dispersal may constrain the variance in AFR.

2.6 Demographic Variation and Flexibility

Many primates exhibit high levels of variance in their group and population sizes (e.g. red lemurs, Erhart and Overdorff 2008; red colobus, Struhsaker 2008, Chapman and Chapman 2000, Gogarten et al. 2014; red howlers, Rudran and Fernandez-Duque 2003; white-faced capuchins, Fedigan and Jack 2012). Indeed, analyses of both longitudinal and cross-sectional data show that the coefficient of variation in group size increases with the number of observation points over time (Fig. 2.5; Strier et al. 2014). This variance may be attributable to any type of change in group or population size, including growth, decline, fluctuations, or shifts between periods of directional change and stability. For example, the capuchin monkey population at Santa Rosa National Park increased by more than 33% over an 8-year period from 1984 to 1992 (Fedigan et al. 1996) and continued to grow until the early 2000s (Fedigan and Jack 2012), but since then, it has remained comparatively stable in size (Campos et al. 2015). By contrast, one of the northern muriqui groups at the RPPN-FMA that has been monitored intensively since 1983 increased more than sevenfold over a 34-year period from 1982 to 2016 (Strier 2017), with corresponding growth across the rest of the population inhabiting this 1000 ha forest fragment (Strier and Ives 2012; Strier and Mendes 2012).

Demographic variation can have significant implications for both behavioural and demographic flexibility. For example, although the initial twofold increase in group size coincided with a corresponding increase in the muriqui's home range area and a shift from cohesive to fission-fusion grouping patterns (Strier et al. 1993), subsequent expansions of home range area in response to continued group size

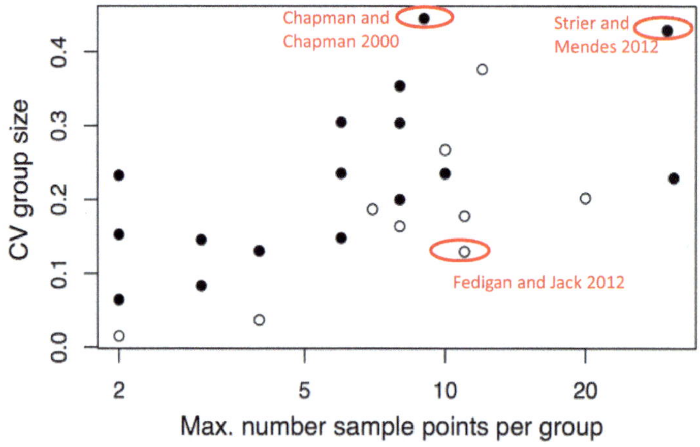

Fig. 2.5 Demographic variation of diverse primates. Variance in group size increases significantly with the number of data points (usually corresponding to duration of the study). Solid circles refer to study groups ($N = 16$ species); open circles refer to study populations ($N = 9$ species). Note:This figure is reproduced from Strier et al. (2014). Studies by authors also contributing to the present volume are highlighted with the reference to the data shown

increases would have been constrained by habitat saturation from comparable increases in the size of other muriqui groups in this isolated forest fragment and may have stimulated the expansion of their vertical niche.

Previous comparative analyses have shown that the demographic threshold for behavioural shifts from cohesive to fluid grouping patterns differs depending on dispersal regime, with female-biased dispersal, such as that of muriquis, and bisexual dispersal permitting greater social flexibility in response to demographic variation than female philopatry (Strier et al. 2014). However, whether dispersal regimes also differentially constrain demographic flexibility merits closer scrutiny, particularly for assessing both the resilience of populations and the factors influencing their potential for growth.

In their consideration of capuchin monkey population growth at Santa Rosa National Park, Fedigan and colleagues (1996) considered a number of ecological, demographic, and social factors (Fig. 2.6). These factors also frequently emerge in comparisons of interpopulation differences in primate group size (e.g. Struhsaker 2008) and in the ways in which factors such as food availability might affect changes in group sizes within a single population over time (Gogarten et al. 2015). Whereas intergroup variation provides insights into the correlated variables that affect demographic variation, intragroup variation provides insights into demographic flexibility and thus illustrates an important limitation of the more common "space-for-time" substitutions used in comparative analyses (reviewed in Strier 2017).

Additional distinctions among the factors affecting population growth can be made (Fig. 2.6). By definition, population growth is an example of demographic flexibility. Populations tend to grow in response to favourable ecological conditions, which can be considered as extrinsic stimuli that elicit the flexible, growth response. Demographic conditions, by contrast, may be both extrinsic and intrinsic, exerting

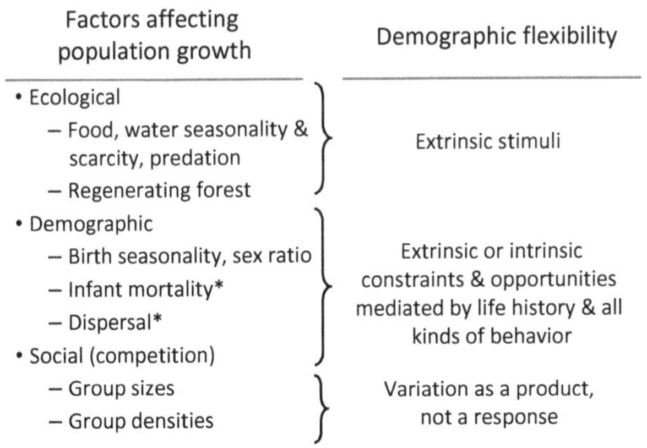

Fig. 2.6 Factors affecting population growth and demographic flexibility. Demographic variation is the product of flexibility in responses to extrinsic and intrinsic stimuli. The list of factors affecting growth is adapted from Fedigan et al. (1996); asterisks include male takeovers, which can contribute to infant mortality and thus affect birth intervals and reproductive rates. See Chapter by Jack and Fedigan (2018) in this volume

both constraints and opportunities on demographic flexibility depending on both behavioural flexibility and the extent to which components of life histories are free to vary in response to biological and ecological constraints. The dynamic interactions between ecological variables, such as food availability, and demographic variables, such as group size and densities, will affect levels of competition and cooperation and, thus, the demographic variation we observe. However, detecting variation in levels of demographic flexibility can only be done along a temporal scale.

2.7 Implications for Conservation

As Fedigan and colleagues (1996) wrote more than two decades ago, "the study of population dynamics of long-lived species in regenerating habitats can offer important insight into conservation research and behavioral ecology." Indeed, their discussion of the effects of habitat quality (including essential access to water) and forest regeneration anticipated the findings from their own long-term studies (Fedigan and Jack 2012; Campos et al. 2015; see also Chapter by Campos (2018) in this Volume) and from other long-term studies of other forest-dwelling primates around the world (e.g. red colobus monkeys, Uganda, Chapman and Chapman 2000, Gogarten et al. 2014; northern muriquis, Brazil, Strier 2000, Strier and Mendes 2012). Considering these comparative perspectives, the demographic variance we have documented in northern muriquis is only unusual because it has been caused mainly by group and population expansion, instead of by fluctuations between periods of growth, decline, and stability (Strier et al. 2006; Strier and Boubli 2006; Strier and Mendes 2012). Nonetheless, the population growth rate has been declining with increases in male-biased birth sex ratios and mortality, and it would have stopped growing entirely if fertility rates had not increased (Strier and Ives 2012).

A similar transition from population growth to stability has also been documented in analyses of demographic responses to climatic oscillations of the Santa Rosa capuchin monkey population (Campos et al. 2015).Comparative analyses of these and six other primate populations indicate that although climate change may have little effect on survival, warmer temperatures are projected to negatively impact fertility in seasonal breeders, which include the northern muriqui (Campos et al. 2017). Whether they possess sufficient behavioural and demographic flexibility to respond to these new challenges remains to be seen.

Acknowledgements I thank Jeremy Hogan and other members of the planning committee for inviting me to participate in the Festschrift honouring Linda Fedigan in December 2016, where an earlier version of this paper was presented. I am also grateful to the editors of this volume and especially my editor, Kathy Jack, and to Louise Barrett and Phyllis Lee for their helpful comments as external reviewers. The muriqui data presented in this chapter were collected with permission from the Brazilian government (CNPq) and the Sociedade Preserve Muriqui, in collaboration with Dr. Sérgio Lucena Mendes, and with support from Conservation International and CI-Brasil. Many funding sources have financed parts of the fieldwork, and many people have contributed to the long-term demographic data; please see these acknowledgements at my website, https://strierlab.anthropology.wisc.edu/muriqui-project-of-caratinga-acknowledgements/.

References

Alberts SC, Altmann J, Brockman DK et al (2013) Reproductive aging patterns in primates reveal that humans are distinct. Proc Natl Acad Sci U S A 110(33):13440–13445

Aureli F, Schaffner CM, Boesch C et al (2008) Fission-fusion dynamics: new research frameworks. Curr Anthropol 49:627–654

Barrett L, Halliday J, Henzi SP (2006a) The ecology of motherhood: the structuring of lactation costs by chacma baboons. J Anim Ecol 75:875–886

Barrett L, Henzi SP, Lycett JE (2006b) Whose life is it anyway? Maternal investment, developmental trajectories, and life history strategies in baboons. In: Swedell L, Leigh SR (eds) Reproduction and fitness in baboons: behavioral, ecological, and life history perspectives. Springer, New York, pp 199–224

Boubli JP, Couto-Santos F, Strier KB (2011) Structure and floristic composition of one of the last forest fragments containing the critically endangered northern muriqui (*Brachyteles hypoxanthus*, primates). Ecotropica 17(2):53–69

Brasington LF, Wikberg EC, Kawamura S et al (2017) Infant mortality in white-faced capuchins: the impact of alpha male replacements. Am J Primatol. https://doi.org/10.1002/ajp.22725

Brockman DK, Van Schaik CP (2005) Seasonality and reproductive function. In: Brockman DK, Van Schaik CP (eds) Seasonality in primates: studies of living and extinct human and non-human primates. Cambridge University Press, Cambridge, pp 269–309

Bronikowski AM, Cords M, Alberts SC et al (2016) Female and male life tables for seven wild primate species. Sci Data 3:160006

Campos FA (2018, this volume) A synthesis of long-term environmental change in Santa Rosa, Costa Rica. In: Kalbitzer U, Jack KM (eds) Primate life histories, sex roles, and adaptability - essays in honour of Linda M. Fedigan. Developments in primatology: progress and prospects. Springer, New York, pp 331–354

Campos FA, Jack KM, Fedigan LM (2015) Climate oscillations and conservation measures regulate white-faced population growth and demography in a regenerating tropical dry forest in Costa Rica. Biol Conserv 186:204–221

Campos FA, Morris WF, Alberts SC et al (2017) Does climate variability influence the demography of wild primates? Evidence from long-term life-history data in seven species. Glob Change Biol. https://doi.org/10.1111/gcb.13754

Chapman CA, Chapman LJ (2000) Constraints on group size in redtail monkeys and red colobus: testing the generality of the ecological constraints model. Int J Primatol 21:565–585

Chapman CA, Rothman JM (2009) Within-species differences in primate social structure: evolution of plasticity and phylogenetic constraints. Primates 50(1):12–22

Chaves PB, Alvarenga CS, Possamai CB et al (2011) Genetic diversity and population history of a critically endangered primate, the northern muriqui (*Brachyteles hypoxanthus*). PLoS ONE 6(6):e20722

Di Fiore A, Rendall D (1994) Evolution of social organization: a reappraisal for primates by using phylogenetic methods. Proc Natl Acad Sci U S A 91:9941–9945

Dias LG, Strier KB (2003) Effects of group size on ranging patterns in *Brachyteles arachnoides hypoxanthus*. Int J Primatol 24:209–221

Dingemanse NJ, Wolf M (2013) Between-individual differences in behavioural plasticity within populations: causes and consequences. Anim Behav 85(5):1031–1039

Dobson AP, Lyles AM (1989) The population dynamics and conservation of primate populations. Conserv Biol 3:362–380

Emery Thompson M, Muller MN, Wrangham RW et al (2009) Urinary C-peptide tracks seasonal and individual variation in energy balance in wild chimpanzees. Horm Behav 55(2):299–305

Emery Thompson M, Muller MN, Wrangham RW (2012) The energetics of lactation and the return to fecundity in wild chimpanzees. Behav Ecol 23(6):1234–1241

Erhart EM, Overdorff DJ (2008) Population demography and social structure changes in *Eulemur fulvus rufus* from 1988 to 2003. Am J Phys Anthropol 136:183–193

Fedigan LM, Jack KM (2012) Tracking neotropical monkeys in Santa Rosa: lessons from a regenerating Costa Rican dry forest. In: Kappeler PM, Watts DP (eds) Long-term field studies of primates. Springer-Verlag, Berlin, pp 165–184

Fedigan LM, Rose LM (1995) Interbirth interval variation in three sympatric species of neotropical monkey. Am J Primatol 37:9–24

Fedigan LM, Rose LM, Avila RM (1996) See how they grow: tracking capuchin monkey (*Cebus capucinus*) populations in a regenerating Costa Rican dry forest. In: Norconk MA, Rosenberger AL, Garber PA (eds) Adaptive radiations of neotropical primates. Plenum Press, New York, pp 289–307

Fedigan LM, Carnegie SD, Jack K (2008) Predictors of reproductive success in female white-faced capuchins (*Cebus capucinus*). Am J Phys Anthropol 137:82–90

Forsman A (2015) Rethinking phenotypic plasticity and its consequences for individuals, populations and species. Heredity 115(4):276–284

Franklin IR (1980) Evolutionary change in small populations. In: Soule ME, Wilcox BA (eds) Conservation biology, an evolutionary-ecological perspective. Sinauer Associates, Sunderland, pp 135–149

Gillespie TR, Chapman CA (2001) Determinants of group size in the red colobus monkey (*Procolobus badius*): an evaluation of the generality of the ecological-constraints model. Behav Ecol Sociobiol 50:329–338

Gilpin ME, Soule ME (1986) Minimum viable populations: processes of species extinction. In: Soule ME (ed) Conservation biology: the science of scarcity and diversity. Sinauer Associates, Sunderland, pp 19–34

Gogarten JF, Brown LM, Chapman CA et al (2012) Seasonal mortality patterns in non-human primates: implications for variation in selection pressures across environments. Evolution 66(10):3252–3266

Gogarten JF, Bonnell TR, Brown LM et al (2014) Increasing group size alters behavior of a folivorous primate. Int J Primatol 35(2):590–608

Gogarten JF, Jacob AJ, Ghai RR et al (2015) Group size dynamics over 15+ years in an African forest primate community. Biotropica 47:101–112

Guedes D, Young RJ, Strier KB (2008) Energetic costs of reproduction in female northern muriquis, *Brachyteles hypoxanthus* (Primates: Platyrrhini: Atelidae). Rev Brasil Zool 25:587–593

Isbell LA (1991) Contest and scramble competition: patterns of female aggression and ranging behavior among primates. Behav Ecol 2:143–155

Jack KM, Fedigan LM (2018, this volume) Alpha male capuchins (*Cebus capucinus imitator*) as keystone individuals in: Kalbitzer U, Jack KM (eds) Primate life histories, sex roles, and adaptability - essays in honour of Linda M. Fedigan. Developments in primatology: progress and prospects. Springer, New York, pp 91–109

Kalbitzer U, Bergstrom ML, Carnegie SD et al (2017) Female sociality and sexual conflict shape offspring survival in a neotropical primate. Proc Natl Acad Sci U S A 114(8):1892–1897

Kamilar JM, Baden AL (2014) What drives flexibility in primate social organization? Behav Ecol Sociobiol 68:1677–1692

Kappeler PM, Kraus C (2010) Levels and mechanisms of behavioural variability. In: Kappeler PM (ed) Animal behaviour: evolution and mechanisms. Springer, New York, pp 655–684

Knott CD (1998) Changes in orangutan caloric intake, energy balance, and ketones in response to fluctuating fruit availability. Int J Primatol 19(6):1061–1079

Lee PC (1996) The meanings of weaning: growth, lactation and life history. Evol Anthropol 5:87–96
Lee PC (1999) Comparative ecology of postnatal growth and weaning among haplorhine primates. In: Lee PC (ed) Comparative primate socioecology. Cambridge University Press, Cambridge, pp 111–136
Lee PC, Kappeler P (2003) Socio-ecological correlates of phenotypic plasticity in primate life history. In: Kappeler PM, Pereira M (eds) Primate life histories. University of Chicago Press, Chicago, pp 41–65
Lee PC, Strier KB (2015) Complexities of understanding female dispersal in primates. In: Furuichi T, Aureli F, Yamagiwa J (eds) Dispersing primate females. Springer, Tokyo, pp 215–230
Maldonado-Chaparro AA, Read DW, Blumstein DT (2017) Can individual variation in phenotypic plasticity enhance population viability? Ecol Model 352:19–30
Mendes SL, Melo FR, Boubli JP et al (2005) Directives for the conservation of the northern muriqui, *Brachyteles hypoxanthus* (Primates, Atelidae). Neotrop Primates 13(Supplement):7–18
Moore J (1999) Population density, social pathology, and behavioral ecology. Primates 40:5–26
Morris WF, Altmann J, Brockman DK et al (2011) Low demographic variability in wild primate populations: fitness impacts of variation, covariation, and serial correlation in vital rates. Am Nat 177(1):E14–E28
Myers N, Mittermeier RA, Mittermeier CG et al (2000) Biodiversity hotspots for conservation priorities. Nature 403:853–858
Printes RC, Strier KB (1999) Behavioral correlates of dispersal in female muriquis (*Brachyteles arachnoides*). Int J Primatol 20:941–960
Rudran R, Fernandez-Duque E (2003) Demographic changes over thirty years in a red howler population in Venezuela. Int J Primatol 24:925–947
Saccheri I, Hanski I (2006) Natural selection and population dynamics. Trends Ecol Evol 21:341–347
Saether BE, Coulson T, Grotan V et al (2013) How life history influences population dynamics in fluctuating environments. Am Nat 182(6):743–759
Shaffer ML (1981) Minimum population sizes for species conservation. Bioscience 31:131–134
Snell-Rood EC (2013) An overview of the evolutionary causes and consequences of behavioural plasticity. Anim Behav 85(5):1004–1011
Strier KB (1999a) Faces in the forest: the endangered muriqui monkeys of Brazil. Harvard University Press, Cambridge, MA
Strier KB (1999b) Predicting primate responses to "stochastic" demographic events. Primates 40:131–142
Strier KB (2000) Population viabilities and conservation implications for muriquis (*Brachyteles arachnoides*) in Brazil's Atlantic forest. Biotropica 32:903–913
Strier KB (2003) Demography and the temporal scale of sexual selection. In: Jones CB (ed) Sexual selection and reproductive competition in primates: new perspectives and directions. American Society of Primatologists, Norman, pp 45–63
Strier KB (2004) Reproductive strategies of New World primates: interbirth intervals and reproductive rates. In: Mendes SL, Chiarello AG (eds) A primatologia no Brasil-8. IPEMA/Sociedade Brasileiro de Primatologia, Vitoria/Espirito Santo, pp 53–63
Strier KB (2008) The effects of kin on primate life histories. Annu Rev Anthropol 37:21–36
Strier KB (2009) Seeing the forest through the seeds: mechanisms of primate behavioral diversity from individuals to populations and beyond. Curr Anthropol 50:213–228
Strier KB (2011) Social plasticity and demographic variation in primates. In: Sussman RW, Cloninger CR (eds) The origins and nature of cooperation and altruism in non-human and human primates. Springer, New York, pp 179–192
Strier KB (2017) What does variation in primate behavior mean? Am J Phys Anthropol 162(S63):4–14

Strier KB, Boubli JP (2006) A history of long-term research and conservation of northern muriquis (*Brachyteles hypoxanthus*) at the Estação Biológica de Caratinga/RPPN-FMA. Primate Conserv 20:53–63

Strier KB, Ives AR (2012) Unexpected demography in the recovery of an endangered primate population. PLoS One 7(9):e44407

Strier KB, Mendes SL (2012) The northern muriqui (*Brachyteles hypoxanthus*): lessons on behavioral plasticity and population dynamics from a critically endangered primate. In: Kappeler PM, Watts DP (eds) Long-term field studies of primates. Springer-Verlag, Berlin, pp 125–140

Strier KB, Ziegler TE (1997) Behavioral and endocrine characteristics of the reproductive cycle in wild muriqui monkeys, *Brachyteles arachnoides*. Am J Primatol 42(4):299–310

Strier KB, Ziegler TE (2000) Lack of pubertal influences on female dispersal in muriqui monkeys (*Brachyteles arachnoides*). Anim Behav 59:849–860

Strier KB, Mendes FDC, Rímoli J et al (1993) Demography and social structure in one group of muriquis (*Brachyteles arachnoides*). Int J Primatol 14:513–526

Strier KB, Ziegler TE, Wittwer DJ (1999) Seasonal and social correlates of fecal testosterone and cortisol levels in wild male muriquis (*Brachyteles arachnoides*). Horm Behav 35(2):125–134

Strier KB, Carvalho DS, Bejar NO (2001a) Prescription for peacefulness. In: Aureli F, de Waal FBM (eds) Natural conflict resolution. Harvard University Press, Cambridge, MA, pp 315–317

Strier KB, Mendes SL, Santos RR (2001b) Timing of births in sympatric brown howler monkeys (*Alouatta fusca clamitans*) and northern muriquis (*Brachyteles arachnoides hypoxanthus*). Am J Primatol 55(2):87–100

Strier KB, Boubli JP, Possamai CB et al (2006) Population demography of northern muriquis (*Brachyteles hypoxanthus*) at the Estação Biológica de Caratinga/Reserva particular do Patrimônio natural-Feliciano Miguel Abdala, Minas Gerais, Brazil. Am J Phys Anthropol 130:227–237

Strier KB, Altmann J, Brockman DK et al (2010) The primate life history database: a unique shared ecological data resource. Methods Ecol Evol 1:199–210

Strier KB, Chaves PB, Mendes SL et al (2011) Low paternity skew and the influence of maternal kin in an egalitarian, patrilocal primate. Proc Natl Acad Sci U S A 108(47):18915–18919

Strier KB, Lee PC, Ives AR (2014) Behavioral flexibility and the evolution of primate social states. PLoS One 9(12):e114099

Strier KB, Possamai CB, Mendes SL (2015) Dispersal patterns of female northern muriquis: implications for social dynamics, life history, and conservation. In: Furuichi T, Aureli F, Yamagiwa J (eds) Dispersing primate females. Springer, Tokyo, pp 3–22

Struhsaker TT (2008) Demographic variability in monkeys: implications for theory and conservation. Int J Primatol 29:19–34

Tabacow FP, Mendes SL, Strier KB (2009) Spread of a terrestrial tradition in an arboreal primate. Am Anthropol 111:238–249

Tokuda M, Boubli JP, Mourthé I et al (2014) Males follow females during fissioning of a group of northern muriquis. Am J Primatol 76:529–538

van Noordwijk MA, van Schaik CP (2005) Development of ecological competence in Sumatran orangutans. Am J Phys Anthropol 127:79–94

Wong BBM, Candolin U (2015) Behavioral responses to changing environments. Behav Ecol 26(3):665–673

Xiang Z, Yang W, Qi X et al (2017) An examination of factors potentially influencing birth distributions in golden snub-nosed monkeys (*Rhinopithecus roxellana*). PeerJ 5:e2892

Chapter 3
Post-Fertile Lifespan in Female Primates and Cetaceans

Mary S. M. Pavelka, Lauren J. N. Brent, D. P. Croft, and Linda M. Fedigan

Abstract Popular and scientific interest in menopause in humans has led to an increased interest in the extent of post-fertile life in other animals, particularly in long-lived social species such as other primates and cetaceans. Information on maximum lifespan achieved and age at last birth are available from long-term observations of known individuals from 11 primate species in the wild. Comparable information from wild cetaceans are more difficult to obtain; however there are relevant fisheries data, as well as a small number of long-term individual-based studies. Using post-reproductive representation (PrR) as a population measure of post-fertile lifespan that allows comparisons across populations and species, this review confirms that among primates, only humans have a maximum lifespan significantly longer than 50 years, and only human female life history includes a significant post-fertile stage of life. We conclude that although a prolonged post-fertile stage of life is very rare in mammals, it does occur in some exceptionally long-lived taxa, such as humans and resident killer and short-finned pilot whales. Thus menopause evolved independently at least three times in mammals, and the reasons for its evolution may differ in different lineages.

Keywords Evolution of menopause · Whale menopause · Post-fertile lifespan primates

M. S. M. Pavelka (✉) · L. M. Fedigan
Department of Anthropology and Archaeology, University of Calgary, Calgary, AB, Canada
e-mail: pavelka@ucalgary.ca; fedigan@ucalgary.ca

L. J. N. Brent · D. P. Croft
Department of Psychology, Washington Singer Laboratories, University of Exeter, Exeter, UK
e-mail: L.J.N.Brent@exeter.ac.uk; D.P.Croft@exeter.ac.uk

U. Kalbitzer, K. M. Jack (eds.), *Primate Life Histories, Sex Roles, and Adaptability*, Developments in Primatology: Progress and Prospects,
https://doi.org/10.1007/978-3-319-98285-4_3

3.1 Introduction

Popular and scientific interest in menopause in humans has led to increased interest in the extent of post-fertile life in other animals, particularly other primates and whales, as whales are another group for whom menopause has been reported (Croft et al. 2015). The focus on these taxonomic groups has arisen, in part, because both contain species in which females are long lived and because of the increasing availability of robust life history data from a range of wild populations. From an evolutionary life history perspective, the important features of menopause are its universality (e.g. all women who live into their late 50s will experience it) and its timing relative to the maximum lifespan of the species (Pavelka and Fedigan 1991; Croft et al. 2015). Regardless of local average life expectancy values, which is highly variable depending on local conditions, *Homo sapiens* are known to be able to live to past 100 years, and thus the longest-lived women will live 50% of their lives in a post-fertile state, and many will live two to four decades in this "sterile" state. Menopause is sometimes erroneously assumed to be an artefact of modernization; however this is based on confusing average lifespan (aka life expectancy), which is highly variable at the population level, with maximum lifespan, which is a species-level ability to maintain life and set by the genome (Pavelka and Fedigan 1991).

At first glance, menopause in captive nonhuman primates appears to be widespread; however on closer examination, these studies simply describe reproductive senescence (a variable decline in reproduction found mostly in very old individuals), not menopause (a species-wide cessation of ovarian cycling that occurs well before somatic senescence). Studies of captive primates, most commonly of macaques (*Macaca*) and chimpanzees (*Pan*), have indicated a decline in ovarian function in individuals who have lived to close to the maximum known lifespan for their species. While there may be important differences in ovarian ageing between Neotropical and Old World primates (Tardif and Ziegler 1992; Tardif et al. 2008), most studies suggest physiological parallels between the hormonal events associated with menopause in women and the reproductive function of older captive primates: failure to ovulate, low oestrogen levels, prolonged follicular phases, and high plasma LH concentrations, for example (e.g. Tardif and Ziegler 1992; Nozaki et al. 1995; Gilardi et al. 1997; Bellino and Wise 2003; Atsalis et al. 2008). However, these laboratory studies generally feature a very small number of very old individuals that may live far longer than individuals in the wild and cannot be considered to describe population-wide traits of evolutionary relevance. Data from wild populations of nonhuman primates reveal the distinctiveness of human female reproductive ageing, with the vast majority of females in wild populations of nonhuman primates dying before they reach the point of reproductive senescence (see below).

Based on research on wild populations to date, two species of toothed whale exhibit a prolonged period of post-reproductive life comparable to humans – the resident ecotype of the killer whale (*Orcinus orca*) and short-finned pilot whale (*Globicephala macrorhynchus*). In the wild, resident killer whale females breed between the ages of 10 and 40 but can live past 90, while short-finned pilot whales breed from 7 to 35 years of age and can live for over 60 years (Marsh and Kasuya 1986; Olesiuk et al. 1990; Croft et al. 2015). Resident killer whales are just one of a number of killer whale ecotypes that differ in their prey specializations, morphology, and behaviour and represent genetically and phenotypically distinct populations. Short- and long-finned pilots are more closely related to each other than either are to killer whales, with the three last sharing a common ancestor in the Miocene.

The current absence of detailed life history information on the other killer whale ecotypes precludes our ability to determine whether menopause is a species-wide trait or one that is confined to the resident killer whales.

In this chapter, we review what is known about reproductive senescence in female primates and cetaceans, focusing on field studies of reproductive history to explore the proportion of the lifespan that is lived after the production of the final infant. Individual cases of old female animals that have ceased to produce infants close to the end of their lifespans are not evidence of a species-wide life history characteristic – for the latter, population-level data are needed. Unfortunately, only a small number of studies on long-lived species with indivdiuals of known age and reproductive history are available for wild animal populations. Life history data for nonhuman primates and cetaceans are especially difficult to obtain given their longevity relative to the age of most research projects. Hormonal profiles, ovarian histology, and direct measures of menstrual activity are not available for most free-ranging primates and cetaceans, although this might well be expected to change as noninvasive methods for sampling are further developed and enhanced. But even so, to address the question of reproductive senescence, these physiological data need to be collected from populations (not just individuals) of known age. Additionally, the very reason that these animals are of particular interest – their long lifespans – means that aged animals followed by researchers from birth are rare. Fortunately, the number of field-based studies that have endured for enough time to be classified as "long term" has increased in recent years (Alberts et al. 2013). Here, we harness information from these studies and ask: just how prevalent is a post-fertile stage of life in primates and cetaceans?

3.2 Reproductive Data from Free-Ranging Individuals

3.2.1 Nonhuman Primates

Preliminary assessments of the life history patterns in free-ranging primates and cetaceans have tended to focus on descriptive values regarding the maximum lifespan achieved, as well as age at last birth. These data allow researchers to describe how close to the end of the *maximum* lifespan final births occur in a species – keeping in mind that the oldest known individuals are almost certainly outliers and that most females die well before this age. Below we describe some of the data that are available for wild primates and cetaceans.

Information on maximum lifespan achieved, as well as age at last birth, are available from individual-based observations of wild populations from at least 11 primate species spanning the taxonomic range (summarized in Fedigan and Pavelka 2011). Although not much has been published on the age at last birth and maximum lifespan for prosimian species, we do have data from the Beza Mahafaly Sifaka Research Consortium indicating that in a population of 219 wild *Propithecus verreauxi*, the oldest known female died at an estimated age of 31 (Bronikowski et al. 2011) and the maximum age at which a female gave birth was 28 years (Alberts et al. 2013). Thus female sifakas can continue to be reproductive up until the age at which time they have completed 90% of their lifespan. As an example from New World monkeys, LMF's 30-year life history study of Costa Rican capuchins (*Cebus capucinus*) also provides evidence of continued reproduction throughout the lifespan. In a sample of 45 females, the oldest individual lived to be 27 years of age, and the maximum age at which a female gave birth was 25, indicating these monkeys can produce infants up until they have completed 93% of their lifespans. And a study sample of 69 Brazilian muriquis revealed that females can live to an estimated 41 years and produce infants until they are 38 (thus up to 93% of their lifespans) (Bronikowski et al. 2011; Alberts et al. 2013; Stier 2018, this volume).

For Old World monkeys, Sommer et al. (1992), Borries et al. (1991), and Koenig et al. (1997) report that in a wild langur group (*Presbytis entellus*) at Jodhpur, India, the oldest female is estimated to have lived to 34 years and the oldest birth occurred in a female aged 32, suggesting that some females can produce infants until they have completed 94% of the species lifespan. These langur researchers also describe a small number of females who lived between 3 and 9 years before death without producing a new infant, suggesting that some idiosyncratic, as opposed to population-wide, termination of reproductive ability may be occurring in this population.

Other Old World monkeys for whom such data are available are baboons and macaques. Alberts and Altmann's data on the Amboseli baboons show that the oldest female in their sample of 211 *Papio cynocephalus* lived to be 28 years and the oldest recorded birth was to a female 24 years of age. Thus, female yellow baboons can continue to produce infants up to the point at which they have completed 86% of their lifespan (Alberts et al. 2013). A population of 207 semi-free-ranging provisioned Barbary macaques (*Macaca sylvanus*), studied over an 11-year period, also showed idiosyncratic variation in the age at which individuals appeared to stop reproducing, but from a species life history perspective, the oldest birth was to a 28-year-old female, and the maximum lifespan in this colony was age 30, thus up to 93% of the lifespan (Paul et al. 1993; Paul 2005).

Japanese monkeys (*Macaca fuscata*) are perhaps the most studied with respect to old age reproduction. Based on demographic data collected since 1954 on the Arashiyama population near Kyoto, the oldest birth was in a 25-year-old female, and the longest-lived female was 33 (Takahata et al. 1995). Pavelka and Fedigan (1999) conducted an extensive analysis of the reproductive histories of 95 females from the Arashiyama West population that had been documented from birth to death. Using Caro et al.'s (1995) interbirth interval criterion for reproductive termination (when the time lag between the last parturition and the death of the mother exceeds two standard deviations of the female's own mean lifetime interbirth interval), they identified that 20 of 70 females eligible for the criterion test were reproductively terminated. While these results could be interpreted as evidence of a phenomenon analogous to human menopause occurring in Japanese monkeys, closer examination of the data revealed that continued reproduction to death is characteristic of the vast majority of females of all ages. This is because a few females of all ages "tested positive" for being post-fertile according to Caro et al.'s (1995) criterion, but these cases were rare at every age. Even between the ages of 20 and 25 (an age category uniformly regarded as old for Japanese macaques), 81% of females were still reproductive. There are no records of any female Japanese macaque ever giving birth after the age of 25. These findings parallel those of Takahata et al. (1995) for Japanese macaques and Walker (1995), Walker and Herndon (2008), and Johnson and Kapsalis (1998, 2008) for the closely related rhesus macaque (*Macaca mulatta*). It is important to note that most macaque females are dead long before they have an opportunity to reach the population-wide cessation of infant production at age 25. Even though they live with almost no predation, less than 3% of the Arashiyama population of provisioned females that survived to reproductive age went on to live to age 25. Using the equation SpeciesA Age X SpeciesA Maximum Lifespan/SpeciesB Age X SpeciesB Maximum Lifespan, a 25-year-old monkey is between 83 and 95 years of age in human terms, depending on whether the human maximum lifespan used is 122 years (https://en.wikipedia.org/wiki/Jeanne_Calment) or the more conservative and commonly used round number of 100 years (Pavelka and Fedigan 1999). As one would expect, those female macaques that did survive to 25 years were visibly aged and showed clear outward signs of advanced age deterioration (somatic senescence).

Great apes are the longest lived of the primates, and here too the evidence strongly suggests that reproductive lifespan is very close to maximum lifespan, especially given the long interbirth intervals in apes. Wich et al. (2004) reported life history data for a wild Sumatran orangutan (*Pongo abelii*) population studied for a 32-year period and found no evidence for reproductive termination and a post-fertile life stage. Maximum lifespan for females was estimated to be 53 years of age, and the maximum age at which a birth was recorded was 50, meaning that almost 95% of the lifespan was completed before the time of last birth. Given that the interbirth interval in this population is greater than 9 years, a female would need to live significantly longer than 9 years after her last offspring before she could be assumed to be post-fertile.

Mountain gorillas (*Gorilla beringei beringei*) have a shorter lifespan, with the maximum lifespan of 43.7 years based on a sample of 66 females from the Virunga Volcano region studied from 1967 to 2004 (Robbins et al. 2006). The oldest female to give birth did so at 40.6 years of age, 93% of the way through the maximum lifespan. The average interbirth interval for these females was 4 years. Using a refinement of the Caro criterion, Robbins et al. identified five females who may have been reproductively terminated; however the post-reproductive lifespan represented only 1–3% of the total lifespan.

Chimpanzees (*Pan troglodytes*) are one of the most common laboratory primates for whom a post-fertile life stage has been reported in a small number of individuals (e.g. Graham 1979; Gould et al. 1981; Jones et al. 2007; Videan et al. 2006). However, none of the reported cases meet the criterion for a population-level cessation of reproductive function that exceeds the pace of general ageing. Thompson et al. (2007) analysed long-term fertility and mortality data in six wild chimpanzee populations and reported that chimpanzees are estimated to be able to live into their early 60s (64 years at Kanyawara and 66 at Ngogo; Wood et al. 2017) and reproduce until approximately 55 years of age. They reported that chimpanzee fertility decline with age is related to overall decline in survivorship and that healthy females maintain high birth rates late into life. Thus, they concluded that, in contrast to other claims, there was no evidence that a post-fertile phase is characteristic of chimpanzee female life history. Included in this data set was Jane Goodall's favourite, the long-lived and very famous female chimpanzee, Flo, from the Gombe Stream National Park in Tanzania. Flo died as an elderly matriarch at an estimated 53 years of age, having given birth to infant Flame at age 49.

3.2.2 Cetaceans

Compared to primates, life history data are arguably even more difficult to obtain from cetaceans due not only to their very long lifespans but also because they live in vast oceanic habitats that make them difficult to track and observe. As a result, and although there are some exceptions that we discuss in greater detail below, much of what we know about cetacean life history unfortunately come from fisheries-based sources or mass stranding events, rather than long-term observations of known individuals. These are not ideal sources of data for a number of reasons, including that maximum lifespans are probably underestimates because many of the oldest/largest animals will have been killed during the years of intensive whaling and because the ages of the animals are not based on longitudinal recordings of known individuals but are estimated from features such as body length or annual growth layers in dentine or cementum (Hamilton et al. 1998). Nevertheless, we can draw useful information regarding the broad patterns of reproduction in whales from these reports. The infraorder Cetacea is divided into two parvorders: the baleen whales (Mysticeti), best known for their baleen filter-feeding system, and the carnivorous toothed whales (Odontoceti). Below we describe the life history parameters for some species from both of these groups, starting with baleen whales and focusing on populations with some of the best described data.

In baleen whales, fin whales are known to survive into their 90s (Lockyer et al. 1979) with 1 female estimated to be as old as 111 (Mizroch 1981). Although the mean age at reproduction of fin whale mothers ranges between 23 and 37 years (Arrigoni et al. 2011), the oldest pregnant female observed to date was 76 years old (Mizroch 1981) or between 68% and 84% of the range of maximum lifespans. Bowhead whales are very long lived, with female maximum lifespan estimated to be over 100 years (George et al. 2004). Moreover, there is physiological evidence (corpora number) that females continue to ovulate in these very late years (George et al. 2004), suggesting that bowhead females are reproductively viable throughout their long lives. The North Atlantic right whale (*Eubalaena glacialis*) is one of the few species of baleen whale for which there is individual-based data, which has been generated using photo identification of known individuals. Seven known females were observed for over 30 years (Hamilton et al. 1998). The oldest of these females was first sighted in 1935 with a calf, and because the mean age at reproductive maturity in this species is 9.6 years, her age in 1995 was estimated as 69. This particular female did not have a calf in any follow-up sightings, which is probably because the sporadic timing of her sightings missed calving events, not because she was reproductively senescent (Hamilton et al. 1998). Of the other six individuals tracked in this population, the oldest female to give birth to a calf was 34 years old, or 52% of the maximum lifespan of this species.

Finer-grained details are known about the life history of toothed whales, including a greater number of individual-based studies and occasional data on the cessation of ovulation from necropsy reports. For example, fisheries data from the long-finned pilot whale, *Globicephala melaena*, demonstrate that females can live upwards of 59 years (Foote 2008; Bloch et al. 1993) with the oldest documented pregnancy occurring in a 55-year-old female (Foote 2008; Martin and Rothery 1993). Thus females of this species can give birth 93% of the way through their maximum lifespan. Necropsy data from fisheries show that less than 4% of mature females had ceased ovulation (Foote 2008; Martin and Rothery 1993), suggesting that somatic and reproductive senescence may be closely aligned in this species. These findings are in stark contrast to those from the closely related short-finned pilot whale, *Globicephala macrorhynchus. Females from this species cease* breeding at a maximum of 36 years of age but can live for over 60 years (Kasuya and Marsh 1984; Marsh and Kasuya 1986). The oldest female to give birth to a calf therefore lived 57.1% of the maximum lifespan of this species. Inspection of corpses from Japanese drive fisheries revealed that 25% of mature females had ceased ovulation, suggesting that short-finned pilot whales undergo physiological changes that preclude reproduction, similar to that which occurs in humans (Kasuya and Marsh 1984).

Evidence for a prolonged post-fertile lifespan in wild killer whales comes from long-term individual-based studies of two populations of resident killer whales in the Pacific waters off North America. These studies started in 1974 and consist of individual demographic data for more than 600 individuals (Olesiuk et al. 1990; Towers et al. 2015). The ages of individuals already alive in 1974 were necessarily estimated, with dates of birth for adult females estimated by subtracting 15 years (the mean age at first reproduction) from the estimated year of birth of her oldest offspring (offspring determined based on a behavioural association and, in some cases, genetic parentage assignment), which was assumed to be her first viable calf (Bigg et al. 1990). Using this method, many females estimated to live between 70 and 80 years (Franks et al. 2016). In these populations, the maximum documented age at parturition is 43 years old. As such, there are currently two mature females in the study populations who are believed to be over 70 years old, and to have last given birth prior to the start of the study in 1974, but who had not been observed to be pregnant or to give birth in the intervening 42 years. It is unknown if there are physiological changes that preclude reproduction in older female resident killer whales (Whitehead and Rendell 2014).

Although the life histories of killer and pilot whales have been documented in the greatest detail, information from a broader range of toothed whales exists. For example, female bottlenose dolphins (*Tursiops truncatus*) can live past 35 years, and pregnancies in these later years have been documented (Stolen and Barlow 2003; Kasuya et al. 1997). In sperm whales, the best data currently available on pregnancy come from fisheries data, where 22 females aged 40–61 years were examined. Of these, only one was pregnant (aged 41) and six were lactating. None of the remaining females were pregnant or ovulating (Best et al. 1984). Age estimates for sperm whales come from tooth layering techniques, and estimated maximum lifespan at between 60 and 70 years (Rice 1989), which has led to the suggestion that the sperm whale may exhibit menopause (Whitehead 2003).

3.3 Measuring Menopause as a Population-Wide Trait

Data on maximum lifespan and age of last birth presented above demonstrate that in most of the species for which information is available, the ability to reproduce is maintained until very late in the lifespan of the species. Thus, these field reports support what has long been asserted by a number of researchers, such as Gosden (1985), who described the "postmenopausal" phase of life in other primates as being short and confined to a few individuals of exceptional longevity, and Small (1982, 1984) who argued that most macaque females continue to cycle until death and that there is great variability in the age of those females that cease to cycle. Also, in 1991, after a careful analysis of all available field and captive data, Pavelka and Fedigan concluded that while a small number of individual monkeys and apes had indeed ceased reproducing or lost reproductive capacity, often in extreme old age, nothing comparable to the universal termination of female fertility halfway through the human lifespan had yet been found in other primates. Menopause in humans, resident killer whales, and short-finned pilot whales occurs in healthy middle age and is distinct from the overall senescence of the organism, and all females who live into the second half of their lifespan will experience it. Women beyond their late 50s (Hill and Hurtado 1991) and female resident killer whales beyond their early 40s are *all* post-fertile (Franks et al. 2016); there are no individuals continuing to reproduce for 70, 80, or 90% of the maximum lifespan, which, as we saw above, is true for virtually all other primates and cetaceans for whom such data exist. Reproductive cessation in nonhuman primates and in many cetaceans, when it occurs, is idiosyncratic, is not population-wide, and mostly occurs in aged individuals such that it is not distinct from the overall somatic senescence of the organism.

There is a tendency among some animal researchers to conclude that they have demonstrated something comparable to menopause when in fact they have demonstrated quite the opposite, from a species life history point of view. For example, Graham (1979:299) concluded that "chimpanzees aged 35 years of age and more show evidence of. .. reproductive senescence closely comparable to that seen during the human climacteric." However, most of the animals he reported upon were very old and in a state of advanced biological senescence, yet *most* of them continued to have primary follicles and to ovulate and cycle until death in old age. The interpretation of physiological changes in individual old females as evidence of menopause, despite continued reproduction until death in the majority of the study subjects, may be driven by the search for a biomedical laboratory model for studying menopause and the treatment of its symptoms in humans. A number of researchers have pointed out that the conflicting claims regarding menopause often rest on the significance attached to individual animals who appear to have ceased producing infants early enough to spend some portion of their lives in a post-fertile state (e.g. Pavelka and Fedigan 1991; Caro et al. 1995).

In an attempt to move beyond the confusion in the literature and to provide an independent assessment of age-related changes in reproductive function across species, Caro et al. (1995) introduced the interbirth interval criterion for identifying post-fertile females. Using demographic data, this method considers a female to be post-reproductive only when she has lived significantly longer (more than two standard deviations longer) than her own average lifetime interbirth interval without giving birth again. It helps to avoid considering females to be post-reproductive simply because they died in a normal interbirth interval. Caro et al. (1995) applied their method to the reproductive records of 14 nonhuman primates from captive breeding colonies in the USA and Europe. They argued that captive colonies were, like most contemporary humans, buffered to some extent from disease and from predation-induced mortality and thus may be more appropriate than comparison with wild primate populations. Additionally, appropriate longitudinal reproductive data on wild female primates were available for few, if any, primate species at the time. Caro et al. (1995) also examined human data from an eighteenth and nineteenth century rural peasant population in Northern Germany. The comparison of age-specific survivorship, fertility, and changes in interbirth intervals in these 15 primates (including humans) showed that in most species, individual females continue to reproduce throughout their lives, while other individuals terminate reproduction well before death. Humans were the exception. The average time between last birth and death values for nonhuman primates ranged from 2 to 6 years in the lemurs and monkeys, 4.5 years for gorillas, 7.1 years for orangutans, and 9.3 years for chimpanzees. However, these may be overestimates as the authors did not subtract the normal interbirth interval from the time period between the last parturition and death, meaning that these values include time during which a female was nursing a new infant, a time in which she would be better considered reproductive than post-reproductive.

Of course, from an evolutionary perspective, we need to know not just whether post-reproductive females are observed in a population but also with what frequency they occur. Measuring post-fertile periods as the time between the end of fertility and the end of life is inherently biased because it includes only individuals that survive after their last reproductive event and does not allow for generalizations to be made regarding the importance of post-fertile lifespan for the population/species (Croft et al. 2015; Levitis and Lackey 2011). Moreover, many measures of the length of post-fertile life tend to be correlated with longevity, making comparisons across species problematic. Recently, Levitis and Lackey (2011) proposed a new measure: post-reproductive representation (PrR) as a population measure of post-fertile lifespan that allows comparisons across populations and species. PrR is an unbiased population measure of reproductive lifespan and quantifies the proportion of females that are post-reproductive under idealized demographic conditions. PrR is calculated from life tables using data from two ages: Age B, the beginning of adulthood, defined as the point at which 5% of lifetime fecundity has been realized, on average, independent of mortality, and Age M, the end of the fecund lifespan, defined as the point at which 95% of lifetime fecundity has been realized, on average, independent of mortality. PrR represents the number of years an average newborn can expect to live as a post-reproductive adult divided by the number of years an average newborn can expect to live as an adult (Levitis and Lackey 2011).

In recent years, a group of ten researchers has collaborated to synthesize their long-term life history data on populations of wild primates that have been studied continuously for decades (e.g. Strier et al. 2010; Morris et al. 2011; Bronikowski et al. 2011). As a result, Alberts et al. (2013) were able to carry out a uniquely extensive comparative study of reproductive senescence in seven natural primate populations, including calculating PrR values (Table 3.1). They compiled individual mortality and reproduction data for 700 individually recognized adult female nonhuman primates that had been studied for more than 250 combined observation years. Human data from the! Kung population of the Kalahari Desert, who at the time of study practised no agriculture and had little contact with agricultural communities, were used to compare the rate of increase in the probability of death with the rate of increase in the probability of reproductive senescence. Thus they were able to test whether reproductive senescence occurs at the same or at a different pace than does general senescence in all seven species. Their results provided strong evidence that reproductive ageing in women is quite distinct from the pattern of human somatic senescence and that it is also distinct from the patterns of reproductive senescence found in a wide range of other primate species. In the nonhuman primates, PrR values ranged from 1% in baboons to 8% in muriquis (Table 3.1). Gorillas and chimpanzees were at 4% and 2%, respectively. In contrast, in the natural human population of Kalahari! Kung, 43% of adult female years are lived by post-fertile individuals. This cross-species analysis stands as the largest and longest-term study to date reporting on reproductive senescence and post-fertile lifespan in humans and other primates.

PrR values in cetaceans are more difficult to calculate due to a general absence of life tables based on observed individuals. However, the PrR is 28% in both short-finned pilot whales (Levitis et al. 2011) and in resident killer whales (Franks

Table 3.1 Published values of female post-reproductive representations (PrRs)

Species	Population	PrR	Reference
Homo sapiens	!Kung	0.425	Levitis et al. (2013)
Homo sapiens	Ache	0.439	Levitis et al. (2013)
Homo sapiens	Haiti 2002	0.460	Levitis et al. (2013)
Homo sapiens	Sweden 1751	0.477	Levitis et al. (2013)
Homo sapiens	Hadza	0.481	Levitis et al. (2013)
Blue monkey (*Cercopithecus mitis*)	Wild	0.02	Alberts et al. (2013)
Blue monkey (*Cercopithecus mitis*)	Wild	0.041	Levitis et al. (2013)
Chimpanzee (*Pan troglodyte*)	Wild	0.018	Levitis et al. (2013)
Chimpanzee (*Pan troglodyte*)	Wild	0.02	Alberts et al. (2013)
Gorilla (*Gorilla beringei*)	Wild	0.04	Alberts et al. (2013)
Hamadryas baboon (*Papio hamadryas*)	Wild	0.005	Levitis et al. (2013)
Muriquis (*Brachyteles hypoxanthus*)	Wild	0.06	Alberts et al. (2013)
Rhesus macaque (*Macaca mulatta*)	Wild	0.007	Levitis et al. (2013)
Sifaka (*Propithecus verreauxii*)	Wild	0.02	Alberts et al. (2013)
White-faced capuchin (*Cebus capucinus*)	Wild	0.04	Alberts et al. (2013)
Yellow baboon (*Papio cynocephalus*)	Wild	0.01	Alberts et al. (2013)
Short-finned pilot whale (*Globicephala macrorhynchus*)	Wild	0.28	Levitis and Lackey (2011)
Resident killer whales (*Orcinus orca*; *based on current age estimates*)	Wild	0.282, $P = 0.001$	

Also shown are the PrRs for the resident killer whales (*Orcinus orca*) calculated using unpublished data (Foster et al. 2012)
The table is adapted from Croft et al. (2015)

et al. 2016) (Table 3.1). These values are well above those reported for any wild nonhuman primate population (Alberts et al. 2013) and indicate that prolonged period of post-reproductive life is a population-level trait in these species.

3.4 The Evolution of Menopause

Menopause is a feature of the life history of all modern humans (*Homo sapiens*) and is not an artefact of modernization (Pavelka and Fedigan 1991; Caro et al. 1995; Fedigan and Pavelka 2011). This population-wide termination of female reproduction decades earlier than the maximum lifespan for a species is, at first pass, very difficult to explain from an evolutionary perspective. Quite simply it should be inherently disadvantageous for individuals to stop reproducing at midlife. Both theoretically and empirically, lifespan and reproductive lifespan are highly correlated in living organisms. It should be no surprise that outside of humans and the two species of toothed whales, no other animals are known to experience early cessation of fertility; instead they do what is predicted by evolutionary theory – they reproduce until close to the end of their lifespans. In the vast majority of species, natural selection clearly favours continued reproduction to the end of the lifespan. Why then has a substantial post-reproductive lifespan evolved in humans and two species of toothed whales?

One possibility is that menopause is a byproduct of the evolution of increasing lifespan in taxa in which the reproductive system could not "keep up" (Pavelka and Fedigan 1991; Fedigan and Pavelka 2011). Over the course of human evolution, the development of a postmenopausal stage of life in humans may be a byproduct of increases in our lifespan beyond the age of 50. Fifty is approximately the same age at which great ape females (chimpanzees, gorillas, orangutans) in the wild cease to give birth (e.g. Nishida et al. 2003; Wich et al. 2004; Robbins et al. 2006; Thompson et al. 2007) and is also close to the maximum lifespan of chimpanzees, our closest living relative. Cutler (1975) looked at the genetic complexity governing rate of ageing and showed that the "maximum lifespan potential" of humans doubled sometime over the past 3 million years. The evolution of the much longer lifespan of human is most likely tied to parallel increases in brain size in our lineage (Rose and Mueller 1998; Kaplan et al. 2000); however, other explanations have been offered. For example, Hawkes (2003) suggested that a novel helping role for post-reproductive grandmothers may explain the increased human lifespan, and Tuljapurkar et al. (2007) suggested that continued old-age male fertility provides a selective force increasing the human lifespan beyond the theoretically expected "wall of death". More recently, drawing from neuroscience, anthropology, and brain imaging research, Raichlen and Alexander (2014) proposed that the evolution of increased physical activity approximately 2 million years ago served to relax genetic constraints on ageing in human ancestors.

The failure of the reproductive system to maintain functionality along with this increased lifespan may be due to a phylogenetic constraint. As female mammals, women have a reproductive pattern referred to as semelgametogenesis: they are born with approximately 1 million oocytes and primordial follicles, which is all they will ever possess. The number of potential egg cells is fixed at birth and steadily declines thereafter, many to degeneration (von Saal et al. 1994), some spent in ovulation. Unlike male mammals who continue to produce new gametes throughout life, females do not. Oocytes and follicles are depleted throughout a woman's life until they reach a minimum threshold below which hormonal signals, and then cycling, begin to fail (Armstrong 2001). Oocyte depletion is widely regarded to be the starting point for a series of changes that underlie the menopausal transition in women (Pavelka and Fedigan 1991; Fedigan and Pavelka 2011). The reason for the semelgametogenesis pattern in female mammals may be antagonistic pleiotropy (e.g. Williams 1957; Rose 1991; Gosden and Faddy 1998): it may have high adaptive value early in the life course (intense reproductive output) and was selected for even if it results in reduced fitness (follicular depletion) later in the life course. In the wider research on somatic ageing, it is commonly accepted that antagonistic pleiotropy is a good explanation for the evolution of senescence in general (Wood et al. 2001). Additionally, in the case of oocyte depletion later in life, this effect would only have been experienced in the very longest-lived mammals. However, new data suggest that the number of gametes and/or the rate of their depletion is not fixed across all mammals. For example, female bowhead whales breed throughout their 100-year lifespan, and fin whale females have been observed breeding past the age of 70. It is also reported that African elephant females continue to reproduce until the end of their natural lifespans, which can be upwards of 65 years (Lee et al. 2016). These data support the suggestion by Gosden and Telfer (1987) that oocyte stocks may be variable, and if they are, we must again consider the possibility that rather than not "keeping up" with the extended somatic lifespan, direct opposing selection may have held back the female reproductive lifespan in some taxa (Cant et al. 2009).

The other possibility then is that menopause evolved because direct selection favoured females who stopped reproducing over those who continued, in the absence of a phylogenetic constraint on continued reproduction. The most common adaptive explanation is that menopause evolved due to the increase in inclusive fitness benefits accruing to those who did not produce any new offspring but instead invested in the care and provisioning of the close relatives they already had: existing offspring and grandoffspring. While some studies have found no evidence of inclusive fitness benefits for post-fertile mothers and grandmothers (Hill and Hurtado 1996; Rogers 1993), a number of others do show that females can positively impact the survival of offspring or grandoffspring (Shanley et al. 2007; Hawkes 2003; Hawkes and Coxworth 2013; Lahdenpera et al. 2004; Foster et al. 2012). However, inclusive fitness models have generated conflicting results with respect to whether the benefits older females gain by helping their relatives can outweigh the costs they incur by ceasing to reproduce (summarized in Croft et al. 2015). Moreover, while "helping" can select for longer lifespans, it cannot explain the cessation of reproduction, unless helping necessarily precludes females from reproducing (Croft et al. 2015). Additionally, evidence that post-fertile individuals currently provide benefits to their descendants is not automatically evidence for the origins of menopause (see Gould and Lewontin 1979). Being post-fertile in conjunction with living in societies that afford them the opportunity to direct fitness-enhancing behaviours to their descendants, human females may have been able to take advantage of their post-menopausal stage of life and create secondarily adaptive benefits from the phenomenon of midlife reproductive senescence.

This uncertainty in the literature has led some researchers to suggest that, in addition to the benefits of helping, the costs of continued reproduction must also be considered in adaptive scenarios for the evolution of menopause. Pavelka and Fedigan (2012) investigated the costs of continued reproduction in old Japanese monkey females and found no difference in body weight between females who ceased to reproduce and those who continued, and older females were no more likely to die in the 12 months after giving birth than were younger females. Additionally, a regression of interbirth interval on age at death revealed no tendency for females who produce infants in closer succession to die any earlier. However, of particular relevance to the evolution of menopause in both humans and whales, Cant and Johnstone (2008) and Johnstone and Cant (2010) have suggested that conflict over resources between breeding females is a critical missing term in previous models and that, together, the inclusive fitness benefits older females gain by helping their relatives combined with the costs of reproducing at the same time as younger female group members (the "reproductive conflict hypothesis") can explain the evolution of menopause. These costs are dependent on the kinship dynamics that are displayed by a species. In resident killer whales, for example, neither males nor females disperse from their natal group. As a result, mean level of relatedness to other group members increases as females age, driven by the fact that females start out with few related males in their groups but start to add closely related males as they produce sons (new related females are added as well but these simply replace older generations of females as they die). With relatedness to the group increasing with age, theoretical models show that older females not only benefit more from helping their group mates but also suffer greater costs from competing against them. That is, once you reach a certain age, continued reproduction leads to costly competition between your new offspring, your daughters, and your grandoffspring. This pattern is in direct contrast to most mammals, where males disperse away from their natal group and females' mean relatedness to the group remains constant, but is similar to what is predicted to occur in populations of humans with female-biased dispersal (which is arguably the human ancestral state): mothers are related to the offspring of their daughter-in-laws, and thus local relatedness increases with age for these females (summarized in Croft et al. 2015). Indeed, recent empirical results support this reproductive conflict hypothesis in humans and resident killers whales. Older females have been shown to have reduced fitness when they reproduce in conflict, i.e. the new offspring of older females are less likely to survive when they are born at the same time as the new offspring of younger (related) females (Croft et al. 2017; Lahdenpera et al. 2012; Strassman et al. 2011).

3.5 Summary

In this paper, investigating the evidence for post-fertile lifespans in primates and cetaceans, we have highlighted a number of important developments in our efforts to understand the evolution of menopause. First it is clear from the available data on, for example, the baleen fin and Bowhead whales, that reproduction well past the age of 50 is possible in some mammals and that there must be a more variable oocyte supply in different mammalian taxa than previously argued. The second important realization from our review is that although a post-fertile stage of life is very rare in mammals, it does occur in some exceptionally long-lived taxa, such as humans and two species of toothed whales. Short- and long-finned pilots are more closely related to each other than either are to killer whales, with the three last sharing a common ancestor in the Miocene, suggesting that menopause arose independently in killer and short-finned pilot whales. This means that menopause likely evolved independently at least three times in mammals. With this recognition, we must consider that the reasons for the evolution of it may differ in different lineages. It may be a byproduct of increased longevity or the result of direct selection in humans.

However, the evidence for continued reproduction into very old age in a number of whales, combined with mathematical models taking both costs and benefits into account, are showing that direct selection to cease reproduction is a strong possibility in species such as the resident killer whales.

Acknowledgements M.S.M. Pavelka's research is supported by ongoing Discovery Grants from the Natural Sciences and Engineering Research Council of Canada (NSERC). L.J.N.Brent is supported by an Early Career Fellowship from the Leverhulme Trust. D.P. Croft acknowledges funding from the NERC (NE/K01286X/1). L.M. Fedigan's research has been supported since 1983 by ongoing Discovery Grants from the Natural Sciences and Engineering Research Council of Canada (NSERC) and from 2002 to 2016 by the Canada Research Chairs Program. We are thankful to Dr. Katharine Jack and two anonymous reviewers for their helpful comments and suggestions on earlier versions of this manuscript.

References

Alberts SC, Altmann J, Brockman DK, Cords M, Fedigan LM, Pusey A et al (2013) Reproductive aging patterns in primates reveal that humans are distinct. Proc Natl Acad Sci USA 110(33):13440–13445

Armstrong DT (2001) Effects of maternal age on oocyte developmental competence. Theriogeneology 55(6):1303–1322

Arrigoni M, Manfredi P, Panigada S, Bramanti L, Santangelo G (2011) Life-history tables of the Mediterranean fin whale from stranding data. Mar Ecol 32:1–9

Atsalis S, Margulis SW, Hof PR (2008) Primate reproductive aging: cross-taxon perspectives. Karger Press, Basel

Bellino FL, Wise PM (2003) Nonhuman primate models of menopause workshop. Biol Reprod 68(1):10–18

Best PB, Canham PAS, Macleod N (1984) Patterns of reproduction in sperm whales, Physeter macrocephalus. Rep Int Whal Commn Spec 6:51–57

Bigg MA, Olesiuk PF, Ellis GM, Ford JKB, Balcomb KC (1990) Social organization and genealogy of resident killer whales (Orcinus orca) in the coastal waters of British Columbia and Washington state. Rep Int Whal Comm Spec 12:383–405

Bloch D, Lockyer C, Zachariassen M (1993) Age and growth parameters of the long-finned pilot whale off the Faroe Islands. Rep Int Whal Comm Spec 14:163–207

Borries C, Sommer V, Srivastave A (1991) Dominance, age, and reproductive success in free-ranging female Hanuman langurs (*Presbytis entellus*). Int J Primatol 12(3):231–257

Bronikowski AM, Altmann J, Brockman DK, Cords M, Fedigan LM, Pusey A et al (2011) Aging in the natural world: comparative data reveal similar mortality patterns across primates. Science 331(6022):1325–1328

Cant MA, Johnstone RA (2008) Reproductive conflict and the separation of reproductive generations in humans. Proc Natl Acad Sci USA 105(14):5332–5336

Cant MA, Johnstone RA, Russell AF (2009) Reproductive conflict and the evolution of menopause. In: Hager R, Jones CB (eds) Reproductive skew in vertebrates: proximate and ultimate causes. Cambridge Universtiy Press, Cambridge, pp 24–50

Caro TM, Sellen DW, Parish A, Frank R, Brown DM, Voland E, Borgerhoff Mulder M (1995) Termination of reproduction in nonhuman and human female primates. Int J Primatol 16(2):205–220

Croft DP, Brent LJN, Franks DW, Cant MA (2015) The evolution of prolonged life after reproduction. Trends Ecol Evol 30(7):407–416

Croft DP, Johnstone RA, Ellis S, Nattrass S, Franks DW, Brent LJ, Mazzi S, Balcomb KC, Ford JK, Cant MA (2017) Reproductive conflict and the evolution of menopause in killer whales. Curr Biol 27(2):298–304
Cutler RG (1975) Evolution of human longevity and the genetic complexity governing aging rate. Proc Natl Acad Sci 72(11):4664–4668
Fedigan LM, Pavelka MSM (2011) Menopause: interspecific comparisons of reproductive termination in female primates. In: Campbell CJ, Fuentes A, MacKinnon KC, Bearder SK, Stumpf RM (eds) Primates in perspective, 2nd edn. Oxford University Press, New York, pp 488–498
Foote AD (2008) Mortality rate acceleration and post-reproductive lifespan in matrilineal whale species. Biol Lett 4:189–191
Foster EA, Franks DW, Mazzi S, Darden SK, Balcomb KC, Ford JKB, Croft DP (2012) Adaptive prolonged postreproductive life span in killer whales. Science 337(6100):1313
Franks DW, Nattrass S, Brent LJ, Whitehead H, Foote AD, Mazzi S, Ford JK, Balcomb KC, Cant MA, Croft DP (2016) The significance of postreproductive lifespans in killer whales: a comment on Robeck et al. J Mammal 97(3):906–909
George J, Follmann E, Zeh J, Sousa M, Tarpley R, Suydam R (2004) Inferences from bowhead whale ovarian and pregnancy data: age estimates, length at sexual maturity and ovulation rates. International Whaling Commission scientific paper 56
Gilardi KG, Shideler SE, Valverde C, Roberts JA, Lasley BL (1997) Characterization of the onset of menopause in the rhesus macaque. Biol Reprod 57(2):335–340
Gosden RG (1985) Biology of menopause: the causes and consequences of ovarian ageing. Academic Press, London
Gosden RG, Faddy MJ (1998) Biological bases of premature ovarian failure. Reprod Fertil Dev 10(1):73–78
Gosden R, Telfer E (1987) Numbers of follicles and oocytes in mammalian ovaries and their allometric relationships. J Zool 211(1):169–175
Gould KG, Flint M, Graham CE (1981) Chimpanzee reproductive senescence: a possible model for evolution of the menopause. Maturitas 3(2):157–166
Gould SJ, Lewontin RC (1979) The spandrels of san Marco and the Panglossian paradigm: a critique of the adaptationist programme. Proc R Soc Lond B 205(1161):581–598
Graham CE (1979) Reproductive function in aged female chimpanzees. Am J Phys Anthropol 50(3):291–300
Hamilton PK, Knowlton AR, Marx MK, Kraus SD (1998) Age structure and longevity in North Atlantic right whales Eubalaena glacialis and their relation to reproduction. Mar Ecol Prog Ser 171:285–292
Hawkes K (2003) Grandmothers and the evolution of human longevity. Am J Hum Biol 15(3):380–400
Hawkes K, Coxworth JE (2013) Grandmothers and the evolution of human longevity: a review of findings and future directions. Evol Anthropol 22(6):294–302
Hill K, Hurtado AM (1991) The evolution of premature reproductive senescence and menopause in human females: an evolution of the "grandmother hypothesis". Hum Nat 2(4):313–350
Hill K, Hurtado AM (1996) Ache life history: the ecology and demography of a foraging people. Aldine de Gruyter, New York
Johnstone RA, Cant MA (2010) The evolution of menopause in cetaceans and humans: the role of demography. Proc R Soc Lond B 277(1701):3765–3771
Johnson RL, Kapsalis E (1998) Menopause in free-ranging rhesus macaques: estimated incidence, relation to body condition, and adaptive significance. Int J Primatol 19(4):751–765
Johnson RL, Kapsalis E (2008) Heterogeneity of reproductive aging in free-ranging female rhesus macaques. In: Atsalis S, Margulis SW, Hof PR (eds) Primate reproductive aging: cross-taxon perspectives. Karger Press, Basel, pp 62–79
Jones K, Walker L, Anderson D, Lacreuse A, Robson SL, Hawkes K (2007) Depletion of ovarian follicles with age in chimpanzees: similarities to humans. Biol Reprod 77(2):247–251

Kaplan H, Hill K, Lancaster J, Hurtado AM (2000) A theory of human life history evolution: diet, intelligence, and longevity. Evol Anthropol 9(4):156–185

Kasuya T, Marsh H (1984) Life history and reproductive biology of the short-finned pilot whale, *Globicephala macrorhynchus*, off the Pacific coast of Japan. Rep Int Whal Comm Spec 6:259–310

Kasuya T, Izumisawa Y, Komyo Y, Ishino Y, Maejima Y (1997) Life history parameters of bottle-nose dolphins off Japan. IBI Rep 7:71–107

Koenig A, Borries C, Chalise MK, Winkler P (1997) Ecology, nutrition, and timing of reproductive events in an Asian primate, the Hanuman langur *(Presbytis entellus)*. J Zool 243(2):215–235

Lahdenpera M, Lummaa V, Helle S, Tremblay M, Russell AF (2004) Fitness benefits of prolonged post-reproductive lifespan in women. Nature 428(6979):178–181

Lee PC, Fishlock V, Webber CE, Moss CJ (2016) The reproductive advantages of a long life: longevity and senescence in wild female African elephants. Behav Ecol Sociobiol 70(3):337–345

Lahdenpera M, Gillespie DOS, Lummaa V, Russell AF (2012) Severe intergenerational reproductive conflict and the evolution of menopause. Ecol Lett 15(11):1283–1290

Levitis DA, Lackey LB (2011) A measure for describing and comparing postreproductive life span as a population trait. Methods Ecol Evol 2(5):446–453

Levitis DA, Burger O, Lackey LB (2013) The human post-fertile lifespan in comparative evolutionary context. Evol Anth 22: 66–79.

Lockyer C, Brown S (1979) A review of recent biological data for the fin whale population off Iceland. Rep Int Whal Comm Spec 29:185–190

Marsh H, Kasuya T (1986) Evidence for reproductive senescence in female cetaceans. Rep Int Whal Comm Spec 8:57–74

Martin AR, Rothery P (1993) Reproductive parameters of female long-finned pilot whales (*Globicephala melas*) around the Faroe Islands. Rep Int Whal Comm Spec 14:263–304

Mizroch SA (1981) Analyses of some biological parameters in the Antarctic fin whale. Rep Int Whal Comm Spec 31:425–434

Morris W, Altmann J, Brockman D, Cords M, Fedigan L, Pusey A et al (2011) Low demographic variability in wild primate populations: fitness impacts of variation, covariation, and serial correlation in vital rates. Am Nat 177(1):E14–E28

Nishida T, Corp N, Hamai M, Hasegawa T, Hiraiwa-Hasegawa M, Hosaka K et al (2003) Demography, female life history, and reproductive profiles among the chimpanzees of Mahale. Am J Primtol 59(3):99–121

Nozaki M, Mitsunaga F, Shimizu K (1995) Reproductive senescence in female Japanese monkeys (*Macaca fuscata*): age- and season-related changes in hypothalamic-pituitary-ovarian functions and fecundity rates. Biol Reprod 52(6):1250–1257

Olesiuk PF, Bigg MA, Ellis GM (1990) Life history and population dynamics of resident killer whales (*Orcinus orca*) in the coastal waters of British Columbia and Washington state. Rep Int Whal Comm Spec 12:209–243

Paul A (2005) Primate predispositions for human grandmaternal behavior. In: Voland E, Chasiotis A, Schiefenhovel W (eds) Grandmotherhood: the evolutionary significance of the second half of female life. Rutgers University Press, New Brunswick, pp 21–37

Paul A, Kuester J, Podzuweit D (1993) Reproductive senescence and terminal investment in female barbary macaques *(Macaca sylvanus)* at Salem. Int J Primatol 14(1):105–124

Pavelka MSM, Fedigan LM (1991) Menopause: a comparative life history perspective. Yearb Phys Anthropol 34:13–38

Pavelka MSM, Fedigan LM (1999) Reproductive termination in female Japanese monkeys: a comparative life history perspective. Am J Phys Anthropol 109(4):455–464

Pavelka MSM, Fedigan LM (2012) Costs and benefits of old age reproduction in the Arashiyama west female Japanese macaques. In: Leca JB, Huffman MA, Vasey PL (eds) The monkeys of Stormy Mountain: 60 years of primatological research on the Japanese macaques of Arashiyama. Cambridge University Press, Cambridge, pp 13–27

Raichlen DA, Alexander GE (2014) Exercise, APOE genotype, and the evolution of the human lifespan. Trends Neurosci 37(5):247–255

Rice DW (1989) Sperm whale *Physeter macrocephalus* Linnaeus, 1758. In: Ridgeway SH, Harrison R (eds) Handbook of marine mammals, vol 4. Academic Press, London, pp 177–233

Robbins AM, Robbins MM, Gerald-Steklis N, Steklis HD (2006) Age-related patterns of reproductive success among female mountain gorillas. Am J Phys Anthropol 131(4):511–521

Rogers AR (1993) Why menopause? Evol Ecol 7(4):406–420

Rose MR (1991) Evolutionary biology of aging. Oxford University Press, Oxford

Rose MR, Mueller LD (1998) Evolution of human lifespan: past, future, and present. Am J Hum Biol 10:409–420

Shanley DP, Sear R, Mace R, TBL K (2007) Testing evolutionary theories of menopause. Proc R Soc Lond B 274(1628):2943–2949

Small MF (1982) Reproducive failure in macaques. Am J Primatol 2:137–147

Small MP (1984) Aging and reproductive success in female *Macaca mulatta*. In: Small MF (ed) Female primates: studies by women primatologists. Alan R Liss Inc., New York, pp 249–259

Sommer V, Srivastava A, Borries C (1992) Cycles, sexuality, and conception in free-ranging langurs *(Presbytis entellus)*. Am J Primtol 28(1):1–27

Stolen MK, Barlow J (2003) A model life table for bottlenose dolphins (*Tursiops truncatus*) from the Indian River lagoon system, Florida, USA. Mar Mamm Sci 19(4):630–649

Strier K, Altmann J, Brockman D, Bronikowski A, Cords M, Fedigan L et al (2010) The primate life history database: a unique shared ecological data resource. Methods Ecol Evol 1(2):199–211

Strier K (2018, this volume). The temporal scale of behavioral and demographic flexibility: implications for comparative analyses and conservation. In: Kalbitzer U, Jack KM (eds) Primate life histories, sex roles, and adaptability - essays in honour of Linda M. Fedigan. Developments in primatology: progress and prospects. Springer, New York, pp 19–33

Strassmann BI, Garrard WM (2011) Alternatives to the grandmother hypothesis. Hum Nat 22:201

Takahata Y, Koyama N, Suzuki S (1995) Do old aged females experience a long post-reproductive life span?: the cases of Japanese macaques and chimpanzees. Primates 36(2):169–180

Tardif SD, Ziegler TE (1992) Features of female reproductive senescence in tamarins (*Saguinus* spp), a New World primate. J Reprod Fertil 94(2):411–421

Tardif SD, Araujo A, Arruda MF, French JA, Sousa MBC, Yamamoto ME (2008) Reproduction and aging in marmosets and tamarins. In: Atsalis S, Margulis SW, Hof PR (eds) Primate reproductive aging: cross-taxon perspectives. Karger Press, Basel, pp 29–48

Thompson ME, Jones JH, Pusey AE, Brewer-Marsden S, Goodall J, Marsden D et al (2007) Aging and fertility patterns in wild chimpanzees provide insights into the evolution of menopause. Curr Biol 17(24):2150–2156

Towers J, Ellis G, Ford J (2015) Photo-identification catalogue and status of the northern resident killer whale population in 2014. Canadian technical report of fisheries and aquatic sciences 3139, 75vi

Tuljapurkar SD, Puleston CO, Gurven MD (2007) Why men matter: mating patterns drive evolution of human lifespan. PLoS One 2(8):e785

Videan EN, Fritz J, Heward CB, Murphy J (2006) The effects of aging on hormone and reproductive cycles in female chimpanzees *(Pan troglodytes)*. Comp Med 56(4):291–299

von Saal FS, Finch CE, Nelson JF (1994) Natural history and mechanisms of reproductive aging in humans, laboratory rodents, and other selected vertebrates. In: Knobil E, Neill J (eds) The physiology of reproduction. Raven Press, New York, pp 1213–1314

Walker ML (1995) Menopause in female rhesus monkeys. Am J Primtol 35(1):59–71

Walker ML, Herndon JG (2008) Menopause in nonhuman primates? Biol Reprod 79(3):398–406

Whitehead H (2003) Sperm whales: social evolution in the ocean. University of Chicago press, Chicago

Whitehead H, Rendell L (2014) The cultural lives of whales and dolphins. University of Chicago Press, Chicago

Wich SA, Utami-Atmoko SS, Mitra Setia T, Rijksen HD, Schurmann C, van Hooff JARAM, van Schaik CP (2004) Life history of wild Sumatran orangutans (*Pongo abelii*). J Hum Evol 47(6):385–398

Williams GC (1957) Pleiotropy, natural selection, and the evolution of senescence. Evolution 11(4):398–411

Wood JW, Holman DJ, O'Connor KA (2001) Did menopause evolve by antagonistic pleiotropy? In: Schultz M (ed) *Homo*: Unsere Herkunft und Zukunft. Cuvillier Verlag, Gottingen, pp 483–490

Wood BM, Watts DP, Mitani JC (2017) Favorable ecological circumstances promote life expectancy in chimpanzees similar to that of human hunter-gatherers. J Hum Evol 105:41–56

Chapter 4
Seasonal Patterns of Infant Mortality in Wild Sanje Mangabeys, *Cercocebus sanjei*

Gráinne M. McCabe and David Fernández

Abstract Infant mortality is a fundamental factor influencing population growth. This study examines the distribution of infant mortality in relation to seasonality for the endangered Sanje mangabey, *Cercocebus sanjei*, endemic to the Udzungwa Mountains, Tanzania. Data were collected over 40 continuous months (August 2008 to November 2011). Censuses conducted during monthly group follows recorded 35 live births, plus 6 births just prior to the start of the study data (23 females, 16 males and 2 of undetermined sex). The survival probability for these infants combined was 60.2% (27 survived and 14 died before 1 year of age). Mean annual infant mortality (39.8 ± 4.1%) was at the higher end of results reported for other wild cercopithecines. All infant deaths occurred within the dry season (June to October), although most (64%, 9 of 14) were highly clustered within the early dry season (June to August), a period coinciding with an initial drop in fruit production and a peak in tick infestation among adult females. Six of the 14 infants that died were conceived in the previous dry season and born in the wet season; our previous work has demonstrated that this is considered 'out-of-phase' with optimal conceptive timing for the species. Previous studies in this species have demonstrated that timing conception with peak food abundance can influence infant survival. We also suggest that tick-borne disease, prevalent in the region, should be further studied as it could be a factor impacting infant mortality and may be interacting with decreasing nutritional intake during the early dry season.

Keywords Infant survivorship · Disease · Ectoparasites · Seasonality

---

G. M. McCabe (✉)
Department of Field Conservation and Science, Bristol Zoological Society, Bristol, UK
e-mail: gmccabe@bristolzoo.org.uk

D. Fernández
Department of Applied Sciences, University of the West of England, Bristol, UK

U. Kalbitzer, K. M. Jack (eds.), *Primate Life Histories, Sex Roles, and Adaptability*, Developments in Primatology: Progress and Prospects,
https://doi.org/10.1007/978-3-319-98285-4_4

4.1 Introduction

A key factor in population growth rate is the number of offspring reaching reproductive age. As such, infant mortality rate has been recognized as a key population health indicator (Lacy and Clark 1990). In order to understand how this factor is impacting a population, it is first necessary to investigate the pattern of infant mortality in a given species and determine what variables may be underlying that pattern.

Infant mortality is often influenced by food availability and disease (e.g. Milton 1980; Lee 1987; Cheney et al. 2004; Gogarten et al. 2012). These are not independent variables, and as such, they may interact to adversely affect infant survival in wild primates. Within their first year of life, infants require a significant amount of energy and some specific nutrients in greater quantity than at other life stages, in order to achieve normal cognitive and physical development (Rasmussen et al. 1980; Lee 1987; Ullrey et al. 2003). For example, primates fed a diet deficient in iron often develop anaemia, which has been reported to contribute significantly to maternal mortality and maternal and foetal morbidity and is associated with adverse behavioural and cognitive development in human infants (van den Broek and Letsky 2000). As much of an infant's nourishment is acquired through breast milk, nutritionally stressed mothers may have offspring with reduced neonatal growth rates, increasing their risk of mortality, as observed in vervets (*Cercopithecus aethiops*) (Lee et al. 1991). Finally, evidence of periods of increased mortality in relation to low food abundance in the environment may indicate that seasonality of food resources impacts the survival of infant primates (Altmann et al. 1985), though disentangling the range of ecological factors that also vary seasonally (e.g. disease, mating competition) often proves difficult.

Sanje mangabeys (*Cercocebus sanjei*) are an endangered Old World cercopithecine endemic to the Udzungwa Mountains of Tanzania (Ehardt et al. 2005; McCabe et al. 2018). Systematic long-term data collection on a habituated group of this species began in 2008 (McCabe and Emery Thompson 2013). The purpose of this study is to examine the rate of infant survivorship and annual distribution of infant mortality in our wild group of Sanje mangabeys and investigate the impact of seasonality on the observed distribution of deaths.

McCabe and Emery Thompson (2013) demonstrated that while births can occur year-round, approximately 78% of Sanje mangabey births occur in the dry season. This study also found that mangabey infant mortality can be influenced in part by the timing of births and conceptions. Infants conceived, rather than born, in the dry season, when food is less abundant and maternal energy balance is low, and subsequently born in the wet season, are eight times more likely to die within the first year of life, than those conceived in the wet season, when high-quality resources are more plentiful for their mothers (McCabe and Emery Thompson 2013). Conceiving in the wet season, when fruit abundance is significantly higher (McCabe et al. 2013), allows mothers to improve their energy balance during the critical period of early gestation, which appears to play a role in infant survivorship.

However, not all infants that die within the first year of life are conceived outside of the identified 'optimal period' (i.e. wet season) for mangabeys. Thus, it is likely that other factors are also contributing to infant death in this species. Other possible causes of primate infant mortality include infanticide, incidental death due to aggression, predation and disease (Gogarten et al. 2012). The first of these two causes of infant mortality has never been observed in the mangabeys since the onset of systematic data collection, even with a pattern of male influxes during the peak conceptive season and regular occurrence of male takeovers (Fernández 2017). Predation has been observed in the area (Jones et al. 2006), and raptors and leopards have been seen stalking the group on occasion (personal observation). Though predation is a potential cause of infant mortality, we do not have any evidence of infants having been predated upon in the study group. We have, however, anecdotal evidence of infants displaying illness prior to their death or disappearance and have observed all group members suffering from seasonal tick infestations during the dry season (McCabe and Fernández, unpublished). In the current study, we conduct a preliminary investigation into possible explanations for the observed patterns of survivorship and mortality in infant Sanje mangabeys in relation to disease, specifically tick infestation.

4.2 Methods

4.2.1 Study Site

This study was conducted in the Mwanihana Forest of the Udzungwa Mountains National Park (UMNP), Tanzania, part of the Eastern Afromontane Biodiversity Hotspot (Mittermeier et al. 2004). The UMNP is a mosaic environment comprised of tropical submontane and montane evergreen primary and secondary forest, with areas of elephant disturbance (Struhsaker et al. 2004). The UMNP was formed in 1992 to protect the Mwanihana Forest (7°40′–7°57′S, 36°46′–36°56′E) of the eastern escarpment (Dinesen 2001). In addition to the Sanje mangabeys, UMNP contains five species of monkeys, including the endemic Udzungwa red colobus (*Procolobus gordonorum*) and up to five species of galago (Ehardt et al. 2005). This region receives an annual rainfall of ca. 1750 mm, with the majority (1650 mm) falling in a wet season between November and May (range: 97–377 mm/wet season month) (Lovett 1996; McCabe et al. 2013). Food availability is influenced by rainfall at this site, and significantly more fruit is produced in the wet season compared to the dry season (based on 47 mangabey food species; for details of phenology, see McCabe et al. 2013).

4.2.2 Study Subjects

First described in 1981 (Homewood and Rodgers 1981), Sanje mangabeys are endemic to the Udzungwa Mountains of Tanzania. Their distribution is limited to two fragments of the Udzungwa Mountains: the Mwanihana Forest, within the

UMNP, and the Udzungwa Scarp Nature Reserve (USNR), more than 120 km southwest and separated by fire-maintained grassland. The Mwanihana Forest harbours approximately 60% of the estimated remaining 1300–3500 mangabeys (Ehardt et al. 2005; Rovero et al. 2009). They are omnivorous, consuming predominantly fruit, seeds/nuts and invertebrates and live in multi-male/multi-female social groups with female philopatry and male dispersal (McCabe et al. 2013). They are semiterrestrial, spending more than 50% of their time on the forest floor, though the mangabeys use the entire canopy (Ehardt et al. 2005). Our study group's home range (ca. 200 ha) lies within the boundaries of UMNP and overlaps with at least three neighbouring Sanje mangabey groups and three other diurnal species of primate.

The study group has been observed since 2006, and systematic data collection began in 2008. No other groups of this species have been habituated for research; thus, we have limited available information on much of their life history and behaviour. We do know that their gestation period is approximately 5.5 months and the interbirth interval after a surviving infant is 24.1 months (Fernández et al. 2018).

At the time of this study, group composition ranged from approximately 63–70 individuals. Data included here are from 41 infants born to 19 adult females. All adult mangabeys were individually identified. Infants were identified based on interactions with their mothers; i.e. nursing, carrying and resting in contact. An infant was defined as an individual less than 1 year of age (Cheney et al. 2004). Of the 41 infants, 6 were present in the group at the start of the project, and their age was estimated to be less than 1 month based on size, pelage colour and behaviour (Fernández and McCabe, unpublished data). The remaining 35 infants were born during the study period.

4.2.3 Data Collection

Reproductive Data Data on infant births and deaths were collected ad libitum from August 2008 through November 2011. On average, the group was followed all day for 12 days per month. Each day that the group was followed, a census of adult females and infants was conducted, and the presence of new infants was recorded. Infant health and development were also noted, including nursing behaviour, injuries and wounds. If an infant was not observed with the mother for several days, the infant was recorded as 'disappeared' and presumed dead (Beehner and Bergman 2008).

Tick Prevalence Over 12 months between July 2009 and 2010, we performed 1267 30-min focal follows on the adult females as part of a larger project on female reproductive ecology. During this time, we noted the presence or absence of ticks on the face and neck, as these were the easiest areas to view from a distance. Both ticks and wounds from tick bites were recorded. Given the proximity of infants to their mothers, it is reasonable to assume that infants were also infected at the same time as their mothers. It should be noted that level of infection was not recorded.

4.2.4 Data Analysis

We used a survival analysis to determine the overall infant survival to year 1 and a log-rank test to examine differences in survival probability of infants born in the wet versus the dry seasons. Rao's spacing test of uniformity was used to examine the distribution of infant deaths across the year and age at death. A Kolmogorov-Smirnov analysis was used to examine possible skew of infant loss and the distribution of tick infections, across mothers in the study. A Mann-Whitney U test was used to examine seasonal differences in the proportion of focal follows in which mothers were observed with ticks. Finally, we used an independent samples t-test to examine sex differences in infant age at death and investigate differences in the presence of tick infection between mothers that lost infants versus those whose infants survived the first year of life.

Survival analysis and the log-rank test, as well as Rao's spacing test of uniformity, were conducted in R version 3.3.2 for Windows (R Core Team 2016) using the survival (Therneau 2015) and circular (Agostinelli and Lund 2013) packages, respectively. The remaining statistical tests were run using SPSS 23.0. The alpha level for all tests was $p < 0.05$.

4.3 Results

4.3.1 Infant Survival Probability

Censuses conducted during monthly group follows between August 2008 and November 2011 recorded 41 births: 23 females, 16 males and 2 infants of undetermined sex (as 1 died before the sex could be confirmed, and the other was not confirmed before the end of the study period). Of these infants, 27 survived and 14 died within the first year of life. The overall survival probability of infants to year 1 was 60.2% (Kaplan-Meier estimator: median = 0.602, 95% CI:0.457–0.793; Fig. 4.1).

4.3.2 Reproductive Timing and Infant Survival to Year 1

We examined infant survivorship to year 1 in relation to timing of the infant's birth. Of 41 infants, 32 were born in the dry season and 9 in the wet season. Individuals born during the dry season had a significantly higher survival probability to year 1 of 70.0 ± 9.1% (Kaplan-Meier estimator: median = 0.700; 95% CI: 0.541–0.903; $X^2 = 5.8$, df = 1, $p = 0.016$) compared to those born in the wet season with 33.0 ± 15.7% (Kaplan-Meier estimator: median = 0.330; 95% CI: 0.132–0.804) (Fig. 4.2).

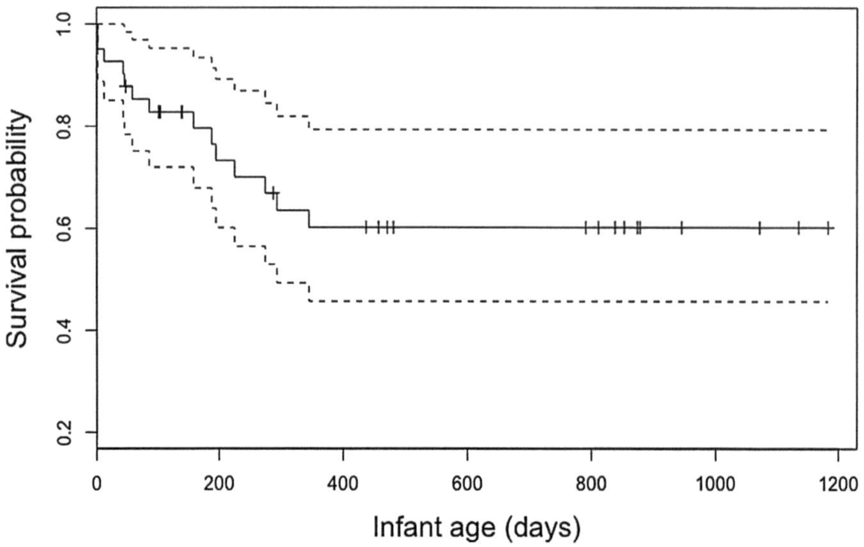

Fig. 4.1 Survival curve for all infants included in the study ($n = 41$, age at death or age at the end of the study for each infant is indicated by vertical ticks) indicating the median survival probability with ± 95% confidence intervals. As some infants born during the study were older than 1 year at the end of the study period, the graph extends beyond 1 year

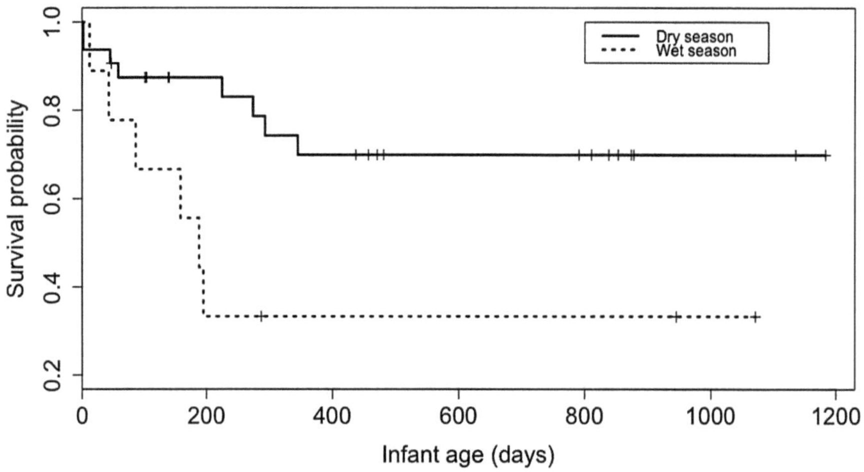

Fig. 4.2 Survival curves for infants born during the dry and wet season. Age at death or age at the end of the study for each infant is indicated by vertical ticks. As some infants born during the study were older than 1 year at the end of the study period, the graph extends beyond 1 year

Table 4.1 Number of infant births, deaths and subsequent annual infant mortality rate over a 3-year period from 2009 to 2011

Year	No. births	No. deaths	Annual infant mortality (%)
2009	11	4	36.4
2010	9	4	44.4
2011	13	5	38.5

4.3.3 Mean Infant Mortality Rate

Mean annual infant mortality over a 3-year period (2009–2011) was 39.8 ± 4.1% (Table 4.1). Infant death was not uniformly distributed across the year (Rao's spacing test of uniformity: $U = 162.5$, $p < 0.05$), as all occurred within the dry season (June to October). The majority of deaths (9 of 14) were clustered within the early dry season (June to August), with a mean date of death of 13 July (Fig. 4.3).

4.3.4 Age at Death and Sex of Infants

There was no discernable pattern in infant age at death. Infant deaths were uniformly distributed throughout the first year of life (Rao's spacing test of uniformity: $U = 132.56$, $p > 0.05$), with deaths ranging between 2 and 345 days of age and a mean of 137.6 ± 117.3 days or approximately 4.6 ± 3.9 months (Fig. 4.4).

Of the infants that died, 9 were female, 4 were male and 1 was undetermined at the time of death. We found no significant difference in mean age at death (independent samples t-test: $t = 1.099$, $df = 11$, $p = 0.295$) between females (160.6 ± 117.1 days) and males (80.8 ± 130.4 days).

4.3.5 Infant Loss Across Females

In this study, 41 infants were born to 19 reproductively active females; however, only 6 females lost infants, representing significant skew of infant loss in this group (Kolmogorov-Smirnov: $D = 0.42$, $p < 0.001$). Five of these females lost over 67.0% of their infants born during this time (mean number of infants lost per female = 2; range = 1–4). Given the small sample size, it was not possible to statistically analyse the impact of maternal rank on the level of infant loss; however, three of these females were low ranking, while three were high ranking; one of the latter was the alpha female (Table 4.2).

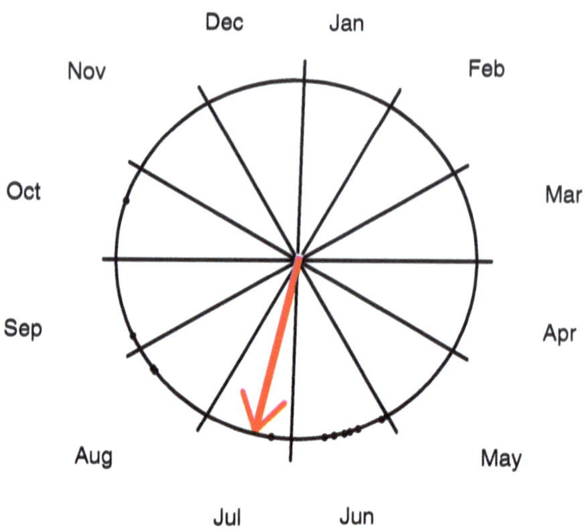

Fig. 4.3 Distribution of infant deaths across the year. Dots indicate the date of an infant death ($n = 14$); the red arrow indicates the mean death date (13 July) over a 3-year period between 2009 and 2011

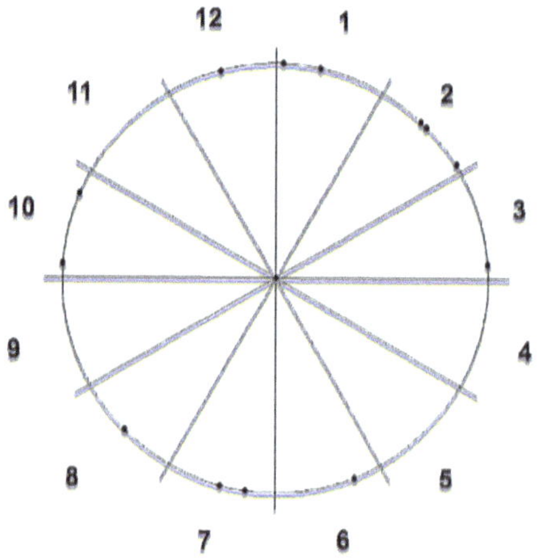

Fig. 4.4 Infant age at death in months. The spokes on the circular graph represent the transition between the end of 1 month and beginning of the next, up to the end of the first year of life. Infant deaths were uniformly distributed (Rao's spacing test of uniformity: $U = 132.56$, $p > 0.05$)

4.3.6 *Maternal Tick Infection and Infant Loss*

All mothers were observed to be affected by ticks at some point in the study. The proportion of follows in which a female was observed with ticks, however, was significantly different between the wet and dry season (Mann-Whitney U: $Z = -2.986$, $p < 0.01$). During the period of transition to the wet season (November 2009), 8.5% (7 out of 82) of focal follows were of females with ticks or tick wounds; however, we found no evidence of tick infestation during the remaining wet season months (Dec to May). In contrast, we saw a large variation in monthly tick prevalence in the dry season months (Jun to Oct) (Fig. 4.5). Within the dry season, the highest number of recorded infections occurred at the beginning, in June, and tapered off towards the end of the dry season.

Table 4.2 Rank of females that lost offspring during the study period, with infant birth season and age at death for infants that did not survive the first year of life

Female	Rank	No. infants born	No. infants died	Birth season	Death season	Infant age at death
Uvimbe (alpha)	High	2	1	Late dry	Early dry	11.5 mon
Madoa	High	3	2	Late dry Early dry	Late dry Late dry	1.5 mon 2 mon
Kisura	High	3	2	Late dry Early dry	Early dry Early dry	10 mon 2 days[a]
Bado	Low	3	3	Late dry Early dry Early wet	Early dry Early dry Early dry	9 mon 2 days 6.5 mon
Pua	Low	4	3	Late dry Early wet Early dry	Early dry Early dry Early dry	7.5 mon 6 mon 2 days
Yeyoo	Low	3	3	Late wet Late wet Early wet	Early dry Early dry Early dry	1 mon 2.5 mon 5 mon

[a]Infant was *ca.* 4 weeks premature and born with visible limb malformations

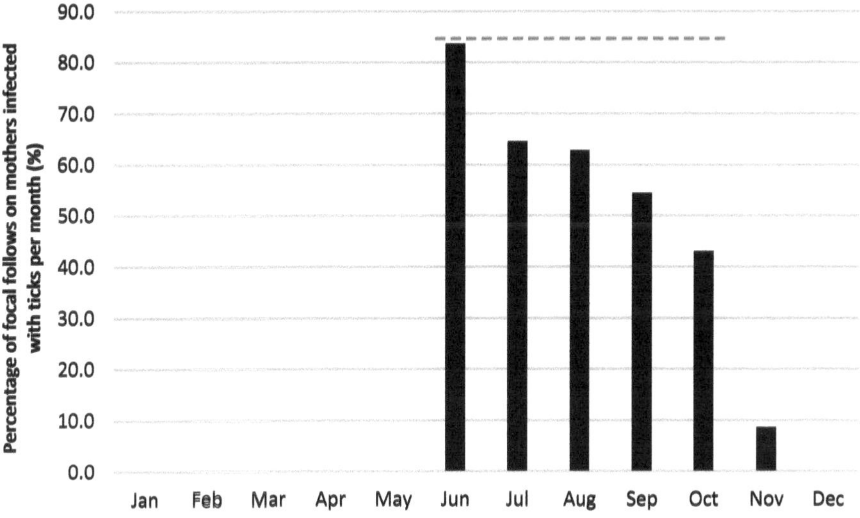

Fig. 4.5 Percentage of focal follows of mothers that were observed with ticks or wounds from tick bites per month. The wet season is from November to May, while the dry season is from June to October (indicated by dashed line)

The distribution of tick infections across females in the study was normal (Kolmogorov-Smirnov: $D = 0.144$, $p = 0.2$). This indicates that all females were equally infected based on a recording of the presence or absence of ticks or tick wounds during focal follows ($n = 1267$ focals, mean per female = 66.7 ± 1.83, range = 63–69). There was no significant difference in the presence of tick infection for those mothers that lost infants (mean = 17.9 ± 2.9 focals) versus those whose infants survived the first year of life (mean = 18.2 ± 4.6 focals) (independent samples t-test: $t = 0.137$, df = 17, $p = 0.893$).

4.4 Discussion

Over a 3-year period, 41 infant mangabeys were born, of which 14 died within the first year of life. Infants that were conceived in the wet season and born in the dry season had a significantly higher survival probability compared to those born in the wet season. More than one-third of infants died on average each year (39.8%), which is at the higher end of the range for other wild populations of similar-sized cercopithecines (5.4–47.0%; mean = 25.7 ± 13.0%; Table 4.3). No discernable pattern is observed in infant age, as deaths occurred throughout the first year of life. All deaths occurred in the dry season, with most clustered in the early dry season between June and August. Anecdotally, this is the time of year when ticks seem to first appear in the forest. This time of year also coincides with a high prevalence of tick infections among mothers, which may be correlated with tick infections among infants, that could be influencing mortality.

We found a link between timing of infant deaths and reproductive timing. Six of the 14 infants that died were conceived in the previous dry season, considered 'out-of-phase' with the optimal conceptive timing for the species, which occurs when food abundance is highest (McCabe and Emery Thompson 2013). Sanje mangabey females that conceive when food abundance is low appear unable to increase their energy balance during the critical phase of early gestation (McCabe and Emery Thompson 2013). This phase is also known as the embryonic phase, when organ formation occurs, in utero growth is the fastest and the embryo's oxygen and nutritional needs are the highest per unit mass (Rosso 1990; Houdijk et al. 2001; Carrai et al. 2003; Ullrey et al. 2003). However, this leaves eight infants whose deaths were unlikely to be explained by issues of inadequate maternal energy or nutrition during their foetal development, as their conceptions occurred during periods associated with high fruit abundance.

Our results indicate that infant mortality is clustered at the onset of the dry season. These results are contrary to what was found in a review of patterns of infant mortality in primates by Gogarten et al. (2012). They found that across the range of prosimian, monkey and ape species examined, only half exhibited a peak in seasonal mortality, yet of those, all had deaths occurring throughout the year. They also found deaths to be correlated with the wet season and to be the most seasonal among folivores. All such findings are in opposition to our results. The highly clustered dry season pattern found among infant deaths in the mangabeys suggests that an extrinsic factor occurring at this time is impacting their survival. While there is an initial drop in preferred mangabey food availability (i.e. fruits) at the end of the wet season in May (McCabe et al. 2013), it is unlikely that this decrease was substantial enough

Table 4.3 Infant mortality rate for wild populations of similar-sized cercopithecine species from the literature

Species	Infant mortality (%)	Reference
Cercocebus sanjei	39.8	This study
Cercopithecus aethiops	40.9	Isbell et al. (2009)
Erythrocebus patas	26.0	Isbell et al. (2009)
Lophocebus albigena	39.5	Arlet et al. (2014)
Macaca cyclopis	16.0	Hsu et al. (2006)
Macaca fuscata	29.0	Sugiyama and Ohsawa (1982)
Papio anubis	22.0 47.0	Smuts and Nicolson (1989), Higham et al. (2009)
Papio cynocephalus	12.0 25.0	Bentley-Condit and Smith (1997), Altmann et al. (1977, 1985)
Papio hamadryas	18.0 12.9	Sigg et al. (1982) Swedell et al. (2014)
Theropithecus gelada	5.4	Beehner and Bergman (2008)

Mean infant mortality across species (excluding *C. sanjei*) is 25.7 ± 13.0%

to severely impact maternal condition and infant survivability alone. If food availability was the primary factor influencing infant mortality, deaths should peak in the late dry season after the mangabeys have endured several months of lowered food availability and mothers may no longer be able to rely on reserve energy to produce nutritious milk for infants (McCabe and Emery Thompson 2013).

It is possible that we do not see a peak in infant mortality in the late dry season because of the availability of the fruit tree, *Tabernaemontana pachysiphon*. *T. pachysiphon* is an important mangabey food species that ripens in late August and early September. It is a large, pulpy fruit that is high in readily available nutrients (total nonstructural carbohydrates: 78.77%) (McCabe, unpublished data) and is abundant within the home range of our study group. This fruit may be improving dietary quality in the mid to late dry season for both mothers and their increasingly independent infants, which may lessen the risk of infant mortality at this time (i.e. only two infants died in the late dry season in this study).

The prevalence of tick infestation is, however, correlated with the timing of infant deaths, potentially indicating disease as a causative factor for the highly clustered pattern of deaths observed in our study group. Mangabeys often scratch open tick bites and a large open wound results. These wounds may then be followed by secondary infection at the bite site and a noticeable bloody swelling (Fig. 4.6). As many of the tick bites are on the face and neck, swelling at the bite site can lead to problems suckling for infant primates (Brain and Bohrmann 1992).

The impact of parasitic arthropods is under-represented in the primate-parasite literature, likely due to the difficulty in studying them without capturing the animals. Many species of ticks are vectors of infectious disease that can be debilitating

Fig. 4.6 Infant mangabey with bloody wounds from tick bites on the neck. (Photo taken at the start of the dry season in June 2010. Photo credit G. McCabe)

or fatal in domesticated animals and wildlife (e.g. Nunn and Altizer 2006; Parola 2006; Maina et al. 2014). In the Udzungwa Mountains, it is possible we are dealing with ticks from the genus *Amblyomma*, which have been found on baboons in Kenya (Kuntz and Myers 1967). These ticks are known to carry *Rickettsia africae*, which is responsible for African tick bite fever in the region (Jensenius et al. 2004). Kalter et al. (1968) suggest that baboons in Kenya may be affected by rickettsial diseases, as evidenced by antibodies found in serological analyses. In UMNP, both tourists and researchers have reported infection with *Rickettsia*, which can lead to severe symptoms if untreated. We recommend that further study into the possible impact of tick infection on infant mortality in the mangabeys be undertaken to better elucidate this situation.

Given that disease and nutritional status are not mutually exclusive variables, it is possible that the observed pattern of infant mortality is a result of the interaction between the observed increase in tick infestation and the initial drop in preferred food availability in the early dry season in the Udzungwa Mountains. A number of studies have demonstrated an increased risk of mortality due to the co-occurrence of disease and lowered nutrition (Gulland 1992; Beisel 1996; Chapman et al. 2005). Dietary stress can adversely affect resistance to disease by reducing the effectiveness of the immune system. As a result, shortages of high-quality food result in a decreased ability to fight off disease (Gulland 1992). This can in turn increase nutritional demands on the host and accentuate effects of food shortages. In other words, nutritional status and disease could have synergistic effects on the host (Chapman et al. 2005). For example, red colobus monkeys at Kibale National Park, Uganda, were found to have more gastrointestinal parasites in forest fragments with lowered food availability (Chapman et al. 2006). The authors' explanation is that parasites are taking advantage of a decline in the animal's immune system that is associated with poor nutrition.

This situation may be amplified in lactating females that already have a high energy burden, as they must provide enough energy for their own daily metabolic activities, quality milk production and infant care. If a lactating female begins to experience nutritional stress and becomes infected with disease, her infant may be at a greater risk of mortality. Beisel (1996) explains that a deficiency in any single essential nutrient can produce a dysfunction in the immune system or other host defence mechanisms. An infant, therefore, may suffer an immunological problem and face an increased risk of mortality even in the absence of severe energetic stress.

While tick infection was found to be uniform across the mothers in this study, all infant loss was highly skewed to a subset of six females. There are a number of possible explanations for this pattern. For example, rank among female primates has been demonstrated to impact reproductive success in a number of species (Fedigan 1983; Bercovitch 1991; Pusey et al. 1997; van Noordwijk and van Schaik 1999; Altmann and Alberts 2003), and three of the six females were of low rank. However, the remaining three females were of high rank, and one was the alpha female. Alternatively, as previously mentioned, energy balance during gestation is a potential driver, but without energy balance data for the complete 3 years of this study, it is not possible to determine if this could explain the observed skew.

Physiological abnormalities may also be impacting reproductive success for some females in this population. Five of the six females that suffered infant mortality lost more than two-thirds of their infants born during this study. For two females, both low ranking, this meant losing all three infants born over the 3-year period (see Table 4.2). Human infertility issues and reproductive loss are well established in the literature (for review see Hull et al. 1985; Miller et al. 2017), and similar physiological problems may be affecting a proportion of mangabey females as well. Without invasive studies, however, we are unable to verify this potential explanation.

In conclusion, our results suggest that infant mortality in Sanje mangabeys is highly seasonal. The pattern of seasonality could be a product of tick infestation and/or tick-borne disease or the result of secondary infection from tick bites. It is possible that the effects of this infection may be amplified by decreasing fruit availability at the start of the dry season leading to a synergistic interaction between infection and nutritional status for infant mangabeys. Further study is needed to elucidate the precise mechanism underlying infant loss in this species and the role of tick infection in the pattern observed.

Acknowledgements We would like to thank the Udzungwa Ecological Monitoring Centre and the Conservation Resource Center for providing logistical support for this project. Permission to carry out research has been granted by the Udzungwa Mountains National Park, Tanzania National Parks, Tanzanian Wildlife Research Institute and Tanzanian Commission for Science and

Technology. Valuable assistance was also provided by the wardens and staff of the Udzungwa Mountains National Park, as well as the Sanje Mangabey Project field assistants. Funding for this study was provided by the National Science Foundation Doctoral Dissertation Improvement Grant (DF and GMM), Leakey Foundation (DF and GMM), Primate Conservation, Inc. (GMM), Margot Marsh Biodiversity Foundation (DF), Conservation International Primate Action Fund (GMM) and IDEA Wild, as well as the University of Texas at San Antonio (UTSA) International Education Fund and Department of Anthropology Doctoral Research Grant (GMM) and Stony Brook University (SBU) Dean's Fellowship for Professional Development (DF). This study was approved by the IACUC regulations committee for both UTSA and SBU.

References

Agostinelli C, Lund U (2013) R package 'circular': circular statistics

Altmann J, Alberts S (2003) Variability in reproductive success viewed from a life-history perspective in baboons. Am J Hum Biol 15:401–409

Altmann J, Altmann S, Hausfater G, McCuskey S (1977) Life history of yellow baboons: physical development, reproductive parameters and infant mortality. Primates 18:315–330

Altmann J, Hausfater G, Altmann S (1985) Demography of Amboseli baboons, 1963–1983. Am J Primatol 8:113–125

Arlet M, Isbell L, Molleman F, Kaasik A, Chancellor R, Chapman C, Mänd R, Carey J (2014) Maternal investment and infant survival in gray-cheeked mangabeys (*Lophocebus albigena*). Int J Primatol 35:476–490

Beehner J, Bergman T (2008) Infant mortality following male takeovers in wild geladas. Am J Primatol 70:1152–1159

Beisel W (1996) Nutrition and immune function: overview. J Nutr 126(suppl_10):2611–2615

Bentley-Condit V, Smith E (1997) Female reproductive parameters of Tana River yellow baboons. Int J Primatol 18:581–596

Bercovitch F (1991) Social stratification, social strategies, and reproductive success in primates. Ethol Sociobiol 12:315–333

Brain C, Bohrmann R (1992) Tick infestation of baboons (*Papio ursinus*) in the Namib Desert. J Wildl Dis 28:188–191

Carrai V, Borgognini-Tarli S, Huffman M, Bardi M (2003) Increase in tannin consumption by sifaka (*Propithecus verreauxi verreauxi*) females during the birth season: a case for self-medication in prosimians? Primates 44:61–66

Chapman C, Gillespie T, Goldberg T (2005) Primates and the ecology of their infectious diseases: how will anthropogenic change affect host-parasite interactions? Evol Anthropol 14:134–144

Chapman C, Speirs M, Gillespie T, Holland T, Austad K (2006) Life on the edge: gastrointestinal parasites from the forest edge and interior primate groups. Am J Primatol 68:397–409

Cheney D, Seyfarth R, Fischer J, Beehner J, Bergman T, Johnson S, Kitchen D, Palombit R, Rendall D, Silk J (2004) Factors affecting reproduction and mortality among baboons in the Okavango Delta, Botswana. Int J Primatol 25:401–428

Dinesen L (2001) Conservation priorities for the forests of the Udzungwa Mountains, Tanzania, based on primates, duikers and birds. Biol Conserv 99:223–236

Ehardt C, Jones T, Butynski T (2005) Protective status, ecology and strategies for improving conservation of *Cercocebus sanjei* in the Udzungwa Mountains, Tanzania. Int J Primatol 26:557–583

Fedigan L (1983) Dominance and reproductive success in primates. Am J Phys Anthropol 26(S1):91–129

Fernández D (2017) Consequences of a male takeover on mating skew in wild Sanje mangabeys. Am J Primatol 79:e22532

Fernández D, Ehardt C, McCabe G (2018) Monitoring the Sanje mangabey population in Tanzania while engaging the local community. In: Behie AM, Teichroeb JA, Malone N (eds) Primate research and conservation in the Anthropocene. Cambridge University Press, Cambridge (in press)

Gogarten JF, Brown LM, Chapman CA, Cords M, Doran-Sheehy D, Fedigan LM, Grine FE, Perry S, Pusey AE, Sterck EH, Wich SA (2012) Seasonal mortality patterns in non-human primates: implications for variation in selection pressures across environments. Evolution 66:3252–3266

Gulland F (1992) The role of nematode parasites in Soay sheep (*Ovis aries*) mortality during a population crash. Parasitology 105:493–503

Higham J, Warren Y, Adanu J, Umaru B, MacLarnon A, Sommer V, Ross C (2009) Living on the edge: life-history of olive baboons at Gashaka-Gumti National Park, Nigeria. Am J Primatol 71:293–304

Homewood K, Rodgers W (1981) A previously undescribed mangabey from southern Tanzania. Int J Primatol 2:47–55

Houdijk J, Jessop N, Kyriazakis I (2001) Nutrient partitioning between reproductive and immune functions in animals. Proc Nutr Soc 60:515–525

Hsu M, Lin J, Agoramoorthy G (2006) Effects of group size on birth rate, infant mortality and social interactions in Formosan macaques at Mt longevity, Taiwan. Ethol Ecol Evol 18:3–17

Hull M, Glazener C, Kelly N, Conway D, Foster P, Hinton R, Coulson C, Lambert P, Watt E, Desai K (1985) Population study of causes, treatment, and outcome of infertility. Br Med J (Clin Res Ed) 291:1693–1697

Isbell L, Young T, Jaffe K, Carlson A, Chancellor R (2009) Demography and life histories of sympatric patas monkeys, *Erythrocebus patas*, and vervets, *Cercopithecus aethiops*, in Laikipia, Kenya. Int J Primatol 30:103–124

Jensenius M, Fournier P, Raoult D (2004) Rickettsioses and the international traveler. Clin Infect Dis 39:1493–1499

Jones T, Laurent S, Mselewa F, Mtui A (2006) Sanje mangabey Cercocebus sanei kills an African crowned eagle Stephanoaetus coronatus. Folia Primatologica, 77:359–363.

Kalter S, Kuntz R, Myers B, Eugster A, Rodriguez A, Benke M, Kalter G (1968) The collection of biomedical specimens from baboons (*Papio sp.*) Kenya, 1966. Primates 9:123–139

Kuntz R, Myers B (1967) Parasites of the Kenya baboon: arthropods, blood protozoa and helminths (Kenya, 1966). Primates 8:75–82

Lacy RC, Clark TW (1990) Population viability assessment of eastern barred bandicoot. In: Clark TW, Seebeck JH (eds) The management and conservation of small populations. Chicago Zoological Society, Brookfield, pp 131–146

Lee P (1987) Nutrition, fertility, and maternal investment in primates. J Zool 213:409–422

Lee P, Majluf P, Gordon I (1991) Growth, weaning and maternal investment from a comparative perspective. J Zool 225:99–114

Lovett JC (1996) Elevational and latitudinal changes in tree associations and diversity in the Eastern Arc Mountains of Tanzania. J Trop Ecol 12:629–650

Maina A, Jiang J, Omulo S, Cutler S, Ade F, Ogola E, Feikin D, Njenga M, Cleaveland S, Mpoke S, Ng'ang'a Z (2014) High prevalence of *Rickettsia africae* variants in *Amblyomma variegatum* ticks from domestic mammals in rural western Kenya: implications for human health. Vector Borne Zoonotic Dis 14:693–702

McCabe G, Emery Thompson M (2013) Reproductive seasonality in wild Sanje mangabeys (*Cercocebus sanjei*), Tanzania: relationship between the capital breeding strategy and infant survival. Behaviour 150:1399–1429

McCabe G, Fernández D, Ehardt C (2013) Ecology of reproduction in Sanje mangabeys (*Cercocebus sanjei*): dietary strategies and energetic condition during a high fruit period. Am J Primatol 75:1196–1208

McCabe G, Fernández D, Butynski T, Struhsaker T, Rovero F (2018) *Cercocebus sanjei* The IUCN Red List of Threatened Species 2016 (in press)

Miller J, Ahn S, Monsanto S, Khalaj K, Koti M, Tayade C (2017) Implications of immune dysfunction on endometriosis associated infertility. Oncotarget 8:7138–7147
Milton K (1980) The foraging strategy of howler monkeys: a study in primate economics. Columbia University Press, New York
Mittermeier R, Robles Gil P, Hoffmann M, Pilgrim J, Brooks T, Mittermeier C, Lamoreux J, da Fonseca G (2004) Hotspots revisited. CEMEX, Mexico
Nunn C, Altizer S (2006) Infectious diseases in primates: behavior, ecology and evolution. Oxford University Press, Oxford
Parola P (2006) Rickettsioses in sub-Saharan Africa. Ann N Y Acad Sci 1078:42–47
Pusey A, Williams J, Goodall J (1997) The influence of dominance rank on the reproductive success of female chimpanzees. Science 277:828–831
R Core Team (2016) R: a language and environment for statistical computing. R Foundation for Statistical Computing, Vienna
Rasmussen K, Thenen S, Hayes K (1980) Effect of folic acid supplementation on pregnancy in the squirrel monkey (*Saimiri sciureus*). J Med Primatol 9:169–184
Rosso P (1990) Nutrition and metabolism in pregnancy. Oxford University Press, New York
Rovero F, Marshall A, Jones T, Perkin A (2009) The primates of the Udzungwa Mountains: diversity, ecology and conservation. J Anthropol Sci 87:93–126
Sigg H, Stolba A, Abegglen J, Dasser V (1982) Life history of hamadryas baboons: physical development, infant mortality, reproductive parameters and family relationships. Primates 23:473–487
Smuts B, Nicolson N (1989) Reproduction in wild female olive baboons. Am J Primatol 19:229–246
Struhsaker T, Marshall A, Detwiler K, Siex K, Ehardt C, Lisbjerg D, Butynski T (2004) Demographic variation among Udzungwa red colobus in relation to gross ecological and sociological parameters. Int J Primatol 25:615–658
Sugiyama Y, Ohsawa H (1982) Population dynamics of Japanese monkeys with special reference to the effect of artificial feeding. Folia Primatol 39:238–263
Swedell L, Leedom L, Saunders J, Pines M (2014) Sexual conflict in a polygynous primate: costs and benefits of a male-imposed mating system. Behav Ecol Sociobiol 68:263–273
Therneau T (2015) A package for survival analysis in S. R package version 2.38
Ullrey D, Allen M, Ausman L, Conklin-Brittain N, Edwards M, Erwin J, Holick M, Hopkins D, Lewis S, Lonnerdal B, Rudel L (2003) Nutritional requirements of non-human primates, 2nd edn. National Research Council: Committee on Animal Nutrition. The National Academies Press, Washington, D.C
van den Broek N, Letsky E (2000) Etiology of anemia in pregnancy in South Malawi. Am J Clin Nutr 72:247–256
van Noordwijk M, van Schaik C (1999) The effects of dominance rank and group size on female lifetime reproductive success in wild long-tailed macaques, *Macaca fascicularis*. Primates 40:105–130

Part II
Sex Roles, Gender, and Science

Louise Barrett

> "It is essential that we not base our image of ourselves on false foundations. What is involved here is not simply the understanding of human nature, but also the image of humanity that grows out of that understanding."
> ~ Ashley Montagu

I have always been struck by the truth and importance of this quotation by the anthropologist, Ashley Montagu. No doubt this is because, as anthropologists, biologists, and primatologists, we bear some responsibility for constructing the foundations of humanity's understanding of itself. There is also no doubt that, on occasion, attempts to build such foundations have simply projected our preconceptions of what is naturally the case onto our primate cousins. Sex differences are a case in point, and investigations of whether and why the sexes differ are often fraught and contentious; positions can become polarised and tensions mount. It takes a brave person to wade into these waters, especially if one wishes to tackle the bigger question of how science itself is conducted, and not simply question the results of our scientific endeavours. Linda Fedigan is one such person, and in this part, we recognise not only Linda Fedigan's contribution to the scientific study of sex roles within and across primate species but also the importance of her work on gender and primatology in science studies.

Right from the beginning of her career, Linda was among a select few practicing scientists who approached the topic of sex and gender from the point of view of feminist epistemology and worked to show how the production of scientific knowledge often reflects the unconsidered assumptions and biases of scientists. This critical approach was, and still is, met with a hostile reception from some scientists; there is a suspicion and dislike of any suggestion that scientific facts are produced or constructed, rather than simply discovered in pure, value-free fashion. To some

L. Barrett (✉)
Canada Research Chair in Cognition, Evolution and Behaviour, Department of Psychology, University of Lethbridge, Lethbridge, AB, Canada
e-mail: louise.barrett@uleth.ca

scientific ears, talk of the "construction" of knowledge sounds almost like an accusation of fraud: that scientists are engaged in "making things up". Talk of the social construction of science also gets conflated with extreme post-modernist views, which deny the very possibility of an external reality and a single overarching truth, and anyone critical of scientific theories and practice is then labelled as being anti-science. As Pamela Asquith makes clear (Chap. 5, this volume), Linda has traversed this dangerous territory with extreme skill, calling into question biased perspectives, and offering her students and colleagues an alternative means by which we can theorise and study primate lives, without veering either to the scientific or post-modernist extreme. As a result, we now have both better science and better scientists.

As just one example of her work in this area, Linda, in collaboration with Kathy Jack (Fedigan and Jack 2013), offered an excellent analysis of the ways in which our use of language can canalise thinking in ways that privilege a male-centred viewpoint: why is it, they ask, that we generally consider female behaviour as a "counter-adaptation" to male behaviour? After all, it is just as evolutionary likely (and often more plausible in certain instances) that male behaviour evolves in response to selection pressures exerted by females. An insistence on male adaptations as primary, with females as following in their wake means we have already dictated the rules of the evolutionary game and thus limited the lines of investigation open to pursue. Recognising where bias creeps into our language—however innocently—and taking steps to counter it serves to open up new avenues of thought: it gives us new ways to conceptualise the inter-play of male and female strategies, and new ways to build an understanding of sex roles from first principles, rather than via the projection of the human status quo onto our primate cousins. As Asquith's chapter makes abundantly clear, Linda's nuanced, thoughtful position on this and other gendered issues, and her unwavering support of women scientists throughout her career, marks her out as a true pioneer of feminist science.

The other chapters in this part are authored by a number of Linda's former students, and all tackle issues related to the social and sexual roles of males. This was not by design, but it works superbly well to make the point that attention to gender issues in science, and an explicitly feminist viewpoint, does not mean an exclusive focus on females at the expense of males. Instead, as the chapters here demonstrate, it entails constructing a view of males that recognises their status as socially engaged beings, rather than simply as competitors for mates: a male's life does not reduce simply to a battle for mating access, any more than a female's reduces to gaining access to the food resources needed to fuel reproduction.

In the chapter co-authored by Katharine Jack and Linda, they use data from their long-term study of white-faced capuchins to test the idea that alpha-males in this species are keystone individuals (i.e., individuals who have a profound and lasting effect on the behaviour of others and the social structure of their groups). This is an idea that has gained prominence in behavioural ecology but has yet to be applied widely to the study of primates. (This in turn points to another feature of Linda's work that deserves mention: her desire to connect her work to broader theoretical and empirical concerns that encompass other non-primate species, and in so doing to draw primatology and behavioural ecology closer together.) Jack and Fedigan conclude that alpha-males do indeed fulfil the criteria for keystone individuals, but their analysis also reveals the limits of our ability to explain why some males achieve this unique position while others do not. Do all males have the capacity to become alpha, they ask, or is it only a certain "type" of male? Are there potential alpha-males who never experience the right combination of circumstances that lets them achieve their potential? Conversely, are there males that lack "classic" alpha traits but nevertheless become an alpha-male, through chance events and demographic accidents? What combination of environmental factors must be in place in order for a potential alpha-male to rise to the top and dominate others so comprehensively? Do males have control over their lives, or is it simply a case of being in the right place at the right time?

The next chapter in this part, by Eva Wikberg and colleagues, continues the capuchin theme and deals with some of the consequences of a system in which alpha-males hold such sway over the lives of others. Specifically, they consider how high levels of reproductive skew influence patterns of male kinship and dispersal, and how this in turn influences patterns of male-male cooperation. Using long-term demographic and genetic data, Wikberg et al.'s analyses reveal that only the alpha-male successfully reproduced in 36 group-years, while in another 4 group-years, only a single subordinate male reproduced. Even in years where multiple males reproduced, the level of skew remained high. As a result, natal males had a strong tendency to reside in groups with their paternal brothers of the same age-cohort. Partly as a result of this, males had a tendency to disperse from their natal group in parallel with their half-brothers of a similar age. Although parallel dispersal was not as common for males engaging in secondary dispersal, males that did disperse together in these circumstances were more closely related to one another than they were to other co-resident males.

Interestingly, given the potential benefits of kin-based coalitions to males, kinship predicted co-residence times of males only in one of three study groups. That is, while young related males may migrate together to a new group, they do not necessarily co-reside in those groups for extended periods of time. Why kinship is such a poor predictor of young males co-residency and the maintenance of male coalitions is a question that remains to be answered. For young males, the short co-residence times are rather puzzling: young males are not able to gain the alpha position in their new troops, and so are not in direct competition with each other for reproductive opportunities, but they can and do reap kin-related benefits of cooperation from living in such groups. Remaining together would therefore seem to be a high benefit-low cost endeavour: it makes sense for such males to support an incumbent alpha-male while biding their time until they are competitive for an alpha position (should this be possible). It is also the case that, once an alpha-male's daughters reach sexual maturity, subordinate males often gain opportunities to reproduce, as daughters avoid mating with their fathers.

One possible answer to the puzzle is that fully adult males who disperse together face a different set of circumstances that make co-residence less beneficial. This is because they can compete directly for the alpha-male position, and hence are rivals.

It is also less beneficial for an adult male to remain in subordinate position to a closely related alpha-male, as adult daughters who avoid breeding with their fathers may also avoid breeding with a closely related beta-male. Thus, inbreeding avoidance may exert a pressure for adult non-kin to reside together, rather than remain with kin. One possibility then is that this pressure on adult males may drive the system as a whole, leading to short co-residence times by all male kin, even when the costs seem trivial and the benefits high. While these ideas remain speculative at present, it is apparent that paternal relatedness and reproductive patterns underlie the patterns of cooperation seen among capuchin males. One cannot predict likely patterns of cooperation by reference to the costs and benefits of particular cooperative strategies alone.

In the final chapter of the part, Sarah Turner and colleagues take up another of Jack and Fedigan's themes, namely that dominance is not a property of the individual, but is it a social construct that depends on how given individuals act and react to others in particular situations. Turner et al. (Chap. 8) present a detailed analysis of the impact of disability on dominance relation among the Japanese macaques of Awaji Island. The macaques of Awaji show a high prevalence of congenital limb malformations (CLM), and Turner et al. ask whether the presence of CLM influences an individual's ability to achieve high rank and whether this differs between the sexes. Their analysis reveals that the effects of disability are less clear-cut than one might imagine. Among females, disability had the effect of lowering female rank acquisition, but only once a female's matrilineal kinship ties were taken into account. This suggests that the extrinsic power that females accrue through their associations with kin can ameliorate any CLM-related reduction in resource holding potential (RHP) that might place them at a disadvantage against members of other matrilines. Within their own families, however, CLM clearly was disadvantageous and led to lower rank positions than might otherwise have been expected.

For males, by contrast, CLM appeared to have no effect on their ability to achieve a certain rank position: males with CLM were distributed throughout the dominance hierarchy, and CLM was not a predictor of lower than average rank. Instead, age-class was a more important predictor. One possibility is that, as males with CLM are less likely to migrate, they have longer tenure lengths within the group and so form the social relationships and gain the social knowledge needed to achieve a given rank. As with the capuchins, there is still much to be unravelled. What is clear, however, is that dominance is associated with more than just age and resource-holding potential and reflects complex social dynamics. In the case of males, in particular, understanding how males achieve high rank requires much a reduced emphasis on physical attributes as the major determinants of resource-holding potential. Turner et al.'s study also illustrates the importance of focusing on what disabled individuals can and do achieve, rather than opting for a default disability-as-deficit model.

Taken together, the chapters in this part illustrate the value of adopting a richer assessment of male behaviour: understanding male mating strategies requires an appreciation of the effects of chance and circumstance, the importance of social factors in addition to physical resource holding potential, and a recognition that patterns of cooperation between kin and non-kin may be intertwined with reproductive patterns in complex and far-reaching ways.

Reference

Fedigan LM, Jack KM (2013) Sexual conflict in white-faced capuchins: it's not whether you win or lose. In: Fisher ML, Garcia JR, Chang RS (eds) Evolution's empress: Darwinian perspectives on the nature of women. Oxford University Press, New York, pp 281–303

Chapter 5
A Woman of Science: Sorting Fact and Illusion in Gender and Primatology

Pamela J. Asquith

Abstract Linda Fedigan's contributions to gender and primatology are discussed in relation to two books that, from a science studies perspective, were key turning points in this aspect of primatology's history. One is Donna Haraway's (1989) *Primate Visions*, and the other is Shirley Strum and Linda Fedigan's (2000) *Primate Encounters*. Fedigan's publications about gender and primatology span years before, during and after these important intellectual syntheses and call to action. Her work is considered here as an object of study by science studies practitioners, for the unique insights she provided to this field, and as a toolkit for testing possible real outcomes of current gender imbalances in primatology and other sciences.

Keywords Gender · Primatology · Science studies · Linda Fedigan

5.1 Introduction

> "…The science of nonhuman primates, primatology, may be a source of insights or a source of illusions. The issue rests on our skills in the construction of mirrors" *(Fedigan [1982] quoting and agreeing with Haraway [1978: 37] in her introduction to the first edition of Primate Paradigms).*

Linda Fedigan's contributions to gender and primatology form a significant corpus of books and papers that are interspersed through nearly 40 years of publications on old and new world primate field research. Her first book, *Primate Paradigms* (Fedigan 1982), written after her doctoral thesis monograph, more or less defined the field of gender and primatology with sustained examination of the neglected role of females, both human and nonhuman, in evolution. The book was a *tour de force* for both primate research and for its early contribution to women's studies and science. It was followed by several important papers that defined, tested and sharpened

P. J. Asquith (✉)
School of Environmental Studies, University of Victoria, Victoria, BC, Canada

Linacre College, Oxford, UK
e-mail: pasquith@ualberta.ca

U. Kalbitzer, K. M. Jack (eds.), *Primate Life Histories, Sex Roles, and Adaptability*, Developments in Primatology: Progress and Prospects,
https://doi.org/10.1007/978-3-319-98285-4_5

the discourse about women, gender and feminist and science studies perspectives on primatology and its related fields. This corpus of work can be more fully appreciated in relation to two books that, from a science studies perspective, were key turning points in this aspect of primatology's history. One is Donna Haraway's *Primate Visions*, published in 1989, and the other is Shirley Strum and Linda Fedigan's edited collection of papers, *Primate Encounters*, published in 2000. Fedigan's publications about gender and primatology span years before, during and after these important intellectual syntheses and call to action. The focus of this review is Linda's influence on, and response to, the issues raised at these critical points in the history of primatology.

I chose Haraway's book for the obvious reason of its prime focus but also because she contributes to the second book, along with other science studies scholars, in the form of a kind of confessional (which is her term) about writing *Primate Visions*. We are thus privy to a very intimate analysis of developments in the debates and reflections about gender and primatology. Strum and Fedigan's *Primate Encounters* was chosen for its representation of nearly all the aspects that Linda has addressed in gender and primatology: changes in the role and presence of females (human and nonhuman), comparisons of gender balance in primatology with its parent disciplines, science studies scholarship with regard to primate studies and comparisons regarding apparently 'feminist' approaches with those of different national traditions. The meeting on which the book was based also represented, I think, the most direct challenge to the very question of gender and primatology. One would be forgiven for wondering why Linda continued with this line of questioning after those debates, but she did, and she continues to reinforce the importance of an equal focus on the females of a species (e.g. Fedigan and Jack 2013).

5.2 Gender and Primatology

Linda's thoughts on the term and concept of 'gender' are discussed later in this paper, but for present purposes, 'gender and primatology' comprises three areas of inquiry. One is sex roles in primates, the second is the influence of feminism on primatological thinking, and the third is the influence of primatology on feminist thought. The second and third are variously broken down into specific questions about how the gender of the scientist may affect research on sex differences, the role of women in making models of human evolution, how women have been portrayed by male and female researchers in those models and the effects of gender on changing views of life history research. Linda is, I believe, the only scholar who has contributed explicitly and significantly to all of these areas, and she is the acknowledged first scholar to examine both images of women in theories of human evolution and the role of female scientists in constructing those images (see Fedigan 1986).

Linda's major early contribution to highlighting the female role in primate societies was her book *Primate Paradigms, Sex Roles and Social Bonds* published in 1982 with a second edition in 1992. Linda wrote the book to examine the neglected perspective of lives of female primates from a Darwinian perspective and to demolish old sex stereotypes. Haraway (1989) drew heavily on *Primate Paradigms* in her analysis of what Linda, as a female primatologist, was bringing to the, until then, male-dominated field. Keeping firmly in mind Linda's statement to me that she hardly recognized herself in Haraway's commentary about her work at the time, we can nonetheless reflect back to contextualize the story as it was then told. The subtitle of *Primate Visions* is *Gender, Race and Nature in the World of Modern Science*. Haraway's aims were both academically and politically motivated to reveal the deep androcentric and ethnocentric biases in natural science generally. That she chose to focus on primate studies was a function of the importance attached to understanding human behaviour through them as well as the happy fact that here was a science now with several leading female researchers, perhaps nearly in equal numbers to their male counterparts. That simple statement alone, regarding relative numbers of male and female researchers, was addressed in Linda's important paper 'Science and the Successful Female: Why There Are So Many Women Primatologists' published in *American Anthropologist* in 1994. No one had actually calculated the numbers, nor taken the further step to contextualize the findings by comparing primatology with other sciences and with its parent disciplines. Linda found that, as of 1992, while there *was* a significantly higher proportion of women in primatology than in analogous single-order biological sciences such as ornithology, benthology or mammalogy, there were *not* significantly more women primatologists than there were women anthropologists, psychologists and animal behaviourists, the 'parent disciplines' of primatology. As she stated in a later paper that while 'the perception that there is a critical mass of women in primatology is likely valid in a comparison across the biological sciences, it is not particularly striking from the perspective of the behavioral sciences, such as anthropology and psychology' (Fedigan 1997:64). She further reflected on reasons why females might be attracted to primatology, among them the fact that it was a relatively young and in some ways 'maverick' discipline; that there had been some very influential male mentors such as Louis Leakey and Sherwood Washburn; that it was in many ways a social science to which women were attracted, though it can equally be considered a biological science these days; and lastly, that at the time, the media coverage of the Trimates (Jane Goodall, Dian Fossey and Birute Galdikas) was very compelling.

Most readers of *Primate Paradigms* probably considered it to be a significant contribution to considering the roles and evolutionary strategies of the females of various species, as well as her examination of the use of primates in models of human evolution. However, Linda's thinking even at that early date included important contributions to the study of science and females' roles in it. British primatologist John H. Crook's (1983) review of *Primate Paradigms* in the journal *Nature* described it as *the* feminist critique of theories of primate and human evolution. Linda herself remarked that she hoped her book would be read by specialists both in primatology and women's studies (Fedigan 1982: xxxiii, 1992 ed.).

Linda's contributions in the first area of sex roles in primates were ground breaking and stood out among other authors. In the introduction to the second edition of *Primate Paradigms*, Linda listed various books that had in the intervening decade documented the lives and roles of female animals, especially those of primates. One was Meredith Small's *Female Primates* published in 1984 reviewed by Linda (Fedigan 1984) for the *American Journal of Primatology*. There, she noted that by the early 1980s although women were responsible for much and varied primate research, their numbers had not yet had the impact they warranted in the realm of theory making and testing. They had gone far, however, in filling in the female part of the hitherto overemphasized male part of the picture.

Turning to the second area of the influence of feminism on primatology, one of her most important early syntheses on gender, written with Larry Fedigan, was *Gender and the Study of Primates* published in a collection on *Gender and Anthropology, Critical Reviews for Research and Teaching* (Fedigan and Fedigan 1989). There, they demonstrated how female primatologists had changed perspectives and contributed to new theory, in contrast to Linda's previous observation in 1984 about the absence of this effect. These developments had been triggered by the recognition that evolutionary forces act directly upon females as well as males, by methodological improvements in the discipline, as well as an impetus from feminist studies. Linda and Larry diligently and clearly reviewed how people, such as Shirley Strum, Barbara Smuts, Thelma Rowell, Jeanne Altmann and Sarah Hrdy, among others, had accomplished these changes partly through letting the data and the animals speak for themselves and partly through improvements in observational sampling methods.

In the review, they considered also the work of scholars in the social studies of science who were discussing whether there might be a feminist epistemology. Evelyn Fox Keller (1985) at the time suggested that Western women scientists tend to be holistic and integrative, perhaps more attuned to social complexities and subtleties and less satisfied with reductionist principles of analysis. In reviewing the work of the above-mentioned primatologists, Linda and Larry tackled the question of the feminist critique of the dichotomy between reason and feeling and scientists' fears of anthropomorphism. Ethologists, primatologists and science studies scholars have argued for the importance of 'intelligent empathy' (projecting our feelings onto our subjects in order to better understand them and assuming they have feelings at least somewhat like our own), which is not the same thing. The main change that had happened in primate studies was a move beyond simply critiquing past androcentrism and cataloguing behaviours. Rather, the questions and insights about the animals had come to be based more on observing their behaviour outside of the context of a pre-existing Western set of folk beliefs about how men and women should behave in society, as well as through framing evolutionary questions about female primates differently.

Linda tackled the question explicitly in her paper 'Is primatology a feminist science?' (Fedigan 1997) published in Lori Hager's edited volume *Women in Human Evolution*. There, she carefully considered the features that have been described as representing feminist models of science. She concluded with a qualified 'yes' to the question but with the caveats that the correlation between feminist science and trends in primatology could just as well be attributed to other, alternative approaches to science, such as African, Japanese, Indian and Marxist perspectives on natural knowledge with their sense of holism and humans as part of nature. Or it could equally be attributed to the maturation of the discipline, having gained what feminists claimed for a feminist science, such as the move from reductionism and dualisms to more complex, sophisticated explanatory models and the development of a female as well as a male perspective.

Harking back to where the feminist critique of science first focussed mainly on primatology, Linda thought that although a good circumstantial case might be made for primatology having been influenced, perhaps transformed, by feminist perspectives and objectives, the almost universally negative reaction by practicing scientists and primatologists to Haraway's (1989) rather positive views of primatology was likely due to her portraying primatology as politics. Another reason many primatologists disliked *Primate Visions* was the postmodern language, the 'multilayered meanings', the trope-laden prose about the 'production of science', the very characterization of which offended many scientists. By contrast, 'multilayered' for Linda was based on layers of data gathered over many years and the interplay of people and of their ideas. Linda was as aware of the story-telling aspect of primatology and human evolution in Landau's (1984, 1991) sense as was Haraway. But in Linda's view, 'the story' never undermined the value or relevance of the hypotheses it generated or the theories that were tested.

At about the same time as the publication of her 1997 paper, Linda returned to her first principles and rationale for the trajectory of her primate studies. Her paper *Changing Views of Female Life Histories* (Fedigan 1996) reviewed the use and meanings of the concept of 'life history' over time and across disciplines and demonstrated its empirical application to primate studies through her own research on Japanese macaques. She noted that female reproductive success and longevity were not related to dominance or access to food in her study group, the transplanted fissioned group from Japan called Arashiyama West, but that this noncorrelation may have been an artefact of provisioning. She therefore sought a nonprovisioned and stable site where she could do a long-term study of factors affecting female reproductive success, settling on Costa Rica. In her retrospective on her 30-year study in Costa Rica (Fedigan 2014), Linda said that what impelled her from the beginning of her research when she received the genealogical records and shared behavioural database of the Japanese macaque Arashiyama site was the richness of the data and the kinds of questions that could be tackled on this basis. For Linda, what was especially interesting was how and why some females experience greater reproductive success than others. In a rather moving conclusion to the paper, Linda reflected on the lives of members of a particular capuchin family she had observed in Costa Rica over two decades and how she had every intention of continuing observations to find out the eventual fates of their offspring.

In considering the third area of the influence of primatology on feminist thought, Linda again has demonstrated a clear and fair evaluation of the issues despite considerable divisions among feminists, primatologists and science studies practitioners. This opportunity arose in a meeting hosted in 1996 by Wenner-Gren and held in the mountains near Teresopolis, Brazil, where participants were invited to discuss, among other topics, how gender might relate to the practice and ideals of science and what role gender played in the history of ideas about primate societies. Never before had representatives from every area of gender and primatology discussed by Linda been gathered in one place. Imagine, then, the excitement of a meeting organized by Linda and Shirley Strum that brought together influential and founding primatologists, primatologists representing different national traditions, science studies scholars and representatives from parent and sister disciplines of primatology such as psychology, anthropology, archaeology and ethology for a week-long discussion. The group of 22 invited participants included science studies and history of science scholars such as Evelyn Fox Keller, Donna Haraway, Bruno Latour, Gregg Mitman, Alison Wylie, Brian Noble and myself; some of the senior founders of primatology such as Robert Hinde, Thelma Rowell, Alison Jolly, and Robert Sussman; primatologists who represented different national traditions such as Hiroyuki Takasaki from Japan and Emilia Yamamoto from Brazil; North American and European primatologists who studied great ape and Old and New World monkeys such as Craig Stanford, Shirley Strum, Linda Fedigan, Sarah Hrdy and Karen Strier; as well as researchers from the sister and parent disciplines of primate studies such as Richard Byrne, Naomi Quinn, Zuleyma Tang-Martinez and Stephen Glickman. The results were published in *Primate Encounters*, edited by Strum and Fedigan (2000). For several months after the meeting, we continued to discuss issues around gender and science, among other topics, in an email forum. Uniquely, the publisher included excerpts from those exchanges in the final publication.

In her chapter 'Gender Encounters', which summarized results of the workshop, Linda remarked that to initiate a discussion about what role gender played in the history of ideas about primate societies was '…like striking a match to check on a gas leak. After the resultant explosion…you pick yourself up and wonder what happened' (Fedigan 2000a: 498). Bruno Latour, renowned anthropologist of science, had recently received a birthday gift of a small wind-up gorilla toy. When faces grew grim during our discussions around the table, he would wind it up and set it off on its bipedal progress to the other side. Even without a white flag, it unfailingly broke the tension. What people were defending so ardently was no less than their presuppositions, about which, as the philosopher R.G. Collingwood (1946) famously remarked, people get angry when they bump up against them. Thus, I consider the meeting and publication to mark something of a watershed in Linda's long encounter with the topic.

In her paper for Strum and Fedigan's *Primate Encounters*, Haraway (2000) elaborated on her approach to assessing primatology. She explained her genuine love of language and its power to make us see things differently or more clearly. In a self-described 'slightly confessional account of…writing about primatology and primatologists', she admitted 'I am in love with words themselves…' (ibid: 399). In a neat co-option of biological terminology, Haraway wrote '…science studies is about the behavioral ecology and optimal foraging strategies of scientists and their subjects; and primatology seems to me to be about the historically dynamic, material-semiotic webs where important kinds of knowledge are at stake' (ibid: 399). Although she has always been concerned with many problems in addition to gender imbalances in science, Haraway (1989) had foreshadowed Teresopolis discussions on gender and science in her chapter that discussed Linda's (Fedigan 1982) research in *Primate*

Paradigms. There, she noted how Linda consistently emphasized agency in female primates rather than seeing them as a resource for males and how, within that frame, a more complex picture of power and knowledge (among both the animals and the scientists) could emerge (Haraway 1989: 320).

5.3 Fedigan on Gender

Readers may wonder why they have waited until now for a statement regarding the use of the term 'gender' in the current context. It is because Linda herself waited until her chapter summary (Fedigan 2000a) to review the many and varied ways that the concept of gender has been used through history and across disciplines and to state her own views of the use of the concept. In a sense she likely felt impelled to, given the vast gulf with regard to 'gender and science' between and among the science studies scholars and scientists that was so apparent in the 1996 meetings on which the book was based. She noted that although for some at the meetings to even ask whether women do science differently is unproductive or even dangerous, she felt that if this is a question in which scientists themselves and the public are interested, then it is important to find ways to modify and adapt the question to meet the objections and to break it down into more specific questions that are more contextualized and multifactorial.

From her own standpoint, Linda stated that 'I learned that all the different human cultures take the basic clay of biological sex differences and construct this clay into quite variable models of how men and women should behave. Thus, I think of gender as the cultural transformation of biological sex differences into stereotyped and dichotomous expectations of the attributes of men and women' (Fedigan 2000a: 500). But of course, she had long since discovered that it was not so straightforward as that. After pointing out that grammarians likely have the original claim to the use of 'gender' (such as in feminine, masculine or neuter nouns), having nothing to do with biological sex or cultural attributions, she went on to portray the various understandings of gender by social psychologists and feminists, among others. With reference to humans, ideas about gender cover a spectrum from some feminists considering that it is not the property of an individual and that one cannot have a gender since gender-appropriate behaviour is constantly changing at both the societal and individual levels and hence cannot be measured to social psychologists who argue that gender is a property of individuals and can be measured for a given culture and time period. For the purposes of her analysis, Linda adhered to Sandra Harding's (1986) understanding of gender as composed of three components (identity, roles and symbols), which, she felt, allowed for much plasticity and diversity. An important part of Linda's reflections on the concept is that gender should not be used synonymously with 'women' and 'femininity' inasmuch as gender refers as much to cultural constructions of masculinity (Fedigan 2000a: 502–503). In all of Linda's papers regarding gender and primatology, she is careful to emphasize that she is discussing men and women in primatology and never assumes that so-called feminist approaches to science are necessarily exclusive to females.

On the basis of her very nuanced and balanced review of the literature, together with what participants at the meeting had discussed, Linda concluded her chapter 'Gender Encounters' with a series of reformulated questions about women and men in primatology, the primatology of gender and gender in primatology. These are worth repeating here as a toolkit for future analyses of gender in primatology and other sciences.

Women and Men in Primatology What difference does it make if few, many or no women/men are present and practicing in primatology and related disciplines? If women are spread evenly across the specialties or concentrated in a few areas? If individual women, or individual men, have acted as influential role models for younger primatologists?

The Primatology of Gender What difference does it make if Western cultural understandings of masculinity and femininity have both influenced and been influenced by primatological constructions of sex differences and of gender? If the work of particular women or men primatologists has changed our ideas about sex and gender? If the feminist critique of science has changed our ideas about sex and gender?

Gender in Primatology What difference does it make if the practices and institutional aspects of primatology are gendered masculine or feminine? If power in primatology is gendered? If, in the past two decades, primatology itself has come to be gendered feminine by outside observers? (Fedigan 2000a: 519).

Linda further asked what we can generalize from this case study of one science and how we can apply our findings to make science more inclusive in all senses of that term. Ultimately, Linda found common ground among the participants 'in their respect for science and their desire to comprehend how science functions in society and how to make it better' and that, in her view, '...one way to improve science is to better understand its multifaceted interactions with gender identities, roles, and symbolic systems' (ibid: 519). A few years further on, while reconsidering 'The Paradox of Feminist Primatology' (Fedigan 2009; see also Fedigan 2001), Linda remarked that the considerable difficulties experienced by scientists and science analysts to discuss gender and primatology were due to the very different models of science they adhered to. In an insightful comparison between scientists and science analysts formulated in discussion with Shirley Strum, she observed that for scientists, the model of science is 'outcome and norms', while for science analysts it is 'process and practice'. Linda noted: 'Often we could not even begin our discussion of the history of ideas in primatology because the participants had such different views of what science is and how it works. We finally realized that we were enacting a local battle in the larger science wars and that C.P. Snow's [1959] model of two cultures still holds sway' (Fedigan 2009: 267).

5.4 Future Research

Linda has not yet applied her specific questions raised in 2000 to further assessment of men and women's position and influence in primatology or science more generally. Two articles that address women's positions in primatology published a dozen years after her article 'Gender Encounters' provides a snapshot of where we are, albeit in more specific terms of career trajectories and representation at professional meetings. One study regarding gender bias in primatology premised its findings on the assumption that participation at professional meetings in symposia is more prestigious than in poster sessions or giving individual talks. Isbell et al. (2012) found that although women are the numerical majority in primatology and have substantial peer recognition in the discipline, they nonetheless are underrepresented in professional meetings in that male-organized symposia have half the numbers of female first authors than symposia organized by women or by both men and women and half that of female participation in talks and posters. The authors make the interesting observation that female representation in areas of primate biology and evolution as compared with primate behaviour/sociology is much less and so omitted those areas from their survey of professional meetings. Linda had remarked many years before that one of the attractions of females to primatology may have been the existence of a social science as well as biological niche for researchers (Fedigan 1997: 64).

In another study, Addessi et al. (2012) found that there still seems to be a glass ceiling effect in which fewer women than would be expected given their numbers and published contributions in the discipline have attained full professor rank (although their scientific impact, as measured by their h-index, may indicate that their fewer articles have higher impact than those of their male colleagues). These are precisely the kind of observations that could be pursued further in terms of Linda's questions (Fedigan 2000a: 519), i.e. what contributions to theory and method are those articles and books making; are females pursuing some approaches or questions that are directing research in a significant way?

On a final note, Linda had raised the importance of bridging another gap within anthropology's subfields, between those who focus on sociocultural aspects of humanness and those who concentrate on its biological phenomena (Fedigan 2000b). In her 'A view on the science: Physical anthropology at the millennium' published in the *American Journal of Physical Anthropology*, she traced her own trajectory at the beginning of the third millennium. There, she concluded that even with our highly varied specialties within physical or biological anthropology, all subareas use cross-disciplinary research and the comparative method, and all have a contribution to make to understand the human place in nature. The same can be said for any varied viewpoints that may be contributed as a result of gender. Linda has laid a coherent and cogent foundation for future studies that address evidence for, and give definition to, contributions by men and women to theory and method and which could provide insight into possible real outcomes of gender imbalances in primatology.

Acknowledgements I wish to thank the organizers for the privilege and pleasure of the opportunity to reflect on Linda Fedigan's work at the Festschrift conference in the magnificent surroundings of the Banff Centre. Many hands are involved in the organization of a conference, but special thanks to Jeremy Hogan and the editors of the Festschrift. Thanks to Mary Pavelka for getting this particular ball rolling on singling out Linda's contributions to gender and primatology. Thanks to two reviewers of the paper for helpful comments. Thanks to Linda Fedigan for responding to questions so rapidly and with good humour as I sought the background to her thoughts on gender and primatology. I thank her too for being such a fine scholar, teacher, colleague and friend over many years.

References

Addessi E, Borgi M, Palagi E (2012) Is primatology an equal-opportunity discipline? PLoS One 7(1):1–6 www.plosone.org/e30458

Collingwood RG (1946) The idea of history. Oxford University Press, Oxford

Crook JH (1983) A matter of sex. Nature 302:359

Fedigan LM (1982) Primate paradigms: sex roles and social bonds, 2nd edn. Montreal, Eden Press, 1992

Fedigan LM (1984) Sex ratios and sex differences in primatology (a review of female primates: studies by women primatologists). Amer J Primatol 7:305–308

Fedigan LM (1986) The changing role of women in models of human evolution. Ann Rev in Anthropol 15:25–66

Fedigan LM (1994) Science and the successful female: why there are so many female primatologists. Am Anthropol 96(3):529–540

Fedigan LM (1996) Changing views of female life histories. In: Morbeck ME, Galloway A, Zihlman AL (eds) The evolving female: a life-history perspective. Princeton University Press, Princeton, NJ, pp 15–26

Fedigan LM (1997) Is primatology a feminist science? In: Hager L (ed) Women in human evolution. Routledge, London, pp 56–75

Fedigan LM (2000a) Gender encounters. In: Strum S, Fedigan LM (eds) Primate encounters. Models of science, gender and society. University of Chicago Press, Chicago, pp 498–520

Fedigan LM (2000b) A view on the science: physical anthropology at the millennium. Amer J Phys Anthropol 113:451–454

Fedigan LM (2001) The paradox of feminist primatology - the goddess's discipline? In: Creager ANH, Lunbeck E, Schiebinger L (eds) Feminism in twentieth century science, technology and medicine. University of Chicago Press, Chicago, pp 46–72

Fedigan LM (2009) The paradox of feminist primatology. In: Wyer M et al (eds) Women, science and technology: a reader in feminist science studies, 2nd edn. Routledge, New York, pp 256–270

Fedigan LM (2014) Questions my mother asked me: an inside view of a thirty-year primate project in a Costa Rican National Park. In: Strier KB (ed) Primate ethnographies. Upper Saddle River, Bergen, pp 186–195

Fedigan LM, Fedigan L (1989) Gender and the study of primates. In: Morgan S (ed) Gender and anthropology. Critical reviews for research and teaching. American Anthropological Association, Washington, DC, pp 41–64

Fedigan LM, Jack KM (2013) Sexual conflict in white-faced capuchins: It's not whether you win or lose. In: Fisher M, Garcia JR, Chang RS (eds) Evolution's empress: Darwinian perspectives on women. Oxford University Press, New York, pp 281–303

Haraway D (1978) Animal sociology and a natural economy of the body politic, part II: the past is the contested zone: human nature and theories of production and reproduction in primate behavior studies. Signs 4:37–60

Haraway D (1989) Primate visions: gender, race and nature in the world of modern science. Routledge, New York

Haraway D (2000) Morphing in the order: flexible strategies, feminist science studies, and primate revisions. In: Strum S, Fedigan LM (eds) Primate encounters. Models of science, gender and society. University of Chicago Press, Chicago, pp 398–420

Harding S (1986) The science question in feminism. Cornell University Press, Ithaca

Isbell LA, Truman PY, Harcourt AH (2012) Stag parties linger: continued gender bias in a female-rich scientific discipline. PLoS One 7(11):1–4 www.plosone.org/e49682

Keller EF (1985) Reflections on gender and science. Yale University Press, New Haven

Landau M (1984) Human evolution as narrative. Amer Scientist 72:262–268

Landau M (1991) Narratives of human evolution. Yale University Press, New Haven

Snow CP (1959) Two cultures and the scientific revolution. The American Library, New York

Strum S, Fedigan LM (eds) (2000) Primate encounters. Models of science, gender and society. University of Chicago Press, Chicago

Chapter 6
Alpha Male Capuchins (*Cebus capucinus imitator*) as Keystone Individuals

Katharine M. Jack and Linda M. Fedigan

Abstract A keystone individual is defined as an individual that has a disproportionate impact on group dynamics relative to its representation in the population. Here we use over 30 years of behavioural, physiological, paternity, and demographic data collected on the Santa Rosa, Costa Rica, capuchin population to address the question of whether or not alpha male white-faced capuchins (*Cebus capucinus imitator*) are keystone individuals. Within groups of white-faced capuchins, the alpha male is easily distinguished from other adult males. He is usually of prime age (10–15 years) and often the largest male due to his pronounced secondary sexual characteristics. He is the most central adult male and the recipient of the highest rates of grooming. He is also the most active participant during encounters with predators and extragroup individuals. Using naturally occurring dispersal events, we assess the impact of the *removal* of an alpha versus subordinate adult male on group dynamics, specifically infant mortality. We found that infant mortality following the removal of an alpha male was more than double the rate observed following the removal of a subordinate adult male. The removal of an alpha male has additional consequences for group success, individual life histories, and population conservation that extend far beyond the immediate aftermath of an alpha male replacement. Based on these findings, we conclude that alpha male white-faced capuchins are keystone individuals, and future research should focus on identifying the factors that enable some males to attain alpha status while others live out their lives as subordinates.

Keywords Keystone individual · Alpha male · Dominance · Demography · Infant mortality

K. M. Jack (✉)
Department of Anthropology, Tulane University, New Orleans, LA, USA
e-mail: kjack@tulane.edu

L. M. Fedigan
Department of Anthropology and Archaeology, University of Calgary, Calgary, AB, Canada
e-mail: fedigan@ucalgary.ca

U. Kalbitzer, K. M. Jack (eds.), *Primate Life Histories, Sex Roles, and Adaptability*, Developments in Primatology: Progress and Prospects,
https://doi.org/10.1007/978-3-319-98285-4_6

6.1 Introduction

> We suggest that the alpha animal in primate groups may not be simply one rank higher than the beta animal. Instead, the alpha animal may behave in a distinctive pattern made up of a variety of related behaviors pertaining to control of the interactions within the group and management of the relationship between the group and external factors. (Gould et al. 1997, p.411)

More than 20 years ago, in their comparative analysis of alpha male ring-tailed lemurs (*Lemur catta*) and white-faced capuchins (*Cebus capucinus imitator*), Gould et al. (1997) recognized that the behaviour of alpha males in these two species was discernably different from that of subordinate males. While their study focused on male vigilance behaviour, they proposed that there are a number of behaviours that enable us and, more importantly, other group members to discriminate the difference between alpha and non-alpha individuals. Earlier studies of primates also recognized alpha males as distinct from subordinate group males. For example, based on his experimental studies of male capuchins (*Cebus albifrons*) and macaques (*Macaca mulatta*), Bernstein (1964 and 1966) suggested that alpha males exhibit a "constellation of behaviours" and when/if this male is removed by either death or experimentation, another male will quickly begin to display these same behaviours.

The existence of *highly influential individuals* that have a large impact on the behaviour and/or success of conspecifics, such as the alpha males described above, has been since noted in many species of social animals (reviewed by Modlmeier et al. 2014). In social insects engaging in collective behaviours, for example, the "behavioural type" of certain individuals (e.g. highly aggressive) has been shown to cause behavioural responses in other group members; these triggered responses can impact prey capture rates in social spiders (*Stegodyphus dumicola:* Pinter-Wollman et al. 2017) or colony movement rates in ants (*Diacamma indicum*; Sumana and Sona 2013), both of which can profoundly affect colony fitness. In species of social mammals that form dominance hierarchies, the alpha male and/or female in a social group will have preferential access to resources that can not only impact social relationships within the group, but can affect the physiology and fitness of other group members (Sih and Watters 2005). In other taxa, influential individuals may have different labels and different roles. In pig-tailed macaques (*Macaca nemestrina*), for example, certain individuals are referred to as *conflict mangers* as they will "police" the group to maintain social order (Flack et al. 2006). In other species some individuals are considered *superspreaders* because of their inordinately high rates of disease transmission (*Oleander aphids*, *Aphis nerii*: Harrison and Mondor 2011), while in bottlenose dolphins (*Tursiops truncatus*), *brokers* are individuals that play a crucial role in maintaining community cohesion (Lusseau and Newman 2004).

Many more terms exist and definitions vary (reviewed by Modlmeier et al. 2014; Pinter-Wollman et al. 2017), though the common feature is that these individuals differentially impact other group members (Modlmeier et al. 2014). In an effort to establish an organizational framework for future research, and to synthesize this topic across fields, Modlmeier et al. (2014) suggest referring to these highly influential individuals as "keystone individuals". Derived from the keystone species concept, which argues that certain species play a unique role crucial to the success of a particular ecosystem (Paine 1969), Modlmeier et al. (2014, p. 55; see also Sih and Watters 2005) define a keystone individual as an "individual that has a disproportionally large, irreplaceable effect on other group members and/or overall group dynamics relative to its abundance".

While the fitness consequences of interindividual trait variation (Bolnick et al. 2003) and the existence of highly influential individuals are not debated, quantitatively documenting the impact of these individuals on a group or population can be problematic, even though it is imperative to informing and expanding this conceptual framework. As such, Modlmeier et al. (2014) outlined a series of methods for detecting the existence of keystone individuals using observational data, controlled manipulations, and/or social network analysis. While the authors state a clear preference for controlled social group manipulation, these methods are not ethical or feasible for most studies of wild primates (Fedigan 2010; MacKinnon and Riley 2010). When using observational data to identify keystone individuals, Modlmeier et al. (2014) suggest collecting descriptions of the behavioural patterns of potential keystone individuals and documenting their effects on group dynamics. If keystone individuals are easily identified, researchers can compare their traits across groups to explore how these traits themselves might be associated with the success of a group (McComb et al. 2011). In some cases, natural experiments wherein the keystone individual is "removed" from the group (e.g. the deaths of aggressive dominant male olive baboons, *Papio anubis*; Sapolsky and Share 2004) can inform our understanding of the impact keystone individuals have on group dynamics, though confounding factors (e.g. change in group size with the removal of the alpha male) may remain an issue.

The concept of keystone individuals was suggested over a decade ago (Sih and Watters 2005) but has seen surprisingly little direct investigation. Interestingly, although Modlmeier et al. (2014) acknowledge the existence of dominant or alpha individuals in social animals as being among the most noticeable examples of keystone individuals, the use of this concept remains largely restricted to the study of collective behaviour in social insects (e.g. Pinter-Wollman et al. 2017 and Pruitt and Pinter-Wolman 2015 on social spiders *Stegodyphus dumicola*). Recognizing the existence of keystone individuals across diverse systems and documenting their impact on group dynamics and success can inform our understanding of the evolution of sociality and cooperative behaviour. In addition, like the keystone species concept, which recognizes particular species whose existence has a far-reaching impact on the functions and survival of the entire biological communities, the keystone individual concept may also help us better understand the impact that the removal of a single, highly influential, individual can have on their group and possibly the population.

In this study, we examine the keystone individual concept using data collected over a 30-year period on the Santa Rosa population of wild white-faced capuchins (*Cebus capucinus imitator*) in Costa Rica. White-faced capuchins are excellent candidates for exploring the keystone individual concept because in this species alpha males are easily recognizable, they behave in a unique way (Gould et al. 1997), and they represent a distinctive life-history stage not experienced by all males (Jack et al. 2014). Following the organizational framework proposed by Modlmeier et al. (2014), we begin with a description of relevant features of white-faced capuchin behavioural ecology (Sect. 6.2) followed by a presentation of our findings to date on the features of alpha male capuchins that distinguish them from other adult males in the population (Sect. 6.3). We then use regularly occurring alpha male replacements and the dispersal of subordinate adult males as natural "removal" experiments to examine the differential impact of these two types of adult males on group dynamics and success, namely, infant mortality (Sect. 6.4). In considering these issues, we explore the significance of alphas as keystone individuals in this species and examine the conservation implications of these findings (Sect. 6.5). Finally, we examine the relevance of the keystone individual concept to our understanding of dominance and make suggestions for future research on this topic (Sect. 6.6).

6.2 General Features of White-Faced Capuchins

Our studies in the Santa Rosa sector of the Área de Conservación Guanacaste, Costa Rica, began in 1983 with demographic, behavioural, ecological, and life-history data collected near-continuously since that time (see Fedigan and Jack 2001, 2012 for detailed history of the project). Santa Rosa is comprised of 108 km^2 of dry deciduous forest and has been federally recognized and protected since 1972. The region experiences extreme seasonal effects, with little to no rain falling during the dry season that extends from mid-December through mid-May and an average of nearly 1500 mm of rain falling during the wet season months (Janzen and Hallwachs 1995; see also Campos 2018, this volume). The Sector Santa Rosa is home to just under 700 capuchins residing in 45 groups (Jack and Fedigan unpublished data), and our long-term research has focused on between two and five groups residing in and around the central administrative area of the park. In addition to our team's research on this species, Susan Perry and her team have been studying a nearby population of white-faced capuchins at Lomas Barbudal, Costa Rica, since 1990 (e.g. Perry et al. 2012), making this one of the most intensively studied species of neotropical primate.

White-faced capuchins are female philopatric and reside in groups comprised of multiple immigrant subadult (6–9 years) and adult males (≥10 years), related adult females, and their offspring (Fedigan 1993). The ratio of adult males to adult females within groups is nearly equal, though a slight bias towards males has been recorded across the Santa Rosa capuchin population (Fedigan and Jack 2011; Fedigan and Jack 2012). Female white-faced capuchins give birth to their first infant at a mean of 6.5 years of age (Fedigan and Jack 2012). The interbirth interval is relatively long (mean = 2.25 years when the prior infant survives) (Fedigan 2003;

Fedigan et al. 2008) in comparison with similarly sized primates, likely because female capuchins continue to nurse their young well into their second year of life (Sargeant et al. 2015). Though births have been recorded during all months, the species is considered a moderately seasonal breeder with most births clustered during the early months of the annual rainy season (May through July) (Carnegie et al. 2011). Females reproduce throughout their lives, and our oldest recorded female to date, followed from birth to death, was just over 23 years of age.

Unlike females who pass directly from the juvenile to adult stage following the birth of their first infant, male capuchins experience a distinctive subadult life-history phase (6 to 9 years of age) during which puberty occurs and they are physically and behaviourally distinct from both juvenile and adult males (Jack et al. 2014). Males do not reach full adult body size until around 10 years of age and only acquire the complete suite of secondary sexual characteristics if and when they attain alpha status within a group (Schoof and Jack 2013; Jack et al. 2014) (see images of adult male capuchins in Fig. 6.1). Male capuchins leave their natal group at approximately 4 years of age (Jack and Fedigan 2004a; Jack et al. 2012) and reside in multiple social groups, at varying dominance ranks, throughout their lives. The oldest male we have recorded from birth is currently 24 years and still an alpha male in one of our study groups (see "Legolas", Fig. 6.1c).

Alpha male replacements (the changeover in a group's alpha male; Teichroeb and Jack 2017) are common in this species, though alpha male tenure length is highly variable, ranging from 2 months to almost 15 years in the Santa Rosa population (mean = 3 years; median = 2.3 years, n = 19 complete alpha male tenures; Jack and Fedigan, unpublished data; see Perry et al. 2012 for similar findings in the Lomas Barbudal population). Intragroup male-male relationships are characterized

Fig. 6.1 Comparative images of adult male capuchins (Photo courtesy of Fernando Campos). Image a: *Weirdo*, a subordinate adult male white-faced capuchin photographed at ~ 18 years of age who, to the best of our knowledge, never attained alpha status during his lifetime. Scars on the bridge of his nose and above his right eye are evidence of his participation in past male-male aggression. Image b: *Legolas*, an alpha male white-faced capuchin photographed at ~15 years of age. This male became alpha via succession following the disappearance of the group's long-term alpha. He sports some large scars on his forehead, one of which was the results of a wound incurred shortly before the photograph was taken. The large black mark on the left side of his chin is natural, but many of the other dark marks are the result of past wounds. Image c: *Marmite*, a recently deposed alpha male white-faced capuchin at ~18 years of age. Note canine punctures under both eyes, scars on the forehead, and almost completely darkened face from past wounds

by mutual tolerance, occasional affiliative interactions, and cooperative group defence against predators and extragroup males (Perry 1998a; Jack 2003; Schoof and Jack 2014). Coresident males in Santa Rosa rarely engage in agonistic interactions making it impossible to reliably determine male dominance rank below the alpha level (Schoof and Jack 2013, 2014). However, capuchins are extremely xenophobic, which leads to intergroup interactions characterized by high levels of aggression, particularly among males from different groups (Perry 1996; Fedigan and Jack 2004). Adult male white-faced capuchins usually immigrate into groups by aggressively and cooperatively evicting resident males (Fedigan and Jack 2004), and immigration is the primary way in which they increase their dominance rank (Jack and Fedigan 2004a; Schaebs et al. 2017). Alpha male replacements often result in the eviction of resident males, infanticide and infant disappearances, and the occasional deaths or disappearances of juveniles, adult males, and adult females (Brasington et al. 2017; Fedigan 2003; Gros-Louis et al. 2003).

6.3 How Do Alpha Male Capuchins Differ from Other Group Males?

Although white-faced capuchins can reside in groups as large as 40 individuals and contain up to 13 adult males (Perry 2012), even the most inexperienced observer of animal behaviour would require no more than an hour of direct observation to discern the group's alpha male. As recognized by Gould et al. (1997), there is a suite of characteristics that enable us, and presumably other group members, to easily discern the alpha male from other males in the group. Since that publication, our long-term research in Santa Rosa, and that of Susan Perry and colleagues at Lomas Barbudal, have greatly increased our knowledge of male behaviour and life histories in this species. The behaviour of an alpha male is conspicuous; he is the most vigilant group member, the most active during encounters with predators and extra-group individuals, the most central, and the most groomed; and he is even the recipient of alpha-specific vocalizations. All in all, there is just no mistaking him. Here we present evidence gathered to date that supports our contention that alpha male capuchins differ from other group males based on their physical, behavioural, physiological, and reproductive characteristics.

6.3.1 Looking Like an Alpha Male

The alpha male in a group of white-faced capuchins is generally in his prime (10–15 years of age), usually the largest male in the group (be this from pure body size differences or his constant state of piloerection), and sports the most pronounced secondary sexual characteristics (e.g. exaggerated brow ridges, exaggerated mandibular girth, and wide shoulder girdle) (Schoof and Jack 2013; Jack et al. 2014; Perry et al. 2017; Schaebs et al. 2017) (Fig. 6.1a, b, and c). While we have not yet quantified our observations that male white-faced capuchins experience an enhancement of their secondary sexual characteristics after attaining alpha status, Fragaszy et al. (2016) recently documented a 20% increase in body mass in male bearded capuchins (*Sapajus libidinosus*) that changed from subordinate to alpha status. The process, they argue, is similar to the rank-dependent weight gain in mandrills (*Mandrillus sphinx*; Setchell and Dixson 2001). In both of these species, weight gain is thought to be triggered by changes in baseline testosterone levels associated with a male's rise to alpha status (Setchell et al. 2008; Mendonça-Furtado et al. 2014), a phenomenon we suspect is also responsible for the changes we have observed in alpha male white-faced capuchins (see Sect. 6.3.4 below; Jack et al. 2014).

6.3.2 Behaving Like an Alpha Male

Within their groups, alpha males are unquestionably the centre of attention, often receiving significantly higher rates of grooming and contact time with other group members compared to subordinate males (Jack 2003). Alpha males (and alpha females) are most frequently found in the centre of their group, though during travel they tend to be in the most forward position (Hall and Fedigan 1997). Being at the front during group movement likely facilitates their priority of access to fruit trees and water since dominance rank has been directly linked to increased energy intake in this species, even in the absence of overt aggression (Vogel 2005).

This forward position in the group may also confer alpha males with the best location for detecting predators or extragroup individuals. Male and female capuchins both engage in vigilance behaviour; however males are more vigilant than females, and alpha males are significantly more vigilant than all other group members (Rose and Fedigan 1995; Perry 1996; Gould et al. 1997; Jack 2001). While some of this vigilance is directed at detecting and avoiding predators, much of it appears to be aimed at detecting conspecifics who might be attempting to take over their alpha position within the group (vigilance is highest in areas of home range overlap where intergroup encounters are most likely to occur) (Rose and Fedigan 1995). Although intragroup relationships among coresident males are tolerant and sometimes affiliative (Jack 2003; Schoof and Jack 2014), severe and sometimes lethal male-male agonism occurs in the context of intergroup interactions and alpha male replacements (Perry 1996; Fedigan and Jack 2004). During intergroup interactions, male white-faced capuchins are the primary participants with alpha males being the first to initiate the encounter and take the lead in the attack, while females only rarely participate and generally flee with immature group members (Jack unpublished data; Perry 1996; Rose and Fedigan 1995; but see Crofoot 2007).

6.3.3 Becoming an Alpha Male

Subadult and juvenile males can quite easily join existing breeding groups as subordinates, often remaining behind with the new group following intergroup interactions or by tagging along with adult males moving between groups (Jack and Fedigan 2004a, b; Jack et al. 2012). However, once males attain adult body size at around 10 years of age, their movement between groups is no longer tolerated, and they must usually fight their way into a bisexual group whether they are attempting to enter as an alpha or a subordinate. To combat this resistance, males often join groups in parallel (dispersing in the company of other males or targeting groups containing familiar males) (Jack and Fedigan 2004a; Perry 2012; see Wikberg et al., 2018, this volume). In cases where multiple males take up residency together, one of them will rise to the position of resident alpha male, often (but not always) with some jockeying among them for this position. We have also observed subordinates dispersing with their group's alpha and continuing to retain their subordinate status through multiple dispersal events.

We have observed male white-faced capuchins to become alpha via all five of the various modes to alpha outlined by Teichroeb and Jack (2017). They are listed here from most to least commonly observed in our study population as reported in Brasington et al. (2017):

1. *Group takeovers*, which usually involve coalitions of extragroup males aggressively attacking and overthrowing the alpha male of a group (33% of AMRs).
2. *Successions*, where a subordinate group male is able to take up the position of alpha male following the death, dispersal, or disappearance of the former alpha male (29% of AMRs). Note that in this case, the departure of the former alpha is not due to aggression received from other males in the group.
3. *Waltz-ins*, where an extragroup male(s) is able to take up residency in a group that is devoid of resident males (often due to their immigration into other groups) (24% of AMRs).
4. *Rank reversals*, where one or more subordinate resident males aggressively challenge and overthrow the existing alpha male and take up this position in the group (10% of AMRs). In both takeovers and rank reversals, the deposed alpha may be killed, remain in the group as a subordinate, or disperse/disappear.
5. *Group fissions*, where an existing group splits into two or more sister groups, with the former alpha retaining his position in one of the subgroups and a subordinate resident male becoming alpha in the other (5% of AMRs).

6.3.4 The Physiology of Alpha Males

Even in the complete absence of observations of how alpha males look and behave, we can discriminate alpha males from subordinate males simply by examining their hormone profiles. Androgens, a group of hormones associated with male secondary sexual characteristics, sexual activity, and aggressive behaviour, have been among the most widely studied hormones with respect to male dominance in mammals. Androgens are associated with the facilitation of aggression in a reproductive context, including aggression that occurs during challenges for access to mates or territory and during the formation of dominance relationships (Wingfield et al. 1990). Therefore, it is not surprising that dominant males in many non-human primate species have higher androgen levels than subordinates when male-male competition is high, such as during the mating season (Bales et al. 2005; Gould and Ziegler 2007), during periods of social instability (Sapolsky 1993; Marshall and Hohmann 2005), or when the timing of challenges for mates or the threat of infanticide is unpredictable (Muller and Wrangham 2004b; Muehlenbein et al. 2004; Beehner et al. 2006; Setchell et al. 2008; Schoof and Jack 2013).

Our cross-sectional analysis found that alpha male fecal androgen (fA) levels in white-faced capuchins were an average of 11 times higher than those of subordinate adult males, whose levels did not differ significantly from subadult males (Jack et al. 2014). A recent longitudinal study of this same species at Lomas Barbudal confirmed these differences (Schaebs et al. 2017). In adult males, elevated fA levels appear to occur independent of male age and be dependent upon attaining alpha status. Schoof et al. (2012) documented a single case of a subordinate male's succession to alpha following the emigration of the group's former alpha male. It took nearly 3 months for this beta male's fA to reach alpha levels. Schaebs et al. (2017) similarly documented increases in fA levels for 15 of the 18 males in their study who rose from subordinate to alpha status. Of the three males who did not experience an increase in fA levels on attaining alpha status, the authors suggest that in one case sampling was an issue (too few samples), while the other two males displayed high androgen levels prior to attaining alpha status due to high levels of competition even as subordinates. Collectively these results indicate that male fA levels may be highly influenced by the competitive situation faced in their groups (Schoof and Jack 2013; Schaebs et al. 2017). Indeed, both of these studies have documented that while fA levels can vary widely across alpha males, within groups, alphas always exhibit the highest fA levels.

Alpha and subordinate male capuchins also differ in the fecal glucocorticoid levels (fGCs), with alpha males generally displaying higher fGCs than subordinate males (Schoof et al. 2012; Jack et al. 2014). GCs are a group of hormones associated with the stress response, which is adaptive in the short term because it focuses energetic resources to areas relevant to immediate survival needs (e.g. energy release) and temporarily suppresses nonvital functions (e.g. growth, reproduction) (Sapolsky 2005). However, long-term or chronic expression of the stress response can be detrimental to individual health (reviewed in Sapolsky 2005; Gesquiere et al. 2011). The relationship between GCs and dominance rank is variable across species and appears to be closely associated with perceived stress by dominant and subordinate individuals (Abbott et al. 2003; Goymann and Wingfield 2004; Sapolsky 2005). For example, in species where threats to male dominance rank are unpredictable, alpha male GC levels will remain elevated (Schoof and Jack 2013), while in other species alpha GCs are only elevated during periods of group instability (Sapolsky 2005).

The higher fGCs in alpha versus subordinate male white-faced capuchins is likely the result of the higher metabolic costs associated with their elevated vigilance and more intensive participation in intergroup encounters and group protection, behaviours which are also undoubtedly influenced by alpha male androgen levels (Schoof and Jack 2013). Given that alpha male capuchins sire the majority of group infants (see Sect. 6.3.5 below), their elevated fGCs are consistent with the idea of perceived risk (Brockman et al. 2009; Schoof and Jack 2013). If their group is targeted for a takeover, alpha males have the most to lose in terms of current reproductive success (i.e. living infants) and future reproductive opportunities, so they should display the highest stress levels. These external threats to male rank are persistent and unpredictable and should lead to consistently higher GC levels in dominant males, as has been reported for chimpanzees (Muller and Wrangham 2004a). However, alpha male GC levels are undoubtedly influenced by other social factors such as the number of coresident males, dependent infants, and/or fertile females in the group (Schoof et al. 2012, 2016; Schoof and Jack 2013). GC levels are also impacted by environmental factors such as rainfall, temperature, or photoperiod (Schoof et al. 2016).

6.3.5 Reproducing Like an Alpha Male

The behavioural, physical, and physiological alpha male traits described above do not come without costs. High-ranked males may experience higher energetic costs (Barrett et al. 2003; Muller and Wrangham 2004b; Muehlenbein et al. 2004; Gesquiere et al. 2011; Mendonça-Furtado et al. 2014), increased rates of injury (Rose 1994; Rose and Fedigan 1995; Drews 1996; but see Archie et al. 2012), and greater health risks (Sapolsky 1993; Muehlenbein 2009; Muehlenbein and Watts 2010; but see Archie et al. 2012). However, in capuchins, as in many other primate species, this assemblage of alpha male traits collectively enables certain males to successfully compete with other males to gain access to breeding groups of females, the benefits of which can be enormous (Di Fiore 2003). This is certainly the case for alpha male white-faced capuchins as they sire between 70 and 100% of all offspring born during their tenure (see Chap. 7 by Wikberg et al. 2018, this volume; Jack and Fedigan 2006; Muniz et al. 2010; Perry 2012).

While the discovery of high reproductive skew towards alpha males can be explained as benefits for the high costs associated with occupying this position, these results were nonetheless curious given the behaviour of subordinate adult male group members. Subordinate adult males are fully integrated into their social group rather than occupying a more peripheral position, as has been reported for some populations of the closely related tufted capuchins (e.g. Janson 1986). For example, subordinate adult male white-faced capuchins also engage in affiliative behaviours with adult females, males, and immature group members (Jack 2003; Schoof and Jack 2014), and, while their vigilance efforts are not at the level of alpha males, they play a key role in group defence from both predators and extragroup males (Rose 1994; Rose and Fedigan 1995; Jack 2001). Because the presence and activities of subordinate males provide benefits, alpha males make a concerted effort to keep them in the group. When subordinate males are separated from the group, alphas will repeatedly lost-call (capuchin long calls) for them, and, once reunited with the group, alphas perform an elaborate reunion display similar to that observed during mating (see description in Perry 1998a). It is also not surprising then that alphas are tolerant of the mating activities of subordinate males and overt intragroup male-male mating competition is rare (Schoof et al. 2014b). While our evidence to date indicates that a female is more likely to mate with a subordinate adult male when she is pregnant or lactating (Carnegie et al. 2005), some subordinates do sire offspring, particularly when alpha males have long tenures and their daughters begin to reproduce (Muniz et al. 2006, 2010; Godoy et al. 2016a, b; Wikberg et al. 2017 and Chap. 7 in this volume).

6.4 Alpha Males as Keystone Individuals: The Impact of Alphas on Group Success and Survival

It is clear from the above review that alpha males embody a distinct category of individual and occupy a unique role within capuchin groups. However, according to Modlmeier et al. (2014), it is critical to differentiate keystone individuals from "generics" the latter of which they argue are replaceable. One could argue that an alpha male capuchin is replaceable in that when he is gone, another male quickly takes up the vacant position in the group and rises to alpha status (e.g. Perry 1998b; Schoof et al. 2012). It is not, however, that simple. The removal of a true keystone individual should differ from the removal of a generic in that the former should have lasting consequences to the group (Modlmeier et al. 2014). To differentiate between the two, therefore, requires removal experiments (Sih et al. 2009) and comparisons between the effects of removing keystone individuals and other, generic, individuals.

All male white-faced capuchins disperse from the natal group at approximately 4 years of age (Jack et al. 2012), and males continue to move between groups throughout their lives at approximately 4-year intervals (Jack and Fedigan 2004a). This pattern of dispersal means that most males reside in three or more distinct social groups during their lifetimes, occupying varying ranks, and some males are alphas in more than one group (Jack and Fedigan 2004a; Schaebs et al. 2017). This regular occurrence of adult male dispersal, and the frequent turnover of group alpha males (alpha male replacements), provides natural removal experiments enabling the examination of their consequences on group dynamics.

The removal (i.e. death, disappearance, dispersal) of any individual from a social group will have some sort of measurable impact on a group. For example, we have documented the impact of subordinate male dispersal on the vigilance behaviour of alpha males (Jack 2001). Here we examine the impacts of male movement on group demographics via an examination of infant mortality in association with changes in male group membership. Increased infant mortality in association with changes in a group's alpha male has been well documented in primates and other social mammals, particularly for those in which male dominance rank is positively associated with male reproductive success (see review by Palombit 2015). In our study groups, we have shown that infant mortality increases significantly during times associated with alpha male replacements compared with periods of stability in the male dominance hierarchy (e.g. Fedigan 1993; Fedigan et al. 2008; Kalbitzer et al. 2017; Brasington et al. 2017). However, we have not yet examined the effects of subordinate adult male dispersal on infant mortality (nor are we aware of these data for other species), a comparison critical to determining whether white-faced capuchin alpha males are indeed true keystone individuals. Following Modlmeier et al. (2014), we address this issue by exploring the following questions:

1. What impact does the "removal" of an alpha male have on infant mortality?
2. What impact does the "removal" of a subordinate adult male have on infant mortality?
3. Are the effects of alpha male replacements (AMRs) on infant mortality long-lasting?

We examined long-term data on infant births and deaths, as well as data on male movement and rank change in five study groups for variable periods between 1986 and 2015 (a total of 85 group years; see Fedigan and Jack 2012 for additional details on data collection). During this period, we documented the birth of 221 infants, 21 alpha male replacements, and the emigration of 53 subordinate adult non-natal males (i.e. all immigrant males aged 10 years and older). Our analysis is restricted to adult males because (a) in all but very rare circumstances (i.e. where there are no other males in the group), only adult males become alphas, (b) we have never observed or suspected a non-adult male to kill an infant, and (c) subadult males only rarely sire offspring (unpublished data; see also Muniz et al. 2010). Based on our prior observation that infants (<1 year of age) are more vulnerable to infanticide in association with alpha male replacements (Fedigan 2003) and following Brasington et al. (2017) and Kalbitzer et al. (2017), we divided our sample of 221 infants into (a) those born in association with an AMR in that the infant was <1 year at the time of the AMR, or was born within 5.5 months following the AMR (i.e. sired prior to but born after the AMR; 5.5 months is the median gestation length for *C. c. imitator*, Carnegie et al. 2011), and (b) those born during periods of group stability.

6.4.1 *What Impact Does the "Removal" of an Alpha Male Have on Infant Mortality?*

In our most recent analysis (Brasington et al. 2017), we documented infant mortality in association with AMRs at 50.9%, a mortality rate more than double the 26.2% recorded during times of group stability (Fig. 6.2) (see Perry 2012 for similar findings in the Lomas Barbudal capuchin population). As mentioned above, all five types of AMRs have been observed in the Santa Rosa capuchins, and each is associated with observed infanticides and higher mortality rates than recorded during periods of group stability (Brasington et al. 2017). Only 4 of the 21 AMRs in our sample were not associated with infant deaths; 2 of these involved groups that did not contain infants, the third involved a group that contained a single infant aged 7 months, and the fourth involved a group that contained a single infant aged 10 months, which puts them at a significantly lower risk of infanticide than infants <6 months of age (Brasington et al. 2017).

Over the years of our study in Santa Rosa, we have witnessed many infanticides and have strong inferential evidence that many more "missing infants" have been the victim of infanticide. While a few cases of infanticide have occurred during periods of social stability (Schoof et al. 2014a), the majority occur during periods associated with alpha male replacements. AMRs are often, but not always, associated with high levels of aggression exchanged between resident and incoming males. However, infant deaths and disappearances most often occur after a new alpha male has established himself in the group, rather than during the initial periods of aggression associated with AMRs (Brasington et al. 2017). In many cases, infants that are born up to 5.5 months after an AMR are targeted, and these infants, who were confirmed or strongly suspected as offspring of the former alpha male, experience the highest observed mortality rates in our sample (63%; Brasington et al. 2017).

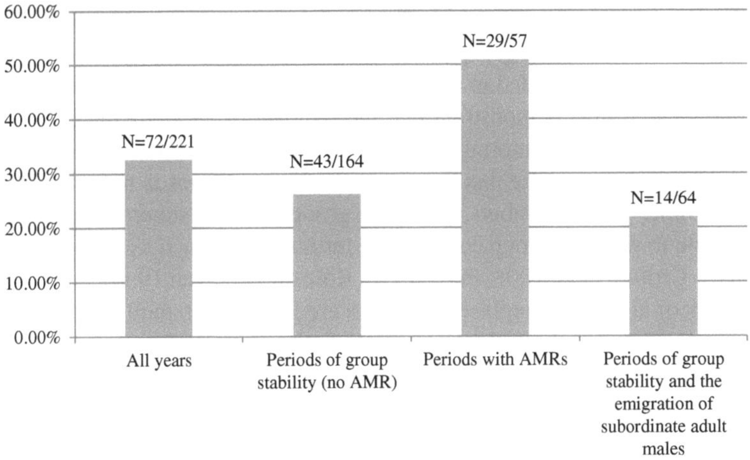

Fig. 6.2 Infant mortality rates across all years and in association with periods of group stability, alpha male replacements, and subordinate male emigrations (N = # of infants died/total number infants born during each period)

6.4.2 What Impact Does the "Removal" of a Subordinate Adult Male Have on Infant Mortality?

When a subordinate adult male dies or disappears, he is not "replaced by another male". Indeed, many of our study groups have spent long periods of time with the "alpha" as the sole resident male following the dispersal of all subordinate males (e.g. Jack 2001, 2003; Jack and Fedigan 2004b). A total of 53 subordinate, non-natal, adult males emigrated or disappeared (referred to as emigrations from herein) from our groups during the study period. Of these males, 22 emigrated in association with AMRs, in that their emigration coincided with the departure of the former alpha male and often other adult and immature males in the group. As mentioned above, male white-faced capuchins often engage in parallel dispersal, and during AMRs it is not uncommon for all males to emigrate simultaneously, resulting in a complete replacement of resident group males (Fedigan and Jack 2004; Jack and Fedigan 2004a; Wikberg et al. 2014). The 22 males that emigrated in association with an AMR were not included in these analyses due to the confounding impact of AMRs on infant mortality.

The remaining 31 subordinate males in the sample emigrated during periods of stability in the group's alpha male. Following the same criteria used above for alpha male removals, 64 of the 164 infants born during periods of group stability were either <1 year at the time of the subordinate male emigration or they were born within 5.5 months following the emigration event (i.e. sired prior to but born after the emigration). Of these 64 infants, 14 died before reaching 1 year of age, for a mortality rate of 21.9% (Fig. 6.1). This is lower than the infant mortality rate recorded during periods of group stability (26.2%) and less than half of the 50.9% infant mortality rate recorded in association with AMRs.

While these results demonstrate that the removal of an alpha male has a much greater impact on infant mortality than does the removal of a subordinate adult male, it is important to note that the removal of subordinates is not without consequence to other group members. Capuchin groups lacking subordinate adult males are less able to win intergroup competition for resources, such as access to water or fruit trees (Crofoot et al. 2008; Perry 1996; Rose and Fedigan 1995), and they are at greater risk of alpha male replacements than are groups with multiple resident adult males (Fedigan and Jack 2011). In addition, in the current data set, there were several instances in which the death/disappearance of infants appeared to be directly related to the departure of the subordinate; 5 of the 14 infants in our sample that died/disappeared in association with a subordinate male leaving the group occurred in close temporal association with the subordinate male's emigration, and 1 of these deaths was an observed infanticide by the group's alpha. The differentiating feature here, however, is that the resulting infant death was due to the action of the resident alpha male, rather than a new male taking up the subordinate's position, as we have documented to occur in the wake of AMRs. Unfortunately, this observation was made in the early years of our study (the 1980s) predating our routine use of genetic paternity data; thus we are not able to determine if the infant killed by the alpha male was sired by the departed subordinate male.

6.4.3 Are the Effects of Alpha Male Removals on Infant Mortality Long-Lasting?

In their conceptual framework, Modlmeier et al. (2014) stress the importance of determining if the removal of keystone individuals has long-lasting effects or if they end after a new individual takes up that particular role or position in the group. Sapolsky and Share (2004), for example, documented the long-lasting effects of the death of all high-ranking aggressive males in a troop of olive baboons (*Papio anubis*) after a tuberculosis outbreak in their study group. Following their "removal" the remaining, more passive, group males rose to the higher ranks, and the entire temperament of the group became more docile in nature, and these changes persisted for more than a decade.

In white-faced capuchins, the impacts of AMRs are very long-lasting. AMRs occur over a period of weeks or months, while the new male(s) integrate themselves into the group, sometimes jockeying for the alpha position when more than one male immigrates together. In addition, female and immature members of the group will flee from and avoid these new males, and it can take many weeks before they begin to tolerate their presence and accept them into the group (Fedigan and Jack 2013). Prolonged female avoidance of these new males is likely a counterstrategy to infanticide, which occurs not only in the immediate context of the replacement itself but continues well after the new alpha male is established in the group. Of the six observed or strongly inferred infanticides in our sample (dead infants were found with canine puncture wounds similar to those in observed cases), the range of time from AMR to infant death was 0–90 days (mean = 46 days). If we extend this to our entire sample of infants that perished in association with AMRs, the mean time between the AMR and infant death was 98 days (range 0–254 days, median 94 days), including numerous cases where infants born up to 5 months following an AMR have perished. These latter infants would have most likely been sired by the group's previous alpha male (median gestation length is 5.5 months; Carnegie et al. 2011). Thus, the impact of alpha male removal on infant survival can last for many months. The impact on female reproduction is even longer lasting. If, for example, a female loses a 3-month-old infant, this represents a minimum of 20.5 months of reproductive time lost (5.5-month gestation +3 months of rearing lost infant +6.5 months post-nursing infertility period [Recabarren et al. 2000] + 5.5-month gestation for the next infant). If a female is very successful and experiences, for example, a 15-year reproductive career, the death of this 3-month-old infant in association with a single AMR represents 11.4% of her reproductive career. It is little wonder that female capuchins actively promote multi-male groups and resist AMRs to the best of their abilities (Fedigan et al. 2008; Fedigan and Jack 2013).

6.5 Discussion

In light of the evidence presented here, white-faced capuchin alpha males appear to be keystone individuals; they have a large impact on group dynamics and success relative to their representation in the population, and their "removal" from a group has a much larger, and long-lasting, impact on group dynamics (infant mortality in particular) than does the removal of a subordinate male. To be sure, changes in alpha males within groups are associated with a suite of consequences extending well beyond increased infant mortality. The demographic impacts of AMRs can continue for years and may indefinitely influence the lives of some individuals. As we have shown elsewhere, changeovers in a group's alpha male impact female reproductive success by altering interbirth intervals (Fedigan et al. 2008), and they can even disrupt the species-specific pattern of female philopatry (Jack and Fedigan 2009). Infants and juveniles that survive an AMR may be nonetheless impacted by the availability (or lack thereof) of play, alliance, and dispersal partners (e.g. see Perry 2012). Infant mortality in association with AMRs will impact the presence of kin in the social group, which may be of particular importance in species where male coalitions are a prime determinant for battles over rank acquisition and maintenance (Bissonnette et al. 2015). In species like white-faced capuchins, long-term coalition partners, who may or may not be kin, are maintained via parallel dispersal that persists even into adulthood (Jack and Fedigan 2004a, b; Perry 2012; Wikberg et al. 2014). These partners can impact a male's success in entering a social group, rise to alpha status, and, ultimately, reproductive success.

Given that alpha males sire the majority of offspring born into a group, AMRs often result in the removal of the father of surviving infants and juveniles. Studies from several primate species have now shown that the presence of the father in the group during immaturity can impact the timing of a number of key life-history events. For example, Charpentier et al. (2008) found that offspring age at sexual maturity decreased for both male and female baboons when the father was present in the group. Perry et al. (2017) have shown that in white-faced capuchins, the occurrence of AMRs during the first 5 years of a male's life influences the timing of natal dispersal and an individual's rise to alpha status.

It is clear that the "removal" of an alpha male from a group has far-reaching, long-lasting impacts. While our discussion has focused on the impact of the "natural removal" of an alpha male (via death, predation, dispersal, or group takeovers), the same impacts will be felt following the "unnatural removal" of an alpha male, such as occurs from poaching. While Costa Rica is well-known for its progressive conservation initiatives (Kappelle et al. 2016), poaching remains an ongoing threat to wildlife. When a capuchin group encounters a poacher, it is the alpha male who takes the lead in defending the group. He will threaten the poacher, often for an extended period of time and at relatively close range, while the rest of the group retreats into the forest. Similar observations have been made where alpha males will threaten oncoming traffic (such as a busload of tourists) while the group crosses the tarmac. The death of an alpha male, no matter whether from poachers or an oncoming vehicle, sets off a chain of events. A new male will quickly try to take up the alpha male position in the group, most likely fighting (sometimes to the death) with other males who also want to assume the alpha position. Group infants will die, adult females will be injured trying to protect their infants, and group demographics will be altered, the impacts of which will remain for years or even lifetimes (e.g. the availability of coalitionary and dispersal partners and the timing of important life-history events). Knowing that killing an alpha male is not simply killing a single monkey, but that these actions will negatively impact that entire group and possibly the community, may help inform community-driven conservation initiatives (e.g. see Jiménez et al. 2017).

While it is clear alpha males play a unique role in group dynamics and success, we are still far from understanding the mechanisms underlying how and why some males become alphas and others do not. Do all males have the capacity to become an alpha male, or is there only a certain "type" of male who is able to attain alpha status? Modlmeier et al. (2014, pp 58–59) suggest that the emergence of keystone individuals occurs (1) via Darwinian selection for their particular "keystone-conferring genotype" and/or (2) due to particular "experiential, state-dependent, or context-dependent phenomenon". We have documented a suite of both behavioural and physiological characteristics associated with alpha male white-faced capuchin, some of which are likely to be under genetic control or influence. Given the strong genetic determination of testosterone levels identified in humans (Ohlsson et al. 2011), it is likely white-faced capuchins are similar in that only some males have the potential to produce the extremely elevated androgen levels characteristic of alpha males (Schoof et al. 2012; Fragaszy et al. 2016; Schaebs et al. 2017) and the immune system to sustain chronically elevated cortisol levels (Sapolsky 2005; Rakotoniaina et al. 2017; but see Beehner and Bergman 2017). If there is a heritable component to an individual's ability to attain high dominance status (e.g. the ability to produce high levels of androgens), it is equally likely that a specific set of environmental variables need to align to enable a particular male to realize this potential. These environmental variables may include (but are not limited to) maternal androgen levels during prenatal development (e.g. Ryckmans et al. 2015), maternal rank (e.g. Meikle and Vessey 1988; Onyango et al. 2008; Surbeck et al. 2011), access to sufficient nutrition during key developmental phases (e.g. Lee et al. 2013), the presence of the father in a group during development (e.g. Lynch et al. 2017), the timing of pubertal onset (e.g. Bercovitch 1993), and/or the possession of particular social skills or personality types that may be acquired or informed during socialization (e.g. Perry et al. 2017).

It is important to remember that dominance is a social construct dependent upon the interactions and reactions of given individuals in particular situations. It is likely that some male capuchins who possess the potential to display the suite of alpha male traits never have the opportunity to become an alpha male. Likewise, it is possible that some males who do not possess these traits will nonetheless be able to take up the role of alpha male within a group as a result of particular circumstances (e.g. being in the right place at the right time). For example, Perry and Manson (2008) describe a situation where multiple adult males fought for the alpha position over several years. Eventually all of these males were so badly wounded that they deserted the group, opening up the opportunity for young, inexperienced males from a neighbouring group to "waltz-in" and take up residency uncontested. We have observed similar cases in our own study groups in Santa Rosa. In one of our smallest study groups, a subadult male (aged 8 years) rose uncontested to alpha status after the group's alpha disappeared following an intergroup encounter. The young male remained alpha for almost 3 years and sired three infants, though none survived their first year of life despite being born during a period of group stability (i.e. their deaths were not associated with AMRs). We have also observed two cases where a series of botched takeover attempts eventually lead to the extinction of the groups, with group females disappearing or emigrating to neighbouring groups (Jack and Fedigan 2009).

6.6 Conclusions and Future Directions

Identifying traits and processes influencing individual variability in reproductive success is fundamental to understanding evolutionary processes. In white-faced capuchins, as in many other socially living mammals including humans, one of the key variables influencing male reproductive success is dominance status (Jack and Fedigan 2006; Muniz et al. 2010). However, we still know very little about how and why some males attain alpha status while others do not. Indeed few studies conducted to date have investigated differences in males prior to their achievement of a particular rank in a dominance hierarchy (see Bercovitch 1993; Setchell and Dixson 2001; Beehner et al. 2006; Schoof et al. 2012; Schaebs et al. 2017). Such data are crucial to furthering our understanding of how dominance impacts the behaviour and reproductive success of individuals and his/her surrounding conspecifics (Chichinadze et al. 2014).

The majority of research on keystone individuals conducted to date has focused on how the addition or removal of keystone individuals displaying particular behavioural types impacts collective behaviour and group success within groups of social insects (e.g. Sih and Watters 2005; Pruitt and Pinter-Wollman 2015; Pinter-Wollman et al. 2017). Due to the constraints on a primatologist's ability to experimentally manipulate group composition, this line of research remains virtually unexplored in wild primates. What little is known does not address the keystone individual concept directly and remains largely descriptive. However, as was the case in the current study, the natural dispersal of keystone individuals from the group enables researchers engaged in long-term behavioural studies to document the resultant effects of the departure of such individuals.

The analysis of primate personalities may provide another arena for exploring and identifying the traits and emergence of keystone individuals. Perry et al. (2017) recently examined the effects of personality types on the age of natal dispersal and the time to attaining an alpha male position in wild white-faced capuchins. They found that more extraverted males (those that were scored as more social, aggressive, assertive, fearless, etc.) attained alpha status sooner than males displaying other personality types. This area of research, and the ongoing long-term studies of wild primates across taxa (see Kappeler et al. 2012), holds promise for future analysis of how keystone individuals longitudinally impact group success. It is time, as Modlmeier et al. (2014, p. 53) suggest, for us to pay more attention to the consequences of trait variation at the individual level, rather than continuing to ignore it and treat it as "*mere statistical noise*".

The study of keystone individuals and their impacts on social groups, populations, and communities is still in its infancy, and we are just beginning to recognize the evolutionary, ecological, and conservation importance of these individuals (Modlmeier et al. 2014). Future research on this topic, particularly in long-lived species such as primates, should explore variation in trait expression across keystone individuals, as well as the impacts of this variation on both individual and group success, and examine the social and ecological factors influencing the development and manifestation of these traits. Further exploration of whether or not alpha individuals, both males and females, in other species may fit the definition of keystone individuals is also warranted.

Acknowledgements We thank the Costa Rican National Park Service for permission to work in Santa Rosa National Park from 1983 to 1989 and the administrators of the Área de Conservación Guanacaste (especially Sr. Roger Blanco Segura) for allowing us to continue our research in the park through the present day. Over the years, many students and local assistants have contributed to the life-history database on the Santa Rosa monkeys, and we are very thankful to all of them. Dr. John Addicott developed the database, which he has since maintained with the help of a series of technicians, to whom we are also grateful. K.M. Jack's research has been supported by the National Geographic Society, Leakey Foundation, Nacey Maggioncalda Foundation, Louisiana Board of Regents, National Science Foundation (grant 0926039), Sigma Xi, Natural Sciences and Engineering Research Council of Canada, Alberta Heritage Scholarship Fund, Killam Foundation, Royal Anthropological Institute, and grants from Tulane University's Research Enhancement Grant, Newcomb Institute, Stone Center for Latin American Studies, Carol Lavin Bernick Grant, and Lurcy Grants. L.M. Fedigan's research has been supported by the Canada Research Chairs Program and since 1983 by ongoing Discovery Grants from the Natural Sciences and Engineering Research Council of Canada (NSERC). We are thankful to Dr. Susan Perry, Urs Kalbitzer, and one anonymous reviewer for their helpful comments and suggestions on earlier versions of this manuscript.

References

Abbott DH, Keverne EB, Bercovitch FB et al (2003) Are subordinates always stressed? A comparative analysis of rank differences in cortisol levels among primates. Horm Behav 43:67–82. https://doi.org/10.1016/S0018-506X(02)00037-5

Archie EA, Altmann J, Alberts SC (2012) Social status predicts wound healing in wild baboons. Proc Natl Acad Sci U S A 109:9017–9022. https://doi.org/10.1073/pnas.1206391109

Bales KL, French JA, Hostetler CM, Dietz JM (2005) Social and reproductive factors affecting cortisol levels in wild female golden lion tamarins (Leontopithecus rosalia). Am J Primatol 67:25–35

Barrett L, Henzi P, Dunbar R (2003) Primate cognition: from "what now?" to "what if ?". Trends Cogn Sci 7(11):494

Beehner JC, Bergman TJ (2017) The next step for stress research in primates: to identify relationships between glucocorticoid secretion and fitness. Horm Behav 91:68–83. https://doi.org/10.1016/J.YHBEH.2017.03.003

Beehner JC, Bergman TJ, Cheney DL et al (2006) Testosterone predicts future dominance rank and mating activity among male chacma baboons. Behav Ecol Sociobiol 59:469–479. https://doi.org/10.1007/s00265-005-0071-2

Bercovitch FB (1993) Dominance rank and reproductive maturation in male rhesus macaques (Macaca mulatta). J Reprod Fertil 99:113–120. https://doi.org/10.1530/JRF.0.0990113

Bernstein IS (1964) Role of the dominant male rhesus monkey to external challenges to the group. J Comp Physiol Psychol 57:404–406

Bernstein I (1966) Analysis of a key role in a capuchin (Cebus albifrons) group

Bissonnette A, Perry S, Barrett L, et al (2015) Coalitions in theory and reality: a review of pertinent variables and processes. Behaviour 152:1–56. https://doi.org/10.1163/1568539X-00003241

Bolnick DI, Svanbäck R, Fordyce JA et al (2003) The ecology of individuals: incidence and implications of individual specialization. Am Nat 161:1–28. https://doi.org/10.1086/343878

Brasington LF, Wikberg EC, Kawamura S et al (2017) Infant mortality in white-faced capuchins: the impact of alpha male replacements. Am J Primatol:e22725. https://doi.org/10.1002/ajp.22725

Brockman DK, Cobden AK, Whitten PL (2009) Birth season glucocorticoids are related to the presence of infants in sifaka (Propithecus verreauxi). Proc Biol Sci 276:1855–1863. https://doi.org/10.1098/rspb.2008.1912

Campos CA (2018, this volume) *A synthesis of long-term environmental change in Santa Rosa, Costa Rica* In: Kalbitzer U, Jack KM (eds) Primate life histories, sex roles, and adaptability – essays in honour of Linda M. Fedigan. Developments in primatology: progress and prospects. Springer, New York, pp 331–354

Carnegie SD, Fedigan LM, Ziegler TE (2005) Behavioral indicators of ovarian phase in white-faced capuchins (*Cebus capucinus*). Am J Primatol 67:51–68. https://doi.org/10.1002/ajp.20169

Carnegie SD, Fedigan LM, Melin AD (2011) Reproductive seasonality in female capuchins (*Cebus capucinus*) in Santa Rosa (Area de Conservación Guanacaste), Costa Rica. Int J Primatol 32:1076–1090. https://doi.org/10.1007/s10764-011-9523-x

Charpentier MJE, Tung J, Altmann J, Alberts SC (2008) Age at maturity in wild baboons: genetic, environmental and demographic influences. Mol Ecol 17:2026–2040. https://doi.org/10.1111/j.1365-294X.2008.03724.x

Chichinadze K, Chichinadze N, Gachechiladze L et al (2014) Physical predictors, behavioural/emotional attributes and neurochemical determinants of dominant behaviour. Biol Rev 89:1005–1020. https://doi.org/10.1111/brv.12091

Crofoot M (2007) Mating and feeding competition in white-faced capuchins (*Cebus capucinus*): the importance of short- and long-term strategies. Behaviour 144:1473–1495. https://doi.org/10.1163/156853907782512119

Crofoot MC, Gilby IC, Wikelski MC, Kays RW (2008) Interaction location outweighs the competitive advantage of numerical superiority in *Cebus capucinus* intergroup contests. Proc Natl Acad Sci U S A 105:577–581. https://doi.org/10.1073/pnas.0707749105

Di Fiore A (2003) Molecular genetic approaches to the study of primate behavior, social organization, and reproduction. Am J Phys Anthropol 122:62–99. https://doi.org/10.1002/ajpa.10382

Drews C (1996) Contexts and Patterns of Injuries in Free-Ranging Male Baboons (Papio Cynocephalus). Behaviour 133:443–474. https://doi.org/10.1163/156853996X00530

Fedigan L (1993) Sex differences and intersexual relations in adult white-faced capuchins (*Cebus capucinus*). Int J Primatol 14:853–877. https://doi.org/10.1007/BF02220256

Fedigan LM (2003) Impact of male takeovers on infant deaths, births and conceptions in *Cebus capucinus* at Santa Rosa, Costa Rica. Int J Primatol 24:723–741. https://doi.org/10.1023/A:1024620620454

Fedigan LM (2010) Ethical issues faced by field primatologists: asking the relevant questions. Am J Primatol 72:754–771. https://doi.org/10.1002/ajp.20814

Fedigan LM, Jack K (2001) Neotropical primates in a regenerating Costa Rican dry forest: a comparison of howler and capuchin population patterns. Int J Primatol. https://doi.org/10.1023/A:1012092115012

Fedigan LM, Jack KM (2004) The demographic and reproductive context of male replacements in *Cebus capucinus*. Behaviour 141:755–775

Fedigan LM, Jack KM (2011) Two girls for every boy: the effects of group size and composition on the reproductive success of male and female white-faced capuchins. Am J Phys Anthropol. https://doi.org/10.1002/ajpa.21414

Fedigan LM, Jack KM (2012) Tracking neotropical monkeys in Santa Rosa: lessons from a regenerating Costa Rican dry forest. In: Long-term field studies of Primates. Springer Berlin Heidelberg, Berlin, Heidelberg, pp 165–184

Fedigan LM, Jack KM (2013) Sexual conflict in white-faced capuchins. In: Evolution's empress. Oxford University Press, pp 281–303

Fedigan LM, Carnegie SD, Jack KM (2008) Predictors of reproductive success in female white-faced capuchins (*Cebus capucinus*). Am J Phys Anthropol 137:82–90. https://doi.org/10.1002/ajpa.20848

Flack JC, Girvan M, de Waal FBM, Krakauer DC (2006) Policing stabilizes construction of social niches in primates. Nature 439:426–429. https://doi.org/10.1038/nature04326

Fragaszy DM, Izar P, Liu Q et al (2016) Body mass in wild bearded capuchins, (*Sapajus libidinosus*): ontogeny and sexual dimorphism. Am J Primatol 78:473–484. https://doi.org/10.1002/ajp.22509

Gesquiere LR, Learn NH, Simao MCM et al (2011) Life at the top: rank and stress in wild male baboons. Science 333:357–360. https://doi.org/10.1126/science.1207120

Godoy I, Vigilant L, Perry SE (2016a) Inbreeding risk, avoidance and costs in a group-living primate, *Cebus capucinus*. Behav Ecol Sociobiol 70:1601–1611. https://doi.org/10.1007/s00265-016-2168-1

Godoy I, Vigilant L, Perry SE (2016b) Cues to kinship and close relatedness during infancy in white-faced capuchin monkeys, *Cebus capucinus*. Anim Behav 116:139–151. https://doi.org/10.1016/j.anbehav.2016.03.031

Gould L, Ziegler TE (2007) Variation in fecal testosterone levels, inter-male aggression, dominance rank and age during mating and post-mating periods in wild adult male ring-tailed lemurs (Lemur catta). Am J Primatol 69:1325–1339. https://doi.org/10.1002/ajp.20438

Gould L, Fedigan LM, Rose LM (1997) Why be vigilant? The case of the alpha animal. Int J Primatol 18:401–414. https://doi.org/10.1023/A:1026338501110

Goymann W, Wingfield JC (2004) Allostatic load, social status and stress hormones: the costs of social status matter. Anim Behav 67:591–602

Gros-Louis J, Perry S, Manson JH (2003) Violent coalitionary attacks and intraspecific killing in wild white-faced capuchin monkeys (*Cebus capucinus*). Primates 44:341–346. https://doi.org/10.1007/s10329-003-0050-z

Hall CL, Fedigan LM (1997) Spatial benefits afforded by high rank in white-faced capuchins. Anim Behav 53:1069–1082. https://doi.org/10.1006/anbe.1996.0392

Harrison JS, Mondor EB (2011) Evidence for an invasive aphid "Superclone": extremely low genetic diversity in oleander aphid (Aphis nerii) populations in the southern United States. PLoS One 6:e17524. https://doi.org/10.1371/journal.pone.0017524

Jack KM (2001) Effect of male emigration on the vigilance behavior of coresident males in white-faced capuchins (*Cebus capucinus*). Int J Primatol. https://doi.org/10.1023/A:1012061231850

Jack KM (2003) Explaining variation in affiliative relationships among male white-faced capuchins. Folia Primatol 74:1–16. https://doi.org/10.1159/000068390

Jack KM, Fedigan L (2004a) Male dispersal patterns in white-faced capuchins, *Cebus capucinus*. Part 2: patterns and causes of secondary dispersal. Anim Behav 67:771–782. https://doi.org/10.1016/j.anbehav.2003.04.015

Jack KM, Fedigan L (2004b) Male dispersal patterns in white-faced capuchins, *Cebus capucinus*. Part 1: patterns and causes of natal emigration. Anim Behav 67:761–769. https://doi.org/10.1016/j.anbehav.2003.04.015

Jack KM, Fedigan LM (2006) Why be alpha male? Dominance and reproductive success in wild white-faced capuchins (*Cebus capuchinus*). New Perspect Study Mesoamerican Primates Distrib Ecol Behav Conserv: 367–386. https://doi.org/10.1007/0-387-25872-8_18

Jack KM, Fedigan LM (2009) Female dispersal in a female-philopatric species, *Cebus capucinus*. Behaviour 146:471–497. https://doi.org/10.1163/156853909X404420

Jack KM, Sheller C, Fedigan LM (2012) Social factors influencing natal dispersal in male white-faced capuchins (*Cebus capucinus*). Am J Primatol 74:359–365. https://doi.org/10.1002/ajp.20974

Jack KM, Schoof VAM, Sheller CR et al (2014) Hormonal correlates of male life history stages in wild white-faced capuchin monkeys (*Cebus capucinus*). Gen Comp Endocrinol 195:58–67. https://doi.org/10.1016/j.ygcen.2013.10.010

Janson CH (1986) The mating system as a determinant of social evolution in capuchin monkeys (*Cebus*). In: Primate ecology and conservation (Else JG, Lee PC, eds). Cambridge: Cambridge University Press, 169–179.

Janzen DH, Hallwachs W (1995) Indicative weather data for the Santa Rosa area and Volcan Cacao. Unpublished ms.

Jiménez A, Monroe MC, Zamora N, Benayas J (2017) Trends in environmental education for biodiversity conservation in Costa Rica. Environ Dev Sustain 19:221–238. https://doi.org/10.1007/s10668-015-9734-y

Kalbitzer U, Bergstrom ML, Carnegie SD et al (2017) Female sociality and sexual conflict shape offspring survival in a Neotropical primate. Proc Natl Acad Sci 114:201608625. https://doi.org/10.1073/pnas.1608625114

Kappeler PM, van Schaik CP, Watts DP (2012) The values and challenges of long-term field studies. In: Long-term field studies of primates. Springer Berlin Heidelberg, Berlin, Heidelberg, pp 3–18

Kappelle M, Lovejoy TE, Gámez Lobo R (2016) Costa Rican ecosystems. University of Chicago Press

Lee PC, Bussière LF, Webber CE, Poole JH, Moss CJ. (2013) Enduring consequences of early experiences: 40 year effects on survival and success among African elephants (Loxodonta africana). Biol Lett 9: 20130011. http://dx.doi.org/10.1098/rsbl.2013.0011

Lusseau D, Newman MEJ (2004) Identifying the role that animals play in their social networks. Proceedings Biol Sci 271(Suppl 6):S477–S481. https://doi.org/10.1098/rsbl.2004.0225

Lynch EC, Di Fiore A, Lynch RF, Palombit RA (2017) Fathers enhance social bonds among paternal half-siblings in immature olive baboons (Papio hamadryas anubis). Behav Ecol Sociobiol 71:120. https://doi.org/10.1007/s00265-017-2336-y

MacKinnon KC, Riley EP (2010) Field primatology of today: current ethical issues. Am J Primatol 72:749–753. https://doi.org/10.1002/ajp.20836

Marshall AJ, Hohmann G (2005) Urinary testosterone levels of wild male bonobos (Pan paniscus) in the Lomako Forest, Democratic Republic of Congo. Am J Primatol 65:87–92. https://doi.org/10.1002/ajp.20099

McComb K, Shannon G, Durant SM et al (2011) Leadership in elephants: the adaptive value of age. Proc Biol Sci 278(1722):3270

Meikle DB, Vessey SH (1988) Maternal dominance rank and lifetime survivorship of male and female rhesus monkeys. Behav Ecol Sociobiol 22:379–383. https://doi.org/10.1007/BF00294974

Mendonça-Furtado O, Edaes M, Palme R et al (2014) Does hierarchy stability influence testosterone and cortisol levels of bearded capuchin monkeys (Sapajus libidinosus) adult males? A comparison between two wild groups. Behav Process 109:79–88. https://doi.org/10.1016/j.beproc.2014.09.010

Modlmeier AP, Keiser CN, Watters JV et al (2014) The keystone individual concept: an ecological and evolutionary overview. Anim Behav 89:53–62

Muehlenbein MP (2009) The application of endocrine measures in primate parasite ecology. Primate Parasite Ecol Dyn. Study Host-Parasite Relationships: 63–81

Muehlenbein MP, Watts DP (2010) The costs of dominance: testosterone, cortisol and intestinal parasites in wild male chimpanzees. Biopsychosoc Med 4:21. https://doi.org/10.1186/1751-0759-4-21

Muehlenbein MP, Watts DP, Whitten PL (2004) Dominance rank and fecal testosterone levels in adult male Chimpanzees (Pan troglodytes schweinfurthii) at Ngogo, Kibale National Park, Uganda. Am J Primatol 64:71–82. https://doi.org/10.1002/ajp.20062

Muller MN, Wrangham RW (2004a) Dominance, cortisol and stress in wild chimpanzees (Pan troglodytes schweinfurthii). Behav Ecol Sociobiol 55:332–340. https://doi.org/10.1007/s00265-003-0713-1

Muller MN, Wrangham RW (2004b) Dominance, aggression and testosterone in wild chimpanzees: a test of the "challenge hypothesis.". Anim Behav 67:113–123. https://doi.org/10.1016/j.anbehav.2003.03.013

Muniz L, Perry S, Manson JH et al (2006) Father–daughter inbreeding avoidance in a wild primate population. Curr Biol 16:R156–R157. https://doi.org/10.1016/j.cub.2006.02.055

Muniz L, Perry S, Manson JH et al (2010) Male dominance and reproductive success in wild white-faced capuchins (*Cebus capucinus*) at Lomas Barbudal, Costa Rica. Am J Primatol 72:1118–1130. https://doi.org/10.1002/ajp.20876

Ohlsson C, Wallaschofski H, Lunetta KL et al (2011) Genetic determinants of serum testosterone concentrations in men. PLoS Genet 7:e1002313. https://doi.org/10.1371/journal.pgen.1002313

Onyango PO, Gesquiere LR, Wango EO et al (2008) Persistence of maternal effects in baboons: Mother's dominance rank at son's conception predicts stress hormone levels in subadult males. Horm Behav 54:319–324. https://doi.org/10.1016/j.yhbeh.2008.03.002

Paine RT (1969) A note on trophic complexity and community stability. Am Nat 103:91–93. https://doi.org/10.1086/282586

Palombit RA (2015) Infanticide as sexual conflict: coevolution of male strategies and female counterstrategies. Cold Spring Harb Perspect Biol 7:1–31. https://doi.org/10.1101/cshperspect.a017640

Perry S (1996) Intergroup encounters in wild white-faced capuchins (*Cebus capucinus*). Int J Primatol 17:309–330. https://doi.org/10.1007/BF02736624

Perry S (1998a) Male-male social relationships in wild white-faced capuchins, *Cebus capucinus*. Behaviour 135:139–172. https://doi.org/10.1163/156853998793066384

Perry S (1998b) A case report of a male rank reversal in a group of wild white-faced capuchins (*Cebus capucinus*). Primates 39:51–70. https://doi.org/10.1007/BF02557743

Perry S (2012) The behavior of wild white-faced capuchins: demography, social relationships, and communication. Adv Study Behav 44:135–181

Perry S, Manson JH (2008) Manipulative monkeys: the capuchins of Lomas Barbudal. Harvard University Press

Perry S, Godoy I, Lammers W (2012) The Lomas Barbudal monkey project: two decades of research on *Cebus capucinus*. In: Long-term field studies of primates. Springer Berlin Heidelberg, Berlin, Heidelberg, pp 141–163

Perry S, Godoy I, Lammers W et al (2017) Impact of personality traits and early life experience on timing of emigration and rise to alpha male status for wild male white-faced capuchin monkeys (*Cebus capucinus*) at Lomas Barbudal Biological Reserve, Costa Rica. Behaviour 154:195–226. https://doi.org/10.1163/1568539X-00003418

Pinter-Wollman N, Mi B, Pruitt JN (2017) Replacing bold individuals has a smaller impact on group performance than replacing shy individuals. Behav Ecol 28:883–889. https://doi.org/10.1093/beheco/arx054

Pruitt JN, Pinter-Wollman N (2015) The legacy effects of keystone individuals on collective behaviour scale to how long they remain within a group. Proc R Soc B Biol Sci 282:20151766. https://doi.org/10.1098/rspb.2015.1766

Rakotoniaina JH, Kappeler PM, Kaesler E et al (2017) Hair cortisol concentrations correlate negatively with survival in a wild primate population. BMC Ecol 17:30. https://doi.org/10.1186/s12898-017-0140-1

Recabarren MP, Vergara M, Martinez MC, et al (2000) Impact of lactation upon fertility in the New World primate capuchin monkey (Cebus apella). J Med Primatol 29:350–360. doi: https://doi.org/10.1034/j.1600-0684.2000.290507

Rose LM (1994) Benefits and costs of resident males to females in white-faced capuchins, *Cebus capucinus*. Am J Primatol 32:235–248. https://doi.org/10.1002/ajp.1350320402

Rose LM, Fedigan LM (1995) Vigilance in white-faced capuchins, *Cebus capucinus*, in Costa Rica. Anim Behav 49:63–70. https://doi.org/10.1016/0003-3472(95)80154-5

Ryckmans J, Millet K, Warlop L et al (2015) The influence of facial characteristics on the relation between male 2D:4D and dominance. PLoS One 10:e0143307. https://doi.org/10.1371/journal.pone.0143307

Sapolsky RM (1993) Endocrinology alfresco: psychoendocrine studies of wild baboons. Recent Prog Horm Res 48:437–468. https://doi.org/10.1016/B978-0-12-571148-7.50020-8

Sapolsky RM (2005) Review: The Influence of Social Hierarchy on Primate Health. Science (80) 308:648–652. https://doi.org/10.1126/science.1106477

Sapolsky RM, Share LJ (2004) A pacific culture among wild baboons: its emergence and transmission. PLoS Biol 2:e106. https://doi.org/10.1371/journal.pbio.0020106

Sargeant EJ, Wikberg EC, Kawamura S, Fedigan LM (2015) Allonursing in white-faced capuchins (*Cebus capucinus*) provides evidence for cooperative care of infants. Behaviour 152:1841–1869. https://doi.org/10.1163/1568539X-00003308

Schaebs FS, Perry SE, Cohen D et al (2017) Social and demographic correlates of male androgen levels in wild white-faced capuchin monkeys (*Cebus capucinus*). Am J Primatol 79:e22653. https://doi.org/10.1002/ajp.22653

Schoof VAM, Jack KM (2013) The Association of Intergroup Encounters, Dominance Status, and Fecal Androgen and Glucocorticoid Profiles in Wild Male White-Faced Capuchins (*Cebus capucinus*). Am J Primatol 75:107–115. https://doi.org/10.1002/ajp.22089

Schoof VAM, Jack KM (2014) Male social bonds: strength and quality among co-resident white-faced capuchin monkeys (*Cebus capucinus*). Behaviour 151:963–992. https://doi.org/10.1163/1568539X-00003179

Schoof VAM, Jack KM, Carnegie SD (2012) Rise to power: a case study of male fecal androgen and cortisol levels before and after a non-aggressive rank change in a group of wild white-faced capuchins (*Cebus capucinus*). Folia Primatol 82:299–307. https://doi.org/10.1159/000337220

Schoof VAM, Wikberg EC, Jack KM et al (2014a) Infanticides during periods of social stability: kinship, resumption of ovarian cycling, and mating access in white-faced capuchins (*Cebus capucinus*). Neotrop Primates 21:191–195. https://doi.org/10.1896/044.021.0206

Schoof VAM, Jack KM, Ziegler TE (2014b) Male response to female ovulation in white-faced capuchins (*Cebus capucinus*): variation in Fecal testosterone, dihydrotestosterone, and glucocorticoids. Int J Primatol 35:643–660. https://doi.org/10.1007/s10764-013-9742-4

Schoof VAM, Bonnell TR, Jack KM et al (2016) Male endocrine response to seasonally varying environmental and social factors in a neotropical primate, *Cebus capucinus*. Am J Phys Anthropol 159:671–682. https://doi.org/10.1002/ajpa.22925

Setchell JM, Dixson AF (2001) Changes in the secondary sexual adornments of male mandrills (Mandrillus sphinx) are associated with gain and loss of alpha status. Horm Behav 39:177–184. https://doi.org/10.1006/hbeh.2000.1628

Setchell JM, Smith T, Wickings EJ, Knapp LA (2008) Social correlates of testosterone and ornamentation in male mandrills. Horm Behav 54:365–372. https://doi.org/10.1016/j.yhbeh.2008.05.004

Sih A, Watters JV (2005) The mix matters: behavioural types and group dynamics in water striders. Behaviour 142:1417–1431

Sih A, Hanser SF, McHugh KA (2009) Social network theory: new insights and issues for behavioral ecologists. Behav Ecol Sociobiol 63:975–988. https://doi.org/10.1007/s00265-009-0725-6

Sumana A, Sona C (2013) Key relocation leaders in an Indian queenless ant. Behav Process 97:84–89. https://doi.org/10.1016/j.beproc.2013.03.006

Surbeck M, Mundry R, Hohmann G (2011) Mothers matter! Maternal support, dominance status and mating success in male bonobos (Pan paniscus)

Teichroeb JA, Jack KM (2017) Alpha male replacements in nonhuman primates: variability in processes, outcomes, and terminology. Am J Primatol. https://doi.org/10.1002/ajp.22674

Vogel ER (2005) Rank differences in energy intake rates in white-faced capuchin monkeys, *Cebus capucinus*: the effects of contest competition. Behav Ecol Sociobiol 58:333–344. https://doi.org/10.1007/s00265-005-0960-4

Wikberg EC, Jack KM, Campos FA et al (2014) The effect of male parallel dispersal on the kin composition of groups in white-faced capuchins. Anim Behav 96:9–17. https://doi.org/10.1016/j.anbehav.2014.07.016

Wikberg EC, Jack KM, Fedigan LM et al (2017) Inbreeding avoidance and female mate choice shape reproductive skew in capuchin monkeys (*Cebus capucinus* imitator). Mol Ecol 26:653–667. https://doi.org/10.1111/mec.13898

Wikberg EC, Jack KM, Fedigan LM, Kawamura S (2018, this volume) The effects of dispersal and reproductive patterns on the evolution of male sociality in white-faced capuchins. In: Kalbitzer U, Jack KM (eds) Primate life histories, sex roles, and adaptability – essays in honour of Linda M. Fedigan. Developments in primatology: progress and prospects. Springer, New York, pp 117–129

Wingfield JC, Hegner RE, Dufty AMJ, Ball GF (1990) The "challenge hypothesis": theoretical implications for patterns of testosterone secretion, mating systems, and breeding strategies. Am Nat 136:829–846. https://doi.org/10.1086/285134

Chapter 7
The Effects of Dispersal and Reproductive Patterns on the Evolution of Male Sociality in White-Faced Capuchins

Eva C. Wikberg, Katharine M. Jack, Linda M. Fedigan, and Shoji Kawamura

Abstract In many mammalian species, philopatric females reside with female kin with whom they form long-lasting cooperative bonds, whereas dispersing males rarely form strong bonds with other males. However, males may have the opportunity to disperse and form long-lasting bonds with paternal male kin in species with high male reproductive skew and parallel male dispersal. We used 54 group-years with demographic and genetic data to investigate how male dispersal and reproductive patterns affected the distribution of male kin in the Santa Rosa white-faced capuchins (*Cebus capucinus imitator*). During 41 of 54 group-years, there was a high degree of reproductive monopolization by either the alpha or a subordinate male. Natal males often co-resided with paternal brothers of similar age. The likelihood of parallel dispersal from the natal group increased with age similarity and paternal sibship. Some males may possibly gain inclusive fitness benefits by engaging in parallel dispersal and forming long-term cooperative bonds with other males. However, only one of three groups displayed a positive association between male-male relatedness and time spent co-resident in the breeding group. Inbreeding avoidance between alpha males and their daughters also gives subordinate males an opportunity to gain direct fitness benefits and may provide a strong incentive for cooperation among males that are not close kin. These findings suggest that cooperation between related and unrelated males likely evolved due to different reasons.

E. C. Wikberg (✉)
Department of Anthropology, University of Texas at San Antonio, San Antonio, TX, USA
e-mail: eva.wikberg@utsa.edu

K. M. Jack
Department of Anthropology, Tulane University, New Orleans, LA, USA

L. M. Fedigan
Department of Anthropology and Archaeology, University of Calgary, Calgary, AB, Canada

S. Kawamura
Department of Integrated Biosciences, University of Tokyo, Kashiwa, Chiba, Japan

U. Kalbitzer, K. M. Jack (eds.), *Primate Life Histories, Sex Roles, and Adaptability*, Developments in Primatology: Progress and Prospects,
https://doi.org/10.1007/978-3-319-98285-4_7

Subordinate males related to the alpha male may gain inclusive fitness benefits, while only unrelated subordinate males gained reproductive opportunities and direct fitness benefits from cooperating.

Keywords Social evolution · Kin cooperation · Kin competition · Inbreeding · Parallel dispersal

7.1 Introduction

Many gregarious mammals show male dispersal and female philopatry (i.e. females remain in their natal group to breed) (Greenwood 1980). As a result, adult females will typically reside with their adult female kin, while adult males will rarely reside with adult male kin after dispersing from their natal group to a breeding group. Having access to related social partners may favour the evolution of social behaviours via kin selection, which occurs when the cost of the behaviour to the actor's own reproductive success is outweighed by an increase in reproductive output by a related recipient (Hamilton 1964a, b). Due to these inclusive fitness benefits, cooperation is more likely to evolve among kin than non-kin (Hamilton 1964a, b). If kin interact with each other more frequently than non-kin, the former may be more reliable allies due to their history of interactions, and long-lasting cooperative relationships with kin may lead to increased reproductive success for both partners (Chapais 2001). Thus, kin cooperation can evolve due to indirect fitness benefits according to Hamilton (1964a, b) or direct fitness benefits according to Chapais (2001).

Several influential socioecological models proposed in the 1980s and 1990s suggested that the sex that relies on cooperative partners to gain high reproductive success should remain in their natal group and cooperate with same-sex kin (Wrangham 1980; van Schaik 1989; Sterck et al. 1997). However, the benefit of kin cooperation does not always compensate the cost of intense competition. Therefore, the sex that experiences high competition for limited resources and does not gain sufficient benefits from forming kin coalitions should disperse from their natal group to avoid kin competition (Griffin and West 2002; West et al. 2002).

These theoretical models explain why many female primates reside with female kin with whom they form long-lasting cooperative bonds, whereas many male primates disperse (Sterck et al. 1997). Female sociality is predicted to have significant fitness benefits, although only a few studies have been able to confirm this prediction in long-lived animals (Silk et al. 2003, 2009; Kalbitzer et al. 2017) due to difficulties in determining fitness outcomes in these populations (Silk et al. 2009). Even though males disperse, they may still form kin-biased affiliative relationships in their natal group. For example, male rhesus macaques (*Macaca mulatta*) form the strongest relationship with their maternal kin, although their relationships were not as strong as those between maternal female kin (Widdig et al. 2016). This difference may be due to mothers biasing affiliation to offspring of the philopatric sex (Kulik et al. 2015; Murray et al. 2016). In contrast, males formed stronger relationships with paternal kin than did females, perhaps because paternal male kin often disperse together (Widdig et al. 2016).

Fedigan (1993) noted early on that although female white-faced capuchins (*Cebus capucinus imitator*) conformed to these expectations, male white-faced capuchins did not. They were unusual in comparison to other male-dispersed primate species, even species in the same genus, by showing frequent affiliation with other males and by forming dominance hierarchies that appeared to be dependent not only on individual traits like body size and strength but also on social factors (Fedigan 1993). Similar reports from other primate species have since appeared. For example, dispersing males benefit from cooperating with each other in Assamese macaques (*Macaca assamensis*) (Schuelke et al. 2010) and Geoffroy's tamarins (*Saguinus geoffroyi*) (Diaz-Munoz and Ribeiro 2014).

Fedigan and Jack (2012) investigated this topic in more detail by using unique longitudinal data sets from the Santa Rosa capuchins. Using decades of behavioural observations from multiple study groups, they found that males benefitted from forming dispersal coalitions (i.e. showing parallel dispersal) because these coalitions were more successful than single males in taking over groups of females (Fedigan and Jack 2004; Jack and Fedigan 2004a, b). From data collected during five park-wide population surveys, they discovered that group productivity (i.e. infant to female ratio) was higher in groups with more males (Fedigan and Jack 2011). These results indicated that it was important for males to cooperate, which can explain why males form tolerant, cooperative, and often affiliative relationships with each other (Jack 2003; Schoof and Jack 2014). What is so remarkable is that male capuchins are able to form cooperative relationships despite dispersing several times during their lifetime (Jack and Fedigan 2004b; Jack and Fedigan (2018), Chap. 6 in this volume) and competing intensely for reproductive opportunities – a combination of traits that is not predicted to co-occur according to theoretical models (Griffin and West 2002; West et al. 2002).

Observational data suggest that white-faced capuchin males may have solved the dilemma of needing to form alliances and dispersing in two ways. First, frequent parallel dispersal leaves open the possibility that males disperse and cooperate with their close kin to gain inclusive fitness benefits (Jack and Fedigan 2004a, b). Following this hypothesis, we predict that male kin will be more likely to disperse in parallel than male non-kin, that immigrant male kin will reside together longer than immigrant male non-kin, and that male reproductive skew will be high (i.e. subordinate males gain indirect rather than direct fitness benefits from cooperating with a related alpha male). Second, cooperating males may share reproductive opportunities in which case all males gain direct fitness benefits from cooperating (Fedigan 1993). Based on this hypothesis, parallel dispersal need not be kin-biased, co-residency times will be longer for co-dispersing males than other males, and reproductive skew will be low. We will evaluate the support for each of the two hypotheses using 20 years of demographic and genetic data from the Santa Rosa white-faced capuchins. Based on the findings of this and previous studies, we will discuss how the costs and benefits associated with kin cooperation and kin competition can explain male sociality during different life history stages.

7.2 Methods

7.2.1 *Study Population*

This study on white-faced capuchins was conducted in Sector Santa Rosa of the Área de Conservación Guanacaste, Costa Rica. Most females are philopatric (Jack and Fedigan 2009), and they often give birth for the first time at 6.5 years of age (Fedigan and Jack 2012). Males typically leave their natal group at around 4 years of age, but the dispersal age varies from 19 months to 11 years in our study population (Jack and Fedigan 2004b; Jack et al. 2012). Although subadult males (7–10 years) are able to copulate with ejaculation, they have not sired any genotyped offspring in our study population to date (Wikberg et al. 2017). Adult males (>10 years) have gained full body size, but it is only upon becoming alpha males that they develop the full complement of secondary sexual characteristics (Jack et al. 2014; also see Chap. 6 by Jack and Fedigan (2018) in this volume).

7.2.2 *Demographic and Behavioural Data Collection*

Demographic and ad libitum (Altmann 1974) behavioural data from four groups (Table 7.1) were collected once a month between 1993 and 2013 under the supervision of Linda Fedigan and Katharine Jack. These data were used to determine group composition, age of individuals, mother-offspring relationships, and alpha male status. Alpha males can easily be distinguished by the direction of submissive interactions,

Table 7.1 Group composition and reproductive skew (Nonacs' B index)

Group	Years	AF[a]	IM[a]	Infants[a]	# years with 1 reproducing male		# years with >1 reproducing male	
					Only alpha	Only subordinate	Skewed/ total[b]	B index
CP	1993–2012	2–10 (14/15)	1–6 (9/11)	0–7 (42/53)	9	4	0/4	−0.25–0.05
EX	2006–2010	3–4 (4/4)	1–4 (6/6)	0–4 (10/16)	3	0	0/1	−0.74
GN	1999–2013	4–11 (12/12)	3–9 (9/13)	0–8 (36/46)	7	0	1/5	−0.13–0.50
LV	1993–2013	4–6 (9/10)	2–10 (19/30)	0–4 (33/49)	12	0	0/4	−0.11–0.11
Total					36/40	4/40	1/14	

[a]The range represents the number of adult females (AF), immigrant males (IM), and infants present in any year. The proportions of genotyped animals are indicated in parentheses. Raw data can be found at University of Calgary's digital repository PRISM

[b]Number of years with reproductive skew/total number of years with two or more reproducing males

such as avoid, fear grin, flee, and supplant (Perry 1998; Jack 2003; Jack et al. 2014). We observed too few submissive interactions between the subordinate males to be able to rank them in relation to each other. Therefore, we will only discuss how alpha versus subordinate status (rather than exact rank number) affects reproductive output.

7.2.3 *Genetic Data Collection*

Two faecal samples were collected from offspring, mothers, and candidate sires (Wikberg et al. 2014a, 2017). In Shoji Kawamura's laboratory at the University of Tokyo, we performed DNA extractions, DNA quantification using real-time quantitative polymerase chain reactions (PCRs), PCR amplifications of up to 20 short tandem repeat loci, capillary electrophoresis, and confirmation of genotypes (Wikberg et al. 2014a, 2017). Sires were assigned at the 95% confidence interval using the software Cervus (Marshall et al. 1998; Kalinowski et al. 2007), and estimated relatedness values (*R*-values) were calculated with ML-Relate (Kalinowski et al. 2006) as previously described (Wikberg et al. 2017).

7.2.4 *Data Analysis*

When multiple resident males reproduced, we assessed whether reproduction was skewed using Nonacs' B index. This index calculates how different the observed reproductive sharing is from being equal while taking into account the number of months each adult male spent resident in the group that year (Nonacs 2000). The index ranges from -1 to 1, and higher values indicate stronger reproductive skew.

We investigated whether dispersal partner choice among natal males was kin-biased using binomial generalized linear mixed models (GLMMs). Each time a natal male dispersed, we coded whether he dispersed with each of the potential dispersal partners (i.e. co-resident natal males that were at least 19 months old). We created one null model that did not contain any fixed effects and three alternative models that included one or two fixed effects: (1) age difference between the dispersing male and his potential partner, (2) kinship (i.e. maternal kin, paternal half-sibling, or non-kin), and (3) age difference and kinship (Table 7.2). We combined maternal half-siblings and distant maternal kin into one category "maternal kin" due to small samples sizes. Our data set did not include any distant paternal kin. The fixed effects were not collinear based on low variance inflation factors (VIF, 1.07–1.20). Each model included the random effects group and male identities. We included random intercepts but not random slopes because the latter did not improve the models. We determined the support for each model based on Akaike information criterion corrected for small sample sizes (AICc) (Akaike 1974). The models were computed with the packages lme4 (Bates et al. 2014) in R version 3.3.3. Several models received similar support, and we took model selection uncertainty into account by

Table 7.2 Models of parallel dispersal partner choice and each model's fixed effects, AICc, and delta and Akaike's weight, residual degrees of freedom, and R-square

Fixed effects	AICc	Delta	Weight	df	R^2
Age difference + kinship	68.49	0.00	0.63	64	73
Age difference	69.58	1.08	0.37	66	95
Kinship	84.97	16.48	0.00	65	45
Null model	96.32	27.83	0.00	67	46

All models contained group and focal identities as random effects

averaging coefficients across models (Burnham and Anderson 2002) using the R package MuMIn (Barton 2013). We also used MuMIn to calculate each model's conditional R-square, which takes both random and fixed effects into account. We used the multiple comparison procedure implemented in the R package multcomp (Hothorn et al. 2014) to investigate differences between all kinship categories.

To investigate whether kin or parallel immigrants resided together longer than other males, we used double-decker semipartialing multiple quadratic assignment procedures in UCINET (Borgatti et al. 2002). This procedure controls for autocorrelation in the data and is therefore well suited for analysing social networks (Hanneman and Riddle 2005). Rows and columns were randomly reshuffled, and R-square and regression coefficients were calculated after each permutation for a total of 10,000 permutations. These permuted values were used to create a sampling distribution to evaluate whether our observed values were significantly different from random. The significance level was set to $p < 0.05$. This procedure regressed the outcome matrix (i.e. number of months immigrant males spent co-resident) against multiple predictor matrices: R-value, parallel immigration status, and whether the dyad included the alpha male. The latter variable was included because immediate inclusive fitness benefits may only be gained by cooperating with a male kin that occupies the alpha position in groups with high reproductive skew. We used R-value instead of kinship categories for this analysis because the mothers and sires remained unknown for the majority of immigrant males. However, several recent studies indicate that marker-based methods reflect genetic similarity more accurately than kinship categories determined from shallow pedigrees (Forstmeier et al. 2012; Robinson et al. 2013), and R-values show a relatively high correlation with actual relatedness in some study populations (Wikberg et al. 2014b, 2017).

7.3 Results

7.3.1 Paternity Assignments

We obtained genotypes for 121 of 164 offspring born in the groups (the remaining infants died at an early age), 40 of 41 adult females (the remaining female died early during the study), 16 of 18 alpha males, and 29 of 43 subordinate immigrant males (the remaining males dispersed before sample collection began or their samples did not amplify). We were able to determine paternity for 111 of 121 infants (Wikberg et al. 2017). For seven of the ten remaining infants, a previous study had genotyped all their candidate sires (including the two alpha males that were not genotyped in this study) and determined their paternity (Jack and Fedigan 2006).

7.3.2 Reproductive Patterns

Male siring success varied from 0 to 24 offspring (mean = 1.76, SD = 3.42), and it was largely dependent on whether the male was alpha or subordinate. Alpha males sired 84%, and subordinate males sired 16% of the offspring. All 18 alpha males sired offspring, but the total number of offspring they sired as alpha varied from 1 to 19 (mean = 5.71, SD = 4.56). Alpha males that only sired 1 offspring had short alpha tenures, sometimes only 1 month (unpublished data). The male that sired the highest number of offspring as alpha did so during his tenure as alpha in two different groups. He transferred voluntarily to a study group with a higher number of females, a process that has been observed a number of times in our study population (e.g. Jack and Fedigan 2004b). Only 10 of 32 subordinate males sired offspring (including subordinate males that were not genotyped but present in the group when all infants were assigned to the alpha male) (mean = 0.42, SD = 1.23). In half of the cases when a subordinate male sired offspring (12/24 offspring), the mother of the offspring was the alpha male's daughter and the subordinate male was not kin with the alpha male or the mother. The other cases when a subordinate male sired offspring occurred when the alpha was past prime age (i.e. >16 years), the alpha had recently gained his position, the alpha was wounded, the alpha ranged in the periphery of the group, or when the mother ranged in the periphery of the group. It is possible that the alpha male failed to monopolize reproduction due to these circumstances. The most successful male, Legolas, sired 24 offspring. He sired seven offspring during his 3 years as a subordinate male in the group, which was the highest number of offspring produced by any subordinate male. After the old alpha male left the group, Legolas inherited the alpha position (also see Jack and Fedigan (2018), Chap. 6 in this volume) and sired 17 offspring during the following 5 years. His tenure as alpha has not yet ended (currently >10 years), and he is likely to have an even higher total reproductive success.

Alpha male monopolization of reproduction decreased sharply after 5 years, which corresponds to the age when their daughters approach sexual maturity (Fig. 7.1). For example, Nose completely monopolized reproduction as a new alpha male, while at the end of his 15-year-long tenure, the subordinate male Legolas sired the majority of offspring that were born to Nose's mature daughters. The proportion of infants sired by the alpha male varied depending on group type and alpha male tenure. Alpha males in uni-male groups sired 100% of the offspring (Fig. 7.2). In multi-male groups, alpha males with relatively short tenures (<5 years) monopolized reproduction, while this was not always the case for alpha males with tenures lasting over 5 years (Fig. 7.2).

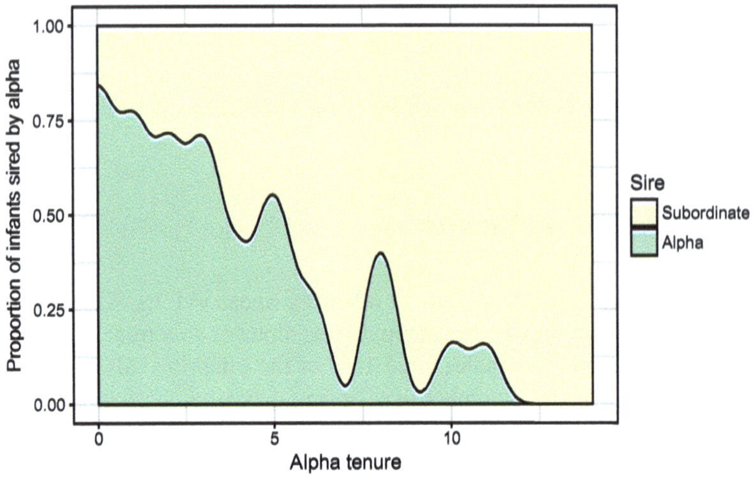

Fig. 7.1 The proportion of infants sired by the alpha male versus the subordinate male(s) during each year of his tenure

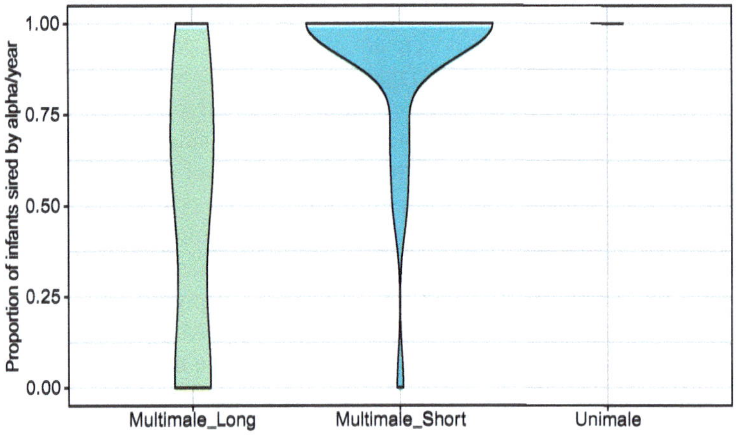

Fig. 7.2 The proportion of infants sired by the alpha male in uni-male groups, multi-male groups with short alpha tenures (<5 years), and multi-male groups with long alpha tenure (>5 years). The length of the violin plot shows the range of values, and the width shows the probability density estimates of values

Alpha males sired all of the infants born during 36 of 54 group-years (Table 7.1). A subordinate male, unrelated to the alpha male Nose, sired all the offspring during 4 of 15 years of the alpha male's tenure (Table 7.1). Six of seven infants born during these 4 years were conceived by the alpha male's mature daughters. Based on Nonacs' B index, reproduction was significantly skewed in 1 of 14 group-years during which the alpha and subordinate males both sired offspring (Table 7.1). Thus, 1 male dominated reproduction during 41 of 54 group-years.

7.3.3 Natal Dispersal Patterns

Of the natal males that dispersed during the study period, 13 dispersed together with other natal males, while 2 dispersed without a partner. However, one of the males that dispersed alone joined a group that contained an unrelated male with whom he had co-resided in his natal group. This dispersal case was therefore classified as delayed parallel dispersal. Most of the dispersing natal males resided with at least one close kin (Fig. 7.3). At the point in time when 1 or more of the 15 natal males dispersed, about half of their potential dispersal partners were close or distant kin ($N = 34$), while the other half were non-kin ($N = 36$). Of these potential dispersal partners, 2/5 maternal half-siblings and 8/12 distant maternal kin, 14/17 paternal half-siblings, and 8/36 non-kin dispersed together. Thus, the category of males that most frequently dispersed together was paternal half-siblings. Our data set did not contain any full siblings or distant paternal kin.

The models that best predicted dispersal partner choice among natal males contained age difference with or without kinship and explained up to 95% of the observed variation, while the null model with only the random effects explained 46% (Table 7.2). The likelihood of parallel dispersal decreased with age difference (Fig. 7.4). Males were more likely to disperse in parallel with paternal kin than with non-kin (Fig. 7.4), while there was no difference between maternal kin and non-kin (Fig. 7.4). The likelihood of dispersing in parallel did not differ between maternal and paternal kin (coefficient = −1.34, 95% CI 4.85–2.15; based on multiple comparison procedures and not visualized in Fig. 7.4 that only shows the effect of the two kinship categories in relation to the baseline level non-kin).

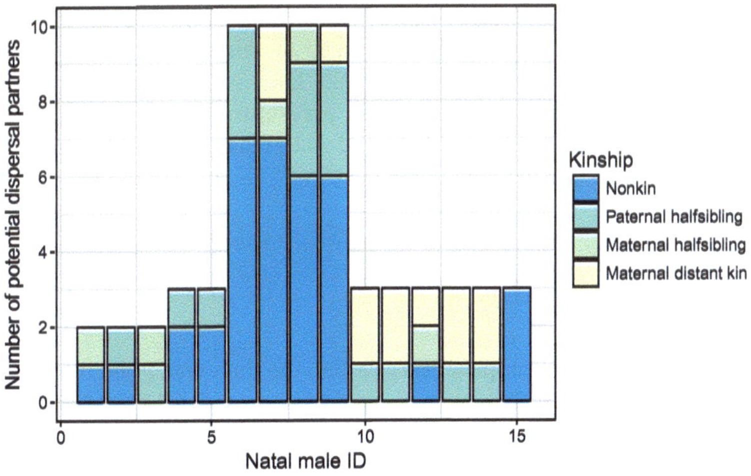

Fig. 7.3 The number of potential dispersal partners of different kinship categories each dispersing natal male had access to

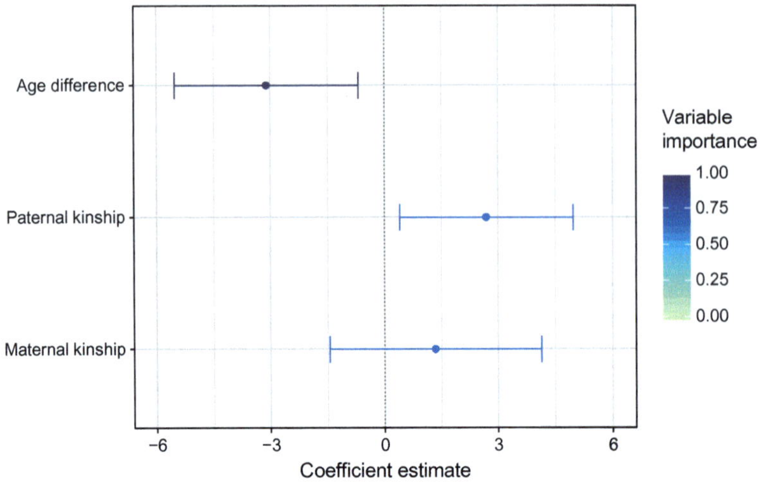

Fig. 7.4 Coefficient estimates and 95% confidence intervals based on averaging across models of parallel dispersal partner choice (see Table 7.2). Coefficient estimates with confidence intervals that do not overlap zero indicate that the variable has a positive or negative effect on the likelihood of parallel dispersal partner choice. Kinship is an ordinal variable with three levels: paternal half-siblings, maternal kin (includes half-siblings and more distant kin), and non-kin (which is the baseline level and not shown in the figure). Age difference is a continuous variable. The variable importance (i.e. the sum of normalized model likelihood over all models that contained the fixed effect) is indicated by colour

7.3.4 Secondary Dispersal Patterns

Of all the cases of secondary dispersal (a.k.a. breeding dispersal) from the study groups when all resident males were genotyped, 11 cases were classified as single emigration or immigration, while 4 cases were classified as parallel emigration. One case of secondary parallel emigration included the alpha male Nose and his two natal sons and one other unrelated natal male. Each of the other three cases consisted of two males that had immigrated in parallel to the study group and had a higher R-value with each other (R: 0.14; 0.45; 0.61) than with the other co-resident immigrant males (R, 0; 0–0.14; 0–0.50).

7.3.5 Co-Residency Times

The groups showed different patterns regarding the time immigrant males spent in co-residency. In LV group, time spent co-resident was predicted by parallel immigration status and R-value (Table 7.3). Males that immigrated in parallel and had lower R-values resided together for longer periods than other males. In GN group, males resided longer with parallel immigrant partners and with the alpha male (Table 7.3). In CP group, males resided longer with the alpha male that had an unusually long tenure (Table 7.3). A small sample size precluded statistical testing in EX group.

Table 7.3 Models of co-residency times between immigrant males

	Model		R-value		Parallel immigrant		Alpha	
Group	R^2	p	Coefficient	p	Coefficient	P	Coefficient	p
CP	0.138	0.026	−0.100	0.343	−0.108	0.168	0.407	0.039
GN	0.319	0.003	0.180	0.107	0.389	0.036	0.274	0.027
LV	0.743	0.034	−0.329	0.042	0.938	0.021	0.051	0.292

R-square and p-values are reported for the overall model, while regression coefficients and their p-values are reported for each of the predictor variables: R-value, parallel immigration partnership, and whether the dyad included the alpha male

7.4 Discussion

These findings show the importance of combining longitudinal demographic and genetic data to fully understand how dispersal and reproductive patterns shape the kin composition of groups and the occurrence of kin cooperation. High male reproductive skew leads to cohorts consisting of paternal siblings, and many natal males dispersed in parallel with their paternal brothers. Surprisingly, kinship did not always have a positive impact on the time immigrant males spent co-resident. It is likely that the costs and benefits of residing with male kin vary with life history stages.

Alpha males with long tenures do not monopolize reproduction to the same degree as relatively new alpha males in white-faced capuchins (Jack and Fedigan 2006; Muniz et al. 2006, 2010; Godoy et al. 2016; Perry 2012; Wikberg et al. 2017), and paternal brothers sired by long-term alpha males may be of dissimilar rather than similar ages. However, most alpha males have relatively short tenures and sire the majority of infants (Jack and Fedigan 2006; Muniz et al. 2006, 2010; Godoy et al. 2016; Perry 2012; Wikberg et al. 2017). This pattern of high reproductive skew leads to most natal animals residing with similar-aged paternal siblings. Our findings indicate that age compatibility and paternal sibship increased the likelihood of parallel dispersal among natal males. Maternal kinship did not have a positive effect on dispersal partner choice even when controlling for age difference, possibly due to this category including both maternal half-siblings and more distant maternal kin. A larger sample size of co-residing maternal brothers is necessary to directly investigate whether males prefer to disperse with their maternal brothers over non-kin.

Parallel secondary dispersal (regardless of kinship) was not common during this study period, but the males that did immigrate in parallel were more closely related to each other than to the other males in the group. Thus, kinship shaped dispersal coalitions both among natal and immigrant males, similar to findings from rhesus macaques (*Macaca mulatta*) (Meikle and Vessey 1981) and long-tailed tits (*Aegithalos caudatus*) (Sharp et al. 2008).

The length of time immigrant males spent co-resident in their breeding group was predicted by parallel immigration status in two of three groups. The reason why the third group did not show a similar pattern may be because it only contained two parallel immigrant males, one of which we believe died during his first year in the study group. In two groups, subordinate males resided longer with the alpha male than with each other. This finding may simply be a by-product of the alpha males having longer tenures than subordinate males (Jack and Fedigan 2004b). However, it may also be a successful strategy for subordinate males to remain with a long-term alpha past the point when the alpha's daughters start to reproduce. Because alpha males and their daughters avoid inbreeding, a subordinate male can gain direct fitness benefits from supporting an alpha male (Muniz et al. 2006; Godoy et al. 2016; Wikberg et al. 2017). Surprisingly, relatedness only predicted time spent co-resident in one of three groups. Although males often immigrate in parallel with their male kin, they do not necessarily reside together for longer time periods than do male non-kin.

These findings beg the question as to whether kinship is a poor predictor of the maintenance of male coalitions because kin co-residency is associated with high costs and few benefits. Males that disperse in parallel are more likely to succeed in immigrating to a new group (Fedigan and Jack 2004). Once they entered a new group, they all benefit from residing in a social group, via reduced predation risk and increased access to food. In this study, the majority of parallel dispersing males were immature natal males. These young males are not able to gain the alpha position and are not in direct competition against each other for reproductive opportunities. Thus, young males incur relatively low fitness costs from supporting an alpha male while biding their time in a subordinate role (van Noordwijk and van Schaik 2001). Because the alpha male benefits in terms of increased reproductive success from having up to two adult or subadult immigrant males in the group (Wikberg et al. 2017), the subordinate male may gain increased inclusive fitness if he is supporting a related alpha male.

In contrast to immature males, adult co-dispersing males will be in direct competition over the alpha position and a monopoly of reproduction once they gain access to a social group. This can take the form of escalated fighting, some of which result in lethal injuries to the fighting males (Gros-Louis et al. 2003; Fedigan and Jack 2004) or infants (Schoof et al. 2015; Brasington et al. 2017). Thus, the benefits of gaining access to a social group is no longer shared equally among adult males, and the benefits of cooperating with kin may not always outweigh the costs incurred from competing with them. For example, it may not be beneficial for two high-quality male kin to co-reside as alpha and beta males, if the alpha male monopolizes reproduction and would do equally well if supported by a male non-kin, and the beta male is likely to become a successful alpha male in another group. As mentioned above, it is particularly beneficial for a subordinate male to cooperate with an alpha male that has resided in the group long enough for his daughters to mature. Mate choice and inbreeding avoidance between the alpha male and his mature daughters give the subordinate male an opportunity to reproduce (Muniz et al. 2006; Godoy et al. 2016; Wikberg et al. 2017). If the alpha and beta males are close kin, females may also avoid breeding with the beta male if they are able to recognize him as paternal kin and they have access to unrelated mates. In all cases where a subordinate male bred with the alpha male's mature daughters in this study, the males were unrelated. Thus, inbreeding avoidance may have favoured the evolution of non-kin rather than kin co-residency among male capuchins. The potential costs coupled with few fitness benefits of kin co-residency between adult male kin can perhaps explain why a greater proportion of secondary dispersers than natal dispersers transferred alone. Alternatively, this pattern may simply be due to secondary dispersers having more limited access to male kin that are compatible dispersal partners.

To fully understand the evolution of male sociality, it is necessary to quantify the direct and indirect fitness consequences of male cooperation and competition. Unfortunately, this is extremely difficult to do in wild populations of dispersing, long-lived animals (Silk et al. 2009). Until this can be done, we propose that immature immigrant males are likely to experience a low cost to high benefit ratio of cooperating with same-sex kin, while this may not be the case for adult males.

The traditional models predicting the evolution of primate social structure highlight the importance of dispersal patterns shaping the kin composition of groups and therefore the occurrence of cooperation. Fedigan (1993) pointed out that white-faced capuchins do not conform to this pattern, and her study has been followed by similar reports from other primate species (Schuelke et al. 2010; Wikberg et al. 2012; Teichroeb et al. 2014; Diaz-Munoz and Ribeiro 2014). Findings from the current study suggest that high reproductive skew coupled with parallel dispersal and female mate choice for unrelated mates may be an overlooked factor that has the potential to affect the evolution of kin and non-kin cooperation (also see Widdig 2013). Furthermore, males are likely to experience different cost-benefit ratios of cooperating versus competing with kin depending on their age and other individual traits. Thus, future modelling efforts will benefit from taking into account how dispersal patterns *and* reproductive patterns shape the composition of groups as well as the fitness outcomes of kin cooperation *and* kin competition during different life history stages.

Acknowledgements This study was approved and supported by the University of Calgary's Life and Environmental Sciences Animal Care Committee, Tulane University's Institutional Animal Care and Use Committee, the Costa Rican Park Service, and Roger Blanco Segura and the administration of the Área de Conservación Guanacaste. The lab work was funded by the Japan Society for Promotion of Science's Postdoctoral Fellowship (ECW) and Grant-in-Aid (SK). The field work was funded by an NSERC Discovery Grant and the Canada Research Chairs Program (LMF) and by the LSB Leakey Foundation, Tulane University's Department of Anthropology, Stone Center for Latin American Studies, Newcomb Institute, and Research Enhancement Fund (KMJ). We would like to thank the many students and research assistants who collected data in the field; Mackenzie Bergstrom, Fernando Campos, Tomohide Hiwatashi, and Akiko Yashima for the help with the lab work and Urs Kalbitzer and two anonymous reviewers for comments that improved earlier versions of this manuscript.

References

Akaike H (1974) A new look at the statistical model identification. IEEE Trans Autom Control 19:716–723. https://doi.org/10.1109/TAC.1974.1100705

Altmann J (1974) Observational study of behavior: sampling methods. Behaviour 49:227–267

Barton K (2013) MuMIn: Multi-model inference. R package version 1.9.5. http://CRAN.R-project.org/package=MuMIn. Accessed 1 May 2017

Bates D, Maechler M, Bolker B, Walker S, Bojesen Christensen RH, Singman H (2014) lme4: linear mixed-effects models using Eigen and S4. http://lme4.r-forge.r-project.org/. Accessed 1 June 2014

Borgatti S, Everett M, Freeman L (2002) Ucinet for windows: software for social network analysis. Analytic Technologies, Harvard

Brasington LF, Wikberg EC, Kawamura S, Fedigan LM, Jack KM (2017) Infant mortality in white-faced capuchins: the impact of alpha male replacements. Am J Primatol 79:e22725. https://doi.org/10.1002/ajp.22725

Burnham KP, Anderson DR (2002) Model selection and multi-model inference: a practical information-theoretic approach. Springer, New York

Chapais B (2001) Primate nepotism: what is the explanatory value of kin selection? Int J Primatol 22:203–229. https://doi.org/10.1023/A:1005619430744

Diaz-Munoz SL, Ribeiro AM (2014) No sex-biased dispersal in a primate with an uncommon social system-cooperative polyandry. PeerJ 2:UNSP e640. https://doi.org/10.7717/peerj.640

Fedigan LM (1993) Sex differences and intersexual relations in adult white-faced capuchins (*Cebus capucinus*). Int J Primatol 14:853–877. https://doi.org/10.1007/BF02220256

Fedigan LM, Jack KM (2004) The demographic and reproductive context of male replacements in *Cebus capucinus*. Behaviour 141:755–775. https://doi.org/10.1163/1568539042245178

Fedigan LM, Jack KM (2011) Two girls for every boy: the effects of group size and composition on the reproductive success of male and female white-faced capuchins. Am J Phys Anthropol 144:317–326. https://doi.org/10.1002/ajpa.21414

Fedigan LM, Jack KM (2012) Tracking neotropical monkeys in Santa Rosa: lessons from a regenerating Costa Rican dry forest. In: Kappeler PM, Watts D (eds) Long-term field studies of primates. Springer, Dordrecht, pp 165–184

Forstmeier W, Schielzeth H, Mueller JC, Ellegren H, Kempenaers B (2012) Heterozygosity-fitness correlations in zebra finches: microsatellite markers can be better than their reputation. Mol Ecol 21:3237–3249. https://doi.org/10.1111/j.1365-294X.2012.05593.x

Godoy I, Vigilant L, Perry S (2016) Inbreeding risk, avoidance and costs in a group-living primate, *Cebus capucinus*. Behav Ecol Sociobiol 70:1601–1611. https://doi.org/10.1007/s00265-016-2168-1

Greenwood P (1980) Mating systems, philopatry and dispersal in birds and mammals. Anim Behav 28:1140–1162. https://doi.org/10.1016/S0003-3472(80)80103-5

Griffin AS, West SA (2002) Kin selection: fact and fiction. Trends Ecol Evol 17:15–21. https://doi.org/10.1016/S0169-5347(01)02355-2

Gros-Louis J, Perry SE, Manson JH (2003) Violent coalitionary attacks and intraspecific killing in wild white-faced capuchin monkeys (*Cebus capucinus*). Primates 44:341–346. https://doi.org/10.1007/s10329-003-0050-z

Hamilton WD (1964a) The genetical evolution of social behaviour. J Theor Biol 7:1–52. https://doi.org/10.1016/0022-5193(64)90038-4

Hamilton WD (1964b) The genetical evolution of social behaviour. II. J Theor Biol 7:17–52

Hanneman RA, Riddle M (2005) Introduction to social network methods. University of California, Riverside

Hothorn T, Bretz F, Westfall P, Heiberger RM, Scheutzenmeister A (2014) Simultaneous Inference in General Parametric Models. http://lme4.r-forge.r-project.org/. Accessed 1 May 2017

Jack KM (2003) Explaining variation in affiliative relationships among male white-faced capuchins (*Cebus capucinus*). Folia Primatol 74:1–16. https://doi.org/10.1159/000068390

Jack KM, Fedigan LM (2004a) Male dispersal patterns in white-faced capuchins, *Cebus capucinus* part 2: patterns and causes of secondary dispersal. Anim Behav 67:771–782. https://doi.org/10.1016/j.anbehav.2003.06.015

Jack KM, Fedigan LM (2004b) Male dispersal patterns in white-faced capuchins, *Cebus capucinus* part 1: patterns and causes of natal emigration. Anim Behav 67:761–769. https://doi.org/10.1016/j.anbehav.2003.04.015

Jack KM, Fedigan LM (2006) Why be alpha male? Dominance and reproductive success in wild white-faced capuchins (*Cebus capucinus*). In: Estrada A, Garber P, Pavelka MSM, Luecke L (eds) New perspectives in the study of Mesoamerican primates: distribution, ecology, behavior, and conservation. Springer, New York, pp 367–386

Jack KM, Fedigan LM (2009) Female dispersal in a female-philopatric species, *Cebus capucinus*. Behaviour 146:471–497. https://doi.org/10.1163/156853909X404420

Jack KM, Fedigan LM (2018, this volume) Alpha male capuchins (*Cebus capucinus imitator*) as keystone individuals. In: Kalbitzer U, Jack KM (eds) Primate life histories, sex roles, and adaptability - essays in honour of Linda M. Fedigan. Developments in primatology: progress and prospects. Springer, New York, pp 91–109

Jack KM, Sheller C, Fedigan LM (2012) Social factors influencing natal dispersal in male white-faced capuchins (*Cebus capucinus*). Am J Primatol 74:359–365. https://doi.org/10.1002/ajp.20974

Jack KM, Schoof VAM, Sheller CR, Rich CI, Klingelhofer PP, Ziegler TE, Fedigan L (2014) Hormonal correlates of male life history stages in wild white-faced capuchin monkeys (*Cebus capucinus*). Gen Comp Endocrinol 195:58–67. https://doi.org/10.1016/j.ygcen.2013.10.010

Kalbitzer U, Bergstrom ML, Carnegie SD, Wikberg EC, Kawamura S, Campos FA, Jack KM, Fedigan LM (2017) Female sociality and sexual conflict shape offspring survival in a Neotropical primate. Proc Natl Acad Sci U S A 114:1892–1897. https://doi.org/10.1073/pnas.1608625114

Kalinowski ST, Wagner AP, Taper ML (2006) ML-RELATE: a computer program for maximum likelihood estimation of relatedness and relationship. Mol Ecol Notes 6:576–579. https://doi.org/10.1111/j.1471-8286.2006.01256.x

Kalinowski ST, Taper ML, Marshall TC (2007) Revising how the computer program CERVUS accommodates genotyping error increases success in paternity assignment. Mol Ecol 16:1099–1106. https://doi.org/10.1111/j.1365-294X.2007.03089.x

Kulik L, Amici F, Langos D, Widdig A (2015) Sex differences in the development of social relationships in rhesus macaques (*Macaca mulatta*). Int J Primatol 36:353–376. https://doi.org/10.1007/s10764-015-9826-4

Marshall TC, Slate J, Kruuk LEB, Pemberton JM (1998) Statistical confidence for likelihood-based paternity inference in natural populations. Mol Ecol 7:639–655. https://doi.org/10.1046/j.1365-294x.1998.00374.x

Meikle DB, Vessey SH (1981) Nepotism among rhesus monkey brothers. Nature 294:160–161. https://doi.org/10.1038/294160a0

Muniz L, Perry S, Manson JH, Gilkenson H, Gros-Louis J, Vigilant L (2006) Father-daughter inbreeding avoidance in a wild primate population. Curr Biol 16:R156–R157. https://doi.org/10.1016/j.cub.2006.02.055

Muniz L, Perry S, Manson JH, Gilkenson H, Gros-Louis J, Vigilant L (2010) Male dominance and reproductive success in wild white-faced capuchins (*Cebus capucinus*) at Lomas Barbudal, Costa Rica. Am J Primatol 72:1118–1130. https://doi.org/10.1002/ajp.20876

Murray CM, Lonsdorf EV, Stanton MA, Wellens KR, Miller JA, Goodall J, Pusey AE (2016) Early social exposure in wild chimpanzees: mothers with sons are more gregarious than mothers with daughters. Proc Natl Acad Sci U S A 111:18189–18194

Nonacs P (2000) Measuring and using skew in the study of social behavior and evolution. Am Nat 156:577–589. https://doi.org/10.1086/316995

Perry S (1998) Male-male social relationships in wild white-faced capuchins, *Cebus capucinus*. Behaviour 135:139–172

Perry S (2012) The behavior of wild white-faced capuchins: demography, life history, social relationships, and communication. In: Brockmann HJ, Naguib M, Mitani JC, Simmons LW (eds) Advances in the study of behavior. Academic, Burlington, pp 135–181

Robinson SP, Simmons LW, Kennington WJ (2013) Estimating relatedness and inbreeding using molecular markers and pedigrees: the effect of demographic history. Mol Ecol 22:5779–5792. https://doi.org/10.1111/mec.12529

Schoof VAM, Jack KM (2014) Male social bonds: strength and quality among co-resident white-faced capuchin monkeys (*Cebus capucinus*). Behaviour 151:963–992. https://doi.org/10.1163/1568539X-00003179

Schoof VAM, Wikberg EC, Fedigan LM, Jack KM, Ziegler TE, Kawamura S (2015) Infanticides during periods of social stability: kinship, resumption of ovarian cycling, and mating access in white-faced capuchins (*Cebus capucinus*). Neotropical Primates 21:192–196

Schuelke O, Bhagavatula J, Vigilant L, Ostner J (2010) Social bonds enhance reproductive success in male macaques. Curr Biol 20:2207–2210. https://doi.org/10.1016/j.cub.2010.10.058

Sharp SP, Simeoni M, Hatchwell BJ (2008) Dispersal of sibling coalitions promotes helping among immigrants in a cooperatively breeding bird. Proc R Soc B Biol Sci 275:2125–2130. https://doi.org/10.1098/rspb.2008.0398

Silk JB, Alberts SC, Altmann J (2003) Social bonds of female baboons enhance infant survival. Science 302:1231–1234. https://doi.org/10.1126/science.1088580

Silk JB, Beehner JC, Bergman TJ, Crockford C, Engh AL, Moscovice LR, Wittig RM, Seyfarth RM, Cheney DL (2009) The benefits of social capital: close social bonds among female baboons enhance offspring survival. Proc R Soc B-Biol Sci 276:3099–3104. https://doi.org/10.1098/rspb.2009.0681

Sterck EHM, Watts DP, van Schaik CP (1997) The evolution of female social relationships in nonhuman primates. Behav Ecol Sociobiol 41:291–309. https://doi.org/10.1007/s002650050390

Teichroeb JA, Wikberg EC, Sicotte P, Ting N (2014) Factors influencing male affiliation and coalitions in a species with male dispersal and intense male-male competition, *Colobus vellerosus*. Behaviour 151:1045–1066. https://doi.org/10.1163/1568539X-00003089

van Noordwijk MA, van Schaik CP (2001) Career moves: transfer and rank challenge decisions by male long-tailed macaques. Behaviour 138:359–395. https://doi.org/10.1163/15685390152032505

van Schaik CP (1989) Social evolution in primates: the role of ecological factors and male behaviour. Proc Br Acad 88:9–31

West SA, Pen I, Griffin AS (2002) Conflict and cooperation - cooperation and competition between relatives. Science 296:72–75. https://doi.org/10.1126/science.1065507

Widdig A (2013) The impact of male reproductive skew on kin structure and sociality in multimale groups. Evol Anthropol 22:239–250

Widdig A, Langos D, Kulik L (2016) Sex differences in kin bias at maturation: male rhesus macaques prefer paternal kin prior to natal dispersal. Int J Primatol 78:78–91

Wikberg EC, Sicotte P, Campos FA, Ting N (2012) Between-group variation in female dispersal, kin composition of groups, and proximity patterns in a black-and-white Colobus monkey (*Colobus vellerosus*). PLoS One 7:e48740. https://doi.org/10.1371/journal.pone.0048740

Wikberg EC, Jack KM, Campos FA, Fedigan LM, Sato A, Bergstrom ML, Hiwatashi T, Kawamura S (2014a) The effect of male parallel dispersal on the kin composition of groups in white-faced capuchins. Anim Behav 96:9–17. https://doi.org/10.1016/j.anbehav.2014.07.016

Wikberg EC, Ting N, Sicotte P (2014b) Familiarity is more important than phenotypic similarity in shaping social relationships in a facultative female dispersed primate, *Colobus vellerosus*. Behav Process 106:27–35. https://doi.org/10.1016/j.beproc.2014.04.002

Wikberg EC, Jack KM, Fedigan LM, Campos FA, Yashima AS, Bergstrom ML, Hiwatashi T, Kawamura S (2017) Inbreeding avoidance and female mate choice shape reproductive skew in capuchin monkeys (*Cebus capucinus imitator*). Mol Ecol 26:653–667. https://doi.org/10.1111/mec.13898

Wrangham R (1980) An ecological model of female-bonded primate groups. Behaviour 75:262–300. https://doi.org/10.1163/156853980X00447

Chapter 8
Disability and Dominance Rank in Adult Female and Male Japanese Macaques (*Macaca fuscata*)

Sarah E. Turner, Masayuki Nakamichi, Toshikazu Nobuhara, Hisami Nobuhara, and Simon M. Reader

Abstract More than 30 years ago, Fedigan (Primate paradigms: sex roles and social bonds, 2nd, 1992 edn. Eden Press, Montreal, 1982) recognized and emphasized that social dominance was not just a reflection of physical size, strength and aggression in nonhuman primates but rather involved a multitude of complex factors and social dynamics. In the free-ranging group of Japanese macaques (*Macaca fuscata*) on Awaji Island, Japan, physical disabilities in the form of congenital limb malformations (CLMs) affect around 17% of the population. This group provides an opportunity to examine dominance rank in relation to physical impairment. Here, we present an analysis of dominance rank in the adult female and adult male Japanese macaques in the Awajishima group using behavioural data collected in 2005, 2006 and 2007 on 38 adult females (12 with CLMs) and in 2015 on 22 adult males (7 with CLMs). Once we controlled for matrilineal kinship, we found that disabled females tended to hold somewhat lower than expected dominance ranks. In contrast, disabled males were interspersed throughout the rank order, and disability was not associated significantly with dominance rank. However, age class had a statistically significant effect on dominance rank in males, such that older males tended to rank higher than younger males. Our results underscore Fedigan's insights by demonstrating some of the complexities of dominance relationships in the context of the extensive physical variation shown in this group of Japanese macaques.

Keywords Dominance rank · Physical impairment · Resource holding potential · Japanese monkey · Physical disability

S. E. Turner (✉) · S. M. Reader
Department of Biology, McGill University, Montreal, QC, Canada
e-mail: sarah.turner@concordia.ca

M. Nakamichi
Osaka University, Osaka, Japan

T. Nobuhara · H. Nobuhara
Awajishima Monkey Center, Hyogo Prefecture, Japan

U. Kalbitzer, K. M. Jack (eds.), *Primate Life Histories, Sex Roles, and Adaptability*, Developments in Primatology: Progress and Prospects,
https://doi.org/10.1007/978-3-319-98285-4_8

8.1 Introduction

Dominance is often used to describe the relationship outcome of the pattern of agonistic interactions between pairs of individuals (Drews 1993). When dyadic agonistic encounters between individuals consistently result in one individual taking physical precedence, obtaining a sought-after food item or otherwise "winning" a contest, fight or interaction, this individual is described as being dominant over the other, subordinate, individual (Drews 1993). In a diverse array of animal species, it is possible to rank individuals in a linear fashion according to their dominance relationships (Shizuka and McDonald 2015). Dominance rank is an emergent property of patterns in social interactions, and as such it should not be taken as a proxy for other variables, such as higher fitness, without clear evidence to suggest a direct connection (Fedigan 1982). However in a wide variety of species, an individual's dominance rank has been shown to influence fitness-associated processes and outcomes, such as access to food and mates, growth patterns, parental care and investment, as well as health measures such as stress hormones and parasite transmission (Abbott et al. 2003; Berard 1999; Dey et al. 2015; Drews 1993; Jack and Fedigan 2006; Majolo et al. 2012; Manson 1992; Sapolsky 2005). A recent meta-analysis of dominance rank and fitness in primates suggested that there are often fitness benefits associated with high dominance rank (Majolo et al. 2012). In particular, higher dominance rank was associated with having a higher proportion of infants surviving their first year of life for females and higher fecundity and mating success in males (Majolo et al. 2012). However, there is considerable inter- and intraspecific variation across primates; there can be costs (such as stress, e.g. Abbott et al. 2003; Sapolsky 2005) and benefits (such as increased mating and paternity success, e.g. Alberts et al. 2006, 2002) to high and low dominance rank that depend on many behavioural, physiological and ecological factors (Abbott et al. 2003; Majolo et al. 2012).

Dominance has been studied extensively in many animal species (Drews 1993; McDonald and Shizuka 2013; Shizuka and McDonald 2015) and has been of particular interest in the study of nonhuman primates (Bergstrom and Fedigan 2010; Drews 1993; Fedigan 1982; Majolo et al. 2012; McDonald and Shizuka 2013; Nakamichi et al. 1995a). More than 30 years ago, Linda Fedigan (1982) provided an in-depth critique and analysis of studies on dominance rank in nonhuman primates. She pointed out that studies of dominance had focused on how animals "organize themselves according to the ability to physically defeat, or intimidate others in conflict" (p. 93), while they often overlooked the many other life history and behavioural factors associated with animal social organization. Fedigan (1982) pointed out that many studies emphasized males and the role of strength and fighting ability in shaping dominance interactions. She argued that the determinants of dominance are complicated, context dependent, only part of an animal's social organization and that "studies of primates generally show that dominance is not… determined primarily by the physical attributes of size and strength" (Fedigan 1982, p. 101).

There are numerous definitions of dominance in the scientific literature; however, dominance is most often measured using contests between individuals (or models of contests between individuals) (Kokko 2013). Individuals vary in terms of their resource holding potential (RHP) (or resource holding power), which is often defined as an individual's ability to win a dyadic all-out physical fight if such a fight were to occur (Kokko 2013; Parker 1974). Fighting ability has been described as "the primary determinant of rank in primates", at least for males, although the authors also wrote that "there is certainly sexual selection on traits other than fighting ability, and there are other sources of variance in male mating and paternity success" (Alberts et al. 2006, p. 1177).

Physical impairment would be expected to have a negative influence on RHP, such that we might expect that the more extensive an individual's physical disabilities, the lower their RHP and the lower their position in the dominance rank order. However, an individual's actual ability to hold resources may not always reflect, or not strongly reflect, their ability to win in a dyadic all-out fight. There are many examples reported of individuals who are very old, injured, toothless or physically disabled living within primate groups and holding high dominance rank positions, as well as individuals employing apparently intelligent strategies to improve their rank or circumvent dominant individuals (Cuozzo and Sauther 2004; Fedigan 1982; Kummer and Goodall 1985; Millette et al. 2009; Nakamichi et al. 1995b; Reader and Laland 2001; Schaller 1963; Stokes and Byrne 2001).

Since the 1960s, ~17% of infant Japanese macaques (*Macaca fuscata*) in the free-ranging group of at the Awajishima Monkey Center on Awaji Island, Japan, have been born with malformations of the limbs and digits (Nakamichi et al. 1997; Turner et al. 2008; Yoshihiro et al. 1979). Many of these disabled individuals survive to adulthood, so that within the group at Awajishima, there are numerous adult females and males with physical impairments. The extent of individual congenital limb malformations (CLMs) varies considerably, ranging from relatively mild malformations of digits or absence of a finger or two to combinations of extensive malformation and absence affecting the hands, feet and limbs or portions thereof (Fig. 8.1). Research has shown that adult females, at least, are able to compensate behaviourally for CLMs; in most respects their behaviour does not differ from that of nondisabled females; however, they perform certain behaviours in unique ways such as using a two-handed/two-arm pinch for grooming and employing individually appropriate styles of locomotion and infant carrying (Turner et al. 2012). Mothers of disabled infants tend to provide compensatory care as well, using an arm to support an infant who is unable to cling to its mother during locomotion and nursing (Turner et al. 2005); otherwise, however, there is little evidence for conspecific care associated with disability in the group (Nakamichi et al. 1983; Turner et al. 2014). However there is also little evidence to suggest social selection against disabled monkeys in the Awajishima group, and disability generally elicits an undifferentiated social response from conspecifics (Turner et al. 2014).

Japanese macaques are known for their formation of stable linear dominance hierarchies, with higher-ranking individuals often having priority of access to contestable food resources and also tending to receive more and give less grooming

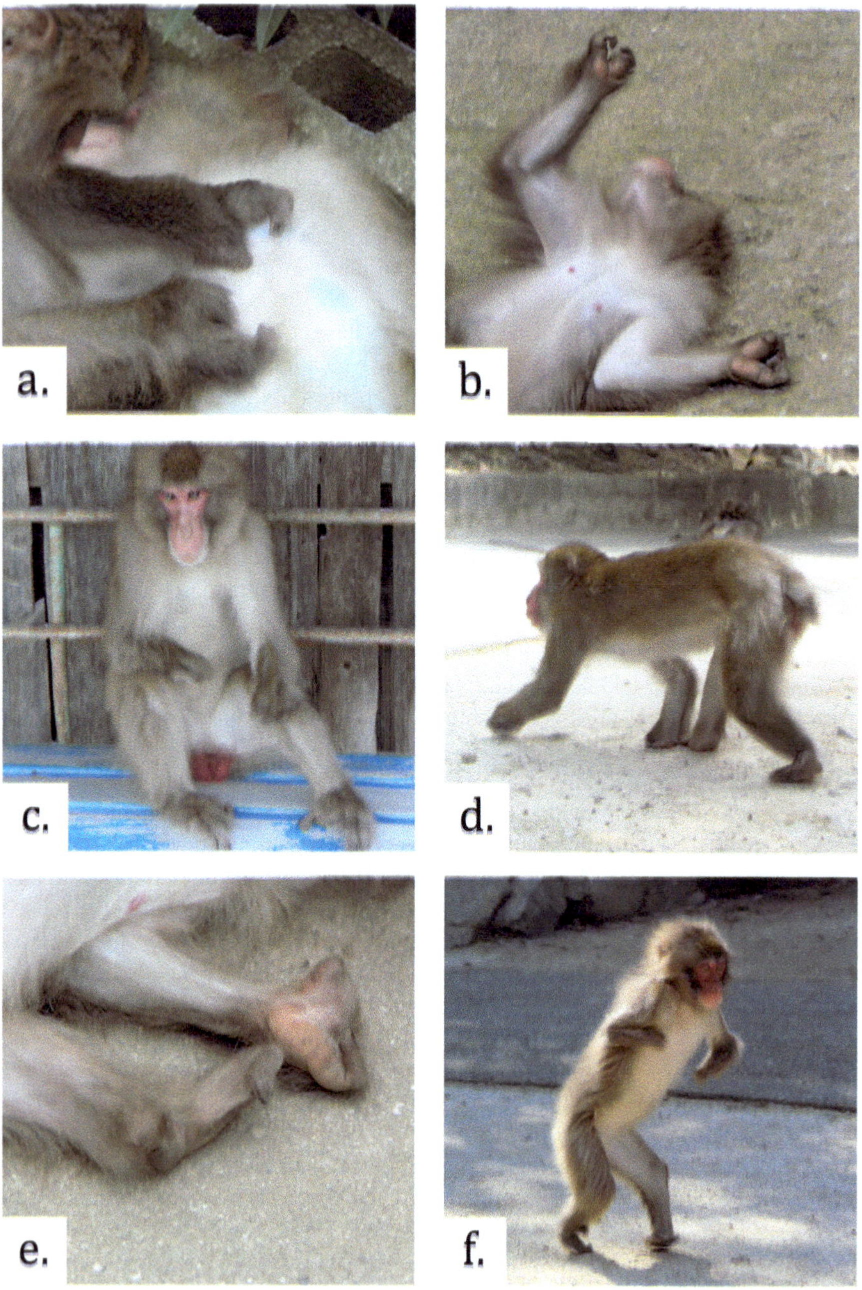

Fig. 8.1 Examples of individuals with congenital limb malformations at Awajishima. (**a**) and (**b**) show top and bottom views of the hands of two young males with similar hands: (**a**) Mochi (born ~2005, index of disability 0.35) and (**b**) Chonan (born 2001, index of disability 0.28). Their hands both show the absence of digits 2, 3 and 4 as well as associated metacarpal bones; (**c**) Gatsu, (born 2005, index of disability 0.47) a young male with the absence of hands and malformations on his

than lower-ranking individuals (Nakamichi et al. 1995a; Nakamichi and Shizawa 2003). Japanese macaques are female philopatric, and kinship is known to be a strong predictor of female dominance rank. In his 1958 study of the provisioned Minoo group of Japanese macaques, Kawamura (1958) discovered two patterns related to female kinship and dominance rank. He observed that (1) mothers outranked their daughters and (2) younger sisters outranked their older sisters. From these observations, he proposed two rules to predict dominance relationships in female Japanese macaques: that females acquire a dominance rank just below that of their mother and the idea of youngest ascendancy (that sisters hold dominance rank inversely ordered by age). These two patterns are now often called "Kawamura's rules" and have been found to be a strong and consistent predictor of female dominance relationships in Japanese macaques (especially in provisioned groups), as well as in other primate species (Bergstrom and Fedigan 2010, 2013; Fedigan 1982; Koyama 1967, 1970; Nakamichi et al. 1995a; Nakamichi and Yamada 2010). Kinship has such a strong influence on dominance rank in female Japanese macaques that other factors such as age and physical characteristics generally have less influence on rank (Nakamichi et al. 1995a). However, not all individual Japanese macaque female ranks follow Kawamura's rules. In a detailed study of female rank in the provisioned Katsuyama group, for example, Nakamichi et al. (1995a) found that about 25% of female-female dyads did not follow Kawamura's rules (Nakamichi et al. 1995a). Deviation from Kawamura's rules was more common in females from medium- and low-ranking matrilines (Nakamichi et al. 1995a). To date, it is unknown if or how physical disabilities influence dominance rank in females.

In male Japanese macaques, who usually immigrate and do not remain in their natal group, kin relationships are expected to be much less important in relation to rank than for females. For males, RHP is expected to be an important predictor of dominance rank (Alberts et al. 2006; Kokko 2013; Parker 1974). In Japanese macaques, age has also been found to be an important predictor of male dominance rank for male (Sprague 1992; Sprague et al. 1998; Suzuki et al. 1998; Takahata et al. 1998), so much so that Sprague characterized dominance rank in terms of the "age-rank structure of troops" (Sprague 1992, p. 437), although the effects of age do vary across populations (Fedigan 1982). For instance, Suzuki et al. (1998) found a linear age-dominance rank relationship in the wild Japanese macaques on Yakushima, with older males tending to be higher ranking, while in the wild Kinkazan group, the relationship was characterized by a humped curve with prime-aged males tending to have higher ranks than younger and old males (Takahashi 2002).

In this study, we examined the influence of physical disability on social dominance rank in adult female and adult male Japanese macaques by addressing the following questions:

Fig. 8.1 (continued) feet. Gatsu is habitually bipedal in his locomotion; (**d**) and (**e**) Steven (born 1997, index of disability 0.66), an adult male with extensive limb malformations. Steven moves with an uneven quadrupedal gait; (**f**) Monmo (born 2012, index of disability 0.74), a young female with extensive limb malformations on all four limbs, who employs both bipedal and quadrupedal locomotion. She uses a jumping movement in her bipedal locomotion (shown here) and uses her elbows and forearms as weight-bearing surfaces during quadrupedal movement

1. Does the dominance rank of disabled females differ from that of nondisabled females after controlling for expected dominance rank within and among matrilines?
2. How does disability relate to age and dominance rank in adult males in the Awajishima group?

8.2 Methods

8.2.1 Study Site, Species and Group

Data were collected at the Awajishima Monkey Center (AMC), on Awaji Island, Japan. The AMC is a privately operated tourist and education centre, where one large group of free-ranging Japanese macaques (*Macaca fuscata*) is systematically provisioned twice a day with wheat, soy, peanuts, fruit and sweet potatoes. Tourists also feed the monkeys through an enclosure. Provision feeding occurs every day that the monkeys arrive at the AMC (about 10 months of the year); during peak fruiting seasons, the monkeys survive on wild foods almost exclusively and generally do not come to the AMC.

The monkeys at the AMC live in one large group (currently~400 individuals). The group is geographically fairly isolated, so males may leave the group at sexual maturity, live in peripheral all-male groups or disperse individually in the surrounding forest and then reintegrate back into the group as full-grown adults. It is sometimes possible to identify these natal males as returning adults, particularly when they have distinctive limb malformations or when they have remained in peripheral all-male groups that avoid the main group but stay in the general vicinity of the AMC and share at least some of the same home range as the main group. However, for most males in the group, even if it is suspected that they are natal males, kin relationships are unknown. There may be other wild groups in the region; however, these populations are undocumented. Some males, particularly a few extensively disabled males, have remained in the Awajishima provisioned group throughout their lives, although even these males lived around the periphery of the group during their transition to adulthood, only returning to the core of the group as adults (H. Nobuhara and T. Nobuhara, unpublished data). Demographic data, including birth records, as well as kinship and matrilineal information for many identified females, have been collected since the 1960s (H. Nobuhara and T. Nobuhara, unpublished data).

8.2.2 Data Collection

Data on female dominance rank were collected on May–August 2005, 2006 and 2007, as part of a study on adult females and disability at Awajishima (Turner 2010). Data on male dominance rank were collected on June–August 2015, as part of a similar investigation of adult male behaviour and disability. ST recorded all instances of agonistic interactions during 30-min continuous focal animal follows, for both females and males. We analysed only dyadic interactions here. Dyadic agonistic interactions were also recorded ad libitum for focal animals and other individuals who were consistently and reliably recognizable to ST during data collection. We used the following behaviours to estimate dominance relationships: bite, chase, cuff, push, displace, grab, lunge, open-mouth threat, avoid and submissive grimace (Turner 2010).

Our final female data set included dyadic interaction data on 38 females, 12 with CLMs. All disabled females with a disability index score of 0.1 or higher (11 of 12 disabled females; see below) were included as focal animals, and each disabled focal animal was pair-matched in the sample with a female relative who shared as many life history similarities as possible. When a female's kin were unknown or if a female did not have close relatives living, she was pair-matched with a female of similar age and parity in the group (for further information on the focal animal sample, see Turner 2010; Turner et al. 2012). Of the 38 females, ST conducted focal animal sampling on 26, and ad libitum observations were collected on the remaining 12 females in the sample. We analysed 376 dyadic interactions among 195 pairs of females from both ad libitum (~240 field days) and focal data (279 h of focal sample data). There were some changes in group composition over the data collection period; however, there were no major changes in the higher-ranking male or female group members over the period of the study. The overall dominance rank order appeared to be stable during data collection. One female focal animal died during the study, and a different individual was added as a focal animal after her death. Data from both individuals were included separately in this analysis. For two females, we had only two seasons of data, and for two others, we had only one season of data. Because we were interested in the relative rank of individuals, using a sample of females from the population (rather than every adult female in the group) should not greatly influence the rank order outcome. That is, even if the actual order in the group is A > B > C > D > E > F and we only have data on A, B, D and F, in a context such as this where rank is linear, the order will still be A > B > D > F among the known individuals. Using a sample of individuals from the group introduces some unavoidable error into the data, as does having pairs with no recorded interactions. In a large group like that at Awajishima, which is known for being more tolerant and less despotic than usual for Japanese macaques, many individuals may simply avoid interacting with one another rather than engaging in direct contests (Fedigan 1982; Kaigaishi et al. 2016; Nakamichi and Yamada 2010). Our data show less certainty in relation to the specific order of females who are closely ranked in the dominance order compared to the certainty associated with the overall dominance order; this aspect of our results likely reflects a combination of behavioural factors (a less despotic, large group where individuals can often avoid conflict) and having individuals in the group missing from the analysis.

We collected data on all males who were central to the group during the period of data collection. Our final male data set included dyadic interaction data on 22 adult males, 8 of which had CLMs. Of the 22 males, ST conducted focal animal sampling on all but 1 (140 h of focal sampling data from ~85 field days, with ad libitum sampling on the remaining individual). Behavioural data collection yielded

95 dyadic interactions among 62 pairs of individuals. For the males, TN and HN also conducted 136 structured food tests among 82 different pairs of individuals, in which a peanut was placed between two adult males[1] and we recorded which monkey took the peanut and other details of their interactions. For a subset of pairs (23 pairs of individuals), we had data from both "peanut tests" and behavioural observations (with a total of 82 associated interactions, 42 of which were from peanut tests). The direction of winner-loser interactions was consistent between peanut tests and observed behaviour in 100% of cases. We therefore amalgamated the peanut test data and the observational data for the males, for a total of 231 dyadic interactions among 119 different pairs.

8.2.3 Potential Predictors of Dominance Rank

8.2.3.1 Disability

For all individuals with congenital limb malformations, we collected detailed photographic records and written descriptions of their individual hand, foot and limb anatomical structures. We then used an index of disability developed for this population to obtain a 0–1 score to measure the extensiveness of an individual's congenital limb malformations, where 0 corresponds with no limb malformations and 1 corresponds with the absence of all 4 limbs (Turner et al. 2008). The index of disability accounts for both absence and malformation in limbs and digits. The index is weighted to account for disability in the following ways: absence is given more weight than malformation; hands are given more weight than feet and thumbs more than other digits.

8.2.3.2 Age of Males

We used available demographic data from Awajishima (H. Nobuhara and T. Nobuhara, unpublished data) to classify individuals according to age. When exact age was unknown, we estimated the age of adult males by comparing physical characteristics with males of known age at Awajishima and by using published age-related physical markers from various Japanese macaque groups (Hamada and Yamamoto 2010). We used characteristics such as body size and shape (younger males are less robust), facial skin, degree of scarring, scrotal development and colour and signs of old age such as shorter or stiffer gait, tremors and tooth loss (Hamada and Yamamoto 2010; Takahashi 2002). We knew the birth year for 8 of 22 of the males in our sample, with known ages ranging from 10 to 26 years at the time of data collection. There was at least one of these reference males of known age for each age group we created. We grouped the males in our sample into five age classes:

[1] Six of the interactions among the females were also observed during informal "peanut tests" that TN and HN conducted spontaneously during provision feeding.

1. Young adult males (males around 10 years old (9–11 years)): these were the youngest adults, with most males in this category being 10 or 11 years; however, as we were estimating, it is possible that the sample contained a 9-year-old. Ten years is usually the age at which males are considered fully adult in Japanese macaques (Hamada and Yamamoto 2010); however in some studies, males are considered adult from around 7 years of age (Inoue and Takenaka 2008).
2. Young prime-aged males: males around 13 years old (12–14 years).
3. Prime-aged males: males around 17 years old (15–19 years), full-sized adults, both in terms of limb length and apparent body weight.
4. Old males: males around 21 years old (20–23 years).
5. Very old males: males around 25 years old, showing strong signs of aging, with stiff, tottering gait, some apparent loss of muscle mass and often some tooth loss.

8.2.3.3 Kinship and Matrilineal Relationships Among Females

As discussed in the introduction, kinship is known to have a strong influence on female dominance rank in Japanese macaques. Using kinship records from the Awajishima Monkey Center (e.g. grandmother, mother, sister and aunt relationships are known for many females; H. Nobuhara, T. Nobuhara, S. Turner and Y. Kaigaishi, unpublished data), we were able to create an expected rank order within each matriline based on Kawamura's rules (Kawamura 1958). We normalized this expected within-matriline rank on a scale from 0 to 1 and used it as a predictor of rank in our statistical model. Three females without known relatives were given the overall mean position for expected within-matriline rank. It had not been previously confirmed that this population follows Kawamura's rules (Nakamichi and Yamada 2010), as the AMC group is reported as less despotic and more tolerant than other groups of Japanese macaques (Inoue-Murayama et al. 2010; Kaigaishi et al. 2016; Koyama et al. 1981; Nakamichi and Yamada 2010). However, we do know that kinship is important in their relationships and could assume that it was likely that Kawamura's principles did apply to this group.

Kin relationships over time in the group should also lead to an overall rank order among matrilines (Koyama 1970). The resulting matriline rank order should then predict individual rank, such that females within the same matriline are expected to be close together in the overall dominance rank order and individual members of higher-ranking matrilines are expected to outrank members of lower-ranking matrilines. However, historical kin relationships in this group are not known well enough to determine an expected rank order among matrilines. Since we did not have an independent measure of the order among matrilines, we controlled for the effect of among-matriline order on individual rank by creating a blocking variable derived from the data. After establishing the most likely overall dominance rank order, we calculated the mean dominance rank among individuals within each matriline and ordered these means sequentially to create an overall rank order for the matrilines. The resulting among-matriline rank order was also normalized on a scale from 0 to

1 and used as a predictor variable in the statistical model. This derived variable effectively centred each individual rank on its matriline average, to better fit the model to known sources of biological variation.

This resulted in two independent variables in the statistical model that we used to represent the expected effect of kin relationships on individual rank order: (1) among-matriline order (derived from the data and therefore used only to control for this known biological effect) and (2) expected within-matriline order (based on Kawamura's rules and independent AMC records of kin relationships within each matriline).

8.2.4 *Analysis*

8.2.4.1 Assessing Dominance Probabilities Using the Percolation-Conductance Method

We calculated female and male dominance rank order separately using directional dyadic agonistic interactions and inverse submissive interactions between known pairs of same-sex individuals. It is important to note that although individual male Japanese macaques are generally dominant over individual female Japanese macaques, some high-ranking females are individually dominant over some males (Fedigan 1982). In particular the highest-ranking female in the Awajishima group was observed to win dyadic food tests with numerous males, including the second highest-ranking male in the group. However, because the behavioural data sets were collected in conjunction with broader sex-specific studies that were separated in time (females 2005–2007 and males 2015), we were unable to quantify and examine the degree to which the sex-specific dominance hierarchies overlapped.

For each sex, we entered behavioural data into a win/loss interaction matrix and then used the percolation-conductance method to create a matrix of dominance probabilities and estimate the best dominance rank order to represent the data. For this analysis, we used the Perc package in R (Crawley 2007; Fushing Lab and McCowan Lab 2015). The percolation-conductance method uses the direct win/loss interactions to generate indirect pathways among individuals. For example, if we know that A > B, B > C and C > D, through indirect pathways, we can estimate information about A's relationships to C and D. The information gained from the indirect pathways is weighted in the model according to the estimated reliability of that information. This method uses a combination of direct win/loss data and the indirect pathways to estimate dominance probabilities among individuals in the matrix and to model the most probable dominance rank order. Through performing a repeated series of random walks (1000 permutations), the percolation-conductance method estimates a series of hierarchies (10) and determines the most likely hierarchy among the series by quantifying the pathway cost associated with creating each hierarchy (Fushing Lab and McCowan Lab 2015).

In the female data set, there were many missing cells in the original matrix, a constraint which had previously limited our analysis of rank order (Turner et al. 2014). The percolation-conductance method allowed us to make full use of the data we had available for the females, in order to examine the possible influence of disability on dominance rank order among females in this population. This method does not assume linearity of the dominance hierarchy. It also provides a visual assessment of the uncertainties associated with different parts of the dominance hierarchy (Fushing Lab and McCowan Lab 2015).

8.2.4.2 Statistical Analyses

We analyzed the relationship among dominance rank order, disability, and kinship (for the females) and dominance rank order, disability, and age class (for the males), using linear regression. In the female regression models, we used linear rank order as the dependent variable with three independent predictors: (1) disability category (disabled or not; analysis 1) or disability index (analysis 2), (2) expected within-matriline rank predicted by Kawamura's rules, and (3) mean rank order for each matriline (we used this as a model offset to account for differences among matrilines, effectively centring each individual rank on its matriline average). In the male regression models, we used linear rank order as the dependent variable with two independent predictors: (1) disability category (disabled or not; analysis 1) or disability index (analysis 2) and (2) age category.

We visually checked quartile plots for normal distribution of residuals and plotted the residuals against the fitted values to check for equality of variance. We then removed non-significant interaction terms and non-significant factors. We chose the final model using the Akaike information criterion (AIC) and the Bayesian information criterion (BIC) to determine which model was best able to explain the variance in the data.

8.3 Results

8.3.1 Estimates of Female Dominance Relationships

In the original female win/loss conflict matrix, the relationships were very linear (205 transitive triangles and 3 intransitive triangles, yielding a transitivity measure of 0.985 and an alpha of 136.2). We were able to use the maximum length of six indirect steps in the model to bootstrap our data, as the addition of each new level of indirect pathway continued to increase the reliable data available. The percolation-conductance method generated ten versions of the dominance rank order, and although these estimated dominance rank orders varied somewhat, with small permutations in individual rank order in the parts of the matrix that contained the highest levels of uncertainty, the overall order was robust,with low levels of uncertainty (Fig. 8.2).

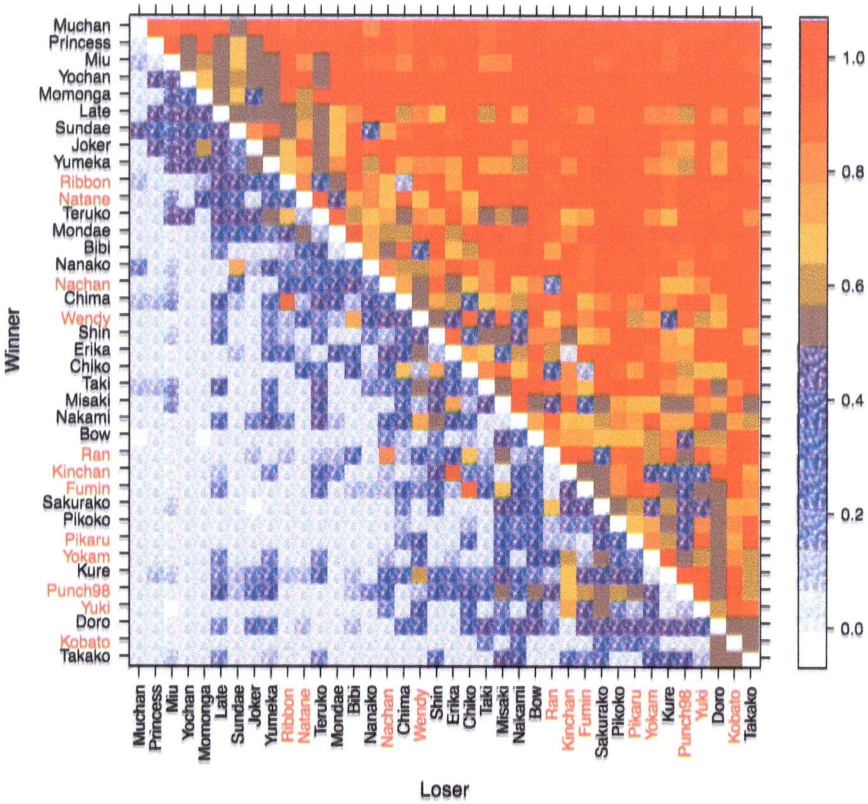

Fig. 8.2 Final female dominance rank probability matrix. The females are ordered from highest (top of Y-axis) to lowest ranking (bottom of Y-axis). Names highlighted are disabled females. The heat map shows the degree of certainty in the dyadic level dominance relationships (individual cells) and the overall pattern of certainty for the dominance rank order. The white squares on the diagonal indicate the intersection of the row and column for the same individual. Values close to 1 indicate a high probability that the individual in the row outranks the individual in the column, whereas values close to 0 indicate a high probability that the individual in the row is subordinate to the individual in the column. Values close to 0.5 indicate a high level of uncertainty with respect to the dominance relationship between two individuals. The degree of certainty is lowest close to the centre line, while further from the centre line, the degree of certainty for relationships increases, suggesting that the overall dominance rank probabilities are quite certain, while there is generally less certainty associated with the exact order of individuals who are closely positioned in the estimated dominance rank order

8.3.2 Predictors of Female Dominance Rank

If disability had a strong negative effect on female dominance rank, we would expect disabled females to hold relatively low-rank positions. While disabled females were distributed throughout most of the dominance rank order, more females occupied positions near the bottom of the dominance rank order (Fig. 8.3). Taking into account matrilineal kinship relationships, both within matrilines and

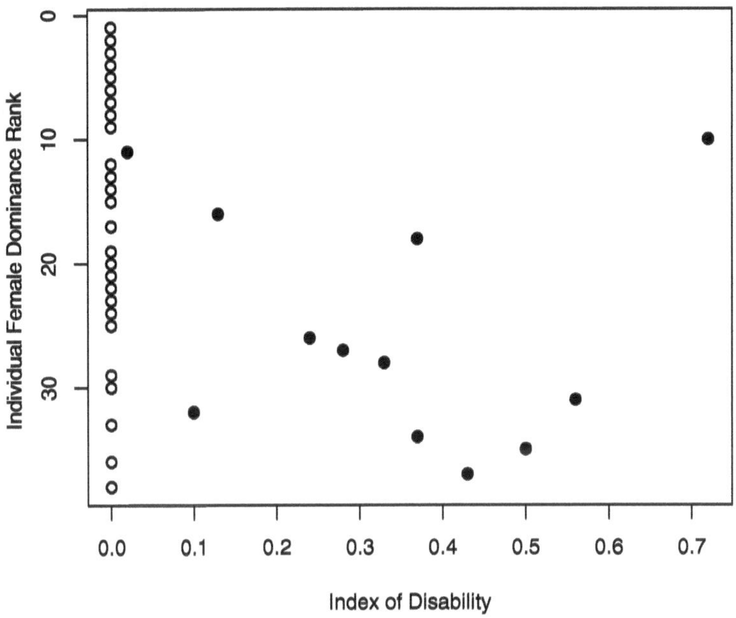

Fig. 8.3 Individual female dominance rank, where 1 and 38 correspond to the highest and lowest ranks, respectively. The open circles represent the dominance rank of individual nondisabled females. The filled circles show the dominance rank order and the extent of physical impairment as measured by our index of disability; for example, the most extensively disabled female in the group holds the tenth position in our estimated dominance rank order

among matrilines, the results of our final linear model confirmed that Kawamura's rules significantly predict female dominance rank order in the Awajishima group (linear regression: $df = 34$, adjusted R-squared = 0.78, $t = -3.52$, $p = 0.001$). Our model also showed that disability had an independent and significant negative influence on dominance rank order for females in the Awajishima group (linear regression (using disability category), $df = 34$, adjusted R-squared = 0.79, $t = 2.89$, $p = 0.007$; (using index of disability), $df = 34$, adjusted R-squared = 0.78, $t = 2.56$, $p = 0.015$; Fig. 8.4).

8.3.3 *Estimates of Male Dominance Relationships*

We were able to estimate a male dominance rank order with a high degree of certainty. Male-male relationships were highly linear (227 transitive triangles and no intransitive triangles, yielding a transitivity measure of 1 and an infinite alpha), and

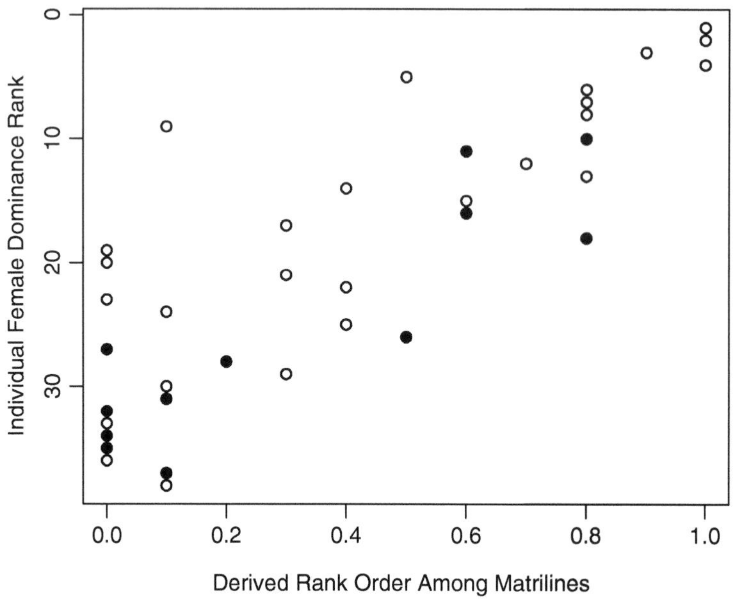

Fig. 8.4 Individual female dominance rank (from 1 to 38) and observed within-matriline rank plotted as a function of the derived matrilineal rank order. Each of the 11 matrilines is separated by 0.1 on the horizontal axis, and the matrilines are ordered according to the mean rank of females within each matriline. Filled circles show disabled females, and open circles show nondisabled females, with the highest ranked adult female in the top right corner of the figure and the lowest ranked adult female near the bottom left corner of the figure. Disabled females tended to be somewhat lower ranking than expected within their matrilines

because the addition of each new level of indirect pathway continued to increase the reliable data available, we were able to use the maximum length of six indirect steps in the model to bootstrap our data in order to increase the information available for deriving dominance estimates. This process resulted in the model producing a consistent most probable dominance rank order for the adult males at AMC (Fig. 8.5).

8.3.4 Predictors of Male Dominance Rank

Disabled males were interspersed throughout the dominance rank order (Fig. 8.6). In our final model, neither the presence/absence of congenital limb malformations nor the severity of disability showed a significant association with dominance rank in males (linear regression (using disability category), $df = 19$, $t = -1.375$, $p = 0.185$; (using index of disability), $df = 19$, $t = -0.928$, $p = 0.365$). Age class, however, was significantly and positively associated with dominance rank such that older males tended to hold higher positions in the dominance rank order (linear regression: $df = 20$, adjusted R-squared $= 0.50$, $t = -4.688$, $p = 0.0001$; Fig. 8.7).

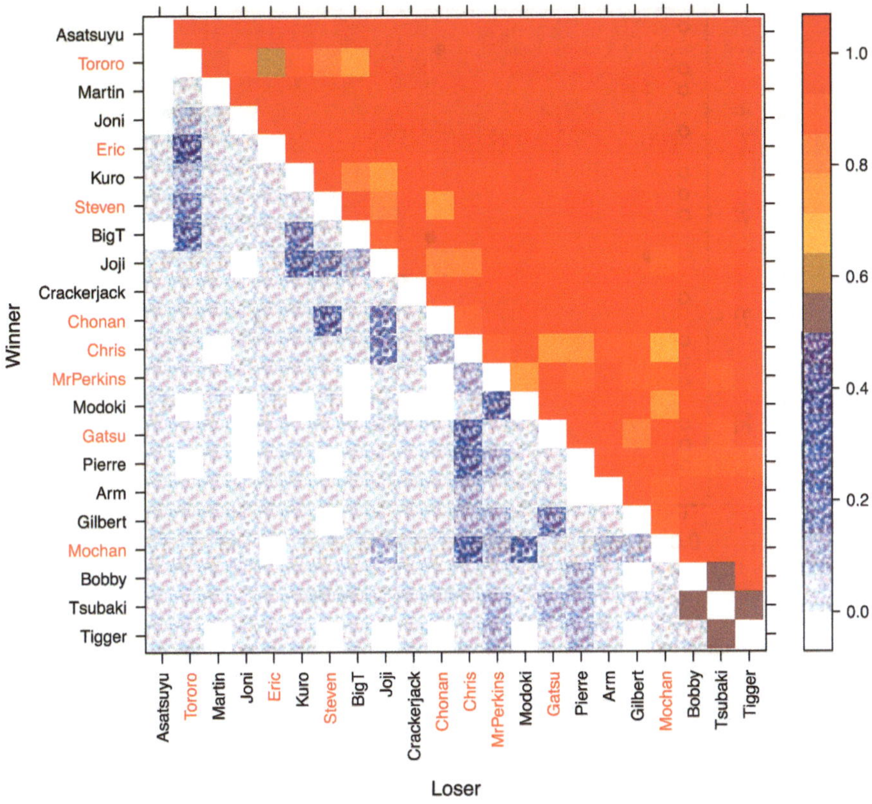

Fig. 8.5 Final male dominance rank probability matrix. The males are ordered from highest (top of the Y-axis) to lowest ranking (bottom of the Y-axis). Names highlighted are disabled males (e.g. Tororo, Eric etc.). The heat map shows the degree of certainty in the dyadic level dominance relationships (individual cells) and the overall pattern of certainty for the dominance rank order. The white squares on the diagonal indicate the intersection of the row and column for the same individual. Values close to 1 indicate a high probability that the individual in the row outranks the individual in the column, whereas values close to 0 indicate a high probability that the individual in the row is subordinate to the individual in the column. Values close to 0.5 indicate a high level of uncertainty with respect to the dominance relationship between two individuals

8.4 Discussion

Among adult female Japanese macaques at Awajishima, disability had a significant negative influence on dominance rank order. However, disabled females were not all low ranking; rather they were spread throughout most of the hierarchy (Fig. 8.3). For example, the most physically disabled female in the group held the tenth highest position in the rank order (Ribbon: index of disability 0.72; Fig. 8.8). Once we controlled for matrilineal kinship using Kawamura's rules, both disability and degree of severity of disability did significantly predict female dominance rank. Although disabled infants are less likely to survive their first year of life compared to

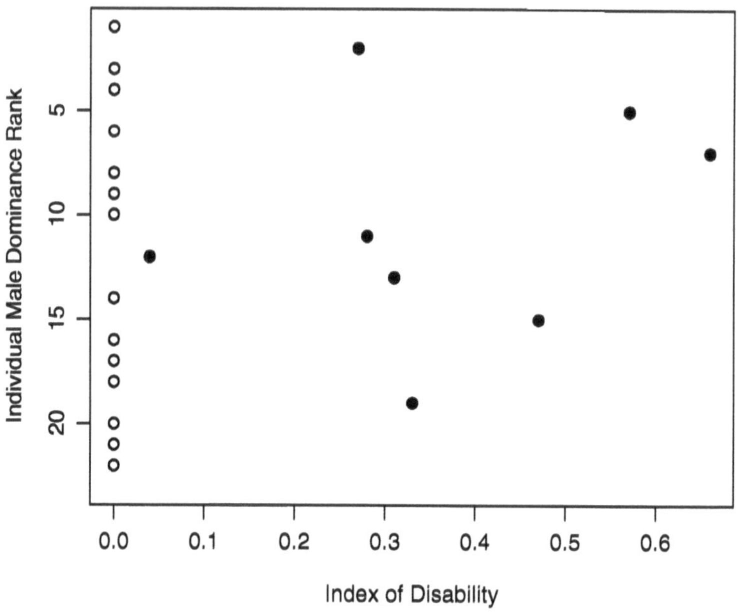

Fig. 8.6 Individual male dominance rank where 1 and 22 correspond to the highest and lowest ranks, respectively. The open circles represent the dominance rank order of individual nondisabled males. The filled circles show the dominance rank order and the extent of physical impairment as measured by our index of disability

nondisabled infants (Nakamichi et al. 1997), we have previously documented few behavioural costs associated with disability in this group (Turner et al. 2012, 2014). The results of this current analysis suggest, however, that lower dominance rank represents such a cost for disabled adult females.

Although disabled females seemed to maintain at least some of the overall rank of their matriline, they were somewhat lower ranking than expected in relation to their own kin. It is possible that while their kin supported them in agonistic interactions with non-kin, nondisabled kin were simultaneously taking advantage of their relatives' lower resource holding potential (RHP) to increase their own dominance rank within their matriline. At no time during the 9 months of data collection did we observe rank reversals, however, and disabled females did not receive more agonism, and received less bite and chase agonism, than nondisabled females (Turner et al. 2014). Consequently, the mechanism and timing of disabled females attaining their individual positions in the rank order remain unclear. Experimental research on Japanese macaque rank acquisition suggests that individual ranks were likely established when females were young and were likely influenced by availability of mothers and close kin for support and alliances (Chapais 1988a, b). It is possible that maternal disability, and/or the disability of close kin, may influence these dynamics at AMC; however, our sample of disabled mother-daughter pairs was too small to examine this possibility statistically.

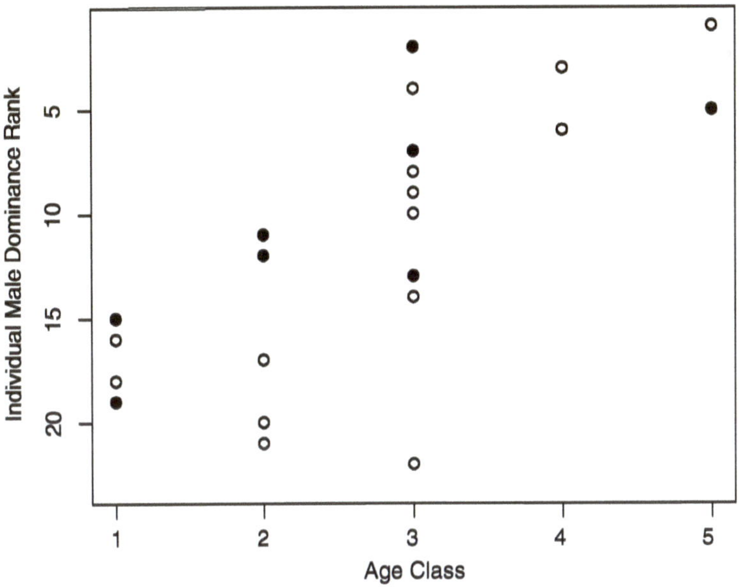

Fig. 8.7 Individual male dominance rank, age class and disability. Males were classified from youngest (1) to oldest (5). Overall, older males held significantly higher dominance ranks than younger males. Disabled males were found across age groups and throughout the dominance rank order. Open circles represent nondisabled males, and filled circles represent disabled males

Fig. 8.8 Disabled monkeys engaged in agonism and dominance-related interactions in the AMC group. Here, Ribbon, the extensively disabled adult female (index of disability 0.72), who held the tenth position in the most probable dominance rank order constructed using the percolation-conductance method for the female data set, lunges and open-mouth threatens another monkey

In previous research, we found that disabled females tended to be less social than nondisabled females across a number of measures, such as number of social partners they affiliated with, and time spent engaged in social behaviours, such as grooming (Turner et al. 2014). We also found that this difference in social behaviour between disabled and nondisabled females was likely not the result of disability-associated bias in the selection of social partners but more likely reflected disabled individuals' need for more resting time (Turner et al. 2014). Another possible explanation for the results of our current study is that lower dominance rank among disabled females emerged from their lower overall engagement in social time and activities. However, though statistically significant, the differences in the social interactions of disabled females were relatively small, and disabled females do engage in agonism and dominance-related interactions (Fig. 8.8), biting and chasing other females at rates that were not statistically different from those of the nondisabled females in our study (Turner et al. 2014).

Among adult males, we found no evidence to suggest that the presence of disability or the severity of disability carried a cost in terms of dominance rank. Disabled males held ranks throughout the dominance hierarchy, and there was no statistically significant relationship between dominance rank and disability. In particular, the second highest-ranking male had moderate manual disabilities (Tororo, index of disability = 0.27), and males holding the fifth rank (Eric, index of disability = 0.57) and seventh rank (Steven, index of disability = 0.66; Fig. 8.1d, e) were both extensively disabled. We found that age class explained a large proportion of the variance associated with rank order among the adult males, with the oldest males holding the highest ranks in the dominance hierarchy. Although we found no significant interaction between age and disability in our model, it is possible that there is an indirect connection between disability and age, via the length of tenure in the group. We did not have an independent measure of length of tenure in the group, but anecdotal evidence suggests that longer tenure was also associated with higher rank and age at Awajishima (H. Nobuhara and T. Nobuhara, unpublished observations). This pattern has been observed in other rhesus and Japanese macaque groups (Manson 1995; Sprague 1992; Sprague et al. 1998). For males, length of tenure in the group may be a key determinant of dominance rank (Suzuki et al. 1998; Takahashi 2002) and if so, likely hinges on the social relationships and knowledge built in the group over time.

It is also possible that disabled males are more likely to remain in the natal group than nondisabled males, leading to longer-than-average tenure. Anecdotally, we know of cases where disabled males never completely left the group. However, it is difficult to compare these cases directly with the emigration and immigration behaviour of nondisabled males because the disabled individuals are more easily individually identifiable over time due to their unique limb configurations, and without further data, it is thus impossible to rule out associated sampling biases (H Nobuhara and T Nobuhara, unpublished data). Tenure in the group and associated social benefits may explain some of the variation in dominance rank among males and may help explain the presence of disabled males throughout the dominance rank order; however, the complexities of these relationships remain to be evaluated.

Our results suggest that RHP may not be a consistent predictor of dominance rank in male Japanese macaques at Awajishima, a suggestion also supported by research on other groups. In wild male Japanese macaques at Yakushima, most rank changes were not the result of fights among males; instead, emigrations and death of high-ranking males were responsible for 95% of male rises in rank, and 81% of drops in rank were the result of males immigrating into the group into a higher position in the hierarchy (Suzuki et al. 1998). In some groups of Japanese macaques, such as the wild group on Kinkazan Island, old males tended to decline in rank, with age showing a humped curve-shaped relationship, such that males in their physical prime held the highest rank positions (Takahashi 2002). However, in the free-ranging Katsuyama group, Nakamichi et al. (1995b) described a case of a very old male who maintained his alpha status through coalitions with high-ranking females, despite deterioration in his ability to walk. In the Awajishima group, three different old males have held the alpha position over the last decade.

How are old and/or disabled males able to hold high rank? Why would younger, nondisabled males not fight to increase their rank over old and/or disabled males? The strength and direction of the relationship between dominance rank and reproductive success are arguably a key element in answering these questions. While high rank can have a strong and positive association with reproductive success (e.g. Alberts et al. 2006; Jack and Fedigan 2006; Majolo et al. 2012), the relationship is variable and not always positive (Berard 1999; Berard et al. 1993; Majolo et al. 2012; Manson 1992, 1995; Widdig et al. 2004). In the free-ranging group of Japanese macaques at Arashiyama, for instance, Inoue and Takenaka (2008) found that length of tenure in the group, which was positively associated with high dominance rank, had a strong but negative influence on paternity rates. When high rank does not carry a benefit in terms of reproductive success, the potential individual costs of attempting to improve rank may not outweigh the benefits (e.g. risk of injury or risk of being chased out of the group by a coalition of conspecifics). Serious one-on-one fights are rare at Awajishima, perhaps reflecting a situation where there is little reproductive benefit to be gained by such encounters.

As disability showed no connection to dominance rank in adult males in our study, our results underscores the importance of examining multiple factors potentially associated with rank and reproductive success when studying dominance rank, e.g. age (as found in this study); ecological and seasonal variables; length of tenure in the group; group behavioural conventions; cognitive performance; personality and formation of coalitions (Fedigan 1982; Nakamichi et al. 1995a; Nakamichi and Yamada 2010; Reader and Laland 2001; Suzuki et al. 1998; Takahashi 2002); as well as female choice, sperm competition, and the relationship between mating success and paternity success in males (Alberts et al. 2006; Alberts et al. 2002; Berard 1999; Berard et al. 1993; Bercovitch and Berard 1993; Manson 1992, 1995; Ostner et al. 2008; Widdig et al. 2004). In particular, as Fedigan (1982) wrote, "adult primates are…strongly influenced and constrained in their behaviour by the behaviour of other group members", and she emphasized that "alliances and coalitions have a strong, even overriding, impact on conflict outcome" (p. 107). When alliances and coalitions occur, individual one-on-one fighting ability becomes less important in determining outcomes. Recent field experiments on cooperative tasks suggest that the monkeys in the Awajishima group may be more tolerant and inclined to cooperate than is typical for Japanese macaques (Kaigaishi et al. 2016). Such cooperation may decrease the importance of RHP as a predictor of rank in this group and increase the likelihood of coalition and alliance formation, although this possibility remains to be investigated.

Finally, recent research has emphasized the importance of endogenous, self-organization properties of dominance relationships (Dey and Quinn 2014). Dey and Quinn (2014) argued that processes that act independently from individual attributes have more influence on the formation and maintenance of dominance hierarchies than have previously been recognized. These processes include such variables as winner and loser effects (Dey and Quinn 2014), priority of "ownership" of a resource (Kasumovic et al. 2010; Kokko 2013) and particular species or group-specific social conventions (Kokko 2013).

Linda Fedigan's analysis of dominance and her application of this concept in primatology is still relevant more than 30 years after the publication of *Primate Paradigms: Sex Roles and Social Bonds* (1982). Early on, she recognized that dominance was not just about physical strength and aggression but involves a multitude of complex factors and social dynamics. As Fedigan described:

> Obviously it helps an individual in a general sense to be physically well and strong, but the power of social learning and social tradition is so pervasive in primates that it often overrides the importance of physical factors in determination of dominance rank. (Fedigan 1982, p. 102)

This was a particularly important insight at a time when many researchers were treating dominance as an immutable trait inherent to the individual, which was more important for males than females, and fundamentally about physical size and strength. The Japanese macaques of Awajishima, many of whom have extensive physical impairments, illustrate the importance of not overemphasizing physical attributes, to the exclusion of other factors, in our investigations of social dominance rank relationships.

Acknowledgements SET would like to thank Linda M. Fedigan for her advising and mentorship over the years, the organizers of the Festschrift and editors of this book, the Nobuhara and Nakahashi family for permission to conduct research at the AMC and for sharing their time and knowledge and conducting peanut tests and K and K Okada and family for their kindness in Japan. SET gratefully acknowledges the funding provided by the Fonds de Récherche de Québec—Nature et Technologies (FRQNT postdoctoral fellowship), McGill University and PhD research funding from The Leakey Foundation, an Izaak Walton Killam Memorial Scholarship, the Animal Behavior Society, Province of Alberta Graduate Scholarships, the University of Calgary, the University of Calgary Dept. of Anthropology and NSERC (post-graduate scholarship B). SMR's research is supported by McGill University, the Natural Sciences and Engineering Research Council of Canada (NSERC grants #418342-2012 and #429385-2012) and the Canada Foundation for Innovation (CFI grant #29433). Thanks to JF Addicott, LJN Brent, CJ Dey, KM Jack, HD Matthews, PO Montiglio, AR Reddon, NJ Turner and two anonymous reviewers for helpful comments on this project and earlier versions of the manuscript and for statistical and data management advice. Thanks for field assistance to Y Kamata, HD Matthews, KL Turner and NJ Turner. Thanks for the support of this project to K Yamada, K Onishi, Y Kaigaishi, K Valenta and the members of Simon Reader's lab at McGill University especially M Cabrera, L Chouinard-Thuly, MF Guigueno, PO Montiglio, AR Reddon, PQ Sims and I Vassileva.

References

Abbott DH, Keverne EB, Bercovitch FD, Shively CA, Medoza SP, Saltzman W, Snowdon CT, Ziegler TE, Banjevic M, Garland T Jr, Sapolsky RM (2003) Are subordinates always stressed? A comparative analysis of rank differences in cortisol levels among primates. Horm Behav 43:67–82

Alberts SC, Watts HE, Altmann J (2002) Queuing and queue-jumping: long-term patterns of reproductive skew in male savannah baboons, *Papio cynocephalus*. Anim Behav 65:821–840

Alberts SC, Buchan JC, Altmann J (2006) Sexual selection in wild baboons: from mating opportunites to paternity success. Anim Behav 72:1177–1196

Berard JD (1999) A four-year study of the association between male dominance rank, residency status, and resproductive activity in rhesus macaques (*Macaca mulatta*). Primates 40(1):159–175

Berard JD, Nürnberg P, Epplen JT, Schmidtke J (1993) Male rank, reproductive behavior, and reproductive success in free-ranging rhesus macaques. Primates 34(4):481–498

Bercovitch FB, Berard JD (1993) Life-history costs and consequences of rapid reproductive maturation in female rhesus macaques. Behav Ecol Sociobiol 32(2):103–109

Bergstrom ML, Fedigan LM (2010) Dominance among female white-faced capuchin monkeys (*Cebus capucinus*): hierarchical linearity, nepotism, strength and stability. Behaviour 147:899–931

Bergstrom ML, Fedigan LM (2013) Dominance style of female white-faced capuchins. Am J Phys Anthropol 150:591–601

Chapais B (1988a) Experimental matrilineal inheritance of rank in female Japanese macaques. Anim Behav 36(4):1025–1037

Chapais B (1988b) Rank maintenance in female Japanese macaques: Experimental evidence for social dependency. Behaviour 104(1):41–58

Crawley MJ (2007) The R book. Wiley, Chichester

Cuozzo FP, Sauther ML (2004) Tooth loss, survival, and resource use in wild ring-tailed lemurs (*Lemur catta*): implications for inferring conspecific care in fossil hominids. J Hum Evol 46:623–631

Dey CJ, Quinn JS (2014) Individual attributes and self-organizational processes affect dominance network structure in pukeko. Behav Ecol 25(6):1402–1408

Dey CJ, Tan QYJ, O'Connor CM, Reddon AR, Caldwell JR, Balshine S (2015) Dominance network structure across reproductive contexts in the cooperatively breeding cichlid fish *Neolamprologus pulcher*. Curr Zool 61(1):45–54

Drews C (1993) The concept and definition of dominance in animal behaviour. Behaviour 125(3/4):283–313

Fedigan LM (1982) Primate paradigms: sex roles and social bonds, 2nd, 1992 edn. Eden Press, Montreal

Fushing Lab and McCowan Lab (2015) Introduction to perc package. University of California, Davis https://cran.r-project.org/web/packages/Perc/vignettes/Perc.html

Hamada Y, Yamamoto A (2010) Morphological characteristics, growth, and aging in Japanse macaques. In: Nakagawa N, Nakamichi M, Sugiura H (eds) The Japanese macaques. Primatology monographs. Springer, Tokyo, pp 27–52

Inoue E, Takenaka O (2008) The effect of male tenure and female mate choice on paternity in free-ranging Japanese macaques. Am J Primatol 70:62–68

Inoue-Murayama M, Inoue E, Watanabe K, Takenaka A, Murayama Y (2010) Behavior-related candidate genes in Japanese macaques. In: Nakagawa N, Nakamichi M, Sugiura H (eds) The Japanese macaques. Primatology Monographs. Springer, Tokyo, pp 293–301

Jack K, Fedigan LM (2006) Why be alpha male? Dominance and resproductive success in wild white-faced capuchins (*Cebus capucinus*). In: Estrada A, Garber P, Pavelka M, Luecke L (eds) New perspectives in the study of Mesoamerican primates: distribution, ecology, behavior and conservation. Kluwer Press, pp 387–386

Kaigaishi Y, Nakamichi M, Yamada K (2016) Cooperative problem solving in Japanese macaques in Awajishima group. Paper presented at the primate research supplement: the 31st congress of Primate Society of Japan. J-Stage:A1
Kasumovic MM, Mason AC, Andrade MCB, Elias DO (2010) The relative importance of RHP and resource quality in contests with ownership asymmetries. Behav Ecol 22(1):39–45
Kawamura S (1958) The matriarchal social order in the Minoo-B group. Primates 1(2):149–156
Kokko H (2013) Dyadic contests: modeling fights between two individuals. In: Animal contests. Cambridge University Press, Cambridge. https://doi.org/10.1017/CBO9781139051248.004
Koyama N (1967) On dominance and kinship of a wild Japanese monkey troop in Arashiyama. Primates 8:189–216
Koyama N (1970) Changes in dominance rank and division of a wild Japanese monkey troop in Arashiyama. Primates 11:335–390
Koyama T, Fujii H, Yonekawa F (1981) Comparative studies of gregariousness and social structure among seven feral *Macaca fuscata* groups. In: Chiarelli AB, Corruccine RS (eds) Primate behavior and sociobiology. Springer-Verlag, Berlin, pp 52–63
Kummer H, Goodall J (1985) Conditions of innovative behaviour in primates. Philos Trans R Soc Lond B 308:203–214
Majolo B, Lehmann J, de Bortoli VA, Schino G (2012) Fitness-related benefits of dominance in primates. Am J Phys Anthropol 147:652–660
Manson JH (1992) Measuring female mate choice in Cayo Santiago rhesus macaques. Anim Behav 44:405–416
Manson JH (1995) Do female Rhesus macaques choose novel males? Am J Primatol 37:285–296
McDonald DB, Shizuka D (2013) Comparative transitive and temporal orderliness in dominance networks. Behav Ecol. https://doi.org/10.1093/beheco/ars192:511-520
Millette JB, Sauther ML, Cuozzo FP (2009) Behavioral responses to tooth loss in wild ring-tailed lemurs (*Lemur catta*) at the Beza Mahafaly Special Reserve, Madagascar. Am J Phys Anthropol 140:120–134
Nakamichi M, Shizawa Y (2003) Distribution of grooming among adult females in a large, free-ranging group of Japanese macaques. Int J Primatol 24(3):607–625
Nakamichi M, Yamada K (2010) Lifetime social development in females. In: Nakagawa N, Nakamichi M, Sugiura H (eds) The Japanese macaques. Primatology monographs. Springer, Tokyo, pp 241–270
Nakamichi M, Fujii H, Koyama T (1983) Development of a congenitally malformed Japanese monkey in a free-ranging group during the first four years of life. Am J Primatol 5:205–210
Nakamichi M, Itoigawa N, Imakawa S, Machida S (1995a) Dominance relations among adult females in a free-ranging group of Japanese monkeys at Katsuyama. Am J Primatol 37:241–251
Nakamichi M, Kojima Y, Itoigawa N, Imakawa S, Machida S (1995b) Interactions among adult males and females before and after the death of the alpha male in a free-ranging troop of Japanese macaques. Primates 36(3):385–396
Nakamichi M, Nobuhara H, Nobuhara T, Nakahashi M, Nigi H (1997) Birth rate and mortality rate of infants with congenital limb malformations of the limbs in the Awajishima free-ranging group of Japanese monkeys (*Macaca fuscata*). Am J Primatol 42:225–234
Ostner J, Nunn CL, Schülke O (2008) Female reproductive synchrony predicts skewed paternity across primates. Behav Ecol 19:1150–1158
Parker GA (1974) Assessment strategy and the evolution of fighting behaviour. J Theor Biol 47:223–243
Reader SM, Laland KN (2001) Primate innovation: sex, age and social rank differences. Int J Primatol 22(5):787–805
Sapolsky RM (2005) The influence of social hierarchy on primate health. Science 308(29 April):648–652
Schaller G (1963) The year of the gorilla. Chicago University Press, Chicago
Shizuka D, McDonald DB (2015) The network motif architecture of dominance hierarchies. J R Soc Interface 12:20150080

Sprague DS (1992) Life history and male intertroop mobility among Japanese macaques (*Macaca fuscata*). Int J Primatol 13(4):437–454

Sprague DS, Suzuki S, Takahashi H, Sato S (1998) Male life history in natural populations of Japanese macaques: migration, dominance rank, and troop participation of males in two habitats. Primates 39(3):351–363

Stokes EJ, Byrne RW (2001) Cognitive capacities for behavioural flexibility in wild chimpanzees (*Pan troglodytes*): the effect of snare injuries on complex manual food processing. Anim Cogn 4:11–28

Suzuki S, Hill DA, Sprague DS (1998) Intertroop transfer and dominance rank structure of non-natal male Japanese macaques in Yakushima, Japan. Int J Primatol 19(4):703–722

Takahashi H (2002) Changes in dominance rank, age, and tenure of wild Japanese macaque males in Kinkazan A Troop during seven years. Primates 43(2):133–138

Takahata Y, Suzuki S, Agetsuma N, Okayasu N, Suguira H, Takahashi H, Yamagiwa J, Izawa K, Furuichi T, Hill D, Maruhashi T, Saito C, Sato S, Sprague DS (1998) Reproduction of wild Japanese macaque females of Yakushima and Kinkazan Islands: a preliminary report. Primates 39(3):339–349

Turner SE (2010) Consequences of congenital limb malformations and disability in adult female Japanese macaques (*Macaca fuscata*). Dissertation, University of Calgary, Calgary

Turner SE, Gould L, Duffus DA (2005) Maternal behavior and infant congenital limb malformation in a free-ranging group of *Macaca fuscata* on Awaji Island, Japan. Int J Primatol 26(6):1435–1457

Turner SE, Fedigan LM, Nobuhara H, Nobuhara T, Matthews HD, Nakamichi M (2008) Monkeys with disabilities: prevalence and severity of congenital limb malformations in *Macaca fuscata* on Awaji Island. Primates 49(3):223–226

Turner SE, Fedigan LM, Matthews HD, Nakamichi M (2012) Disability, compensatory behavior, and innovation in free-ranging adult female Japanese macaques (*Macaca fuscata*). Am J Primatol 74:1–16. https://doi.org/10.1002/ajp.22029

Turner SE, Fedigan LM, Matthews HD, Nakamichi M (2014) Social consequences of disability in a nonhuman primate. J Hum Evol 68:47–57

Widdig A, Bercovitch FB, Streich WJ, Sauermann U, Nürnberg P, Krawczak M (2004) A longitudinal analysis of reproductive skew in male rhesus macaques. Proc R Soc Lond B Biol Sci 271:819–829

Yoshihiro S, Goto S, Minezawa M, Muramatsu M, Saito Y, Sugita H, Nigi H (1979) Frequency of occurrence, morphology, and causes of congenital malformation of limbs in the Japanese monkey. Ecotoxicol Environ Saf 3:458–470

Part III
Primate-Environment Interactions

Michael A. Huffman

In Charles Darwin's 1859 seminal book, *On the Origin of Species*, he wrote, "It is not the strongest species that survive, nor the most intelligent, but the most responsive to change". All animals are highly adapted to their environment, and in that sense are equally "intelligent" when it comes to negotiating their way around their social and ecological environments. Regarding food, animals are obliged to continuously monitor and respond to environmental changes with flexible feeding strategies to maintain health homeostasis and increase reproductive fitness (Schoener 1971; Mangel and Clark 1986).

Primates manipulate their habitat in the search of food to fulfill the basic requirements of maintenance and growth. How they go about this is influenced by both short-term resource availability and long-term seasonal and interannual variability of foods in their environment (Hill 1997; Hladik 1988; Lambert 2011; Lambert and Rothman 2015; Overdorff 1993; Peres 1994; Tutin et al. 1997). Primates inhabit an extremely wide diversity of ecological zones, and not surprisingly therefor, a wide dietary diversity has evolved within the Order, ranging from narrow food type specialists (gumivores, insectivores, folivores) to very broad generalists (omnivores) that consume a wide variety of food types, both plant and animal. It is not uncommon for two or more primate species to live sympatrically in the same habitat, and such interspecies coexistence is made possible by differences in foraging, food processing, and overall dietary strategies (Yamagiwa and Basabose 2006; Hladik 1988; McGraw et al. 2016; Overforff 1993; Tutin et al. 1997).

Plants also manipulate primate behavior by advertising their palatability with positive color, smell, and taste signals, or by repelling them with physical deterrents like hard protective covers and spines or with unpleasant taste-linked chemical deterrents such as alkaloids, tannins saponins, and cyanogenic glycosides (Chapman

M. A. Huffman (✉)
Primate Research Institute, Kyoto University, Kyoto, Japan
e-mail: huffman.michael.8n@kyoto-u.ac.jp

and Chapman 2002; Ganzhorn 1989, Glander 1979, 1982; Overdorff 1993; McGraw et al. 2016). Taken in the proper dosage, such toxins can also promote homeostasis and long-term fitness benefits in the form of self-medication (Forbey et al. 2009; Huffman 1997).

Understanding how these interactions increase individual fitness can tell us much about the ecology and evolution of a species (Lambert and Rothman 2015). In order to understand this big picture of the primate-environment interface, it is important to examine the daily interactions of an animal in detail, from both behavioral and biological perspectives. Long-term monitoring of these interactions and analysis at the appropriate time scales is crucial to our understanding of the nuances and affect of such variation on the individual and how it adjusts to change. The insights gained through the multidisciplinary perspective are expected to be much larger than the sum of its parts, and the research showcased in the chapters of this part nicely illustrate the value of this approach.

This part consists of five chapters, covering a diverse range of perspectives and disciplinary approaches and focusing specifically on the interface between food and the forager; in this case, the focus is on Santa Rosa and how primates maneuver at the interface. This long-term study site is located in the northwest corner of Guanacaste province, Costa Rica, and was established in 1983 by Linda M. Fedigan. These chapters introduce and apply cutting-edge behavioral, ecological, genetic, and theoretical approaches to shed new light on the interactions that occur at the primate-environment interface. Chap. 8 by Melin et al. takes up a seemingly simple but immensely important question; how should we study feeding behavior? The authors provide a detailed discussion of some of the methodological sticking points commonly encountered when initiating field studies of primate dietary ecology, and they suggest possible ways forward. This question is taken to the next level in Chap. 9 by looking at the flexible nature of capuchin foraging in response to changing food resource availability and predictability over short- and long-term periods, providing recommendations for the best time interval to measure and evaluate food availability (Hogan and Melin). Foraging is not just about how much to eat, but how much of what to eat, and its consequences on the nutritional balance provided by all items consumed. In Chap. 10 Bergstrom et al. examine how female capuchins balance food intake relates to estimated individual requirements by measuring variation in food abundance and climatic variables, collecting detailed data on foraging behavior, dietary profiles, nutritional content, and nutrient intake. To better understand the long-term effects of these interactions, in Chap. 11 Valenta and Chapman take a theoretical approach using the concept of "fruit syndromes" to match fruit characteristics (color, odor, size) with consumer behavioral and morphological traits that facilitate seed dispersal. The primate fruit syndrome hypothesis is reviewed, and the authors present a way forward to test this hypothesis in primates using new technology and methodology. This part is aptly finished off by Kawamura in Chap. 12 where he presents a historical review of the groundbreaking studies on the eco-genetics of the visual opsin genes and primate color vision, that trait that determines how primates see the world of food and guides capuchins and other primates at Santa Rosa at the interface of their Neotropical environment.

This part, like the rest of the book showcases an impressive range of perspectives both investigative and theoretical, is drawn from research conducted on a broad range of species and topics. This body of work reflects the academic rigor, broad and equally deep thinking and tireless efforts of their mentor and colleague, Prof. Linda M. Fedigan. It is through Linda's patience, kindness, and generosity of sharing ideas and resources that such profound insights have been gained. Without these traits and the academic environment Linda has nurtured, such an impressive array of collaborative work would not have been, or continue to be, possible.

References

Chapman CA, Chapman LJ (2002) Foraging challenges of red colobus monkeys: influence of nutrients and secondary compounds. Comp Biochem Physiol A Mol Integr Physiol 133:861–875

Forbey J, Harvey A, Huffman MA, Provenza F, Sullivan R, Tasdemir D (2009) Exploitation of secondary metabolites by animals: a behavioral response to homeostatic challenges. Integr Comp Biol 49(3):314–328. https://doi.org/10.1093/icb/icp046

Ganzhorn JU (1989) Primate species separation in relation to secondary plant chemicals. J Hum Evol 4(2):125–132

Glander KE (1979) Howling monkey feeding behavior and plant secondary compounds: a study of strategies. In: Montgomery GG et al (eds) The ecology of arboreal Folivers. Smithsonian Institution Press, Washington DC, pp 561–574

Glander KE (1982) The impact of plant secondary compounds on primate feeding behavior. Yearb Phys Anthropol 25:1–18

Hill D (1997) Seasonal variation in the feeding behavior and diet of Japanese macaques (Macaca fuscata yakui) in lowland forest of Yakushima. Am J Primatol 43:305–322

Hladik CM (1988) Seasonal variations in food supply for wild Primates. In: de Garine I, Harrison GA (eds) Coping with uncertainty in food supply. Clarendon Press, Oxford, pp 1–25

Huffman MA (1997) Current evidence for self-medication in primates: a multidisciplinary perspective. Yearb Phys Anthropol 40:171–200

Lambert JE (2011) Primate nutritional ecology. In: Campbell C, Fuentes A, MacKinnon K, Panger M, Bearder S (eds) Primates in perspective, 2nd edn. University of Oxford Press, Oxford, pp 512–534

Lambert JE, Rothman JM (2015) Fallback foods, optimal diets, and nutritional targets: primate responses to varying food availability and quality. Annu Rev Anthropol 44:493–512

Mangel M, Clark CW (1986) Towards a unified foraging theory. Ecology 67(5):1127–1138

Overdorff DJ (1993) Similarities, differences, and seasonal patterns in the diets of Eulemur rubriventer and Eulemur fulvus rufus in the Ranomanafana National Park. Int J Primatol 14(5):721–753

Peres CA (1994) Primate responses to phenological changes in an Amazonian terre firme forest. Biotropica 26:98–112

McGraw WS, van Casteren A, Kane E, Geissler E, Burrows B, Daegling DJ (2016) Feeding and oral processing behaviors of two colobine monkeys in tai Forest, Ivory Coast. J Hum Evol 98:90–102

Schoener TW (1971) Theory of feeding strategies. Annu Rev Ecol Syst 2:36–404

Tutin CEG, Ham RM, White LJT, Harrison MJS (1997) The primate community of the Lopé reserve, Gabon: diets, responses to fruit scarcity, and effects of biomass. Am J Primatol 42:1–24

Yamagiwa J, Basabose AK (2006) Diet and seasonal changes in sympatric gorillas and chimpanzees at Kahuzi-Biega National Park. Primates 47:74–90

Chapter 9
Data Collection in Field Primatology: A Renewed Look at Measuring Foraging Behaviour

Amanda D. Melin, Shasta E. Webb, Rachel E. Williamson, and Kenneth L. Chiou

Abstract Studies of wild animals reveal how foraging activities unfold in natural contexts and are essential for understanding the evolutionary pressures that shape diet, feeding behaviour, and niche divergence. Yet, observing free-ranging animals presents many challenges including limited visibility and difficulties drawing comparisons across individuals. Here, we review and discuss methods for recording foraging behaviours—in particular, feeding efficiency, activity budgets, and modes of sensory investigation—of diurnal, group-living primates. When the goal of the research is to compare feeding efficiencies, we suggest flexible use of focal animal follows that maximize ability to record the intake rate and food investigation sequences of different individuals in the same food patch. In particular, we suggest observers aim to record a set number of ingest events under conditions of good visibility, while still applying appropriate stopping rules, rather than to complete a focal follow of a set duration. In addition, we consider the benefits and challenges of different methods of data collection, including scan sampling, all occurrence sampling, and focal animal follows, for studying foraging activity budgets and other aspects of feeding ecology. Finally, we discuss strategies for collecting different types of data to integrate goals of long-term (multi-year or multi-decade) studies

A. D. Melin (✉)
Department of Anthropology and Archaeology, University of Calgary, Calgary, AB, Canada

Department of Medical Genetics, University of Calgary, Calgary, AB, Canada

Alberta Children's Hospital Research Institute, University of Calgary, Calgary, AB, Canada
e-mail: amanda.melin@ucalgary.ca

S. E. Webb · R. E. Williamson
Department of Anthropology and Archaeology, University of Calgary, Calgary, AB, Canada

K. L. Chiou
Department of Psychology, University of Washington, Seattle, WA, USA

U. Kalbitzer, K. M. Jack (eds.), *Primate Life Histories, Sex Roles, and Adaptability*, Developments in Primatology: Progress and Prospects,
https://doi.org/10.1007/978-3-319-98285-4_9

with shorter-term projects. Our goal is to provide a critical discussion of some of the methodological sticking points that students commonly encounter when initiating field studies of primate dietary ecology and possible ways forward.

Keywords Foraging ecology · Behavioural observation · Methods in primatology · Focal animal follow, Scan sampling

9.1 Primate Foraging Ecology

The study of foraging ecology focuses on how food-procuring strategies of animals are shaped by their physical surroundings (Kramer 2001; Newman 2007). Primates occupy a wide breadth of ecological and dietary niches. Most primates consume ripe fruits when available, although some taxa specialize in the consumption of leaves (e.g. colobine monkeys), unripe fruits and nuts (e.g. saki monkeys), grass (e.g. geladas), lichen (e.g. snub-nosed monkeys), gums (e.g. pygmy marmosets), and insects (e.g. tarsiers). Others, including baboons and capuchins, are generalists, well-known for eclectic and opportunistic diets (Rowe and Myers 2017). Understanding dietary selectivity, food assessment strategies, and other interactions between animals and the environment provides insight into their adaptive radiation, niche divergence, and present-day morphology, physiology, and behaviour (Schoener 1968; Peres 1993; Barnes et al. 2011; Kusch and Schmitz 2013; Chao et al. 2013; Righini 2017). In primatology, understanding the foraging ecology of species aids researchers in understanding why and how primates diverged from other mammals and evolved into the extant species we see today. Indeed, many hypotheses of primate origins focus on the interrelationships between diet and other ecological variables including activity pattern, for example, the "nocturnal visual predation" hypothesis (e.g. Cartmill 1992).

Study of foraging behaviours, including the investigation of time budgets, dietary selectivity, and food intake efficiency, can also elucidate how individuals are affected by and respond to ecological changes over short time periods and can reveal behavioural plasticity, meaning an organism's ability to adapt to environmental changes by modifying its behaviour (Hine and Martin 2015). For example, behavioural plasticity includes if and how primates respond to seasonal shifts in rainfall, temperature, and food abundance by adjusting their diets and food search behaviours. These responses form a crucial component of the suite of strategies primates can use to adjust to changing habitats (Grueter et al. 2013; Fernandez-Duque and van der Heide 2013; Sengupta and Radhakrishna 2016; Nagy-Reis and Setz 2017). Other research aims to understand the complex ways in which variables such as health, reproductive status, dominance, age, and social group size can affect and be affected by the foraging behaviours of individuals (Murray et al. 2006; Vogel and Janson 2007; Nowell and Fletcher 2008; Grove 2012; Ruivo et al. 2017; Nadia Corp and Byrne 2002; Gunst et al. 2010; Visalberghi et al. 2016). In addition to contributing to a better understanding of species-specific ecology, research in this area contributes to a comparative framework for recognizing the impact of these variables on human adaptation (Luca et al. 2010).

9.2 Goals of the Chapter

Studying the foraging ecology of wild primates can be exceptionally challenging. This is particularly true when the animals are small and/or arboreal, and detailed data on feeding rates or dietary preferences are of interest. Many of the traditional practices for collecting behavioural data are appropriate for generating lists of food species, but are not optimized for recording foraging sequences or comparing food intake rates of multiple individuals, nor was this the type of data collection for which methods in primate behavioural ecology were originally proposed. Our goal is to provide a critical discussion of some of the methodological sticking points that students and other researchers commonly encounter when initiating field studies of primate foraging ecology and possible ways forward. To do so, we use our experience studying the foraging ecology of Costa Rican capuchin monkeys as a case study. Our intended audience is early-level graduate students seeking to design their data collection protocols. We presume that students have carefully read introductory literature on studying primate behaviour (Altmann 1974; Paterson and Vandenbeld 2001; Martin and Bateson 2007). We hope this discussion will be broadly useful to investigators in diverse areas of research for thinking through the aspects of study design, data collection, and analysis.

We begin by (1) listing some of the research questions and the types of variables that are of interest to investigators in this field. We then (2) identify the data needed to calculate the variables and the methods that are commonly used in our field to collect these data. Following this, we (3) overview and highlight some of the strengths and weaknesses of traditional data collection methods, including a detailed discussion of a flexible application of focal animal sampling in the context of comparing food intake rates among co-feeding individuals. We then (4) provide examples of how we are using the data collected and variables analysed to answer different types of research questions. Following this, we (5) provide a brief discussion of a hierarchical data analysis strategy to help maintain independence of nested analyses. Because primates are long-lived animals, and many of the variables needed to assess fitness and adaptation require long-term study, we also (6) discuss the additional challenges of collecting behavioural data in ways that are compatible with longitudinal analysis of behaviour over multiple years and decades. We explain our strategies for integrating short-term behavioural data collection guided by independent (typically student-led) projects, with long-term, standardized data collection that comprises our behavioural and ecological data repository. We hope to spark discussion of data collection methods and analysis in the face of difficult viewing conditions and of integrating short- and long-term research.

9.3 Research Questions Asked and Variables Measured in Studying the Foraging Ecology of Wild Primates

There is a vast array of research that falls under the broad category of primate foraging ecology. Many questions in this area can be addressed by measuring variables pertaining to food encounter, assessment, processing, intake rates, sharing, and diet selectivity. Here we provide several examples of the types of research questions that are commonly asked, overview some of the common variables measured, and show how they are useful for addressing various types of research questions. We also discuss some of the common difficulties inherent in measuring these variables and possible ways through or around these obstacles.

From providing insight into mechanisms of niche divergence, flexibility, and adaptation within and between species to linking behavioural differences with growth, development, and reproduction, the study of primate foraging ecology can address many questions at proximate and ultimate levels. Operationalization of questions in this field may look like this—do age, sex, social dominance, body size, personality, etc. affect:

(a) The ability of individuals to find resources in their environment?
(b) How much time individuals spend in different types of foraging behaviours and/or areas of the habitat?
(c) Which types of foods (e.g. insects, fruit, flowers, leaves, pith, and subcategories within these broad classifications) individuals consume most often?
(d) How efficiently individuals are able to consume different food items?
(e) The method and efficiency of primates to identify edible from inedible food items, high-quality from low-quality foods, and how selective they are about what they will eat?

Below are some common variables we and others have measured in studies of primate foraging ecology. The variables described here are suited to record different stages in, or aspects of, the foraging process. We highlight the research questions (a–e above) that these variables address. We also identify some of the common challenges that we have encountered in measuring these variables in studies of wild monkeys. Some of these challenges are exacerbated for small, arboreal monkeys relative to larger, more terrestrial primates. However, we anticipate that at least some of these challenges are generalizable. In Sect. 9.4, we explain our approaches and methods for dealing with the challenges we identify in Sect. 9.3.

9.3.1 Food Patch Encounter Rate (PER)

Food patch encounter rate (PER) is a measure of how often, per unit time spent in food search behaviours, individuals locate food patches in their habitat (question a). This is an important variable for addressing dietary frequency and, when combined with measures of food abundance in the habitat, dietary selectivity. The definition of a "patch" varies from study to study but is often considered to be a discrete source of food of a given type. Commonly, a single tree or bush, including the area directly underneath its canopy, is considered a patch. This may include fruits, leaves, flowers, or pith (Pruetz and Isbell 2000; Vogel and Janson 2007). Instances of clumped invertebrates (e.g. termite nests, large colonies of embedded ants, caterpillars with high host specificity in a single tree) may also be considered discrete patches (Janson 1988). In the case of lianas, several tree canopies may be considered a single patch if they are all covered with a single type of liana from which a single phenophase (e.g. ripe fruit or flower bud) is consumed (Castellanos and Chanin 1996; Melin et al. 2014).

The food PER for a social group is the number of unique patch visits per unit of observation (group contact) time (Leighton 1993). Questions can be asked regarding whether group PER depends on the composition of the group with respect to the variable(s) of interests, for example, the number of trichromatic individuals (see Veilleux et al. 2016). It may also be possible to note the first monkey to enter a new food patch. In this case, the first monkey to detect a new patch can be examined with respect to variables of interest, including age, sex, and colour vision type (Hogan et al. 2018). Note that patch occupancy duration and number of co-feeders are separate variables that speak to patch profitability and defendability, which could be influenced by fruit biomass-to-primate ratio (Leighton 1993; King et al. 2011).

Challenges of measuring food PER accurately include (a) teasing apart memory of previously visited food patches from new discovery (Janson 1998; Garber et al. 2009; Janmaat et al. 2013), (b) identifying the individual primate who was the first to locate (i.e. the producer) the food patch in group-living primates and controlling for observation effort at the individual level (Hogan et al. 2018), (c) observing all of food patch discoveries without bias towards larger or more conspicuous food patches or bias due to visibility of the habitat or number of observers collecting data (Vogel and Janson 2007), and (d) how to characterize separate food patches (Leighton 1993; Dew 2005).

9.3.2 *Foraging Activity Budget*

We use activity budget to refer to estimates of time spent in different types of activities, in different parts of the forest, and the importance of different foods (research questions b and c). Comparison of activity budgets can reveal niche divergence in foraging activities and individual strategies based on sensory phenotype, sex, dominance, age, or linked variables (e.g. predation risk, Miller 2002). For example, socially dominant individuals may be able to monopolize some foods that are high in energy or limiting macro- or micronutrients or vitamins (Janson 1985; Barton and Whiten 1993; Foerster et al. 2011, Bergstrom et al., Chap. 11, this volume). Differences in activity budget may also occur due to different nutritional requirements (e.g. due to age or reproductive status, Eadie 2015; Ruivo et al. 2017). Among humans, polymorphism in taste perception leads some people to avoid or prefer certain foods (Sandell and Breslin 2006; Hayes et al. 2010; Feeney et al. 2011). We focus on measuring individual activity budgets intraspecifically, but this variable could be assessed at the group level and analysed with respect to group composition. Activity budgets measured at either the group or individual level could also be used to compare among species.

Interpreting the biological significance of differences in activity budgets among individuals or groups can be difficult. For example, individuals may conceivably spend more time in activities for which they are more efficient and thus prefer, or because they are inefficient at these activities and need to expend more time to achieve the same net gain. To begin to tease apart these possibilities, it is often useful to combine interindividual comparisons of activity budget with intake rates of different food types (Schülke et al. 2006, Bergstrom et al., Chap. 11, this volume). This will reveal relative differences in food intake per unit time in different activities and also overall daily intake (i.e. net gain in nutrition from different foods). In addition, differences between individuals may not show up in time allotted to different behaviours. Rather than differences in types of activities, individuals may perform the same activity in different areas of the habitat due to accessibility. For example, smaller individuals can use thinner foraging structures.

Challenges of accurately measuring activity budgets include obtaining even sampling effort across all individuals and environments. This may be particularly challenging for peripheral or difficult-to-identify individuals and/or in parts of the habitat or seasons where visibility is low.

9.3.3 Food Intake Rate

The intake rate is a common measure of feeding efficiency used to estimate the relative foraging ability of individuals, and the consequences for energy balance, health, growth, and development (research question d; (Janson and van Schaik 1988; Johnson and Bock 2004; Nakagawa 2009; O'Mara 2015; Visalberghi et al. 2016)). It is the end result following from the number of food items detected per minute (detection rate), the number of potential foods investigated (investigation rate), the proportion of investigated foods that are accepted (acceptance ratio) of fruits, and the handling time and processing efficiency of food eaten (see Hiramatsu et al. 2008). Ideally, the dry weight consumed per unit time should be estimated. However, whether it is measured at the level of individual food items consumed or of individual bites of a given food type, by calculating food ingestion per unit time, investigators seek to estimate nutritional reward and energy gain from different foods. Accordingly, study of intake rates is perhaps most useful when combined with measures of the available nutrients of different foods of appropriate developmental state, ripeness, and maturity (Rothman et al. 2008, 2012; Aristizabal et al. 2017; Johnson et al. 2017, Bergstrom et al., Chap. 11, this volume).

Challenges of accurately measuring intake rate include (a) ensuring intake rates of different individuals are measured under comparable ecological conditions, (b) seeing monkeys well enough to record the high level of detail necessary, and collecting sufficient data on each monkey for robust analysis.

9.3.4 *Food Processing Efficiency*

For some foods that need to be processed before consumption, such as snails, invertebrates embedded in trees, pulp or aril in thick-husked fruits, or fruits protected by urticating hairs, a major influence on intake rate is the efficiency of extracting edible food components (research question d, i.e. how quickly the edible parts are excised and consumed). This is affected by both food item qualities (e.g. fruit ripeness or invertebrate size and location) and individual-level qualities (e.g. size, strength, experience, personality) (Lambert 1999; Dew 2005; O'Malley and Fedigan 2005; Gunst et al. 2010; Carter et al. 2014). Both food-level and individual-level variables can affect an individual's food choice (i.e. which food types or species to eat, leading to potential for intraspecific niche divergence) (Melin et al. 2008; Eadie 2015).

Challenges of measuring feeding efficiency of protected foods include accurately noting the transitions between food extraction and food consumption behaviours, which can occur quickly and repeatedly. In cases where multiple observers are collecting this data, either contemporaneously or at different periods over a long-term study, ensuring that all observers follow the same rules for switching among states and recording events is critical.

9.3.5 *Diversity and Length of Food Investigation Sequences*

From the time a potential food is initially detected, a primate uses its senses and past experience to assess the item and decide whether to eat or reject it. The food investigation sequence is a chronological record of the types of sensory investigation that a food item is subjected to by the foraging primate and can be used to assess the method and efficiency of identifying edible food species (research question e). The following senses may be employed: *haptic* (manual or buccal; e.g. squeezing or biting to probe softness or texture), *visual* (e.g. close gazing to assess colour, shape, size, or to look for signs of damage), *olfactory* (e.g. sniffing to detect odorants associated with preferred or avoided phenophase), *gustatory* (e.g. licking or biting to detect tastants associated with preferred or avoided phenophase), and/or *auditory* (e.g. exploring sounds made through tapping on foods and listening to reverberations to assess hollowness) (Dominy et al. 2001; Dominy 2004; Melin et al. 2009; Melin and Veilleux 2018). Both the length and composition of the food investigation sequence are informative. It can be short and simple (one to two behaviours) or longer and more complex (three or more different behaviours), and it can reveal the senses that are important during the evaluation of a particular food item. For example, investigation sequences often vary when assessing fruits of different plant species (Valenta et al. 2015).

In general, we predict that foragers that can reach an appropriate decision (i.e. eat versus reject) using fewer senses may be more efficient and able to consume foods at a higher rate and/or expend less energy to reach a decision. Sensory phenotype may also impact the form of assessment used. For example, primates with dichromatic colour vision are more likely to sniff some fruits than are trichromatic group mates (Melin et al. 2009). As monkeys age, they may also learn the most informative forms of assessment for different foods and forage more efficiently (Melin et al. 2009, 2017).

Challenges of measuring food investigation sequences accurately are similar to those linked with intake rates and include (a) collecting comparable data across individuals sampled under different ecological conditions; (b) seeing monkeys well enough to record this level of detail and, relatedly, collecting sufficient data on each monkey for robust analysis; and (c) recording sequences when handfuls of food items are collected and handled together, rather than full evaluation of each fruit prior to proceeding to another fruit.

9.3.6 Quality of Foods Consumed Within Patches

Finally, the ability to make effective choices among different food options (e.g. maturity stages) within food patches is an important component of primate foraging ecology (Gomes and Bicca-Marques 2012; Rushmore et al. 2012). Not all food items within a tree canopy or other patch will have equal nutritive value and energetic reward; some—typically ripe vs. unripe fruits and young vs. mature leaves—differ drastically, although digestibility of different food items varies by primate species (Lambert 1998; Rothman et al. 2012). Intraspecifically, it is likely that individual-level variables impact ability to consume high-quality foods. For example, dominant monkeys may monopolize the best patches within a tree, while juvenile monkeys may be able to access small branches in upper canopies where ripe fruits and/or new leaves tend to first appear (Janson 1985, 1988). Monkeys with red-green colour vision may be more accurate in selecting the most desirable fruits, flowers, or leaves from mature green leaves for foods where reddish colours are informative (Melin et al. 2017). Finally, food choice may be influenced by factors affecting food processing efficiency and physiological tolerance of by-products, including phenols and tannins (Glander 1982; Lambert 1998; O'Malley and Fedigan 2005; Gunst et al. 2010). Sensory phenotype may also be important if there are differences in ability to detect properties of or compounds in foods (Kawamura and Melin 2017).

The challenges of measuring quality of ingested foods in the field are perhaps the greatest among the topics we discuss in this chapter. Data on location of the foraging site within a patch and food selectivity are almost completely lacking from the literature on the study of non-human primates. Often fruit or leaves of multiple phenophases (maturity/ripeness) are simultaneously present in a food patch, and it is often exceedingly difficult to accurately assess the maturity of each food item ingested using binoculars from the forest floor. Occasionally, the colours of handled/ingested fruits can be seen (Melin, pers. obs). Yet, some fruits do not change colour with maturity, and for those that do, there is variation among plant species in how well the timing of colour changes indicates ripeness (Melin et al. 2015). It is also occasionally possible to record the properties of bitten and rejected foods, or incompletely eaten foods, from items recovered from forest floor. However, it is often difficult to accurately measure the properties (e.g. colour, odour, softness, toughness) of partially eaten foods as they can spoil quickly. Additionally, there are a number of assumptions that go into classifying dropped fruits as "rejected", and there are reports that primates are "wasteful" foragers that drop seemingly edible fruits only partially eaten (Chapman 1995; Ferrari and Lopes 2002).

For each of the variables discussed above, the process of generating data by observing, recording, and analysing the behaviours of wild primates can be daunting. Additionally, not all data collection methods are equally suitable for addressing each variable. In the following section, we review some of the most influential methods used by primatologists to guide behavioural data collection. We discuss their uses and limitations in the context of collecting data relevant to the foraging ecology of wild primates.

9.4 Data Collection Methods for Observing Primate Behaviour

Students interested in the study of primate behaviour should thoroughly familiarize themselves with the seminal books and articles in this field, including those cited in this paragraph. Here, we focus on a more nuanced discussion of three sampling methods for collecting standardized data most relevant to foraging ecology: (1) *all occurrences sampling* (Altmann 1974; Paterson and Vandenbeld 2001), also known as behaviour sampling (Martin and Bateson 2007); (2) *scan sampling*; and (3) *focal animal sampling* (Altmann 1974; Paterson and Vandenbeld 2001; Martin and Bateson 2007). After reviewing the three sampling methods, we provide a detailed discussion of practices for handling out-of-sight time and implications of using different practices for data collection and analysis. A fourth method, ad libitum sampling, is not considered in much detail here. Ad libitum sampling is by definition an unstructured practice of recording behaviours and interactants that are deemed relevant or otherwise interesting. Ad libitum observations are useful during pilot observations, for providing context during group census records and recording dominance interactions within and between social groups where the directionality, rather than the frequency or rate, of interactions is important.

9.4.1 All Occurrences Sampling/Behaviour Sampling

During all occurrences sampling, the researcher observes the group of study subjects as a whole and records each occurrence of a particular behaviour, or set of behaviours, of interest (Altmann 1974; Paterson and Vandenbeld 2001; Martin and Bateson 2007). All occurrences sampling is useful for recording conspicuous events that are sufficiently infrequent such that each event can accurately be recorded. Paterson and Vandenbeld (2001) also mention the importance of visibility in the habitat for this form of sampling, to guard against missing or biased data. In contrast to ad libitum sampling, in which the researcher records anything deemed interesting or relevant, all occurrences sampling is a more structured behavioural data collection method with the aim of recording each instance of a specific behaviour.

In the context of foraging ecology, all occurrence sampling can be used to document *visits to new food patches* (patch encounter rate, PER) by a group of primates (i.e. any time any primate of the social group being followed feeds in a food patch, a food patch visit is recorded (Leighton 1993; Spence-Aizenberg et al. 2016; Nagy-Reis and Setz 2017)). These data can be used to calculate measures of *dietary frequency* (one measure of dietary importance) and *dietary selectivity* (e.g. Melin et al. 2014). At minimum, the species and food type eaten (e.g. fruit or flowers), which is used to assess dietary selectivity, should be recorded. To assess individual-level variation in *food patch encounter rates*, the ID of first monkey to arrive should be noted (Hogan et al. 2018). Whenever possible, additional metadata should be recorded to address questions of *food patch profitability*, including patch size (circumference at breast height of the tree, or canopy size metric), patch capacity (maximum number of monkeys in the tree at the same time), phenology score, and total patch residence time. These variables can be analysed to ask a variety of questions, for example, factors influencing how long the group will stay at a given food source (i.e. patch residence time) (Link et al. 2012; Gonzalez et al. 2016; Cabana et al. 2017). Numerous other questions can be addressed by integrating food patch visit data and food trait data with the metadata listed above, including questions in foraging ecology concerning sensory systems and nutrition, nutrition and digestive systems (e.g. do preferred foods have a certain odour/mechanical properties, size, nutrient composition) (Melin et al. 2015; Valenta et al. 2015).

The most prominent difficulty of all occurrence sampling in this context is that it is difficult to observe all food patch visits, and this goal is perhaps unattainable when following a large social group. As a surrogate for this, one solution we take is to record all observed patch visits and in our analyses to control for the number of monkeys in the social group as well as the number of human observers recording food patch visits. An assumption of this approach is that the visits observed are representative of all visits that occur. However, it is important to be aware that short visits to small food patches may be underrepresented by this approach. Another important consideration is ensuring that we control for the likelihood of observing different individuals, as some may be more peripheral or otherwise less conspicuous to researchers.

Finally, the impact of memory on food patch encounter is difficult to address, as there are numerous lines of evidence that primates return to food patches based on recent—or more distant—experience (Garber and Paciulli 1997; Asensio et al. 2011; Janmaat et al. 2013). Ideally, the same group should be followed for many days consecutively, so that repeat visit to trees can be identified and reuse of food trees from year to year with evidence of pre-feeding "monitoring" can be recorded and accounted for. In our analyses of food encounter, we exclude from our analyses records of patch visits when the patch has been visited previously (i.e. a repeat visit). We also explicitly focus on visits to small, ephemeral patches. These are less likely to be memorized and are ecologically interesting due to a high finders' reward as small patches may be quickly depleted (Bunce et al. 2011; Hogan et al. 2018). The cut-off between large "memorable" and "ephemeral/unmemorable" has not been defined and is likely specific to different species of primates and so remains a subjective decision. Further, in addition to patch size, food preference, importance, and rarity in the environment should be considered; foods that are essential during periods of seasonal or interannual food dearth may also be especially relevant for fitness consequences (Melin et al. 2014) and would be interesting to investigate in future research.

9.4.2 Scan Sampling

Scan sampling, as defined by Martin and Bateson (2007), is carried out by rapidly censusing a group of subjects at set time points. The behaviour of each individual in the group (or a subset of interest) is recorded. This is a form of instantaneous scan sampling (Altmann 1974; Paterson and Vandenbeld 2001). Scan sampling requires an ethogram consisting of state behaviours that can be quickly and accurately assessed, and it may be less detailed than one constructed for focal animal sampling. The time required to complete a scan will range from a few seconds (e.g. DeLuycker 2012; Bryson-Morrison et al. 2017) to several minutes (e.g. Mekonnen et al. 2010; Vandercone et al. 2012; Lewis and Justin O'Riain 2017), depending on the level of detail recorded, group size, and how dispersed individuals are. However, the time between scans should be considerably longer than the time it takes to complete one scan to maintain as much independence among data points as possible. It is often not possible to sample all individuals in social groups, and an upper limit on the search duration for scan sampling should be enforced to maintain the independence of scan samples recorded at set time points.

Scan data can be used to gather data about each group member's *activity budget* (Link et al. 2012; Grueter et al. 2013), *diet* (Abreu et al. 2016; Nagy-Reis and Setz 2017), *habitat use* (e.g. height in canopy (DeLuycker 2012) or *type of vegetation occupied* (DeLuycker 2012; Bryson-Morrison et al. 2017)). It can be used to calculate activity budgets for each individual, both at the level of broad behavioural categories (i.e. foraging versus resting) and at the level of assessing the importance of different food types and food species in the diet (Table 9.1). This is complementary

Table 9.1 Ethogram of behaviours recorded during individual scan sampling of white-faced capuchin monkeys in Sector Santa Rosa, Costa Rica

Type of behaviour	Description
Forage: visual	Actively looking for food, including gleaning insects while moving
Forage: fruit (extractive)	Pounding fruit, picking seeds from within
Forage: invertebrate (extractive)	Tearing branches, ripping bark while foraging, exploring tree holes
Forage: flower	Feeding/foraging on flowers
Forage: fruit	Feeding/foraging on fruit
Forage: invertebrates	Feeding/foraging on invertebrates without extractive behaviours
Forage: other	Foraging on bromeliad leaves, pith, vertebrates
Forage: out of sight	Focal individual is foraging but hands, mouths, or other relevant features cannot be seen clearly
Out of sight	Focal individual is completely out of sight
Drink	Intake of water from ground sources or tree holes
Other	Any other behaviour not covered, comment and ad-lib notes taken
Intergroup encounter (IGE)	Interaction with another monkey group
Mobbing	Vocalizing, chasing, or displaying at a predator
Play	Play: Biting, chasing, hitting, bouncing, pushing, non-agonistic
Solitary rest	Resting without touching other monkeys
Social active	Affiliative behaviour; allogrooming
Social aggressive	Chasing, biting conspecifics, agonistic contexts
Self-directed	Grooming self
Social rest	Individual is inactive and is in contact with another individual
Travel	Travel; not foraging or socializing
Excretion	Passing of faeces, urine, or vomit
Vigilant	Scanning intently at a long range (not for food)

to the calculation of dietary importance using all occurrence sampling of visits to food patches. Scan data will capture influences of the duration of visits to food patches and number of monkeys feeding in metrics of resource "importance". Using scan data to calculate activity budgets has numerous practical advantages including relatively even collection of data across study subjects, times of day, weeks, and seasons, which facilitates behavioural and ecological comparisons across individuals (Martin and Bateson 2007). In addition, we can also examine niche divergence among individuals using data collected at the same time point, which helps to control for the influence of weather or forest type in patchy environments. Finally, it is advantageous that scan sampling is compatible with other types of data collection, such as all occurrence sampling, which can be recorded between scan samples. In contrast, it is not possible to record all occurrences of visits to food trees at the group level while full-day focal animal follows are being conducted, for example.

We also record instantaneous scans at the group level as part of the long-term longitudinal data collection component of our research project. Here, we record up to two behaviours in which the majority of the group is engaged. This is useful for looking at *group-level activity budgets* and correlations with changing environmental variables such as temperature, food abundance, or rainfall. This could be calculated from the individual scans, but occasionally researchers are not collecting scan sampling for their projects; in this case, the group-level scans continue as a behavioural record of group activity. The ethogram for group scans at Santa Rosa is nearly identical to the individual ethogram, except with omitting the following categories: (1) out of sight, if the entire group is not in sight, behavioural data cannot be collected; (2) excretion, the majority of the group is never in an excreting behaviour (passing faeces) at the same time; and (3) foraging out of sight, this level of detail is not recorded during group scans.

Potential pitfalls of scan sampling include not being able to capture the level of detail possible with focal animal sampling and the increased potential for bias towards conspicuous behaviours or monkeys. To deal with these limitations, we have worked with a dozen behaviourists to design an ethogram for scan sampling that is detailed enough to meet our needs with respect to activity budget but general enough to be quickly and accurately assessed (Table 9.1). We also make efforts to begin the scan sample with a different individual each time and to keep track of scanned monkeys, such that more cryptic group members can be actively sought out.

9.4.3 Focal Animal Sampling

The application of focal animal sampling (FAS) for studying primate behaviour was standardized decades ago in a landmark paper (Altmann 1974), and FAS is generally considered one of the most satisfactory methods for studying primates (Paterson and Vandenbeld 2001; Martin and Bateson 2007). During FAS, the observer collects data on one individual for a predetermined amount of time ("time rule") and keeps a record of all state and/or event behaviours of interest continuously, at set time points, or within set intervals. Altmann (1974) also suggests that the FAS could be ended after the *N*th iteration of a behaviour, where *N* is predetermined by the researcher. However, she cautions that the stopping rule should be independent of the behavioural parameters of interest and asserts that using predetermined amounts of time is typically more desirable; in practice the "time rule" is almost exclusively used in current primatology (Matsuda et al. 2014; O'Mara 2015; Visalberghi et al. 2016; McGraw et al. 2016; Mallott et al. 2017). The ideal duration will differ by species, habitat, and research goals, but the upper limit should precede observer fatigue (which is study dependent), and the lower limit should be long enough to capture the full durations or sequences of the behaviours of interest (Altmann 1974). A wide range of FAS durations are seen in the literature, from >10 h (full-day or full-night follows; (e.g. Campbell-Smith et al. 2011; Norscia et al. 2012; Matsuda et al. 2014; Aristizabal et al. 2017 to as short as 1 min (e.g. DeLuycker 2012; Johnson et al. 2017; Melin et al. 2017). It is important to note that longer focals (i.e. multi-hour or full-day/full-night follows) decrease the number of data points per study individual, depending on the research question. Using this method makes it more difficult to obtain a robust sample for a number of individuals that is appropriate for statistics. Prior to determining focal length, researchers should consider the specific statistical analyses they intend to use to ensure that sufficient data can be collected.

Once a duration has been selected, one of two methods for timekeeping during focal animal samples can be used (Altmann 1974). These methods vary in how they treat the time when a focal subject is out of sight (OOS) of the observer: (1) focal individuals are followed for a set amount of time, regardless of how long they are out of sight; or (2) the observer keeps track of cumulative "time-in" (i.e. time the focal subject is in sight) (Chancellor and Isbell 2009; Cords and Nikitopoulos 2015; Corewyn 2015). In the latter scenario, once time-in reaches the predetermined duration, the focal is ended. A review of recent literature suggests that most primatologists follow the former method (although this is often not explicitly stated; O'Mara 2015; Abreu et al. 2016; McGraw et al. 2016; Bryson-Morrison et al. 2017; Cabana et al. 2017). Explicit treatment of OOS time is often not mentioned in primatological literature. However, when it is mentioned, researchers typically report the following rules for how much OOS time is allowed before the focal follow data are discarded (Cords and Nikitopoulos 2015; Corewyn 2015; Visalberghi et al. 2016; Bădescu et al. 2016).

The applications of data collected using focal animal sampling to study foraging ecology are widespread, and it is the standard method suggested for observing the behaviour of fruit and seed foraging (Dew 2011; Eadie 2015; Abreu et al. 2016; Visalberghi et al. 2016; Sengupta and Radhakrishna 2016). FAS can generate data used to calculate many of the variables listed above, including (individual-level, rather than group-level) *food patch encounter rates*, *activity budgets* (Campbell-Smith et al. 2011; Mallott et al. 2017), *fruit and insect intake rates* (McCabe and Fedigan 2007; DeLuycker 2012), *food processing efficiency* (Eadie 2015; Visalberghi et al. 2016; McGraw et al. 2016), and *foraging investigation sequences* (Melin et al. 2009). In addition, under cases of exceptional visibility, intake rates of foods of different quality could be assessed using FAS when food quality (e.g. ripeness) can be visually discerned by the observer. Despite numerous advantages, there are downsides to FAS $\geq$ 10 min in duration that follow strict OOS rules and a rigid rotation schedule. The following points are common concerns that have been raised by students and other researchers, specifically in the context of studying feeding rates in food patches using FAS: (1) how much time should be spent looking for a specific monkey to follow before moving to next individual on the rotation?; (2) should focal animal samples be collected during the behaviours of interest (e.g. foraging) or follow a random sampling scheme independent of the behaviour observed?; (3) should focal follow with OOS time be summarily discarded?; and (4) how can FAS be optimized to sample different monkeys under similar ecological conditions?

The answers to these questions depend on the answer to one critical question: *will the data collected during focal animal samples be used to be representative of an overall activity budget?* If yes, then the guidelines outlined by Altmann (1974) and others should be strictly applied, namely, (1) random sampling schedule, (2) sampling regardless of the behaviour at the time the FAS is started, (3) rigorously defined and applied OOS rule, and (4) randomized or predetermined rotation schedule among individuals, with protocols for evenly distributing sampling effort across time of day and season. However, if another method, for example, scan sampling, is being used to calculate activity budgets (and we highlight benefits of this in the previous section), then we suggest that FAS can be optimized and used more flexibly to best capture the data of interest. This practice of flexible application of FAS, occasionally used in conjunction with other data collection methods, has been used in recent primate studies (e.g. (DeLuycker 2012; Eadie 2015; Johnson et al. 2017). Below we summarize some difficulties of using even moderately long (≥10 min) FAS during foraging behaviours to record intake rates and discuss our thoughts on an approach optimized for collecting detailed feeding and foraging data.

First, during arboreal foraging, visibility of study subjects is often low. It is not uncommon for a relatively small set of viewing "windows" through the foliage to be available to observers. To make the best use of these observation spots and avoid wasting large amounts of time searching for specific monkeys, we recommend a modified sampling scheme in which researchers cycle through age-sex categories, rather than search for specific monkeys while still avoiding watching the same monkey repeatedly in the same patch. Aspects of this approach are evident in recent research (Eadie 2015; Gonzalez et al. 2016; Mallott et al. 2017). For recording intake rates, and especially foraging sequences, excellent viewing conditions are paramount. While focal animal samples are often done by just one observer, for intake rate data, we suggest the use of a voice recorder (e.g. dictaphone or similar application) or observer team where one observer calls out the data and a second person records the data. One person simply cannot record the detail needed because momentarily looking away to write or type will result in missing several data points when food investigations happen quickly. The use of video equipment may be another possible solution (Gunst et al. 2010), but stabilization and need to adjust position to follow small movements of the focal animal and post-processing times for videos are not trivial concerns.

Second, if activity budgets will be captured by scan sampling at regular time intervals, this frees up observers to target FAS towards recording behaviours of interest. Again, we reiterate that this approach should not be used if FAS data will address how long or how often a monkey does a behaviour (activity budget questions). Rather, it is suited to ask questions that arise once the focal animal is already doing a behaviour. For example, we could measure the efficiency per unit time, strategy, technique, substrate or tool use, and success rates. If behaviours are relatively infrequent, this approach increases the chances that sufficient data will be collected on the behaviour of interest for all individuals, especially if the study is short or the behaviour is linked to climatic seasonality or other temporally bounded environments or activities (Eadie 2015; Visalberghi et al. 2016; Sengupta and Radhakrishna 2016; Johnson et al. 2017).

Third, when very detailed data on foraging sequences or intake rates are needed, "OOS" needs to be recorded when the mouth (and possibly also hands or feet) is not clearly visible, even if it is still possible to observe that the monkey is in a general foraging state. This results in OOS being a very common occurrence. For these reasons, we advise against strict OOS rules for discarding focals. Rather, we suggest that data from targeted FAS can be used, even if a large proportion of the FAS is OOS. The assumption made here is that observations made while the animal is in view are representatives of those made when the animal is obscured or partially obscured. Finally, the end of the targeted focal may often be determined by the monkey moving completely out of sight or by an end of focal timer.

This raises a linked question: how long is long enough for a targeted focal? Although these may not apply to all studies, we suggest that criteria for using focal data in the context of a feeding in a food patch should be in part guided by the three following questions: (1) *was the visibility good enough that my "time-in" data are reliable?* This involves judgement about the quality of the viewing conditions; (2) *did I observe enough feeding behaviour during the "time-in" that this is an accurate and representative sample in this food patch?* To generate a rough guideline for how many food ingest events need to be observed to estimate a feeding rate that is representative for that monkey in that given food patch, we present an analysis based on repeatedly subsampling a large dataset of foraging behaviour collected by our team (Box 9.1). Our conclusion is that approximately 20 ingest events should be observed, and this holds for fruits of varying sizes (Box 9.1, Fig. 9.1). Ideally, similar analyses will be conducted by researchers studying other primates and plants that will speak to the generalizability of this result. The third and final question to ask is: (3) *was I careful not to end the focal based on the data I am interested in?* For example, "I'll end this focal as soon as this monkey finishes chewing that last bite" is not an appropriate rule. Rather, timers set at 30 s could be used. When the timer alarms, the observer should ask him/herself, have I collected enough data? Yes? Then stop. No? Then continue collecting data until the next 30 s mark (or until monkeys goes completely out of site).

Finally, often patch residence times from entry of first to last group member are 30 min or less (Melin pers. obs, Gonzalez et al. 2016; Johnson et al. 2017). For studying feeding rates of primates, the relatively long durations of traditional FAS (often ≥10 min) reduce the chances of sequentially recording behaviours of many different primates *in the same food patch*. This is problematic because the plant species, phenological state, size, and other characters unique to each food patch affect the food intake rate of primates significantly. In many cases, the effect of phenology or food patch ID has a greater impact than do the individual-level variables, such as age or sensory phenotype (Melin et al. 2017). For comparing food intake rates or foraging sequences rates, it is more powerful (and much more feasible) to sample many monkeys for shorter durations in succession than to watch one monkey per patch for 10 min or longer. While this important point was discussed decades ago (e.g. Janson 1985) and some primatologists use short FAS in food trees (e.g. Vogel 2005; DeLuycker 2012; Eadie 2015; Johnson et al. 2017), we feel it is worth re-emphasizing the point here.

In summary, when the goal is to compare food intake rates among individual primates, for example, to assess the impact of age, sensory phenotype, or social dominance, within and between sympatric species, it is ideal to observe individuals in same food tree and to use FAS short enough to sample multiple individuals that span the levels of variable of interest (e.g. age, rank) while making focals long enough to record >20 food ingest events.

Box 9.1 How much sampling is necessary in order to estimate intake rates accurately?

Elsewhere in this chapter, we discuss challenges inherent in observing feeding behaviours, which presents the critical question: how much sampling is necessary in order to estimate intake rates accurately? To address this question, we analysed a large dataset of capuchin fruit foraging behaviours that was the subject of another recent analysis (Melin et al. 2017). Out of 1603 total foraging bouts encompassing a range of monkey ages, sexes, and fruit species, we analysed a total of 64 bouts that included at least 40 discrete intake events of 9 different fruit species. Fruit species with a mean diameter from 0.05 cm to 1.19 cm were included in this analysis (Fig. 9.1a). We then used resampling techniques to estimate, for each bout, how many intake events must be observed before our permuted intake rate resembled the "best possible" intake rate (i.e. the rate calculated for all intake events recorded) for that bout. To conduct the permutation, we assigned a duration to each intake event as the sum of half the time since the previous intake event and half the time before the subsequent intake event (Formula 9.1). This allowed us to calculate a duration and associated intake rate for our resampled data as the number of resampled intake events divided by the cumulative duration assigned to the same intake events. Because the first and last events in each bout sequence could not be assigned a duration in this manner, we excluded them from analysis and correspondingly adjusted the overall "best possible" duration and intake rate of each bout.

Formula 9.1:

$$D_i = \frac{1}{2}(t_i - t_{i-1}) + \frac{1}{2}(t_{i+1} - t_i)$$

where for each event i, t represents the timestamp and D represents the assigned duration.

Fruit intake rates could potentially vary based on sequential position within a bout due to resource depletion effects. We tested for such an effect, however, using a mixed-effects Poisson regression with fruit species as a random intercept and found no significant relationship between intake event sequence position (i) and the duration (D_i) assigned to the event (slope = −0.002, $z = -1.69$, $p = 0.091$) (Fig. 9.2). After concluding that the depletion effects were sufficiently small, we conducted our permutations as follows.

(continued)

Box 9.1 (continued)

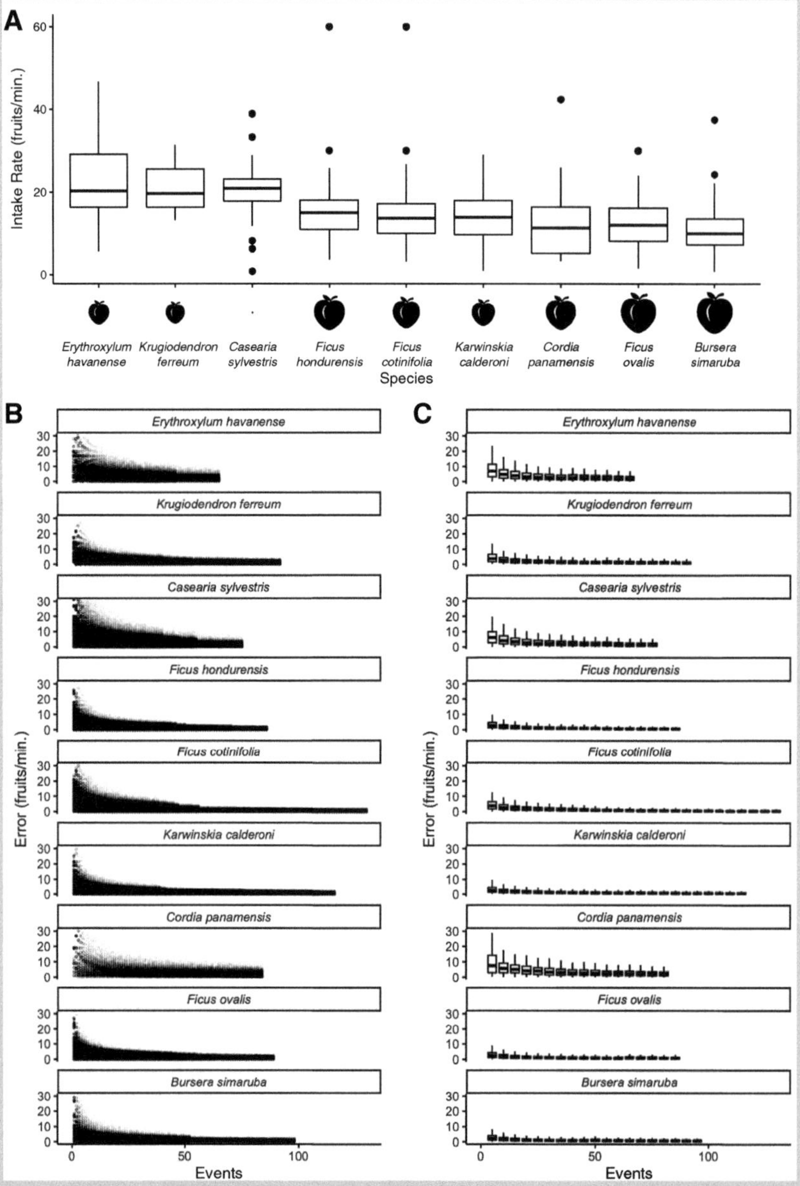

Fig. 9.1 Analysis of capuchin fruit ingest rates, with (**a**) boxplot of intake rates (number of ingest events ≥40 for all bouts analysed) for nine fruit species. Displayed fruit sizes are proportional to the actual size of each fruit; (**b**) distribution of bout intake rate errors, calculated as the absolute value of the difference between the "best possible" intake rate and the permuted intake rate estimated through resampling, with 1000 estimates per number of fruit intake events; (**c**) boxplot of bout intake rate errors estimated through resampling, shown at five-fruit intervals with outliers not shown (see panel B for all points, including outliers)

(continued)

Box 9.1 (continued)

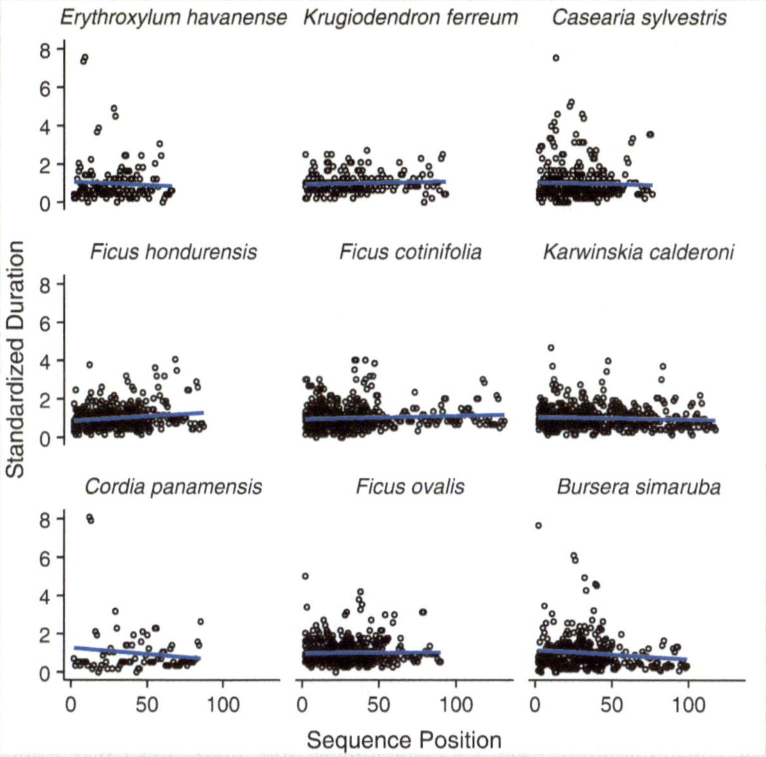

Fig. 9.2 Standardized durations assigned to fruit intake events as a function of position in a bout sequence. Resource depletion effects are expected to decelerate the fruit intake rate and thus increase event durations over time. Our results, however, demonstrate that this effect is either small or nonexistent for feeding rate data on nine species included in our analyses

For each bout, we sampled intake events with replacement 1000 times for each of 1 to the total number of fruits for that bout. In each iteration, we calculated a permuted intake rate as the number of sampled ingest events divided by the total assigned duration of sampled events. We also calculated an error estimate as the magnitude of difference between the permuted intake rate and the best possible intake rate of that bout.

The resulting distribution of error estimates is presented in Fig. 9.1b–c. As expected, we find that the error in estimated intake rate falls as the number of observed intake events increases, eventually stabilizing around 20 observed intake events. Observing more food intake events leads to lower error in estimating the mean rate for that bout (i.e. a certain tree of a certain species in a certain phenophase at a certain time). While increasing the number of observed events continues to decrease the error in intake rate estimates, the returns on that effort are increasingly marginal. By 20 observed events, 95% of intake

(continued)

Box 9.1 (continued)

rate estimates for all but 2 fruit species in our dataset fell within 10 fruits/min of the best possible intake rate, with a range of 3.9 (*Bursera simaruba*) to 12.7 (*Cordia panamensis*). By 40 observed events, the 95% quantile of intake rate error estimates ranged from 2.7 (*Bursera simaruba*) to 8.4 (*Cordia panamensis*). By 80 observed events, for instance, the 95% quantile of intake rate estimates ranged from 1.9 (*Bursera simaruba*) to 5.9 (*Cordia panamensis*). In this case, a quadrupling of sampling effort corresponds roughly to a reduction in error of only one half.

Our results indicate that the distribution of error as a function of sampled intake events differs across species. This is likely in part due to differences in sample size between fruit species. *Cordia panamensis*, for instance, was represented by only one bout, which likely introduced greater stochasticity into our simulations.

Our results suggest guidelines for observation of food intake rates that differ from recommendations for focal follows that traditionally suggest uniform durations, often of 10 min or longer. However, only a portion of this follow may occur in a food patch. When the research goal is to compare intake rates among individuals, we recommend sampling as many different individuals in the same food patch as possible, and watching each individual consume at least 20 fruits (or ideally many more). This could be spread out across several short (1–5 min) focal follows, as visibility permits (but see recommendations for timer use and stop times, previously). The amount of observation time required to observe 20 foraging events will vary based on the food tree characteristics. For white-faced capuchins, mean fruit intake rates vary from 22 fruits/min in 1 tree species (*Erythroxylum havanense*) to 1–3 fruits/min (*Randia monantha*). In the former case, only 1 min of foraging observation time may be needed. In the latter case, which applies to plant species with large or difficult to process fruits, the fruit investigation rate is very low, and "mini-focal" durations may approach the length of traditional focal animal samples. In the case of *Randia monantha*, for example, more than 7 min would be needed. Additionally, individual-level characteristics may be important; young monkeys feed more slowly; and feeding rates are slower when fruits are larger (Melin et al. 2017). Finally, this approach will not likely be useful in small food patches that are quickly depleted or that cannot support multiple monkeys at the same time or in rapid succession. In those cases, these data are best recorded during traditional focal animals follows or other methods (all-occurrence, scan sampling).

9.5 A Brief Comment on Analyses: Using a Hierarchical Structure to Maintain Independence of Multiple Analyses

Analytical approaches to the study of animal behaviour can take many forms depending on data collection methods and research questions. In this section, we focus on using a hierarchy of analyses to maintain statistical independence while asking questions at different levels.

Datasets based on scan sampling or focal animal sampling offer considerable flexibility when it comes to analytical approaches. However, it is important to maintain statistical independence when the same data are used to answer different research questions. For example, if an activity budget has five possible states, each of these states cannot be treated as independent of the others, because the number of records of the fifth state can be calculated by subtracting the total number of records of other states from the total number of overall records. Examining only a subset of activities that are of particular relevance to the research question and creating a hierarchy help to maintain independence when constructing statistical models (Raudenbush and Bryk 2002). Using an example research question from our study site at Santa Rosa, we demonstrate a hierarchy structure to assess the extent of niche divergence among group members. Specifically, we demonstrate a proposed hierarchy for investigating foraging activity time budgets at different levels (Fig. 9.3). Foraging activity could be derived from number of scan records or duration of minutes based on focal animal samples.

In Fig. 9.3, the light grey boxes represent the response variables and the dark grey boxes represent an "offset" term in a linear model. The leftmost column represents variables in a model that could, for example, test whether minutes of time spent in foraging behaviour differs among individuals (response variable),

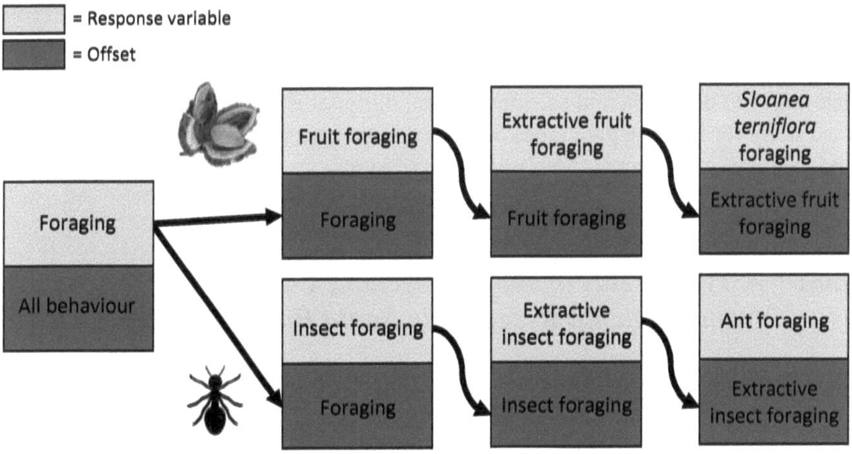

Fig. 9.3 Illustration of a nested hierarchy of variables used to assess activity budget at different levels, from more general (left) to more specific (right). For details see text

controlling for amount of time recorded in any behaviour. As questions become more specific and aimed at behaviours nested in behaviours (e.g. amount of foraging time devoted to fruit foraging, controlling for time spent in any foraging behaviour), the response variable and offset terms change. The black arrows highlight the pattern that response variables can become offset terms in subsequent models, as they test more specific, nested questions.

9.6 Integrating Data Collection Methods for Short- and Long-Term Studies of Behaviour

It is often useful to integrate two or more methods of behavioural sampling into a daily collection strategy. For example, several recent studies have integrated scan sampling to record activity budgets with focal animal sampling and/or all occurrence sampling and/or ad libitum sampling to record more specific behaviours (e.g. Spence-Aizenberg et al. 2016; Sengupta and Radhakrishna 2016; Nagy-Reis and Setz 2017; Cabana et al. 2017). This layered sampling approach has the disadvantage of perhaps being more complex to organize and analyse but has numerous advantages when collecting any one data type is not sufficient to answer the research questions being asked. This tiered sampling is also quite flexible and becomes useful when integrating data that will be collected across multiple years and decades (i.e. for longitudinal study of primate life histories and responses to environmental variation, with data for focused shorter-term projects, such as thesis research or other discrete projects).

In this section, we overview our data collection protocol for studying the behavioural ecology of capuchin monkeys in Sector Santa Rosa, including how we collect standardized longitudinal data that is deposited in the Primates Adapting to Changing Environments (PACE) database, initiated by Dr. Linda Fedigan. At Santa Rosa, our long-term goals in the context of feeding ecology are to understand how sex, reproductive status, age, dominance rank, and sensory phenotype (e.g. colour vision type) impact foraging activity budgets, feeding efficiency, food preferences, and resource detection and selection. Here, our goal is to describe Santa Rosa as a case study for how the data types described in the previous section can be collected and customized during group observation sessions.

9.6.1 A Typical Day of Data Collection: Case Study from Capuchin Research in Santa Rosa

9.6.1.1 Documenting Observers and Study Groups

For each primate observation session, we record the *name of the observer* and the *name of the study group* of primates on Android tablets running a custom-made Android data logging app created by Dr. Scott Johnson. The app records a date-time-location stamp at the start of each new line of text. We additionally keep records of the number and the identities of researchers with the primate group that day to control for observational effort. At first contact with the study group, we record a code for the *type of contact* (e.g. located at sleep tree, join another researcher already with the group, or find a group after they have left sleep tree). This becomes important as only "sleep tree to sleep tree" days of data collection are used for some forms of data analysis, including studies of full-day ranging activities.

9.6.1.2 All Occurrences: Food Patch Visits

When any monkey of the group enters a food patch, we record a food patch visit (FPV) and note the *food species* and, for trees, the *circumference at breast height* (CBH) as a proxy for food patch size. We also record the *phenological state*. We use a 5-point system: 0, absent; 1, 1–24%; 2, 25–49%; 3, 50–74%; and 4, 75–100% to measure the percentage of the canopy covered in fruit, leaves, and flowers, and the proportion of those items in mature versus immature stages. For example, a fruit tree with 60% of its canopy covered in fruit, of which 20% was ripe, we would record "3 and 1" for the fruit cover and ripeness scores, respectively. This can later be multiplied and used together with the CBH to estimate ripe fruit biomass for that food source. Each FPV is also assigned an ID number that is unique to that group's visit on that particular day. This allows us to control for FPV ID, along with the relative amount of ripe fruit (or phenophase) being consumed, during data analysis. In addition, when possible, we record the first monkey into the FPV and note if this is a repeat visit by the group.

9.6.1.3 Scan Sampling

Every 30 min, on the hour and half hour, one researcher per study group records up to two *state behaviours* that are exhibited by the majority of the group. Along with the behaviours and *location*, we also record the *height* of the majority of the group in the forest, the predominant *canopy cover*, and the *weather*. For foraging behaviours, we record the *type of food and taxonomy* whenever possible. These group scan data are continuously collected as part of the long-term records in Santa Rosa. We use these data to examine, for example, the impact of fruit biomass, rainfall, and temperature on primate ranging behaviours, diet selection, and activity budgets at the group level.

Depending on his or her own research goals, investigators may also choose to follow the group scan with scan sampling of individuals of the social group. Additional fine-grained detail, such as size and type of supports used during the behaviours, can also be added to the data collection protocol. These data can be used, among other things, for examining individual-level differences in activity budgets and foraging preferences and the extent of niche divergence within the group.

9.6.1.4 Focal Animal Sampling

Focal animal sampling (FAS) is conducted in between group scan (and optionally individual scan) samples, and FPV records, and/or by another researcher who is not collecting the scan or FPV data. The duration of the FAS, and the choice of behavioural detail to record, is a project-based decision, as is whether to use continuous or instantaneous sampling. We use an extensive, standardized ethogram, and researchers use a core set of codes and then may choose to increase the level of detail to record for different types of behavioural categories. For example, some students are more interested in *social behaviour* (Bergstrom and Fedigan 2013; Schoof and Jack 2014), while others are more interested in *foraging behaviour* (McCabe and Fedigan 2007; Hogan et al. 2018). The more detailed states and events are nested, with each behaviour falling under a larger behavioural category. For example, the more detailed "hand sniffing" behaviour is listed under a broader category of "social active". In this way, different studies can often still be compared using broader behavioural categories. When needed, researchers can add new behaviours, or relevant and associated environmental metadata, and nest this within an appropriate overarching behaviour.

In the context of foraging, researchers will often use FAS to record foraging on invertebrates outside of clumped (e.g. trees full of caterpillars) patches (McCabe and Fedigan 2007; Mosdossy et al. 2015; Melin et al. 2016). Because capture rates of non-clumped invertebrates are far slower than fruit investigation rates, the general protocols for focal animal sampling, including randomization among individuals and durations of 10 or 15 min, work well for studying this type of behaviour. If researchers are exclusively interested in invertebrate foraging, they may target this behaviour and attempt to sample equal amounts of invertebrate foraging for each monkey. For example, the researcher would not start a FAS when a study subject was resting but rather wait until they began to forage. As a reminder, this is only appropriate when FAS is not used to calculate activity budgets (Sect. 9.4).

When the primates are feeding in a clumped patch, and the researchers are interested in comparing feeding rates among individuals, a modified method of focal animal sampling is used to collect data on investigation sequences from detection of food to decision about edibility (e.g. touch, sniff, bite, reject) and intake rates of fruit, flowers, or clumped invertebrates. As explained in Sect. 9.4, the goal here is gathering sufficient data for multiple individuals under the most similar ecological conditions possible. The duration of the modified "mini" focal animal sample should be sufficient to observe ca. 20 ingest events; the duration needed is often much shorter than 10 min and is affected by the species of plant being foraged in because the speed at which monkeys consume fruits varies greatly depending on the size and ripeness conspicuity of the foods (Melin et al. 2017).

9.6.1.5 Ongoing Census Record and Additional Contextual Information

The final type of data we collect each session are ad libitum data for ongoing census, group membership, and dominance records, which are summarized and archived monthly. For *babies born* in our study groups, we use the date of previous contact and size/colour of any new infants to approximate date of birth. We also record *deaths*, including manner of death when observed, or disappearances. Because we individually recognize all individuals, it is also often possible to track *emigrations* or *immigrations* between social groups. We additionally collect data on *physical appearance*, such as new wounds or injuries, and update ID sheets with appearance of birthmarks and scars, which can vary with season. Finally, we record other ad libitum *contextual data*, including observed aggression and submission between individuals, mating behaviour, details around intergroup encounters, comments on the extent of habituation of new immigrants, and concerns about health of study subjects, especially infants. The monthly census information is used to ask questions concerning demography that require long-term data, for example, reproductive success, inter-birth intervals, sex skew at birth, age of first reproduction, cessation of reproduction, and life expectancy.

In sum, our data collection protocol includes (a) documenting the contact with monkey groups for each session; (b) all occurrence data for new food patch visits and other infrequent items of interest to long-term data (e.g. intergroup encounters); (c) instantaneous group (and often individual) scans every 30 min; (d) 10- or 15-min focal animal samples tailored to the goals of the study, which may be done together with "mini" focal animal samples of customized duration when monkeys are foraging together in trees for clumped resources; and (e) census updates each session to record births, deaths or disappearances, wounds, immigrations, and emigrations, along with other contextual ad libitum data for behaviours of interest to long-term study.

9.7 Summary and Final Thoughts

In this chapter, we highlight the relevance of studying foraging behaviours and dietary ecology and pinpoint several key variables that can be measured and analysed to answer research questions in this field. A key message we hope to convey is that researchers should explicitly tailor their methods to the questions they seek to answer, and we draw attention to the pros and cons of different sampling techniques. In particular, we suggest that relatively short, sequential focal animal follows of multiple individuals foraging in the same food patch offer perhaps the best ability to draw comparisons among the food intake rates of individuals linked to individual-level traits (e.g. dominance, sex, age). We suggest that at least 20 ingest events per individual per food patch should be observed to calculate a relatively stable and representative feeding rate. Overall, rather than conducting focal animal samples in food trees for a set duration, it could be productive to observe primates until they have eaten at least 20 (ideally more) fruits. However, we emphasize two points: (1) the stop time for the sample should be independent of the fruit ingest events. We recommend a timer alarm, set for 30 s (or so), be used to signal the end of the focal follow (e.g. stop the focal sample at the next alarm sounding, following the desired number of ingest events; (2) these focal data cannot be used to calculate activity budgets. Rather, it may be helpful to combine this targeted focal sampling with scan sampling to measure activity budgets.

Integration of multiple sampling strategies (scan sampling, focal sampling, all occurrence sampling) allows researchers to answer different questions and also helps facilitate simultaneous collection of standardized data for long-term studies with individualized data for student theses or other shorter research projects. We conclude our chapter by commenting on methods for analysis of behavioural data. There are a wide variety of valid approaches to analysing behaviour of wild animals. Our collective experiences with both long- and short-term datasets have revealed the utility of nested hierarchies in data analysis. This approach helps to maintain independence between analyses when the same overall dataset is used to answer questions at different levels, regardless of the eventual statistical models and tests used.

Throughout the chapter, we use our long-term observations of capuchins in Sector Santa Rosa to illustrate our points and to provide an example of how our research group is approaching the study of primate dietary and sensory ecology. One goal of our chapter is to draw attention to the range of methods being used in primatology currently to study feeding and foraging, and we hope to inspire further discussion about practices and methods in this field.

9.7.1 Outlook for Primate Foraging Ecology

The future of research in primate feeding ecology is bright. We conclude by highlighting a few of the promising avenues of current and future research in this area. By integrating detailed study of behaviour using methods outlined above with the steady advances in the portability and precision of equipment for measuring food properties (e.g. mechanical, nutritional, volatile compounds) and improving methodologies for preservation and analysis, we can ask increasingly refined questions. For example, current and future work will improve our understanding of how the nutritional contents of food—including micronutrients—influence rates of food intake and subsequent food patch choices across individuals and species. Our understanding of the genomic underpinnings of sensory phenotypic variation is also continually improving. By examining sensory investigation sequences and intake rates along with genomic sequencing and gene expression studies, future studies will reveal how differences in genetic variation underlying taste, odour, or hardness perception affect food choice and processing intra- and interspecifically. Studies of the cognitive processes and brain structure combined with observations of ranging behaviour, resource monitoring, and sensory engagement in captive and wild studies will provide new insight into forces shaping the evolution of cognition, including how memory and novel discovery interact during foraging behaviours. Examining the ontogeny of foraging skills in different contexts, improvements in feeding efficiency, and components of the learning process within and across species will also further our understanding of cognitive mechanisms as they are related to food acquisition and problem-solving in an evolutionary context. These research topics, and many more, will benefit from careful selection of observation methods, and the integration of behavioural data with the study of food and habitat properties, morphology, genes, and genomic-level processes.

Dedication We dedicate this chapter to Linda M. Fedigan, whose vision and leadership in areas great and small has made this work possible, and also a terrific experience.

Acknowledgements We thank Urs Kalbitzer and Kathy Jack for inviting this contribution. We also thank Urs and two reviewers for constructive comments on earlier versions of the chapter. We acknowledge the decades-long contributions of Linda M. Fedigan and John Addicott to the Santa Rosa project, which enabled the research we discuss here. We also extend our warmest thanks to Kathy Jack for her ongoing collaboration and contributions to research in Sector Santa Rosa (SSR). We would be remiss if we did not highlight the key role that past and current researchers in SSR played in this process of standardizing behavioural data collection, including generating a shared ethogram. We would like to especially acknowledge Fernando Campos, Jeremy Hogan, Monica Myers, Gillian King-Bailey, Mackenzie Bergstrom, Eva Wikberg, and Valerie Schoof. Conversations with colleagues over many years have contributed to the ideas presented in this chapter, and we thank Shoji Kawamura, Chihiro Hiramatsu, Filippo Aureli, Steig Johnson, Nate Dominy, and Erin Vogel. We thank Roger Blanco Segura and Maria Marta Chavarria and staff from the Área de Conservación Guanacaste and Ministerio de Ambiente y Energía. Warmest thanks also to Nina Beeby, Adrienne Blauel, Saul Cheves Hernandez, Adrian Guadamuz, Brandon Klug, Kelly Kries, Michael Lemmon, Ronald Lopez, Nigel Parr, Emily Walco, and Laura Weckman for assistance in the field.

References

Abreu F, De la Fuente MFC, Schiel N, Souto A (2016) Feeding ecology and behavioral adjustments: flexibility of a small neotropical primate (*Callithrix jacchus*) to survive in a semiarid environment. Mamm Res 61:221–229

Altmann J (1974) Observational study of behavior: sampling methods. Behaviour 49:227–266

Aristizabal JF, Rothman JM, García-Fería LM, Serio-Silva JC (2017) Contrasting time-based and weight-based estimates of protein and energy intake of black howler monkeys (*Alouatta pigra*). Am J Primatol 79:1–8

Asensio N, Brockelman WY, Malaivijitnond S, Reichard UH (2011) Gibbon travel paths are goal oriented. Anim Cogn 14:395–405

Bădescu I, Watts DP, Anne Katzenberg M, Sellen DW (2016) Alloparenting is associated with reduced maternal lactation effort and faster weaning in wild chimpanzees. R Soc Open Sci 3:160577

Barnes LM, Leclerc M, Gray CA, Williamson JE (2011) Dietary niche differentiation of five sympatric species of Platycephalidae. Environ Biol Fish 90:429–441

Barton RA, Whiten A (1993) Feeding competition among female olive baboons, *Papio anubis*. Anim Behav 46:777–789

Bergstrom ML, Fedigan LM (2013) Dominance style of female white-faced capuchins. Am J Phys Anthropol 150:591–601
Bryson-Morrison N, Tzanopoulos J, Matsuzawa T, Humle T (2017) Activity and habitat use of chimpanzees (*Pan troglodytes verus*) in the anthropogenic landscape of Bossou, Guinea, West Africa. Int J Primatol 38:282–302
Bunce JA, Isbell LA, Grote MN, Jacobs GH (2011) Color vision variation and foraging behavior in wild Neotropical titi monkeys (*Callicebus brunneus*): possible mediating roles for spatial memory and reproductive status. Int J Primatol 32:1058–1075
Cabana F, Dierenfeld E, Wirdateti W et al (2017) The seasonal feeding ecology of the javan slow Loris (*Nycticebus javanicus*). Am J Phys Anthropol 162:768–781
Campbell-Smith G, Campbell-Smith M, Singleton I, Linkie M (2011) Raiders of the lost bark: orangutan foraging strategies in a degraded landscape. PLoS One 6:e20962
Cartmill M (1992) New views on primate origins. Evol Anthropol 1:105–111
Carter AJ, Marshall HH, Heinsohn R, Cowlishaw G (2014) Personality predicts the propensity for social learning in a wild primate. PeerJ 2:e283
Castellanos HG, Chanin P (1996) Seasonal differences in food choice and patch preference of long-haired spider monkeys (*Ateles belzebuth*). In: Adaptive radiations of neotropical primates. Springer, Boston, MA, pp 451–466
Chancellor RL, Isbell LA (2009) Food site residence time and female competitive relationships in wild gray-cheeked mangabeys (*Lophocebus albigena*). Behav Ecol Sociobiol 63:1447–1458
Chao A, Simon-Freeman R, Grether G (2013) Patterns of niche partitioning and alternative reproductive strategies in an East African dung beetle assemblage. J Insect Behav 26:525–539
Chapman CA (1995) Primate seed dispersal: coevolution and conservation implications. Evol Anthropol 4:74–82
Corp N, Byrne RW (2002) The ontogeny of manual skill in wild chimpanzees: evidence from feeding on the fruit of *Saba florida*. Behaviour 139:137–168
Cords M, Nikitopoulos E (2015) Maternal kin bias in affiliative behavior among wild adult female blue monkeys. Am J Primatol 77:109–123
Corewyn LC (2015) Dominance, access to females, and mating success among coresident male mantled howlers (*Alouatta palliata*) at La Pacifica, Costa Rica. Am J Primatol 77:388–400
DeLuycker AM (2012) Insect prey foraging strategies in *Callicebus oenanthe* in northern Peru. Am J Primatol 74:450–461
Dew JL (2005) Foraging, food choice, and food processing by sympatric ripe-fruit specialists: *Lagothrix lagotricha poeppigii* and *Ateles belzebuth belzebuth*. Int J Primatol 26:1107–1135
Dew JL (2011) Feeding ecology, frugivory and seed dispersal. In: Field and Laboratory Methods in Primatology. Cambridge University Press, Cambridge, pp 225–236
Dominy NJ (2004) Fruits, fingers, and fermentation: the sensory cues available to foraging primates. Integr Comp Biol 44:295–303
Dominy NJ, Lucas PW, Osorio D, Yamashita N (2001) The sensory ecology of primate food perception. Evol Anthropol 10:171–186
Eadie EC (2015) Ontogeny of foraging competence in capuchin monkeys (*Cebus capucinus*) for easy versus difficult to acquire fruits: a test of the needing to learn hypothesis. PLoS One 10:e0138001
Feeney E, O'Brien S, Scannell A et al (2011) Genetic variation in taste perception: does it have a role in healthy eating? Proc Nutr Soc 70:135–143
Fernandez-Duque E, van der Heide G (2013) Dry season resources and their relationship with owl monkey (*Aotus azarae*) feeding behavior, demography, and life history. Int J Primatol 34:752–769
Ferrari SF, Lopes MA (2002) Fruit rejection by tufted capuchins (*Cebus apella*: Primates, Cebidae) during the predation of *Cariniana micrantha* seeds: suboptimal or just "wasteful" foraging behavior? Revista Etologia 4:3–9

Foerster S, Cords M, Monfort SL (2011) Social behavior, foraging strategies, and fecal glucocorticoids in female blue monkeys (*Cercopithecus mitis*): potential fitness benefits of high rank in a forest guenon. Am J Primatol 73:870–882

Garber PA, Bicca-Marques JC, Azevedo-Lopes MO (2009) Primate cognition: integrating social and ecological information in decision-making. In: Garber PA, Estrada A, Bicca-Marques JC et al (eds) South American primates. Developments in primatology: progress and prospects. Springer, New York

Garber PA, Paciulli LM (1997) Experimental field study of spatial memory and learning in wild capuchin monkeys (*Cebus capucinus*). Folia Primatol 68:236–253

Glander KE (1982) The impact of plant secondary compounds on primate feeding behavior. Am J Phys Anthropol 25:1–18

Gomes DF, Bicca-Marques JC (2012) Capuchin monkeys (*Cebus nigritus*) use spatial and visual information during within-patch foraging. Am J Primatol 74:58–67

Gonzalez M, Clavijo L, Betancur J, Stevenson PR (2016) Fruits eaten by woolly monkeys (*Lagothrix lagothricha*) at local and regional scales. Primates 57:241–251

Grove M (2012) Space, time, and group size: a model of constraints on primate social foraging. Anim Behav 83:411–419

Grueter CC, Li D, Ren B, Li M (2013) Overwintering strategy of Yunnan snub-nosed monkeys: adjustments in activity scheduling and foraging patterns. Primates 54:125–135

Gunst N, Leca J-B, Boinski S, Fragaszy D (2010) The ontogeny of handling hard-to-process food in wild brown capuchins (*Cebus apella apella*): evidence from foraging on the fruit of *Maximiliana maripa*. Am J Primatol 72:960–973

Hayes JE, Sullivan BS, Duffy VB (2010) Explaining variability in sodium intake through oral sensory phenotype, salt sensation and liking. Physiol Behav 100:369–380

Hine R, Martin E (2015) A dictionary of biology. Oxford University Press, New York

Hiramatsu C, Melin AD, Aureli F et al (2008) Importance of achromatic contrast in short-range fruit foraging of primates. PLoS One 3:e3356

Hogan JD, Fedigan LM, Hiramatsu C, Kawamura S, Melin AD (2018) Trichromatic perception of flower colour improves resource detection among New World monkeys. Scientific Reports 8:10883

Janmaat KRL, Ban SD, Boesch C (2013) Chimpanzees use long-term spatial memory to monitor large fruit trees and remember feeding experiences across seasons. Anim Behav 86:1183–1205

Janson C (1985) Aggresive competition and individual food consumption in wild brown capuchin monkeys (*Cebus apella*). Behav Ecol Sociobiol 18:125–138

Janson CH (1988) Intra-specific food competition and primate social structure: a synthesis. Behaviour 105:1–17

Janson CH (1998) Experimental evidence for spatial memory in foraging wild capuchin monkeys, *Cebus apella*. Anim Behav 55:1229–1243

Janson CH, van Schaik CP (1988) Recognizing the many faces of primate food competition: methods. Behaviour 105:165–186

Johnson CA, Raubenheimer D, Chapman CA et al (2017) Macronutrient balancing affects patch departure by guerezas (*Colobus guereza*). Am J Primatol 79:1–9

Johnson SE, Bock J (2004) Trade-offs in skill acquisition and time allocation among juvenile chacma baboons. Hum Nat 15:45–62

Kawamura S, Melin AD (2017) Evolution of genes for color vision and the chemical senses in primates. In: Saitou N (ed) Evolution of the human genome I: the genome and genes. Springer, Tokyo

King AJ, Clark FE, Cowlishaw G (2011) The dining etiquette of desert baboons: the roles of social bonds, kinship, and dominance in co-feeding networks. Am J Primatol 73:768–774

Kramer DL (2001) Foraging behavior. In: Fox CW, Roff DA, Fairbairn DJ (eds) Evolutionary ecology concepts and case studies. Oxford University Press, Oxford/New York, pp 232–246

Kusch J, Schmitz A (2013) Environmental factors affecting the differential use of foraging habitat by three sympatric species of *Pipistrellus*. Acta Chiropt 15:57–67

Lambert JE (1999) Seed handling in chimpanzees (*Pan troglodytes*) and redtail monkeys (*Cercopithecus ascanius*): implications for understanding hominoid and cercopithecine fruit-processing strategies and seed dispersal. Am J Phys Anthropol 109:365–386

Lambert JE (1998) Primate digestion: interactions among anatomy, physiology, and feeding ecology. Evol Anthropol 7:8–20

Leighton M (1993) Modeling dietary selectivity by Bornean orangutans: evidence for integration of multiple criteria in fruit selection. Int J Primatol 14:257–313

Lewis MC, Justin O'Riain M (2017) Foraging profile, activity budget and spatial ecology of exclusively natural-foraging Chacma baboons (*Papio ursinus*) on the Cape Peninsula, South Africa. Int J Primatol 38:751–779

Link A, Galvis N, Marquez M et al (2012) Diet of the critically endangered brown spider monkey (*Ateles hybridus*) in an inter-Andean lowland rainforest in Colombia. Am J Primatol 74:1097–1105

Luca F, Perry GH, Di Rienzo A (2010) Evolutionary adaptations to dietary changes. Annu Rev Nutr 30:291–314

Mallott EK, Garber PA, Malhi RS (2017) Integrating feeding behavior, ecological data, and DNA barcoding to identify developmental differences in invertebrate foraging strategies in wild white-faced capuchins (*Cebus capucinus*). Am J Phys Anthropol 162:241–254

Martin P, Bateson P (2007) Measuring behaviour: an introductory guide. Cambridge University Press, Cambridge

Matsuda I, Akiyama Y, Tuuga A et al (2014) Daily feeding rhythm in proboscis monkeys: a preliminary comparison with other non-human primates. Primates 55:313–326

McCabe GM, Fedigan LM (2007) Effects of reproductive status on energy intake, ingestion rates, and dietary composition of female *Cebus capucinus* at Santa Rosa, Costa Rica. Int J Primatol 28:837–851

McGraw WS, van Casteren A, Kane E et al (2016) Feeding and oral processing behaviors of two colobine monkeys in Tai Forest, Ivory Coast. J Hum Evol 98:90–102

Mekonnen A, Bekele A, Fashing PJ et al (2010) Diet, activity patterns, and ranging ecology of the Bale monkey (*Chlorocebus djamdjamensis*) in Odobullu Forest, Ethiopia. Int J Primatol 31:339–362

Melin AD, Chiou KL, Walco ER et al (2017) Trichromacy increases fruit intake rates of wild capuchins (*Cebus capucinus imitator*). Proc Natl Acad Sci U S A 114:10402–10407

Melin AD, Fedigan LM, Hiramatsu C et al (2009) Fig foraging by dichromatic and trichromatic *Cebus capucinus* in a tropical dry forest. Int J Primatol 30:753

Melin AD, Fedigan LM, Hiramatsu C, Kawamura S (2008) Polymorphic color vision in white-faced capuchins (*Cebus capucinus*): is there foraging niche divergence among phenotypes? Behav Ecol Sociobiol 62:659–670

Melin AD, Hiramatsu C, Parr NA et al (2014) The behavioral ecology of color vision: considering fruit conspicuity, detection distance and dietary importance. Int J Primatol 35:258–287

Melin AD, Shirasu M, Matsushita Y et al (2015) Examining the links among fruit signals, nutritional value, and the sensory behaviors of wild capuchin monkeys (*Cebus capucinus*). Am J Phys Anthropol 156:223–224

Melin AD, Veilleux CC (2018) Primate senses: finding and evaluating food. In: Lambert JE, Rothman JM (eds) Primate diet and nutrition: needing, finding, and using food. University of Chicago Press, Chicago

Melin AD, Wells K, Moritz GL et al (2016) Euarchontan opsin variation brings new focus to primate origins. Mol Biol Evol 33:1029–1041

Miller LE (2002) Eat or be eaten: predator sensitive foraging among primates. Cambridge University Press, Cambridge

Mosdossy KN, Melin AD, Fedigan LM (2015) Quantifying seasonal fallback on invertebrates, pith, and bromeliad leaves by white-faced capuchin monkeys (*Cebus capucinus*) in a tropical dry forest. Am J Phys Anthropol 158:67–77

Murray CM, Eberly LE, Pusey AE (2006) Foraging strategies as a function of season and rank among wild female chimpanzees (*Pan troglodytes*). Behav Ecol 17:1020–1028

Nagy-Reis MB, Setz EZF (2017) Foraging strategies of black-fronted titi monkeys (*Callicebus nigrifrons*) in relation to food availability in a seasonal tropical forest. Primates 58:149–158

Nakagawa N (2009) Feeding rate as valuable information in primate feeding ecology. Primates 50:131–141

Newman J (2007) Herbivory. In: Stephens DW, Brown JS, Ydenberg RC (eds) Foraging: behavior and ecology. University of Chicago Press, Chicago, pp 175–218

Norscia I, Ramanamanjato JB, Ganzhorn JU (2012) Feeding patterns and dietary profile of nocturnal southern woolly lemurs (*Avahi meridionalis*) in Southeast Madagascar. Int J Primatol 33:150–167

Nowell AA, Fletcher AW (2008) The development of feeding behaviour in wild western lowland gorillas (*Gorilla gorilla gorilla*). Behaviour 145:171–193

O'Malley RC, Fedigan L (2005) Variability in food-processing behavior among white-faced capuchins (*Cebus capucinus*) in Santa Rosa National Park, Costa Rica. Am J Phys Anthropol 128:63–73

O'Mara MT (2015) Ecological risk aversion and juvenile ring-tailed lemur feeding and foraging. Folia Primatol 86:96–105

Paterson JD, Vandenbeld D (2001) Primate behavior: an exercise workbook. Prospect Heights, Ill, Waveland

Peres CA (1993) Structure and spatial organization of an Amazonian terra firme forest primate community. J Trop Ecol 9:259–276

Pruetz JD, Isbell LA (2000) Correlations of food distribution and patch size with agonistic interactions in female vervets (*Chlorocebus aethiops*) and patas monkeys (*Erythrocebus patas*) living in simple habitats. Behav Ecol Sociobiol 49:38–47

Raudenbush SW, Bryk AS (2002) Hierarchical linear models: applications and data analysis methods. SAGE, Thousand Oaks

Righini N (2017) Recent advances in primate nutritional ecology. Am J Primatol 79:1–5

Rothman JM, Chapman CA, Van Soest PJ (2012) Methods in primate nutritional ecology: a user's guide. Int J Primatol 33:542–566

Rothman JM, Dierenfeld ES, Hintz HF, Pell AN (2008) Nutritional quality of gorilla diets: consequences of age, sex, and season. Oecologia 155:111–122

Rowe N, Myers M (2017) All the world's primates. In: The international encyclopedia of primatology. Wiley Blackwell, Chichester, pp 1–4

Ruivo LVP, Stone AI, Fienup M (2017) Reproductive status affects the feeding ecology and social association patterns of female squirrel monkeys (*Saimiri collinsi*) in an Amazonian rainforest. Am J Primatol 79:e22657

Rushmore J, Leonhardt SD, Drea CM (2012) Sight or scent: lemur sensory reliance in detecting food quality varies with feeding ecology. PLoS One 7:e41558

Sandell MA, Breslin PAS (2006) Variability in a taste-receptor gene determines whether we taste toxins in food. Curr Biol 16:R792–R794

Schoener TW (1968) The Anolis lizards of Bimini: resource partitioning in a complex fauna. Ecology 49:704–726

Schoof VAM, Jack KM (2014) Male social bonds: strength and quality among co-resident white-faced capuchin monkeys (*Cebus capucinus*). Behaviour 151:963–992

Schülke O, Chalise MK, Koenig A (2006) The importance of ingestion rates for estimating food quality and energy intake. Am J Primatol 68:951–965

Sengupta A, Radhakrishna S (2016) Influence of fruit availability on fruit consumption in a generalist primate, the rhesus macaque *Macaca mulatta*. Int J Primatol 37:703–717

Spence-Aizenberg A, Di Fiore A, Fernandez-Duque E (2016) Social monogamy, male-female relationships, and biparental care in wild titi monkeys (*Callicebus discolor*). Primates 57:103–112

Valenta K, Brown KA, Rafaliarison RR et al (2015) Sensory integration during foraging: the importance of fruit hardness, colour, and odour to brown lemurs. Behav Ecol Sociobiol 69:1855–1865

Vandercone RP, Dinadh C, Wijethunga G et al (2012) Dietary diversity and food selection in hanuman langurs (*Semnopithecus entellus*) and purple-faced Langurs (*Trachypithecus vetulus*) in the Kaludiyapokuna Forest Reserve in the dry zone of Sri Lanka. Int J Primatol 33:1382–1405

Veilleux CC, Scarry CJ, Di Fiore A et al (2016) Group benefit associated with polymorphic trichromacy in a Malagasy primate (*Propithecus verreauxi*). Sci Rep 6:38418

Visalberghi E, Albani A, Ventricelli M et al (2016) Factors affecting cashew processing by wild bearded capuchin monkeys (*Sapajus libidinosus*, Kerr 1792). Am J Primatol 78:799–815

Vogel ER (2005) Rank differences in energy intake rates in white-faced capuchin monkeys, Cebus capucinus: the effects of contest competition. Behav Ecol Sociobiol 58:333–344

Vogel ER, Janson CH (2007) Predicting the frequency of food-related agonism in white-faced capuchin monkeys (*Cebus capucinus*), using a novel focal-tree method. Am J Primatol 69:533–550

Chapter 10
Intra- and Interannual Variation in the Fruit Diet of Wild Capuchins: Impact of Plant Phenology

Jeremy Hogan and Amanda D. Melin

Abstract Understanding primate diet plasticity and what causes foraging variation is crucial to understanding their ecology and evolution. Foraging choices are limited by food availability, and primate diets may shift within and between years. How primates respond to interannual variation and the extent of their dietary flexibility are poorly known. White-faced capuchins (*Cebus capucinus imitator*) have a diverse diet and are informative study subjects for investigating dietary shifts in response to environmental variation. We investigate the phenological characteristics of plant foods and calculate monthly fruit production and probability of fruiting for each dietary species and report on how capuchin foraging varies intra- and interannually. We compare the dietary importance of foods consumed over two 1 year-long sampling periods. Finally, we determine how food use changes seasonally by comparing biweekly diet variation. Most plant food species examined produce fruit seasonally, although two important species are aseasonal. The overall biomass of capuchin fruit foods varies considerably monthly and annually. During scan sampling, capuchins were observed to consume 90 different plant food species across 24 months of study. Almost half of these plant foods were only consumed during one of the two study years, but such foods accounted for only 10% of the plant diet. Within 2-week periods, capuchins were typically reliant on one or two plant foods (i.e., >50% of the diet for that period), but few plant foods were important for multiple periods. This study illustrates the importance of studying both food and foraging characteristics at different scales and demonstrates the necessity of long-term projects for interpreting foraging behaviour.

J. Hogan (✉)
Department of Anthropology and Archaeology, University of Calgary, Calgary, AB, Canada
e-mail: jhogan@ucalgary.ca

A. D. Melin
Department of Anthropology and Archaeology, University of Calgary, Calgary, AB, Canada

Alberta Children's Hospital Research Institute, University of Calgary, Calgary, AB, Canada

Department of Medical Genetics, University of Calgary, Calgary, AB, Canada

U. Kalbitzer, K. M. Jack (eds.), *Primate Life Histories, Sex Roles, and Adaptability*, Developments in Primatology: Progress and Prospects,
https://doi.org/10.1007/978-3-319-98285-4_10

Keywords Phenology · Seasonality · Foraging variation · Food choice

10.1 Introduction

Foraging ecology has long been a focus of studies on wild animals because of the evolutionary implications that diet has for morphology, feeding patterns, social behaviour, and life history characteristics (Lambert and Rothman 2015). Primate diet choices often have important downstream implications for their foods as well. Many primate species positively impact plant reproductive success via seed dispersal (Chapman 1995; Gautier-Hion et al. 1985; Valenta et al. 2016), and some plant species may have evolved fruit characteristics specifically to attract primate consumers, likely through diffuse coevolution (Sussman 1991; Sussman et al. 2013). Importantly, food availability is rarely static, and very few primates will consume one food monotonously or even follow the same broad dietary patterns consistently (Altmann 2009; Marshall et al. 2009).

Intra-annual and spatial dietary variation is well documented throughout the primate order, and much of this variation is a product of seasonal fluctuation in resources (Brugiere et al. 2002; Campera et al. 2014; Marshall et al. 2014; Nagy-Reis and Setz 2016). Many animals attempt to maximize their fitness by timing their life history events to exploit predictable environmental variation (Fenner 1998; Snow and Whigham 1989). The factors that influence a plant's fruit, flower, and leaf production are highly variable and context-dependent, and the phenological patterns of strongly seasonal ecosystems are often driven by climatic events (Wright and van Schaik 1994). Phenological patterns are also scale-dependent: in many cases strikingly different patterns will be observed at the community, species, individual, and even branch levels (Boyle and Bronstein 2012). Furthermore, not all plant species respond in the same way to the same climatic conditions, and factors such as a taxon's evolutionary history and microsite differences within a habitat can have significant effects (Dahlgren et al. 2007). As data from long-term field projects become available, it is becoming apparent that interannual plant phenology patterns are inconsistent and often unpredictable (Chapman et al. 1999; Chapman et al., Chap. 17, this volume). Understanding the extent of fruit production variation at the habitat and species levels is important to fully understand what parameters shape primate food choice and is only possible after many years of continuous data collection.

One consequence of environmental variability and the plasticity of primate behaviour is that too often primate diets are broadly characterized from limited, short-term studies that do not capture the full extent of foraging variation and are likely not representative of longer time periods. For example, white-faced capuchin monkeys (*Cebus capucinus imitator*) are generally referred to as frugivore-insectivores, an accurate classification given that fruit and invertebrates combined comprise over 90% of the annual foraging budget (Bergstrom 2015; Bergstrom et al. 2018). However, such broad characterizations of diet choice risk masking important food choices over short timescales. For example, a multi-year comparison study of capuchin foraging behaviour revealed that flowers were important foods seasonally and that which flower species were important varied between years

(Hogan et al. 2016). Additionally, vertebrate prey can also be important to capuchins (Rose 1997). These results illustrate the value of collecting long-term data across different seasons to better understand foraging behaviour and dietary breadth. White-faced capuchins exhibit food preferences, selecting some foods more frequently than would be expected based solely on that item's abundance in the forest, and underselecting others (Leighton 1993; Melin et al. 2014). However, a plant food's abundance in the forest is subject to natural phenological variation and selectivity and preference may also vary over time. It is necessary, therefore, to explore food use patterns over variable lengths of time, under different phenological conditions, and at a finer scale (i.e., at the food-species level) than the broad food classes often employed. In this study, we aim to provide greater insight into the relationship between ecological variation and foraging flexibility in primate populations by examining phenological variation in capuchin fruit food trees in relation to detailed foraging records at two different timescales.

The specific purpose of our study is to determine the extent to which species-specific fruit production is predictable annually, and to investigate the dietary patterns of a population of wild white-faced capuchins at two different timescales (annually and between 2-week periods). White-faced capuchins are an appropriate study species for investigating dietary variation, as they are known to be omnivorous and highly plastic in their food choices (Fragaszy et al. 2004). Our primary research objectives are to assess (1) the extent of variation in fruit biomass and how predictable individual species' fruiting patterns are annually, (2) how diverse the capuchin diet is and which plant foods are consumed more frequently, and (3) to what extent the dietary importance of specific foods changes over time. To answer these questions, first we compared the monthly fruit production patterns of capuchin foods over a 9-year period to determine how consistent monthly fruit biomass production is, whether plants produced fruit in a predictable seasonal fashion, and to what extent individual conspecifics produced fruit seasonally. Second, we observed capuchin foraging behaviour over two study seasons and determined how diverse the capuchin diet was within each sampling period, which foods were important during both years, and which foods were uniquely consumed during just one sampling period. Finally, we analysed foraging data within smaller (2-week) timeframes to determine to what extent dietary patterns remain constant within a year.

10.2 Methods

10.2.1 Study Site

Data were collected in Sector Santa Rosa, Área de Conservación Guanacaste (hereafter referred to as Santa Rosa), located in the northwest corner of Guanacaste Province, Costa Rica. Santa Rosa's forests are mostly regenerating endangered tropical dry forest (DeGama-Blanchet and Fedigan 2006). Tropical dry forest is characterized by extreme seasonality of precipitation; in Santa Rosa, rainfall averages over 100 mm/month during the wet season (typically May–November), and almost no rainfall occurs during the dry season (December–April; see chapter by Campos 2018, in this volume, for detailed study of Santa Rosa's climate). Temperatures are relatively constant year-round, although the dry season is warmer and has a higher daily temperature range on average. During the dry season, up to 80% of the vegetation sheds leaves (Janzen 1988), resulting in little shade, and most water sources dry up (Campos and Fedigan 2009). Santa Rosa is home to three primate species: white-faced capuchins, mantled howler monkeys (*Alouatta palliata*), and black-handed spider monkeys (*Ateles geoffroyi*).

10.2.2 Phenological Data Collection

To investigate the fruit productivity patterns of capuchin foods, we analysed plant phenological data that has been collected consistently on important fruit foods since 2008. Within 3 days of the first of a month, researchers record the percentage of fruit coverage and maturity of selected food trees using a 0–4 scale (0 = 0%, 1 = 1–25%, 2 = 26–50%, 3 = 51–75%, 4 = 76–100%), resulting in two measurements monthly for each individual plant. Data have been collected from 38 species of capuchin plant foods continuously since 2008, with some species that were only included more recently being excluded from the present analysis. Whenever possible, eight individual representative plants per species are included in phenology data, although some species are underrepresented due to their rarity in the environment. In the case of tree death, a suitable replacement of the same species was added to the dataset as soon as possible. Interobserver reliability was established through a prolonged and rigorous training regime and was tested monthly by ensuring some trees were observed by several researchers.

10.2.3 Biomass Estimates

Capuchin plant food species are not uniformly distributed in Santa Rosa, nor are their fruit crops equal in size. To account for this, we calculated the estimated monthly fruit production (ripe fruit biomass index) for each species in our phenological sample. Plant species density and productivity estimates were obtained via transect sampling projects carried out in 2007–2010 and 2015–2017. Along 151 transects that were 100 m × 2 m (2007–2010) and 273 that were 100 m × 4 m (2015–2017), every individual tree of at least 0.5 cm in diameter and at least 1 m tall was identified to the species level, and its circumference at breast height (CBH) was recorded. In total 48,799 individual trees were identified in 9.06 ha of forest. Although figs are important capuchin foods, *Ficus* species are sufficiently rare in Santa Rosa that they were not captured in sampling transects. For these species, we calculated their biomass by all-occurrences sampling. We defined the sample area as the total area included within 25 m of capuchin group daily travel paths (791 ha; Campos et al. 2014), and we exhaustively searched for and recorded any *Ficus* trees ($n = 316$) in this zone from 2007 to 2014. Fruit productivity scales with tree size and can be estimated for animal-dispersed species from DBH measurements (Peters et al. 1988). For each species, we summed the estimated maximum productivity of every individual tree observed in a transect and divided this by the total transect area, providing an estimate of maximum fruit biomass density per species (kg/ha). This estimate represents an overestimate of fruit productivity that is only achieved if all individual conspecific trees were to fruit to their full potential simultaneously. Ripe fruit scores (i.e., fruit phenology score) were then calculated for each species from the phenology data by determining the average value of (coverage*maturity/4) for a species in a given month for each respective plant part. We then multiplied the species' maximum biomass estimate by its monthly fruit phenology score to calculate a monthly ripe fruit biomass index per species, which were summed to produce annual biomass estimates. To guard against overestimating productivity, we excluded individual transect trees if they were smaller than the smallest tree capuchins that have been observed foraging from during 2 years of behavioural follows (Melin et al. 2014). For biomass assessment, only species that were monitored in our phenology routes for all years were included in monthly and annual estimates. Finally, we also excluded wind-dispersed species because their fruit productivity is not well assessed via DBH (Peters et al. 1988). All calculations used for biomass estimates are contained within customized R scripts written by Fernando Campos and are publicly available at https://github.com/pace-primates/paceR.

10.2.4 Determining Seasonality of Fruiting Patterns

To determine how predictable fruiting patterns are for capuchin food species interannually, we used generalized linear mixed models with sinusoidal terms as predictor variables to determine whether fruit was produced significantly more often in a particular time period (season) yearly. This model system allows for analysis of a binomial dependent variable (i.e., fruit production) against a circular predictor variable (month of the year; Pewsey et al. 2013). Statistical analysis was conducted using R software and the lme4 package (Bates et al. 2015), and significance threshold was considered to be $P < 0.05$ for all analyses. We converted months to circular data by creating a numeric value for each month (e.g., January = 1, December = 12) and then converted these values to radians using the formula (month*2*pi/12). We calculated the number of representative trees producing fruit for each month and used cosinus of month in radian and sinus of month in radian as predictor variables to determine whether fruit production occurred in a predictable seasonal fashion. Models for each species were conducted separately. Since for each individual tree fruit production was a presence/absence observation, we used binomial distribution, and we included study year as a random effect. To test for model significance, we compared the results to null models that did not include the cosinus and sinus terms as predictors, using a chi-square test. Using the model coefficients, we then calculated and plotted predicted probability curves of fruit production per month for each species. These estimate the probability of fruit being produced for each species within any given month (i.e. the predicted proportion of phenology trees in fruit for that species).

10.2.5 Behavioural Data Collection

To determine the diversity of the capuchin plant food diet and to assess how the diet changes over time, we analysed data collected from four capuchin groups in 2007–2008 (LV, CP, EX, GN) and three groups in 2013–2014 (LV, AD, RM). In 2013, CP group fissioned to form groups AD and RM, therefore all three study groups observed in the second sampling period were directly comparable to those observed during the 2007–2008 season, and there is notable home range overlap amongst these groups (Bergstrom et al. 2017). Group sizes fluctuated during both study seasons due to births, deaths, and male transfer between groups (Table 10.1). The individuals living in these groups are well habituated and individually identifiable.

For this study, we analysed two sampling periods of data: 155 full days in 2007–2008 (Season 1) and 107 in 2013–2014 (Season 2). In 2007–2008 data were collected February–April 2007, September 2007–January 2008, and May–August 2008. In 2013–2014, data were collected between May–July 2013 and October 2013–March 2014. During both sampling periods, we followed study groups for 2–4 consecutive days per study "cycle", which was monthly in 2007–2008 and every 2 weeks in 2013–2014. Researchers followed the focal study group from sunrise to sunset (approximately 06:00–18:00 +/– 30 min) and conducted scan sampling every 30 min during this time. Data collection for scan samples lasted for up to 10 min, during which time field personnel attempted to locate and record the behaviour of every individual in the group. Only adults, subadults, and large juveniles were observed for scan sampling. If a monkey was foraging on plant or animal matter, researchers recorded the type of the food item, to the species and part level whenever possible for plant foraging. A total of 29,533 scan records were collected (14,591 of them in 2007–2008 and 14,942 in 2013–2014).

We aimed to determine how diverse the capuchin diet was, which foods were important, and how the diet varied interannually. We defined dietary diversity as the

Table 10.1 The range of the number of individuals living in each study group observed for this study. Groups AD and RM formed following fission of CP in early 2013

Group	2007–2008 population	2013–2014 population
LV	19–22	11–14
GN	27–35	Not observed
EX	8–12	Not observed
CP	20–25	N/A
AD	N/A	19–22
RM	N/A	18

number of plant food items included in the diet for a given study period. We measured the importance of food items as the proportion of scan samples during which at least one monkey was foraging on that food item; foods foraged on more frequently were considered more important. We calculated the total number of plant feeding records and the number of times a specific plant food item was consumed within a year, used these values to calculate the proportion of the overall plant foraging budget comprised of each food item, and ranked them in order of use. To assess interannual variation in food use, we compared the top 10 food items from each sampling period and determined how many foods were consistently used in both seasons and finally calculated how much of the diet consisted of the same foods in both seasons. Because the 2007–2008 data were collected over 12 months, we assumed a priori that the number of different food items observed during the first sampling period would be higher than those consumed over the course of the second field study. To account for this, we only used data from 2007 to 2008 that was collected within 2 weeks of the periods data were collected in 2013–2014.

To examine the variation in food use patterns across shorter timescales, and determine the extent to which foraging patterns in these short periods differ from annual and interannual patterns, we compared food use within 2-week study "cycles" from the 2013–2014 data. For each cycle, we determined the number of plant food items consumed and their relative importance within that cycle. We report the most important plant foods for each foraging cycle and compare these results to annual foraging results to investigate whether important foods remain important year-round or are only consumed within short timeframes (but in high quantities). To determine whether dietary diversity changes over time, we compare the number of plant foods consumed per cycle and compare this to the overall annual pattern.

10.3 Results

10.3.1 Inter- and Intra-Annual Variation in Available Fruit Biomass

The estimated biomass of capuchin fruit foods varied considerably across months and years (Fig. 10.1). The mean annual biomass of capuchin foods was 1179 kg/ha, meaning that if production were to be distributed equally throughout the year, we would expect approximately 100 kg/ha of fruit to be available each month. The interquartile ranges for 5/12 months (June, July, August, November, December) are below this mean, while only 1 month (April) has an interquartile range above this (Fig. 10.1b). June and July have the smallest interquartile ranges, and they, along with December, do not overlap with February's or April's ranges. The most productive year, 2009, was especially productive during the wet season (May–October, except for August), and the single highest monthly production occurred during this

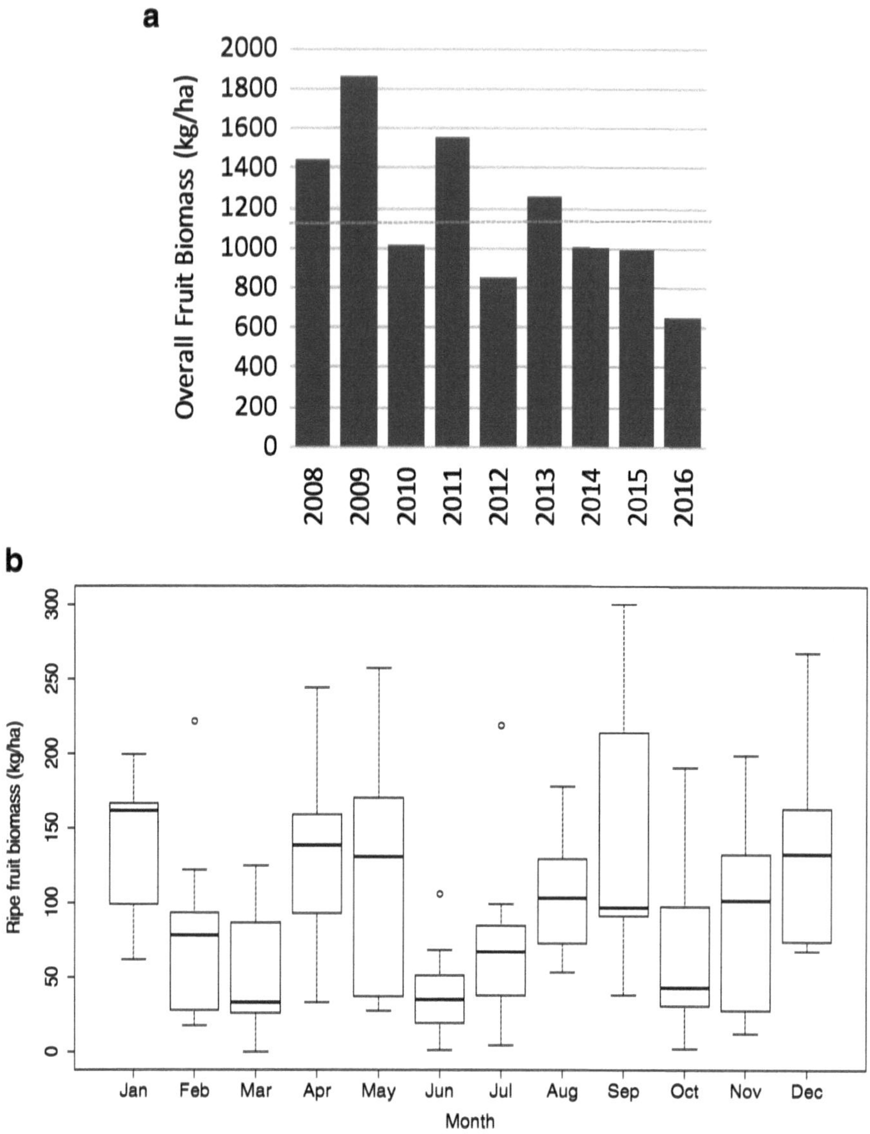

Fig. 10.1 The (**a**) annual and (**b**) monthly biomass estimates (kg/ha) for 38 capuchin food species in 2008–2016. Monthly variation across years is displayed via boxplots: for each month, the bolded central line represents the median biomass estimate for all years, the boxes encapsulate the interquartile range, dashed lines are the maximum and minimum values, and open circles are outliers

same year (300 kg/ha in May). Conversely, 2016, which had the lowest annual production, also had the lowest production for 8/12 months, including the entire wet season. December 2016 is also the only month for any year with no recorded ripe fruit production.

10.3.2 Fruiting Seasonality of Capuchin Food Species

Almost all (36/38) trees in our phenology system were seasonal, as they produced fruit significantly more frequently in one part of the year (Fig. 10.2, Table 10.2). However, the probability of a species producing fruit in a specific month was low: the peak fruiting month (i.e., the month with the highest expected number of individuals in fruit for a given species) for 26 of 38 species had a fruiting probability of <50%, and only two species (*Guazuma ulmifolia* and *Karwinskia calderoni*) were at least 75% likely to produce fruit in their peak months (Table 10.2).

10.3.3 Plant Diet Diversity and the Importance of Specific Plant Food Items

Capuchins were observed to be feeding on an identifiable food item during 15% of scan records. A total of 90 different plant food items were recorded to be consumed during scan sampling over the course of both sampling periods (Fig. 10.3). Nearly half (42/90) of plant food items were unique to one sampling period. These unique foods were typically rarely used, and combined they accounted for 10% of the plant

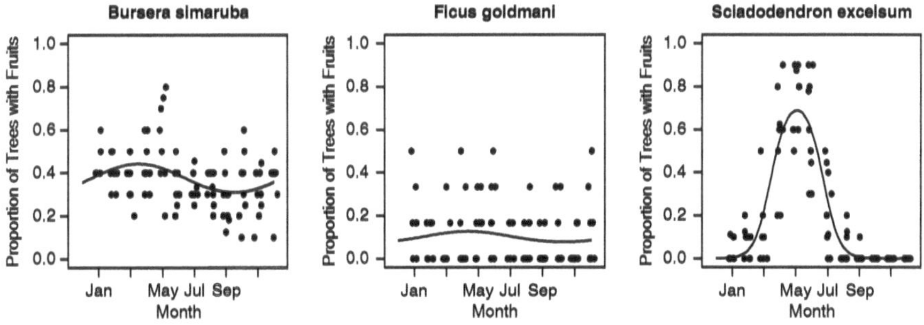

Fig. 10.2 Monthly probability of fruit production for three species of capuchin fruit foods: the weakly seasonal but persistently fruiting *Bursera simaruba*, the aseasonal, asynchronous fig species *Ficus goldmani*, and the highly seasonal, highly synchronous *Sciadodendron excelsum*. Each point represents the proportion of fruiting trees for that species monthly, with monthly data from nine consecutive years. A light horizontal jitter was applied to the figure to increase the visibility of overlapping points

Table 10.2 The degree of seasonality of fruit production, peak fruit month, and the peak probability of fruit production for 38 capuchin fruit foods representing over 90% of the annual capuchin plant diet. Data is derived from generalized linear mixed models with sinusoidal terms of months as predictor variables, using 8 years of phenological data. Chi-square and p values are derived from comparisons between the seasonal and null models; a *p* value <0.05 indicates fruit is more likely to be produced in a particular time period annually

Species	χ^2	*p*	Peak month	Probability of fruiting in peak month (%)
Manilkara chicle	73.4	<0.001	January	25
Guazuma ulmifolia	198.6	<0.001	March	85
Luehea candida	13.8	<0.001	March	15
Malvaviscus arboreus	46.9	<0.001	March	30
Simarouba glauca	270.7	<0.001	March	55
Bursera simaruba	9.9	<0.01	April	40
Dipterodendron costaricense	156.8	<0.001	April	50
Genipa americana	158.5	<0.001	April	55
Sloanea terniflora	94.0	<0.001	April	55
Vachellia collinsii	97.7	<0.001	April	20
Sciadodendron excelsum	431.3	<0.001	May	70
Zuelania guidonia	297.7	<0.001	May	50
Byrsonima crassifolia	172.7	<0.001	June	50
Ficus hondurensis	30.7	<0.001	June	45
Ficus ovalis	24.2	<0.001	June	40
Sebastiana pavoniana	57.1	<0.001	June	25
Tabebuia ochracea	8.3	<0.05	June	10
Cecropia peltata	144.9	<0.001	July	60
Jacquinia nervosa	63.7	<0.001	July	60
Psidium guajava	65.9	<0.001	July	50
Trichilia martiana	164.4	<0.001	July	50
Allophylus occidentalis	146.8	<0.001	August	40
Ficus cotinifolia	54.7	<0.001	August	30
Guettarda macrosperma	190.0	<0.001	August	40
Maclura tinctoria	379.7	<0.001	August	65
Spondias mombin	426.1	<0.001	August	70
Stemmadenia obovata	72.4	<0.001	August	30
Cordia panamensis	26.1	<0.001	September	15
Sapium glandulosum	61.0	<0.001	September	30
Randia thurberi	20.8	<0.001	October	30
Alibertia edulis	137.7	<0.001	November	60
Diospyros salicifolia	82.9	<0.001	November	35
Randia monantha	45.1	<0.001	November	25
Annona reticulata	42.7	<0.001	December	30
Karwinskia calderoni	375.6	<0.001	December	75
Trichilia americana	10.4	<0.01	December	25

(continued)

Table 10.2 (continued)

Species	χ^2	*p*	Peak month	Probability of fruiting in peak month (%)
Ficus goldmani	2.2	0.3306	April	15
Ficus morazaniana	1.6	0.4501	September	15

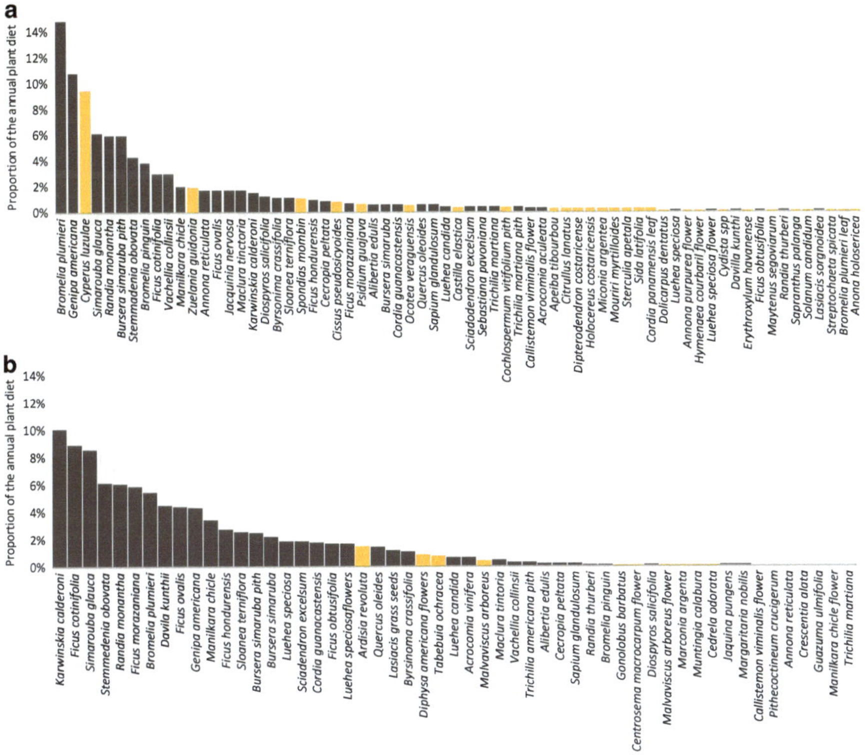

Fig. 10.3 The frequency of plant feeding records for each food item in (**a**) 2007–2008, and (**b**) 2013–2014. Lighter bars indicate foods that were not consumed in the other sampling period ($n = 42$); these "unique" foods account for 10% of the total plant food diet

food diet. Of these foods only used in one sampling period, only one, *Cyperus luzulae*, was amongst the ten most important food species (3rd, 2007–2008). It is important to note that this study does not reflect the total range of plant foods utilized by capuchins and that scan sampling is known to underestimate rare, fleeting behaviours, a shortcoming that can be compensated for via all-occurrences sampling (Leighton 1993; Melin et al. 2014).

Capuchins had high levels of diversity in their plant diet in both sampling periods, with 67 plant foods consumed in 2007–2008 and 53 in 2013–2014. Most food items that were consumed constituted less than 5% of the annual plant diet, and only four foods (*Bromelia plumieri*, 15% in 2007–2008; *Genipa americana*, 11% in 2007–2008; *Cyperus luzulae*, 10% in 2007–2008; *Karwinskia calderoni*, 10% in 2013–2014) accounted for 10% or more of the annual diet during either season.

The most important food items remained relatively static interannually. In both sampling periods, the top 10 foods accounted for approximately 2/3 of the plant food diet, and six of the top 10 foods annually remained consistent during both sampling periods. The 43 species included in phenology transects accounted for over 90% of the plant diet. Of the plant foods not currently included in phenology transects, very few were consumed in large quantities, the exceptions being grass seeds of *Cyperus luzulae* (3rd ranked 2007–2008), pith of *Bursera simaruba* (6th 2007–2008), and the fruits of *Davilla kunthii* (8th 2013–2014), a liana.

10.3.4 How Scale Affects Dietary Diversity and Importance Measures

Diet diversity varied greatly throughout the 2013–2014 sampling period, during which observation periods were broken into equal 2-week cycles. The number of different plant food items consumed by all three capuchin groups within a 2-week cycle varied from a low of 2 to a high of over 20 (Fig. 10.4). However, during 2-week cycles with high levels of diversity, typically only a few plant foods

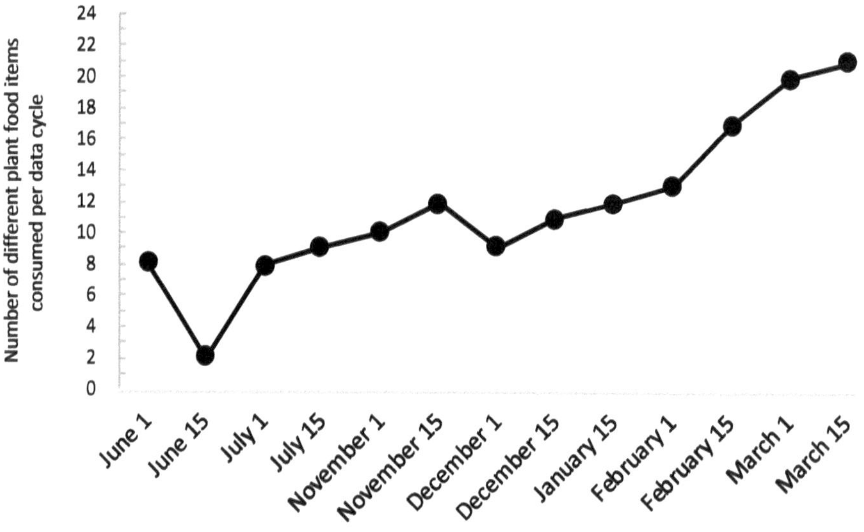

Fig. 10.4 Dietary diversity of white-faced capuchins within 2-week behavioural observation periods ("cycles") during 2013–2014. The date listed on the x-axis is the start of the 2-week window for a given cycle. Data were not collected August–October 2013

were consumed with high frequencies. For example, during 12 of 13 cycles, the top two foods accounted for over half the scan records for that period (Fig. 10.5). The diversity of plant food use by white-faced capuchins was lowest in the early wet season (June) and peaked in the late dry season (March). This pattern appeared to be inversely related to invertebrate foraging: during the period of highest invertebrate reliance, the diversity of the plant diet was lowest (Fig. 10.6). It is important to note that diet diversity is measured at the group level per cycle and is a binary presence/absence response; therefore it does not measure the importance of any food items in the diet or how frequently any item was consumed.

The timescale of the study period affects which foods ranked as the most important in the capuchin diet: several foods were very important for at least one 2-week cycle yet did not show up amongst the most important foods annually (Table 10.3). The timing of a cycle also affected importance: very few foods remained important in the capuchin diet for multiple cycles, and only one species (*Ficus cotinifolia*) was the most important food for multiple cycles. Figs (*Ficus* spp.) in general were extremely important throughout the year and were a top 3 food item in 8/13 cycles.

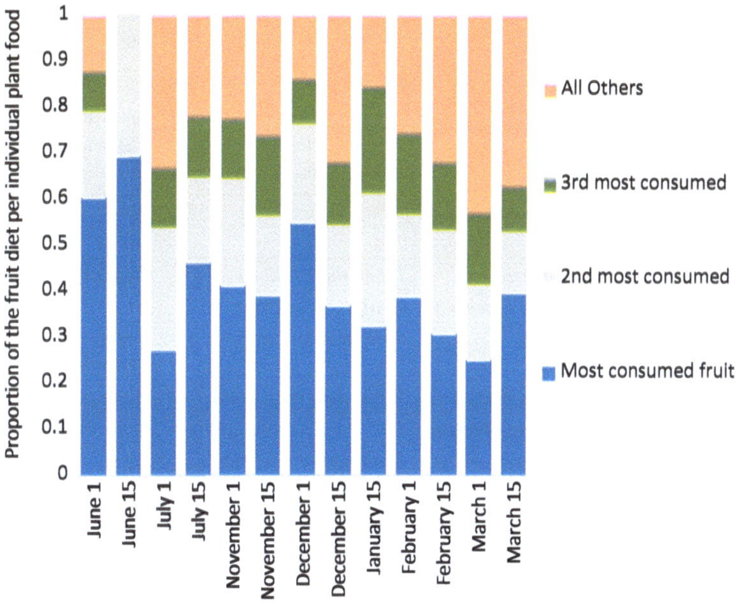

Fig. 10.5 Proportion of the capuchin plant food diet comprised of the three most consumed food items and all other plant foods (i.e., fourth up to 21st) within each 2-week study cycle. At short (2-week) timescales, the capuchin diet is dominated by just a few food items

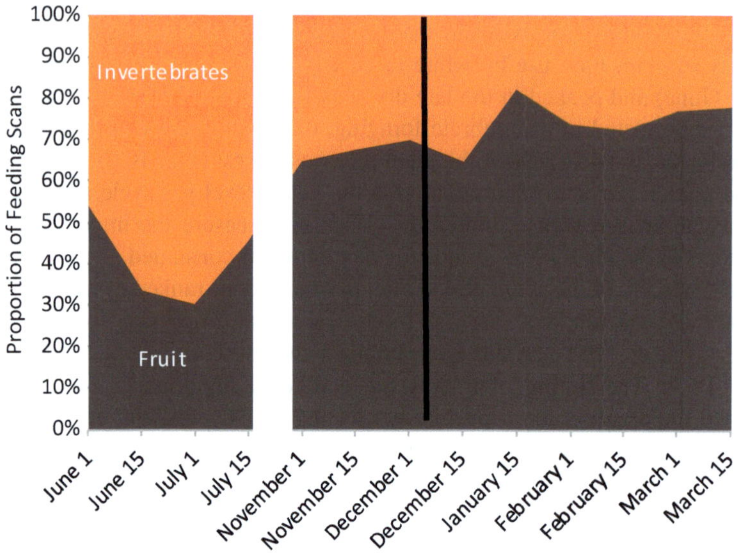

Fig. 10.6 Proportion of scan samples where capuchins were foraging for an identifiable type of invertebrate (light area) or plant food (dark area) during the 2013–2014 sampling period ($n = 2680$ scan samples). The vertical black bar is a rough estimate of the onset of the dry season in late 2013. Data collection did not occur August–October 2013

Table 10.3 The three most important (most frequently consumed) capuchin plant foods observed for each 2-week behavioural observation period ("cycle") in 2013–2014. Species highlighted in bold are foods that were also a top 10 food item in terms of annual importance

	Importance rank of food per cycle		
Cycle midpoint	1st	2nd	3rd
June 1	***Genipa americana***	*Tabebuia ochracea*	*Acrocomia vinifera*
June 15	*Sciadendron excelsum*	***Genipa americana***	n/a
July 1	***Ficus ovalis***	***Genipa americana***	*Quercus oleoides* *Byrsonima crassifolia*
July 15	*Cordia guanacastensis*	*Byrsonima crassifolia*	*Maclura tinctoria*
November 1	***Stemmadenia obovata***	*Bursera simaruba* (pith)	***Ficus cotinifolia***
November 15	*Ficus hondurensis*	***Stemmadenia obovata***	*Lasiacis* species (grass)
December 1	***Ficus cotinifolia***	***Stemmadenia obovata***	*Quercus oleoides*
December 15	***Ficus cotinifolia***	*Luehea speciosa* (flowers)	***Ficus morazaniana***
January 15	***Ficus ovalis***	***Karwinskia calderoni***	***Manilkara chicle***
February 1	***Karwinskia calderoni***	***Ficus morazaniana***	***Randia monantha***
February 15	***Randia monantha***	***Karwinskia calderoni***	***Bromelia plumieri***
March 1	***Davilla kunthii***	***Bromelia plumieri***	***Ficus morazaniana***
March 15	***Simarouba glauca***	*Sloanea ternifolia*	*Bursera simaruba*

10.4 Discussion

10.4.1 How Does Fruit Phenology Vary, and How Does This Variation Affect Capuchin Foraging?

Our study reveals that while most plant species predictably produce fruit around the same annual time period, there is relatively low probability of fruit production for most species for any given month and large interannual variation in the overall monthly and annual fruit abundance. The fruit production patterns observed here for most species (seasonal production but low monthly predictability) are consistent with other field sites (Chapman et al. 1999). Such interannual phenological variation provides a natural experiment, allowing for comparison of food selectivity during periods when several plants are concurrently productive and years in which one or more species are unproductive in the same timeframe.

Many of the top annual food items (e.g. *Karwinskia calderoni, Spondias mombin, Zuelania guidonia*) are very large trees with large, fleshy fruit crops that can hold an entire capuchin group repeatedly over multiple days (Melin et al. 2014; Parr et al. 2011). These important large-crop species that fruit during low-resource abundance period are likely the most crucial individual food item for capuchins, and their absence in a given year is likely the most detrimental. For example, in May, which is the transition period between the dry and wet seasons in Santa Rosa, only *Zuelania guidonia* and *Sciadendron excelsum* are in their peak fruiting windows, and ripe fruit biomass data indicates that May is extremely variable for fruit productivity, with occasional productive years tempered by more typical low-productivity years (median of 100 kg/ha; Fig. 10.1b). Although early May was not included in the sampling period during the 2013–2014 field season, this study also demonstrates that capuchin plant diet diversity is at its lowest immediately after this time, with <10 plant foods consumed in late May (and only two in mid-late June). *Zuelania guidonia* and *Sciadendron excelsum* are large and two of the most predictably seasonal and synchronous fruiting species (Table 10.2). Should both species fail to produce in a particularly bad year, capuchins could be faced with a severe food shortage. While starvation to the point of population crashes is thought to be rare, it has been observed elsewhere, in one instance reducing the local white-faced capuchin population by an estimated 70% (Milton and Giacalone 2014). Future research at this field site will be telling as to how plastic the capuchin diet truly is: Santa Rosa experienced extreme drought in 2015 (Fedigan unpublished data), leading to the lowest recorded biomass estimates for capuchin foods since the inception of phenology data collection. Many fruit crops (e.g. *Spondias mombin, Sloanea terniflora, Zuelania guidonia*) that are typically very important annually (i.e. amongst the ten most consumed foods annually) failed to produce altogether. Knowing which foods capuchins switch to in times of preferred food shortages will provide great insight into the limits of their foraging plasticity and may help to predict their success adapting to escalating climactic and environmental variation.

10.4.2 How Diverse Is the Capuchin Diet, and How Does It Vary at Different Timescales?

Over the course of two sampling periods comparing the same time periods, we observed 90 different plant foods being consumed by capuchins, half of which were only used in one of two periods, and the most important food item accounted for no more than 15% of the annual plant diet. While studies reporting on the types of foods consumed at the species level remain rare, our results are typical for a frugivorous primate, although studies from other field sites inhabited by capuchins (both *Cebus and Sapajus* species) reveal slightly less plant diet diversity (Guillotin et al. 1994; Perry and Ordonez Jimenez 2006; Simmen and Sabatier 1996). Possibly this is an artefact of different study lengths and methods, but an alternative hypothesis is that because Santa Rosa is regenerating tropical dry forest, it is less productive than wetter, more established forests studied elsewhere, leading to reduced consumption of any one plant species and the inclusion of lower-quality foods. One study across three sites in Costa Rica suggests that while the vast majority of foods are consumed by white-faced capuchins at different sites, there are "cultural" differences that result in some edible species being ignored by a population (Panger et al. 2002). As long-term data become available at other field sites, comparative studies of capuchin foraging strategies investigating the differences in the long-term diet and forest structure will be illuminating.

When compared to most primates, the capuchin diet appears to be more diverse than average. Of the eight sympatric primate species inhabiting Gabon rainforest observed over 10 years, only western lowland gorillas (*Gorilla gorilla gorilla*, 188 items) and central chimpanzees (*Pan troglodytes troglodytes*, 161 items) utilized more plant food types than Santa Rosa's capuchins, while primates more similarly sized to capuchins had much less diverse plant diets (Tutin et al. 1997). Furthermore, many of the food items consumed by these two ape species were leaves, likely primarily as a protein source (Felton et al. 2009). Capuchins rarely consume leaves, instead focusing their protein intake efforts on invertebrates. To accurately compare the diversity of diets requires a holistic approach including all food classes and accounting for macro- and micronutrient requirements. However, due to the size of most invertebrate foods, the speed with which capuchins locate and consume them, and their patchiness in the environment, it is currently impossible to accurately assess the diversity of invertebrates consumed by capuchins at the level of detail possible for plant parts. Recent advances in DNA barcoding techniques of faecal samples are promising and may prove to be the key to better understanding the overall dietary diversity of insectivorous primates (Mallott et al. 2016).

The importance of different food items within different timeframes also varied considerably, demonstrating the scale-dependent nature of the diet. For example, in 2013, only two plant foods were included in the diet in early June, and there are strong indications of an inverse relationship between the diversity of the plant diet and the consumption of invertebrates. Typically during this time period, which was near the start of the rainy season, much of the capuchin day is spent gleaning invertebrates off plant surfaces, and a significant proportion of the diet is comprised of Lepidoptera larvae (Bergstrom 2015; Mallott et al. 2016). Fruit biomass estimates also indicate this time period to be a period of relatively low fruit availability for capuchins (Fig. 10.5b), which suggests that caterpillars may be essential to their ability to survive a lean "crunch" period, an interpretation supported by recent work on capuchin metabolism (Bergstrom et al. 2017; Milton and Giacalone 2014). Whether capuchins are consuming less fruit (and fewer species of fruit) because there is a fruit food shortage, or simply because they prefer to eat caterpillars, is a question worthy of future investigation.

Overall, which specific food items are important, and even how important food classes (e.g. fruits) are to the capuchin diet, is variable depending on the length of time investigated as well as the specific time the investigation occurs. This finding highlights the need for longitudinal studies that encompass multiple years and different seasons to reliably capture the full range of foraging behaviours of which this species is capable. Capuchins have a very narrow diet at times and are surprisingly dependent on a small selection of food items over short timescales. Whether this is driven by preference for maximum exploitation of a resource while it is available or a lack of other options needs to be further explored. It is likely that the phenology and abundance patterns of certain important food species are influential.

10.4.3 Conclusion and Future Directions

Perhaps a better question than "what characteristics do capuchin foods share" is "what plant foods are present that capuchins do NOT eat, and what stops them from doing so?" This question need not be limited to phenological differences between foods and non-food plant species. Food quality is determined by many factors beyond the characteristics examined in this study, including accessibility, availability, nutritional quality, and the intrinsic characteristics of the consumer (Felton et al. 2009; Lambert and Rothman 2015). One important question yet to be answered is whether dietary fruits share morphological or chemical characteristics that make them attractive to capuchins and whether this may even vary between study sites due to environmental characteristics. For example, two hard-shelled fruits common to Costa Rican dry forest (*Guazuma ulmifolia* and *Sterculia apetala*) are typically not consumed by Santa Rosa capuchins, whereas both are consumed in high proportions by capuchins living in the nearby Lomas Barbudal Biological Reserve (Panger et al. 2002). Comparing the physical qualities (particularly hardness) of these fruits between sites, as well as the overall forest phenological patterns during the times these species are in fruit at both field sites, could further strengthen our understanding of capuchin food choice. The interaction effects between fruiting patterns and fruit quality are also likely to prove important. For example, capuchins may consume fruits that appear unpredictably and for short time periods only if they are of particularly high quality. Fortunately, vegetation transect work in Santa Rosa has resulted in an extensive tree database, cataloguing nearly 50,000 trees to the species level within capuchin home ranges. From this dataset, future research can be directed towards determining which non-food species are most abundant but not consumed and from there to conduct a comparison of characteristics that are not shared with important dietary plant species.

Acknowledgements We would be remiss to start these acknowledgements any other way than by acknowledging the past and present guidance, support, and friendship of Dr. Linda M. Fedigan; it is an honour to be a part of a festschrift for such a wonderful primatologist and person. We thank the government of Costa Rica, particularly the park administration staff of Santa Rosa, for the opportunity to conduct this research and for their continued support of our project. The data behind this chapter was collected by many dedicated field personnel over many years; thanks to you all. We thank Urs Kalbitzer as well as one anonymous reviewer for their thoughtful suggestions that greatly improved this manuscript, and thanks also to Urs for his patient assistance in compiling R code for creating GLMMs and their graphical output. Funding was provided by the University of Calgary, the Natural Sciences and Engineering Research Committee, and the Canadian Research Chairs Program.

References

Altmann SA (2009) Fallback foods, eclectic omnivores, and the packaging problem. Am J Phys Anthropol 140:615–629. https://doi.org/10.1002/ajpa.21097

Bates D, Mächler M, Bolker B, Walker S (2015) Fitting linear mixed-effects models using lme4. J Stat Softw 67. https://doi.org/10.18637/jss.v067.i01

Bergstrom ML (2015) Seasonal effects on the nutrition and energetic condition of female white-faced Capuchin monkeys. University of Calgary

Bergstrom ML, Emery Thompson M, Melin AD, Fedigan LM (2017) Using urinary parameters to estimate seasonal variation in the physical condition of female white-faced capuchin monkeys (*Cebus capucinus imitator*). Am J Phys Anthropol. https://doi.org/10.1002/ajpa.23239

Bergstrom ML, Melin AD, Myers MS, Fedigan LM (2018) Dietary profile, food composition, and nutritional intake of female white-faced capuchins. In: Kalbitzer U, Jack KM (eds) Primate life histories, sex roles, and adaptability – essays in honour of Linda M. Fedigan., developments in primatology: progress and prospects. Springer, New York, pp 213–238

Boyle WA, Bronstein JL (2012) Phenology of tropical understory trees: patterns and correlates. Rev Biol Trop 60:1415–1430

Brugiere D, Gautier J-P, Moungazi A, Gautier-Hion A (2002) Primate diet and biomass in relation to vegetation composition and fruiting phenology in a rain forest in Gabon. Int J Primatol 23:999–1024

Campera M, Serra V, Balestri M, Barresi M, Ravaolahy M, Randriatafika F, Donati G (2014) Effects of habitat quality and seasonality on ranging patterns of collared Brown Lemur (Eulemur collaris) in littoral Forest fragments. Int J Primatol 35:957–975. https://doi.org/10.1007/s10764-014-9780-6

Campos FA (2018, this volume) A synthesis of long-term environmental change in Santa Rosa, Costa Rica. In: Kalbitzer U, Jack KM (Eds) Primate life histories, sex roles, and adaptability – essays in honour of Linda M. Fedigan, developments in primatology: progress and prospects. Springer, New York, pp 331–354

Campos FA, Fedigan LM (2009) Behavioral adaptations to heat stress and water scarcity in white-faced capuchins (*Cebus capucinus*) in Santa Rosa National Park, Costa Rica. Am J Phys Anthropol 138:101–111. https://doi.org/10.1002/ajpa.20908

Campos FA, Bergstrom ML, Childers A, Hogan JD, Jack KM, Melin AD, Mosdossy KN, Myers MS, Parr NA, Sargeant E, Schoof VAM, Fedigan LM (2014) Drivers of home range characteristics across spatiotemporal scales in a Neotropical primate, Cebus capucinus. Anim Behav 91:93–109. https://doi.org/10.1016/j.anbehav.2014.03.007

Chapman CA (1995) Primate seed dispersal: coevolution and conservation implications. Evol Anthropol 4:74–82

Chapman CA, Wrangham RW, Chapman LJ, Kennard DK, Zanne AE (1999) Fruit and flower phenology at two sites in Kibale National Park, Uganda. J Trop Ecol 15:189–211

Dahlgren JP, von Zeipel H, Ehrlén J (2007) Variation in vegetative and flowering phenology in a forest herb caused by environmental heterogeneity. Am J Bot 94:1570–1576

DeGama-Blanchet HN, Fedigan LM (2006) The effects of forest fragment age, isolation, size, habitat type, and water availability on monkey density in a tropical dry forest. In: New perspectives in the study of Mesoamerican primates. Springer, New York, pp 165–188

Felton AM, Felton A, Lindenmayer DB, Foley WJ (2009) Nutritional goals of wild primates. Funct Ecol 23:70–78. https://doi.org/10.1111/j.1365-2435.2008.01526.x

Fenner M (1998) The phenology of growth and reproduction in plants. Perspect Plant Ecol Evol Syst 1:78–91

Fragaszy DM, Visalberghi E, Fedigan LM (2004) The complete capuchin: the biology of the genus *Cebus*. Cambridge University Press, Cambridge/New York

Gautier-Hion A, Duplantier J-M, Quris R, Feer F, Sourd C, Decoux J-P, Dubost G, Emmons L, Erard C, Hecketsweiler P et al (1985) Fruit characters as a basis of fruit choice and seed dispersal in a tropical forest vertebrate community. Oecologia 65:324–337

Guillotin M, Dubost G, Sabatier D (1994) Food choice and food competition among the three major primate species of French Guiana. J Zool 233:551–579

Hilje B, Calvo-Alvarado J, Jiménez-Rodríguez C, Sánchez-Azofeifa A (2015) Tree species composition, breeding systems, and pollination and dispersal syndromes in three Forest successional stages in a tropical dry Forest in Mesoamerica. Trop Conserv Sci 8:76–94. https://doi.org/10.1177/194008291500800109

Hogan JD, Melin AD, Mosdossy KN, Fedigan LM (2016) Seasonal importance of flowers to Costa Rican capuchins (*Cebus capucinus imitator*): implications for plant and primate. Am J Phys Anthropol 161:591. https://doi.org/10.1002/ajpa.23059

Janzen DH (1988) Tropical dry forests: the most endangered major tropical ecosystem. In: Wilson EO (ed) Biodiversity. National Academy Press, Washington, DC, pp 130–137

Lambert JE, Rothman JM (2015) Fallback foods, optimal diets, and nutritional targets: primate responses to varying food availability and quality. Annu Rev. Anthropol 44:493–512. https://doi.org/10.1146/annurev-anthro-102313-025928

Leighton M (1993) Modeling dietary selectivity by Bornean orangutans: evidence for integration of multiple criteria in fruit selection. Int J Primatol 14:257–313

Mallott EK, Garber PA, Malhi RS (2017) Integrating feeding behavior, ecological data, and DNA barcoding to identify developmental differences in invertebrate foraging strategies in wild white-faced capuchins (Cebus capucinus). Am J Phys Anthropol 162(2):241–254. https://doi.org/10.1002/ajpa.23113

Marshall AJ, Boyko CM, Feilen KL, Boyko RH, Leighton M (2009) Defining fallback foods and assessing their importance in primate ecology and evolution. Am J Phys Anthropol 140:603–614. https://doi.org/10.1002/ajpa.21082

Marshall AJ, Beaudrot L, Wittmer HU (2014) Responses of Primates and other frugivorous vertebrates to plant resource variability over space and time at Gunung Palung National Park. Int J Primatol 35:1178–1201. https://doi.org/10.1007/s10764-014-9774-4

Melin AD, Hiramatsu C, Parr NA, Matsushita Y, Kawamura S, Fedigan LM (2014) The behavioral ecology of color vision: considering fruit conspicuity, detection distance and dietary importance. Int J Primatol 35:258–287. https://doi.org/10.1007/s10764-013-9730-8

Milton K, Giacalone J (2014) Differential effects of unusual climatic stress on capuchin (*Cebus capucinus*) and howler monkey (*Alouatta palliata*) populations on Barro Colorado Island,

Panama: Climatic stress and differential mortality. Am J Primatol 76:249–261. https://doi.org/10.1002/ajp.22229

Nagy-Reis MB, Setz EZF (2016) Foraging strategies of black-fronted titi monkeys (Callicebus nigrifrons) in relation to food availability in a seasonal tropical forest. Primates. https://doi.org/10.1007/s10329-016-0556-9

Panger MA, Perry S, Rose L, Gros-Louis J, Vogel E, Mackinnon KC, Baker M (2002) Cross-site differences in foraging behavior of white-faced capuchins (Cebus capucinus). Am J Phys Anthropol 119:52–66. https://doi.org/10.1002/ajpa.10103

Parr NA, Melin AD, Fedigan LM (2011) Figs are more than fallback foods: the relationship between *Ficus* and *Cebus* in a tropical dry Forest. Int J Zool 2011:1–10. https://doi.org/10.1155/2011/967274

Perry S, Ordonez Jimenez JC (2006) The effects of food size, rarity, and processing complexity on white-faced capuchins' visual attention to foraging conspecifics. In: Hohmann G, Robbins MM, Boesch C (eds) Feeding ecology in apes and other Primates. Ecological, physical and behavioral aspects. Cambridge University Press, Cambridge, pp 203–236

Peters RH, Cloutier S, Dube D, Evans A, Hastings P, Kaiser H, Kohn D, Sarwer-Foner B (1988) The allometry of the weight of fruit on trees and shrubs in Barbados. Oecologia 74:612–616

Pewsey A, Neuhauser M, Ruxton G (2013) Circular statistics in R, 1st edn. Oxford University Press, New York

Rose L (1997) Vertebrate predation and food-sharing in *Cebus* and *Pan*. Int J Primatol 18:727–765

Simmen B, Sabatier D (1996) Diets of some French Guianan primates: food composition and food choices. Int J Primatol 17:661–693

Snow AA, Whigham DF (1989) Costs of flower and fruit production in *Tipularia discolor* (Orchidaceae). Ecology 70:1286–1293

Sussman RW (1991) Primate origins and the evolution of angiosperms. Am J Primatol 23:209–223

Sussman RW, Rasmussen DT, Raven PH (2013) Rethinking primate origins again: primate origins again. Am J Primatol 75:95–106. https://doi.org/10.1002/ajp.22096

Tutin CE, Ham RM, White LJ, Harrison MJ (1997) The primate community of the Lopé reserve, Gabon: diets, responses to fruit scarcity, and effects on biomass. Am J Primatol 42:1–24

Valenta K, Nevo O, Martel C, Chapman CA (2016) Plant attractants: integrating insights from pollination and seed dispersal ecology. Evol Ecol. https://doi.org/10.1007/s10682-016-9870-3

Wright SJ, van Schaik CP (1994) Light and the phenology of tropical trees. Am Nat 143:192–199

Chapter 11
Dietary Profile, Food Composition, and Nutritional Intake of Female White-Faced Capuchins

Mackenzie L. Bergstrom, Amanda D. Melin, Monica S. Myers, and Linda M. Fedigan

Abstract Seasonal variation in food availability and nutritional intake can ultimately affect female reproductive success. Although many primate studies have looked at foraging behaviour as a measure of diet, nutritional ecology and associated physiological consequences are a relatively new area of research. We present data on variation in the dietary profiles, foraging behaviour, and nutritional intake of female white-faced capuchins (*Cebus capucinus imitator*) in response to temporal variation in food abundance within the home ranges of three groups in the Área de Conservación Guanacaste, Sector Santa Rosa, Costa Rica. We estimated nutritional requirements based on published laboratory research and nutritional standards to determine whether females meet requirements in the face of seasonal variation in food abundance. Our results show that fruit contributed most to overall energy gain despite females devoting a greater proportion of foraging time to invertebrates; thus time spent foraging does not accurately reflect energy intake. On a dry matter basis, fruits provided the most important source of water-soluble carbohydrates, whereas high proportions of protein intake came from invertebrates, particularly when fruit availability was low. However, the greater weight of fruit items compared to invertebrates on a dry matter basis likely contributed to females' ability to consume macronutrients at higher rates while foraging on fruit due to higher mass intake per unit time. Requirement estimates and observed intake revealed that there are times

Electronic supplementary material: The online version of this chapter (https://doi.org/10.1007/978-3-319-98285-4_11) contains supplementary material, which is available to authorized users.

M. L. Bergstrom (✉) · M. S. Myers · L. M. Fedigan
Department of Anthropology and Archaeology, University of Calgary, Calgary, AB, Canada

A. D. Melin
Department of Anthropology and Archaeology, University of Calgary, Calgary, AB, Canada

Department of Medical Genetics, University of Calgary, Calgary, AB, Canada

Alberta Children's Hospital Research Institute, University of Calgary, Calgary, AB, Canada

U. Kalbitzer, K. M. Jack (eds.), *Primate Life Histories, Sex Roles, and Adaptability*, Developments in Primatology: Progress and Prospects,
https://doi.org/10.1007/978-3-319-98285-4_11

during low-fruit seasons in which females do not appear to meet their minimum nutritional requirements, warranting further investigation into the physiological consequences of this shortfall.

Keywords Nutrition · Diet · Seasonal variation · Food availability · Neotropical primate

11.1 Introduction

Variation in the spatial and temporal abundance of food resources presents several challenges related to energy and nutrient availability that may ultimately affect the reproductive success of mammals living in seasonal habitats (Clutton-Brock and Harvey 1977; Clutton-Brock et al. 1982; Bronson 1985). Gaining access to the food resources necessary for important biological processes such as growth, maintenance of body condition, and reproduction greatly influences the survival and reproductive success of mammals, including primates (Altmann 1998). Therefore, primates may attempt to maximize overall energy intake, maximize specific nutrient intake (e.g. protein), avoid secondary plant metabolites (e.g. tannins), minimize the intake of factors that inhibit efficient digestion (e.g. fibre), and/or balance the intake of nutrients or modify behaviour to reduce energy expenditure to meet their nutritional goals (reviewed by Felton et al. 2009a). If unsuccessful at meeting nutritional requirements, consequences include weight loss, decelerated growth and development, decreased reproductive output, and increased mortality or population-level changes in density and distribution (van Schaik et al. 1993; Kay et al. 1997).

In addition to ecological variation in nutrient abundance, there is considerable variation in nutritional demands related to morphology that impact the growth and development of individuals as well as the care of offspring and reproductive success across an individual's lifespan. Males and females often have dietary differences due to the added metabolic demands of the larger body size of males in sexually dimorphic species and of energetically demanding reproductive states (i.e. late gestation and early lactation) for females (Bell 1971; Clutton-Brock et al. 1982). Nutritional intake and body condition can significantly affect female lifetime reproductive success by influencing the timing of reproductive maturation as well as the persistence of ovulatory cycling at regular intervals. Previous studies of human and non-human primates have shown that females who consume a high-quality diet and maintain better body condition, or time reproduction with food abundance, give birth at an earlier age and/or exhibit higher rates of reproduction and infant survival than do other females (e.g. reviewed in humans and apes, Bentley 1999; in humans, Voland 1998; in marmosets, Tardif and Jaquish 1997; in tamarins, Miller et al. 2006; in mangabeys, McCabe and Emery Thompson 2013; see Chapter by McCabe and Fernandez 2018, in this Volume). Maternal nutrient intake during gestation impacts foetal development (reviewed by Hinde and Milligan 2011), with consequences as extreme as impairment of foetal cerebral development due to maternal nutrient restriction (baboons, *Papio* spp.; Antonow-Schlorke et al. 2011). Nutritional intake by primates during infancy can influence the secretion of hormones and growth factors, also leading to long-term effects on growth, metabolism, and susceptibility to disease (Mott et al. 1990, 1991; Lucas 1998), which may ultimately impact reproductive success (Altmann 1991).

We investigate seasonal variation in foraging behaviour (i.e. actively searching for and consuming food items), dietary profile (i.e. the types of foods consumed by female capuchins), and nutritional intake as it relates to broad-scale estimates of the requirements in a wild population of white-faced capuchin monkeys (*Cebus capucinus imitator*). Capuchins are arboreal monkeys that make use of multiple forest strata to exploit a wide diversity of resources including fruit (also encompassing seeds, grasses, bromeliads, and arils), invertebrates, flowers, pith, and vertebrates (Chapman and Fedigan 1990; Fragaszy et al. 2004). They are broadly categorized as omnivorous in that they consume foods from multiple trophic levels, and they are specifically categorized as frugivore-insectivores since fruit and invertebrates make up the majority of their diet (Fragaszy et al. 2004). Capuchins are sexually dimorphic in body mass (*Cebus* spp. males = 3.1 kg, females = 2.3 kg; Ford and Davis 1992), and sex has been shown to be a strong predictor of differences in foraging activities and of the types of foods ingested. White-faced capuchin males and females differ in foraging strategies in that males spend more time foraging for invertebrates on the ground, whereas females spend more time foraging for embedded invertebrates (Melin et al. 2010). In other capuchin species, males have been reported to forage more on animal matter and females to focus more of their foraging efforts on plant-based food items (e.g. wedge-capped capuchins, *C. olivaceus*, Fragaszy and Boinski 1995).

To better understand the relationship between foraging behaviour and diet, it is important to estimate species-specific nutritional demands for maintenance, reproduction, and growth. Unfortunately, species-specific values and information on metabolizable energy (i.e. gross energy per food item minus the undigested energy lost in the faeces, urine, and combustible gases) are unknown for most non-human primate species (National Research Council [NRC] 2003). The dietary and morphological similarities between capuchins and humans (e.g. relatively large small intestine and small caecum) allow for broad-level cross-species inferences and comparisons regarding digestive processes contributing to nutrition and energy gain (Chivers and Hladik 1980; Milton 1987); however, humans can likely digest fibre to a greater extent due to slower gut passage time (Cummings et al. 1976; Milton 1981).

By estimating nutritional requirements and determining species-specific patterns of temporal variation in foraging behaviour and nutritional intake, we can better understand variation in reproductive success in primates living in seasonal habitats across the primate order. However, a number of factors complicate a simple presentation of how primates use food resources to meet nutritional needs, including seasonality of resources, variation in dietary food types, variation in intake rates as a result of food distribution and required processing, and differences in the energy and macronutrient availability across food types (Oftedal et al. 1991). It is therefore important to consider these confounding factors and directly assess the relationships among diet composition, nutritional intake, and behaviour rather than making assumptions regarding diet based on foraging behaviour alone. Here, we investigate the nutritional ecology of adult female white-faced capuchin monkeys in Sector Santa Rosa, Costa Rica, by measuring ecological variation (i.e. food abundance), foraging behaviour, and dietary profile, as well as the nutritional composition of food items and nutrient intake with respect to estimated nutritional requirements. We address the following sets of questions:

1. Characterization of female foraging budgets, dietary profile, and intake:
   (a) What percentage of total foraging time do females spend on different types of food (fruits, flowers, invertebrates, pith, and vertebrates)?
   (b) What is the dietary profile of female capuchins?
   (c) Is foraging time representative of intake?
2. Nutritional composition of capuchin foods:
   (a) Does the macronutrient composition and energy density of capuchin foods differ across food types (i.e. fruit, seeds, flowers, caterpillars, and other invertebrates)?
   (b) Do females consume macronutrients and energy at different rates depending on the type of food eaten?
3. Evaluation of nutritional requirements and intake:
   (a) What are the estimated energy intake requirements for female capuchins?
   (b) How do foraging behaviour, energy intake, and macronutrient intake differ in accordance with variation in fruit abundance?

By enhancing our understanding of the foraging behaviour and nutritional composition of foods eaten by white-faced capuchins, we can estimate whether they are meeting nutritional requirements, gain species-specific insight into the range of dietary flexibility exhibited by females, and identify limiting factors that may influence how they meet their nutritional goals. Knowledge of foraging behaviour and nutrition in wild populations may also help to inform captive management practices and in situ conservation efforts. More broadly, these data are valuable for making cross-species comparisons linking foraging ecology to reproductive success and patterns of sociality.

11.2 Methods

11.2.1 Study Site

We conducted this study in the highly seasonal tropical dry forest of Sector Santa Rosa (SSR; 10°50′30′′N, 85°37′0′′W), in the Área de Conservación Guanacaste, northwestern Costa Rica. The rainy season extends from mid-May until mid-November (see Chapter by Campos 2018, in this Volume). During the study period (2009–2011), the mean annual rainfall based on daily records was 2304 mm (SD = 737), only 32 mm of which fell during dry season months. The temperature ranged from a mean minimum of 22.9 °C (SD = 0.1) in the wet season to a mean maximum of 29.6 °C (SD = 0.8) during the dry season. We recorded daily temperature using a Kestrel Pocket Weather Tracker (Nielsen-Kellerman Co., Bloomberg, Pennsylvania) and daily rainfall using a standard cylindrical rain gauge.

11.2.2 Study Subjects

White-faced capuchins compete directly over fruit (Vogel 2005), a high-quality, clumped, and monopolizable food resource (Isbell 1991), but also show dietary flexibility by altering foraging behaviour to exploit alternate food resources in their seasonal environment (i.e. distinct rainy and dry seasons) where abundance, distribution, and nutritional quality of resources fluctuate. The SSR population shows considerable variation in diet and ranging patterns during an annual cycle (Chapman and Fedigan 1990; McCabe 2005; Campos et al. 2014; Melin et al. 2014b; Mosdossy et al. 2015). Reproductive events fluctuate with seasonal variation in food availability; although not strictly seasonal breeders, female births increase considerably between April and September (Carnegie et al. 2011a), but the role of nutrition and energy balance in the timing and success of reproduction is unknown.

We observed three study groups, LV, CP, and GN, over a period of 20 months between September 2009 and May 2011. The average annual home range size for capuchin groups in this study population is 197.77 ± 12.52 hectares (Campos et al. 2014). We collected data during three 4-month periods (Sep–Dec 2009, May–Aug 2010, and Jan–Apr 2011) to account for seasonal variation. With the aid of one to two field assistants per season, we followed groups from dawn until dusk on a rotational basis for 4 to 6 days per group per month [in total 575 focal hours: LV = 120 h, CP = 227 h, GN = 228 h); 2124 h of observational contact with the groups]. The size of the study groups ranged from 20 to 37 individually known subjects (LV, 20–23; CP, 26–33; GN, 33–37). Details of the demographics for study groups are reported in Bergstrom (2015).

We focused data collection on adult females because their reproductive success is thought to be more directly constrained by access to food relative to males (Wrangham 1980; Sterck et al. 1997; Koenig 2002). We classified females ≥6 years old as adults based on the average age at first conception (Carnegie et al. 2011b), unless they conceived before that time, in which case they were included as an adult at the beginning of the data collection season during which they gave birth to their first infant. We did not include females that disappeared or died during the study ($N = 3$) due to lack of data on these individuals. The number of female study subjects ranged from 24 to 25 (LV: 5; CP: 10; GN: 9–10) per season due to the maturation and inclusion of an additional female in GN group during the course of the study.

11.2.3 Estimating Capuchin Nutritional Requirements

We estimated the nutritional requirements of capuchins based on the general metabolic demands determined for mammals as well as published studies of the nutritional requirements for maintenance and growth of captive capuchins. Mean energy intake requirements for maintenance and growth can be estimated for animals by measuring energy intake, energy expenditure, and changes in body mass and have been documented for captive primates, including adult male capuchins (*Cebus albifrons*, body weight 1.5–3.0 kg) as 395 kJ/kg/day (range 250–500 kJ/kg/day) (Ausman and Hegsted 1980). Accordingly, the estimated intake requirement for *Cebus capucinus imitator* females, who weigh an average of 2.54 kg (Smith and Jungers 1997), is roughly 1000 kJ per day, assuming intake is proportional to the change in body size. We broadly categorized female reproductive state per month (cycling, gestating, and lactating; see Carnegie et al. 2011b and Bergstrom 2015 for details on reproductive state categorization). We assigned a monthly energy coefficient of 1.00 for cycling females, 1.25 for pregnant females, and 1.50 for lactating females based on the estimated energy demands of each reproductive state (Key and Ross 1999). The overall energy intake requirement of 1000 kJ/day was thus estimated to be 1.25 times higher in pregnant females (1250 kJ/day) and 1.5 times higher in lactating females (1500 kJ/day).

11.2.4 Behavioural Data Collection

We collected behavioural data using 10-min focal animal follows and ad libitum sampling (Altmann 1974). Often, two people worked together to record behavioural data to ensure accuracy of data collection; one person dictated behaviours, while another recorded data on a hand-held computer (Psion Workabout MX). To observe focal females, we followed a random rotation schedule, outlined in Bergstrom and Fedigan (2010) and following Perry (1996) to achieve a representative and as unbiased sample as possible. During follows, we continuously recorded the general state behaviours of the focal female (e.g. travel, forage, rest, feed, social, and solitary) and detailed foraging event behaviours (e.g. feeding rates) (Bergstrom 2015).

We used durations of behaviours gathered during focal follows to calculate time budgets as well as ingestion rates for each female. We calculated the mean percentage of time all females spent in six broad behavioural states [forage, social, rest, travel, and other (which included self-direct, vigilant, and allospecific association)] by summing the amount of time spent in each category, dividing by the total amount of time that all females were observed across the entire study period, and then multiplying this value by 100. To construct foraging time budgets, we determined the percentage of time females spent actively foraging on different types of resources (invertebrates, fruit, flowers, pith, small vertebrates, and water/other) as well as visually foraging (whereby females visually searched trees without continuously handling substrates such as bark or leaves).

To compare the consumption patterns of broad food categories, foraging time was further divided by food type; we calculated ingestion rates as the number of food items eaten per minute of observed feeding time by species (i.e. consistent targeting of and hand-to-mouth ingestion of food items) and then averaged values for all items within the two most commonly consumed food types, fruit (including seeds) and invertebrates to obtain a mean value for these two categories (O'Malley and Fedigan 2005; McCabe and Fedigan 2007). Feeding bouts, which ranged from seconds to the entire 10-min focal follow, were used to calculate ingestion rates per plant species. Fruit species or invertebrate groups for which we observed less than five items, two bouts, or 10 min of total feeding time (across samples) were excluded from the dataset as these items were eaten too infrequently to accurately quantify ingestion rates. If an item was partially eaten during an observation, only the proportion eaten (e.g. 0.1 items) was used in the calculation of total items consumed per minute. Food characteristics (e.g. size and abundance) greatly differ across plant species and affect variation in the rate at which each species can be ingested. For example, larger-sized food items take longer to ingest, on average, and are consequently eaten at slower rates, and species that are more spatially clumped and abundant may be located and ingested more rapidly. Accordingly, we calculated energy and macronutrient intake rates (items/minute) to compare the importance of different food types (i.e. fruit, seeds, flowers, caterpillars, non-caterpillar invertebrates) in terms of the nutrients ingested.

To estimate energy consumption per day, we extrapolated the total energy intake from the energy ingested during 10-min focal follows. Energy ingested by each individual, determined based on observed intake and the energy content of each item, was summed and divided by daily focal time to determine hourly energy intake rates, and these were used to extrapolate the total estimated energy consumed based on the number of active hours. Although there are limitations to extrapolating from 10-min follows in that they are a snapshot of activity, the benefit to using this shorter sampling scheme is that more data points can be collected per individual and across individuals, both throughout the day and over a series of days, that may be more representative of the true variation in activity than fewer, longer focals (e.g. full day) (see Chapter by Melin et al. 2018, in this Volume). Capuchins feed very regularly from dawn until dusk, and we aimed to collect a sufficient number of focal follows to be representative of this variation in activity.

11.2.5 Ecological Data Collection

For nutritional analysis, we collected samples from plant species (including fruit, seeds, flowers, and pith) and broad categories of invertebrates (single species or groups of related species) consumed by female capuchins in the field, and we processed and dried them in the laboratory at the field station. Sample size per species ranged from five to thousands of specimens, depending on size and weight, to achieve a total dry weight of at least 16 grams, which was the minimum mass required by the lab where we sent samples for analysis (Dairy One Forage Laboratory, New York, USA). If sample collection was not possible on the day the species was observed to be eaten, we collected samples later (≤3 days for fruit and flowers) from as many of the locations in which foraging was observed as possible. Invertebrate samples were an exception; we collected those samples during a season in May 2013 dedicated to this analysis. Capuchins often consume specific parts of food items (shell, flesh, and seed); therefore, when processing fruits with multiple parts, we separated the shell from the flesh and cut away the pulp from the seed. We only analysed the part of the food item that was consumed by the capuchins for nutritional content unless separation of components resulted in loss of the item's integrity and composition (e.g. water content). Subsequently, we dehydrated the samples at 30 °C using a food dehydrator (Nesco American Harvest Gardenmaster Pro, Model FD-1020) and stored them in airtight waterproof bags with silica until exported for analysis.

We could not always taxonomically identify invertebrates on a fine scale. Consequently, we grouped invertebrate samples into the following broader categories for nutritional analysis: peppered roaches (*Archimandrita tesselata*), cicadas (*Fidicina mannifera*), shield bugs (Pentatomidae), ants (Hymenoptera), satellite sphinx caterpillars (*Eumorpha satellitia*, mean wet mass = 1.47 g), medium-sized noctuid caterpillars (*Euscirrhopterus poeyi* and *Gerra Hallowach01*, mean wet mass = 0.26 g), medium-sized caterpillars from various families (Lepidoptera), small-sized caterpillars from various families (Lepidoptera, mean wet mass = 0.05 g), jumping bean moth larvae (*Cydia deshaisiana*), crickets (Gryllidae), grasshoppers and katydids (Caelifera), wasp larvae (*Polistes*), scorpions (*Centruroides limbatus*), and a bulk category that included unidentified small non-caterpillar invertebrates. We collected, flash-froze, and dehydrated the invertebrate samples at a temperature of 42 °C using a hot air oven for 2–6 days, depending upon size and density, to avoid moulding that may occur when drying animal matter at slower rates (DSO-3000DF, Digisystem Laboratory Instruments Inc.).

Nutritional samples were consistently monitored for contamination; any food items showing signs of fermentation or moulding were discarded, and new samples were collected (Harborne 1984; Conklin-Brittain et al. 2006). Due to the higher risk of contamination associated with dehydrated animal matter, we stored invertebrate samples in a temperature and humidity-controlled room until exported for analysis. All samples were transported to Dairy One Forage Laboratory to measure the macronutrient composition (e.g. crude protein, crude fat, water-soluble carbohydrates, neutral detergent fibre, total ash, and organic matter; see Appendix A).

Because there can be large differences in the nutritional composition of plant foods of the same species across space and time (Chapman et al. 2003), we collected samples from multiple trees (two to ten individuals per species based on rarity, fruit size and density, and distribution across home ranges), and the collection of these plant samples spanned the period in which each species was eaten. Thus, values should be more representative of a species average, rather than tied to a specific time and location. We consider the nutritional values subsequently obtained as applicable to this study, but caution should be used when using these values for broad application.

For laboratory nutritional sample analysis, specifics regarding analytical procedures performed by Dairy One can be found in Appendix A. Briefly, additional drying was done to accurately calculate total dry mass. Moisture content was determined by first subtracting the final dry matter weight (grams) from the initial weight of the wet fruit sample collected in the field and then dividing that value by the total wet weight to obtain an overall percentage (Conklin-Brittain et al. 2006). Crude protein (CP) was measured; however, this measure does not subtract bound proteins, which are metabolically unavailable, and thus may overestimate the amount of protein intake by capuchins in this study. Crude fat (CF), water-soluble carbohydrates (WSC), and neutral detergent fibre (NDF) were also measured.

We used the per item macronutrient values determined by the laboratory nutritional analyses to calculate the total energy density of food items as described below. Energy density is an estimate of the total amount of digestible energy (in kJ or kcal) per gram of dry matter. Mean gross energy concentrations for macronutrients determined using bomb calorimetry and published by the National Research Council (2003) are known: 4.1 kcal for carbohydrates, 5.6 kcal for protein, and 9.4 kcal for fat. In an attempt to approximate the digestible portion of gross energy, physiologically available energy conversion factors assigned based on early studies of food digestibility (Merrill and Watt 1955) have been published for humans as 4 kcal for carbohydrates, 4 kcal for protein, and 9 kcal for fat. Unfortunately, requirements based on metabolizable energy (gross energy minus the undigested energy lost in faecal matter) for each nutritional component have not yet been quantified for primates (NRC 2003); therefore, we used the values reported for humans. Here, we converted from kilocalories to kilojoules (kJ) using the factor 4.184 (16.74, 16.74, and 37.66, respectively). We used Formula 1 to calculate the energy density (kJ per gram dry matter) of each food item, where CP is the proportion of dry mass as crude protein, WSC is the proportion of dry mass as water-soluble carbohydrates, and CF is the proportion of dry mass as crude fat (Janson 1985; NRC 2003). We multiplied energy density by the dry mass per item to calculate the energy per food item (kJ/item).

Formula 1

$$\text{Energy}\,(\text{kJ}) = \left(16.74 \times (\text{CP} + \text{WSC})\right) + (37.66 \times \text{CF}).$$

We did not include fibre (measured in this study as neutral detergent fibre, NDF) in this calculation for a number of reasons. First, given their frugivorous-insectivorous diet, *Cebus* monkeys have a simple gastrointestinal system and a relatively short gut transit time of approximately 3.5 h, which is comparable to other frugivorous platyrrhines (e.g. *Ateles*, 4.4 h) but much shorter than that of folivorous platyrrhines (e.g. *Alouatta*, 20.4 h) (Milton 1981). Second, because capuchins lack a specialized digestive system, it is unlikely that they are able to extract significant amounts of energy from difficult to digest plant materials such as fibre. Regarding invertebrate consumption, crude protein is likely a more accurate estimate of the energy available from chitin than is fibre for species like capuchins that produce chitinase (Finke 2007).

We collected and analysed a total of 53 plant-based food items and 10 invertebrate categories during this project. For species that we could not collect during this study for nutritional analyses, we used published data on specimens collected from Sector Santa Rosa, Costa Rica, by McCabe (2005); collected from the nearby site of Lomas Barbudal Biological Reserve, Costa Rica, by Vogel (unpublished, 2004, 2005); and information for one fruit species published by the US Department of Agriculture (USDA) (2014). We estimated energy values for unknown plant species or plant species for which we were unable to conduct nutritional analyses or use literature values. Specifically, when possible, we used the energy values for congeners of species with unknown values ($N = 1$) or species with similar size and composition ($N = 2$; Appendix B, C). When using congener species was not possible, we used the median energy value (kJ/item) for fruit, the mean energy value for flowers, and an assigned energy value for pith (1 kJ per 1-inch unit). Similarly, we calculated energy and macronutrient values per item for two groups of invertebrates (shield bugs and crickets), which were not analysed by Dairy One, using the mean macronutrient values for the same order of invertebrates (Hemiptera and Orthoptera, respectively) multiplied by the dry mass per item (obtained during sample collection for this study).

11.2.6 Fruit Abundance: Fruit Biomass (kg/ha) and Ripe Fruit Energy Density (kJ/ha)

To determine the energy density of ripe fruit (kJ/ha) in the study area per monthly round of data collection, we first calculated the fruit abundance in terms of ripe fruit biomass (kg/ha) using phenological data in combination with biological transects and then applied energy values (kJ/gram wet mass) per fruit species.

Tree abundance data were obtained from 151 botanical transects covering a total area of 3.02 hectares and distributed across all group home ranges (for details see Melin et al. 2014a). Following Peters et al. (1988), for each tree in the transect, the biomass was calculated using Formula 2, where F is the estimated grams of fruit produced by a tree of a given DBH and 47 and 1.9 are coefficients based on regressions between plant height and fruit weight generated using a number of tree species.

Formula 2

$$\text{Tree fruit biomass}(\text{F}) = 47 \times \text{DBH}^{1.9}$$

Monthly phenological data are collected as part of a long-term and ongoing collaborative project at SSR. We assessed the monthly fruit coverage and maturity values for data collected from January 2009 through December 2011 for approximately 8 individual trees ($\bar{x} = 7.93$, SD = 2.46) for 30 fruit species. Although it would be ideal to have monthly phenological data for all dietary plants, this is unfeasible, and these 30 species comprise ca. 80% of the capuchin annual diet (Melin et al. 2014a). We used a 5-point index (0 = absent, 0–25% = 1, 25–50% = 2, 50–75% = 3, and 75–100% = 4) to assess the score for the percentage of fruit coverage (*C*) and the score for the percentage of mature fruit (*M*) (Melin Meachem 2011). These species represented 44% (9394 of 21,347) of the fruit ingestion events recorded during focal observations. Many of the capuchin food species missing from this list were either wind-dispersed species (e.g. *Luehea candida* and *L. speciosa*), for which the Peters et al. (1988) equation (Formula 2) is not likely to provide an accurate fruit biomass estimate, or species for which it is difficult to gain accurate phenological information such as lianas, shrubs, palms, or bromeliads. Our overall calculations of fruit abundance underestimate total fruit abundance; however, as the most consumed species, we feel they provide an accurate measure of the relative abundance of ripe fruit in each month of our study. We calculated a combined index score (CI) for each tree (*i*) by multiplying the proportion coverage index (*C*/4), by the proportion maturity index (*M*/4) (Campos et al. 2014). Using this combined index, trees were assigned the minimum score of 0 when the coverage or maturity indices were 0, and trees received the maximum score of 1 when the coverage and maturity was considered 100% (Formula 3).

Formula 3

$$\text{Tree combined index score}\left(\text{CI}_i\right)=\left(\frac{C}{4}\times\frac{M}{4}\right)$$

Then, we calculated the mean monthly index (MI) for each species (*s*) as the mean combined index (CI_i) score for that species in that month. The fruit biomass score per species (B_s) in kilograms per hectare (kg/ha) per month was calculated as the sum of F (from Formula 2) divided by the sampled area (3.02 ha) and multiplied by 1000 to convert grams to kilograms (Formula 4).

Formula 4

$$\text{Species ripe fruit biomass}\left(B_s\right)=\text{MI}_s\times 1000\times\left(\frac{\sum_{i=1}^{n}F_i}{3.02}\right)$$

The monthly total energy availability from ripe fruit (EA) in kilojoules per hectare (kJ/ha) was calculated as the sum of the species-specific fruit biomass score (B_s) multiplied by the species-specific energy content E_s (kJ/kg wet weight) for all fruit species (Formula 5). We categorized data collection months as high- and low-energy density based on these calculations.

Formula 5

$$\text{Total ripe fruit energy availability}\left(\text{EA}\right)=\sum_{s=1}^{n}B_s\times E_s$$

11.2.7 *Statistical Analyses*

11.2.7.1 Nutritional Composition of Capuchin Foods

To compare the dry and wet mass between ripe fruit and invertebrate food items (and thus water content), we performed independent-samples Mann-Whitney U tests ($\alpha = 0.05$, 2-tailed) because sample sizes were unequal, and Kolmogorov-Smirnov tests confirmed that the data were not normally distributed. We performed a non-parametric Kruskal-Wallis test with Dunn's post hoc tests to compare the nutritional composition among five food types (fruit, seed, flower, non-caterpillar invertebrate, and caterpillar), because sample size for some food categories was small and the data were not normally distributed. We report the standardized test statistic and adjusted significance (multiple comparisons) using the Dunn-Bonferroni for post hoc tests.

11.2.7.2 Evaluation of Nutritional Requirements in Light of Fruit Abundance

We ran linear mixed effects models to examine the effects of seasonal variation in fruit abundance on foraging and intake after confirming our data conformed to assumptions of normality of distribution and homoscedasticity visually and through exploratory statistics. In these models, we included ripe fruit energy density as the fixed effect (standardized as a unitless Z-score) and female ID as the random effect. We ran eight separate models, each with one of the following measurements as response variable using a monthly mean value per female: (a) the proportion of time spent foraging, (b) the proportion of energy intake from fruit, (c) mass intake rate (gDM/hr), (d) energy intake (kJ/hr), (e) crude protein intake (g/hr), (f) crude fat intake (g/hr), (g) water-soluble carbohydrates (g/hr), and (h) neutral detergent fibre (g/hr).

All statistical analyses were performed in SPSS 21.0 (IBM Corp., Armonk, NY) using an alpha of 0.05 except in the case of multiple post hoc analyses, where levels were adjusted for running multiple comparisons.

11.3 Results

11.3.1 *What Percentage of Total Foraging Time Do Females Spend on Different Types of Food?*

The mean annual time budget for all general activity indicates that females in LV, CP, and GN groups spent the majority of their time foraging (60.4%) followed by resting (18.1%) and socializing (12.7%). There was considerable variation in general activity patterns across study months. Although females spent the most time foraging relative to other behaviours throughout the annual cycle, they spent roughly the same amount of time foraging as resting in April, the month with the highest maximum temperature (33.6 °C) and high fruit abundance. Of the total time foraging, capuchins spent 70.3% foraging on invertebrates, which represented the majority of the average time spent foraging over the entire study period. They spent 20.0% foraging on fruit, 1.1% foraging on pith, 0.7% foraging on flowers, 0.4% foraging on vertebrates, 1.3% foraging for water, and 6.2% of their time visually foraging. Both the percentage of total time spent foraging and the contribution of fruit and invertebrates to the foraging budget varied across months (Fig. 11.1). Females spent the greatest percentage of time foraging on invertebrates during the early rainy season months of June and July, which coincided with a flush of caterpillars, whereas the peak in the percentage of time spent foraging on fruit occurred during the late dry season in March and April, which corresponded to a peak in fruit abundance.

11.3.2 What Is the Dietary Profile of Female Capuchins?

Female capuchins in Sector Santa Rosa consumed fruit (including arils, bromeliads, seeds, and grasses), flowers, pith, invertebrates, vertebrates, and a small number of other items (e.g. dirt) during the study. We identified 88 plant food items (fruit from multiple ripeness stages, flowers, and pith) from 41 plant families, including fruit, seeds, bromeliads, and grasses from 64 species, flowers from 7 species, and pith from 3 species. The species for a small proportion of ingested fruit (two species), flowers (one species), and pith (one species) could not be identified. However, using the average per item nutritional value of known species, we estimated that these species comprised less than 1% of the energy consumed during this study. The capuchins also consumed 29 identifiable types of invertebrates from 10 orders, including Araneae, Blattodea, Coleoptera, Hemiptera, Hymenoptera, Lepidoptera, Odonata, Orthoptera, Phasmatodea, and Scorpiones. Based on the subset of species analysed for nutritional content (Appendices B and C), fruit items were significantly heavier ($N = 55$, range = 0.01–19.24 gDM, $\bar{x} = 1.51 \pm 0.46$) than invertebrates ($N = 13$, range = 4.72×10^{-3} to 2.29 gDM, $\bar{x} = 0.38 \pm 0.18$) on a dry matter basis (Mann-Whitney; $U = 226.0$, $p = 0.040$). However, the wet mass of ripe fruit was not significantly greater ($N = 55$, range = 0.02 to 99.00 g, $\bar{x} = 6.52 \pm 2.39$) than that of invertebrates ($N = 13$, range = 0.01 to 7.29 g, $\bar{x} = 1.29 \pm 0.56$) (Mann-Whitney; $U = 240.0$, $p = 0.067$).

11.3.3 Is Foraging Time Representative of Food Consumption and Energy Intake?

Females not only spent the largest percentage of their time foraging on invertebrates; they also comprised the largest percentage of the annual diet in terms of the number of items ingested followed by fruit, flowers, and pith (Table 11.1). However, fruit made a notably higher contribution to the percentage of dry matter ingested and the overall energy intake than did invertebrates (Table 11.1).

Table 11.1 Contribution of different food types to the annual foraging profile (Values are displayed as mean ± SE for female white-faced capuchins at Sector Santa Rosa, Costa Rica)

Food type	Foraging time (%)	Items ingested (%)	Dry matter ingested (%)	Energy ingested (%)
Fruit	19.96 ± 0.20	30.19 ± 1.11	61.69 ± 1.93	57.58 ± 2.01
Invertebrate	70.34 ± 0.22	66.56 ± 1.04	35.77 ± 1.87	39.06 ± 1.90
Flower	0.74 ± 0.10	1.84 ± 0.36	1.20 ± 0.36	0.80 ± 0.28
Pith	1.12 ± 0.15	1.49 ± 0.29	Na	1.23 ± 0.27

The calculation of percentage dry matter ingested does not include pith, as we were unable to successfully measure dry mass per unit. Total energy ingested is only the sum of categories for which nutritional processing and estimation were possible and excludes vertebrates and miscellaneous items

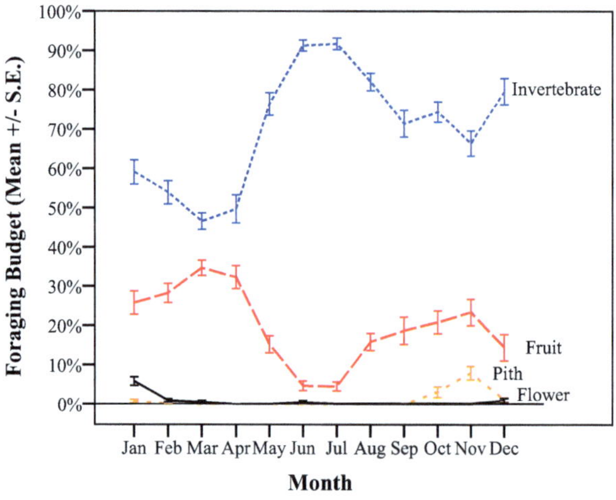

Fig. 11.1 Monthly foraging budget. The line graph depicts the mean ± SE percentage of time spent foraging on the four most common food types (invertebrates, fruit, pith, and flowers) by all three study groups

11.3.4 Does the Macronutrient Composition and Energy Density of Capuchin Foods Differ Across Food Types?

Nutritional composition varied greatly among food types (Table 11.2). Water content significantly differed by food type (Kruskal-Wallis; $N = 81$, $\chi^2(4) = 11.787$, $p = 0.019$). Flowers ($\bar{x} = 80.24\% \pm 3.76$) and caterpillars ($\bar{x} = 79.84\% \pm 5.38$) contained the most moisture, and seeds contained the least amount of moisture ($\bar{x} = 40.83\% \pm 13.35$). The difference in water content between caterpillars and seeds was significant (Dunn's test; $z = 2.957$, $p = 0.031$). Similarly, the percentage of crude protein significantly differed with respect to food type (Kruskal-Wallis; $N = 81$, $\chi^2(4) = 43.704$, $p < 0.001$). Fruit was significantly lower in crude protein compared to caterpillar (Dunn's test; $z = 3.790$, $p = 0.002$) and non-caterpillar invertebrates (Dunn's test; $z = -5.650$, $p < 0.001$). The water-soluble carbohydrates (WSC) significantly differed with respect to food type (Kruskal-Wallis; $N = 81$,

Table 11.2 Summary of macronutrient content per food type for food items eaten by adult females in the study groups

Food item	N species	% H_2O	% CP	% CF	% WSC	% NDF
Ripe fruit, seeds, and grasses	58	66.21 ± 2.47	8.86 ± 0.72	12.46 ± 2.13	35.85 ± 3.20	30.46 ± 2.47
Ripe fruit	53	68.60 ± 2.18	**8.11 ± 0.64**	12.06 ± 2.29	**38.21 ± 3.30**	30.45 ± 2.62
Seeds and grasses	5	**40.83 ± 13.35**	16.89 ± 3.65	16.64 ± 4.95	10.79 ± 3.52	30.62 ± 8.09
Flowers	3	80.24 ± 3.76	13.64 ± 3.13	3.94 ± 2.22	28.61 ± 8.65	21.25 ± 4.07
Invertebrates	13	72.99 ± 2.73	64.92 ± 4.27	16.17 ± 4.01	5.79 ± 1.67	21.65 ± 2.55
Non-caterpillar invertebrates	8	68.71 ± 1.89	**67.25 ± 5.16**	12.56 ± 1.82	**5.50 ± 2.63**	26.40 ± 2.82
Caterpillars	5	**79.84 ± 5.38**	**61.18 ± 7.87**	21.94 ± 10.13	**6.26 ± 1.55**	14.06 ± 2.22

CP crude protein, *CF* crude fat, *WSC* water-soluble carbohydrates, *NDF* neutral detergent fibre. Mean ± SE macronutrient values are listed as the percentage of dry mass. Species with estimated nutritional composition are excluded. Significant differences in composition as calculated by a pairwise comparisons using Dunn's test are highlighted in bold (please see Sect. 11.3.4 for details)

$\chi^2(4) = 33.833$, $p < 0.001$). Fruits contained significantly higher concentrations of water-soluble carbohydrates than did non-caterpillar invertebrates (Dunn's test; $z = 5.006$, $p < 0.001$) or caterpillars (Dunn's test; $z = -2.911$, $p = 0.036$). The differences in fat content across categories were not significant (Kruskal-Wallis; $N = 81$, $\chi^2(4) = 7.719$, $p = 0.102$). There were no significant differences in the percentage of neutral detergent fibre (NDF) across food types (Kruskal-Wallis; $N = 80$, $\chi^2(4) = 5.778$, $p = 0.216$).

Finally, food types differed in their total energy content per gram dry matter (Kruskal-Wallis; $N = 81$, $\chi^2(4) = 18.705$, $p = 0.001$; Fig. 11.2a). Non-caterpillar invertebrates contained significantly more energy than did fruits (Dunn's test; $z = -2.983$, $p = 0.029$). There was no significant difference in energy per gram dry matter between non-caterpillar invertebrates and flowers (Dunn's test; $z = -2.687$, $p = 0.072$) or between caterpillars and fruit (Dunn's test; $z = 2.746$, $p = 0.060$) and flowers (Dunn's test; $z = 2.805$, $p = 0.050$). When energy was assessed per gram wet mass to account for differences in water content, there were also significant differences across food types (Kruskal-Wallis; $N = 81$, $\chi^2(4) = 12.330$, $p = 0.015$; Fig. 11.2b), but none of the differences between pairs significantly differed from each other according to post hoc tests (Dunn's test; all p-values >0.074). When energy was assessed per item, there was a significant difference across food types (Kruskal-Wallis; $N = 81$, $\chi^2(4) = 16.234$, $p = 0.003$; Fig. 11.2c). Specifically, fruit contained significantly more energy (kJ) per item than did seeds (Dunn's test; $z = 3.093$, $p = 0.020$).

11.3.5 Do Females Consume Macronutrients and Energy at Different Rates Depending on the Type of Food Eaten?

The mean annual diet of female capuchins as a percentage of dry matter ingested, including all food types, was comprised of 27.15% (SE = 1.00) crude protein, 9.93% (SE = 0.27) crude fat, 35.75% (SE = 1.48) water-soluble carbohydrates, and 25.54% (SE = 0.07) neutral detergent fibre. As a percentage of metabolizable energy, protein comprised 30.15% (SE = 1.11), fat comprised 25.32% (SE = 0.60), and sugar comprised 43.36% (SE = 1.52).

The item intake rate for fruit and seeds was not significantly different from the intake rate for invertebrates (Mann-Whitney; $U = 164.000$, $p = 1.000$; Table 11.3; for item-specific rates, see Appendix D). However, the energy intake rate (kJ/min) was significantly higher for fruits and seeds than invertebrates (Mann-Whitney; $U = 21.000$, $p < 0.001$). The difference in energy consumption was a result of the significantly higher nutrient intake rate (g/hr) of sugar (Mann-Whitney; $U = 15.000$, $p < 0.001$) and fat (Mann-Whitney; $U = 55.000$, $p = 0.003$) during fruit and seeds versus invertebrate consumption, as the difference in the intake of protein between fruit and seeds and invertebrates was not significant (Mann-Whitney; $U = 122.000$,

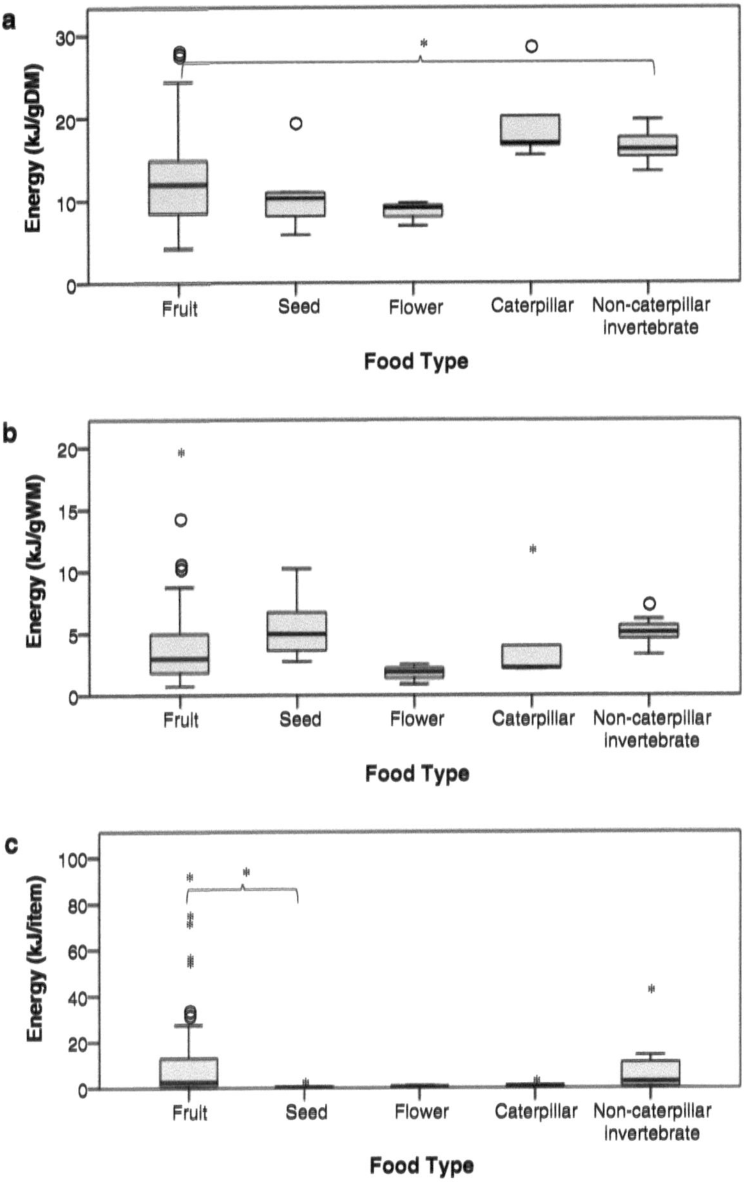

Fig. 11.2 Box plots of the nutritional composition of energy consumed by adult female study subjects. Graphs depict (**a**) energy per dry matter (kilojoules/gram), (**b**) energy per wet mass (kilojoules/gram), and (**c**) energy per item (kilojoules/item). Open circles and stars represent outliers >1.5 times the IQR and >3 times the IQR, respectively. Pairwise comparisons were conducted using Dunn's post hoc tests, and significant results are depicted by brackets and * for $p < 0.05$

Table 11.3 Intake rate and nutritional profitability of two food types (fruit and seeds and invertebrates) eaten by female capuchins

	Food type	
Variable	Fruit and seeds (mean ± SE)	Invertebrates (mean ± SE)
Species (N)	44	8
Item intake rate (N/min)	4.82 ± 0.77	3.89 ± 1.15
Energy intake (kJ/min)	**10.70 ± 1.36**	**1.75 ± 0.57**
Protein (CP)	5.71 ± 0.85	3.58 ± 1.12
Fat (CF)	**7.18 ± 1.34**	**0.84 ± 0.25**
Sugar (WSC)	**28.89 ± 5.13**	**0.77 ± 0.56**
Fibre (NDF)	**29.13 ± 5.68**	**1.27 ± 0.52**

Food items included in calculations are a subset of the larger nutritional dataset for which the targeting of specific species while foraging allowed for measurement of foraging bout length. Macronutrient intake (protein, fat, sugar, and fibre) is measured in grams per hour. Significant differences are highlighted in bold (see Sect. 11.3.5 for details)

$p = 0.256$). There was also significantly higher intake in fibre during fruit and seeds versus invertebrate consumption (Mann-Whitney; $U = 22.000$, $p < 0.001$).

11.3.6 What Are the Estimated Energy Intake Requirements for Female Capuchins?

Based on the number of females that were gestating and lactating during each month and assigning the factors of 1.25 and 1.50, respectively, for increased demands of these two reproductive states, the mean monthly reproductive demand for females during this study was 1.40 (SE = 0.01) times the requirements for a non-cycling female. Thus, the mean daily energy requirement for all females, who spend a disproportionate amount of the annual cycle in a lactational state (mean weaning completion age = 20.50 mo., Fragaszy et al. 2004; mean interbirth interval = 26.36 mo., Fedigan and Rose 1995), was estimated as 1400 kJ/day.

11.3.7 How Does Foraging Behaviour, Energy Intake, and Macronutrient Intake Differ with Variation in Food Abundance?

During this study, biannual peaks in ripe fruit energy density (kJ/ha) occurred from February–March to September–October (Fig. 11.3). Although fruiting patterns are variable interannually, these peaks coincided with the mean annual variation in fruit biomass collected between 2007 and 2013 at this field site (Campos et al. 2014). Monthly mean ripe fruit energy density (kJ/ha) was 557,763 kJ/ha (SEM = 94,514).

The highest energy density occurred during February 2011 (1,066,184 kJ/ha), whereas the lowest occurred during December 2009 (74,030 kJ/ha).

11.3.7.1 Differences in Foraging Behaviour and Energy Intake in Relation to Fruit Abundance

Ripe fruit energy density was a significant predictor of the mean proportion of time spent foraging across study months (Table 11.4). Females spent a lower proportion of their total activity budget foraging when fruit energy density was high compared to periods when fruit energy density was low (Fig. 11.4a). As expected, ripe fruit energy density was also significantly related to the proportion of energy intake from fruit (Table 11.4), whereby the more fruit energy that was available, the higher the proportion of fruit-based energy intake by females (Fig. 11.4b). Females consumed the highest average proportion of energy from fruit during April 2011 ($\bar{x}$ = 0.861 ± 0.036) and the lowest proportion of energy intake from fruit during May 2010 ($\bar{x}$ = 0.132 ± 0.042). Neither the rate of food intake (grams of dry matter per hour) nor the rate of total energy intake was significantly predicted by ripe fruit energy density (Table 11.4, Figs. 11.4c and 11.5d). However, the lowest mean monthly food intake rate ($\bar{x}$ = 3.251 g/hr ± 0.433) and energy intake rate ($\bar{x}$ = 53.097 kJ/hr ± 6.270) did occur during June 2010, a month with very low ripe fruit abundance (256,209 kJ/ha compared to the mean of 557,763 kJ/ha). The

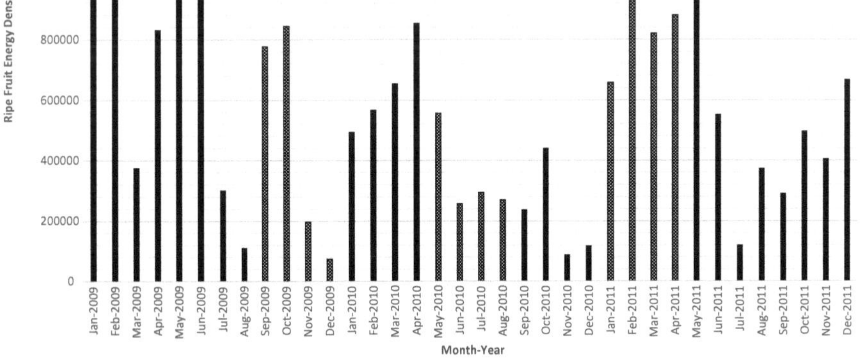

Fig. 11.3 Energy density from ripe fruit (kJ/ha) based on 30 fruiting species important to the diet of white-faced capuchins at Sector Santa Rosa, Costa Rica. Data associated with data collection periods for this study are indicated by patterned bars

Table 11.4 Summary of linear mixed effects models analysing the effect of monthly ripe fruit energy density (kJ/ha) on variation in mean feeding time and nutrient intake

								95% confidence interval	
Response variable	Fixed effect	Intercept	Estimate	Standard error	df	*t*	*p*-value	Lower	Upper
Feeding time (proportion)	**Energy density**	**0.573**	**−0.075**	**0.009**	**270.565**	**−8.065**	**<0.001**	**−0.094**	**−0.057**
Proportion of intake from fruit	**Energy density**	**0.583**	**0.167**	**0.018**	**270.546**	**9.184**	**<0.001**	**0.131**	**0.202**
Intake (gDM/hr)	Energy density	10.856	0.768	0.573	294.000	1.340	0.181	−0.360	1.897
Energy intake (kJ/hr)	Energy density	142.121	7.637	6.446	270.368	1.185	0.237	−5.054	20.329
Crude protein (g/hr)	**Energy density**	**1.832**	**−0.323**	**0.050**	**269.815**	**−6.475**	**<0.001**	**−0.421**	**−0.225**
Crude fat (g/hr)	Energy density	0.851	−0.001	0.046	294.000	−0.020	0.984	−0.091	0.090
Water-soluble carbohydrates (g/hr)	**Energy density**	**4.656**	**0.834**	**0.300**	**270.375**	**2.775**	**<0.01**	**0.242**	**1.425**
Neutral detergent fibre (g/hr)	Energy density	2.820	0.331	0.180	294.000	1.842	0.067	−0.023	0.685

Macronutrient weight is shown as grams dry matter. Female ID was included as a random effect. Significant effects are highlighted in bold

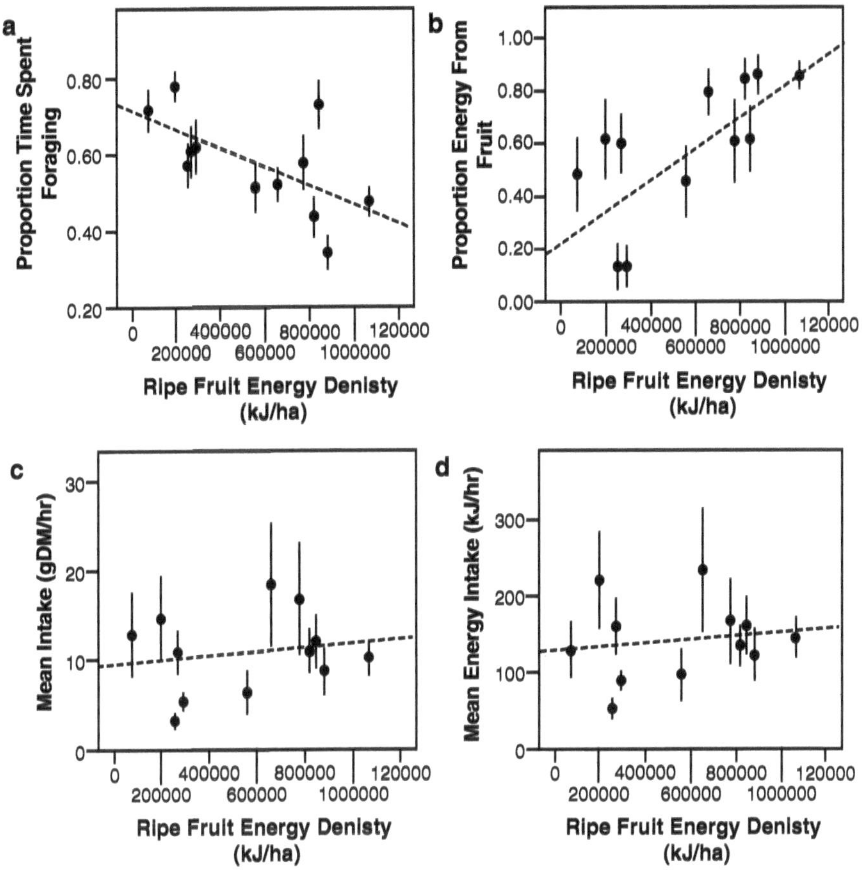

Fig. 11.4 Monthly variation in foraging time and intake by females. Points with error bars show the monthly mean values ± 95% confidence interval for (**a**) foraging time, (**b**) the proportion of energy intake from fruit (kJ/hr), (**c**) mass ingested (grams dry matter per hour), and (**d**) energy intake (kJ/hr) plotted against monthly ripe fruit energy density for 25 white-faced capuchin females at Sector Santa Rosa, Costa Rica. The dashed line depicts the predicted relationship between the variables according to the linear mixed effects models (see Table 11.4)

highest food intake rate (18.451 ± 3.336 g/hr) and energy intake rate (233.329 ± 39.669 kJ/hr) occurred during a month with higher than the annual mean ripe fruit abundance in January 2011 (657,730 kJ/ha). Based on these data, mean daily (12-h) energy intake is estimated to have reached as high as 2800 kJ/day and as low as 637 kJ/day.

11.3.7.2 Differences in Macronutrient Intake According to Fruit Abundance

Ripe fruit energy density significantly predicted protein intake rate (g/hr, Table 11.4). Females consumed protein at lower rates when fruit abundance was high compared to months when fruit abundance was low (Fig. 11.5a). Ripe fruit energy was

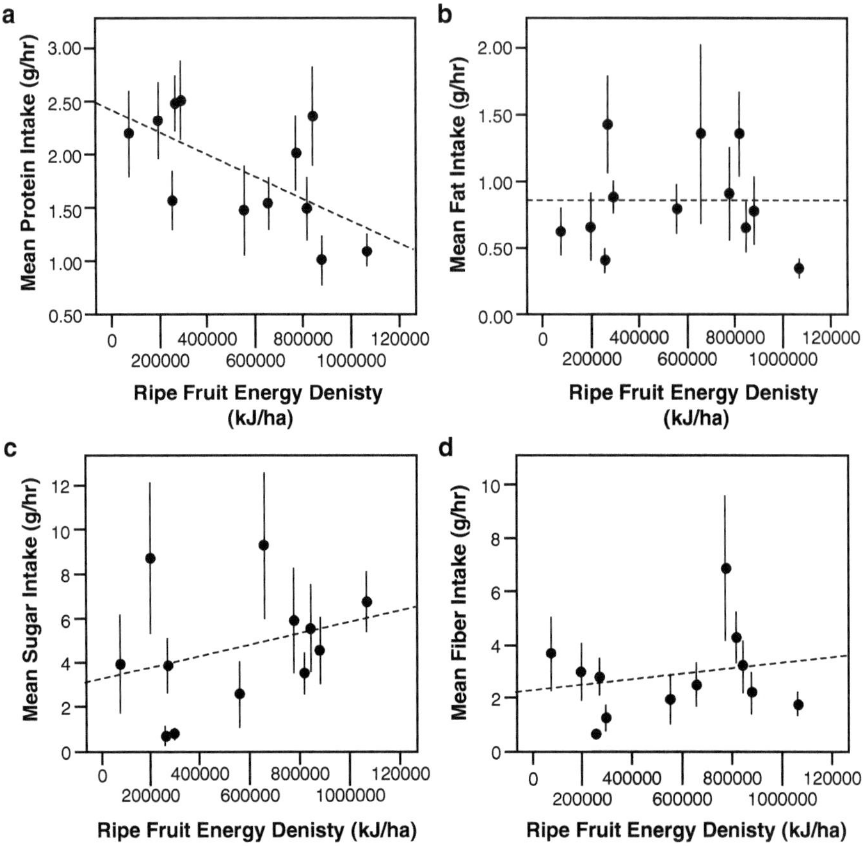

Fig. 11.5 Monthly variation in macronutrient intake by female capuchins. Points with error bars show the mean values ± 95% confidence interval for monthly intake, reported in grams intake per hour, for (**a**) crude protein, (**b**) crude fat, (**c**) water-soluble carbohydrates (WSC), and (**d**) neutral detergent fibre (NDF) plotted against monthly ripe fruit energy density for 25 female white-faced capuchins at Sector Santa Rosa, Costa Rica. The dashed line depicts the predicted relationship between the variables according to the linear mixed effects models (see Table 11.4)

significantly related to the rate of sugar intake (Table 11.4), whereby females consumed sugar at higher rates during months with higher ripe fruit energy density compared to month with lower ripe fruit energy density (Fig. 11.5c). Finally, ripe fruit energy was not significantly related fat intake or fibre intake (Table 11.4, Fig. 11.5b, d).

11.4 Discussion

The objectives of this study were to characterize female capuchin diet through assessment of the foraging budget, dietary profile, and nutritional composition of different food types, as well as to evaluate temporal variation in feeding behaviour, energy intake, and macronutrient intake in light of nutritional requirements. This is important for gaining a better understanding of the behavioural flexibility that capuchins exhibit in the face of seasonal changes in food availability and the physiological consequences (i.e. reproductive output) during seasons with low food abundance.

Female foraging budgets Adult female capuchins spent over half of their time foraging, during which they consumed a diversity of food items from two key dietary categories, plants and invertebrates. Despite interannual variation in food abundance and diet, the food item diversity is comparable to previous studies with documented foraging behaviour conducted at this field site (Chapman and Fedigan 1990; MacKinnon 2006). The observed foraging budget of 60.4% is also comparable to previous studies at this site (e.g. females, 53%, Rose 1994; all individuals, 53%, Melin et al. 2009), as well as the range of 41–58% across *Cebus* and *Sapajus* species (van Schaik and van Noordwijk 1989; Fragaszy 1990; Moura 2004; Matthews 2009; Izar et al. 2012).

Dietary profile Although there is variation in diet across the annual cycle, fruit and invertebrates make the most significant contribution to capuchin foraging time, number of items ingested, and overall energy gain across all months, as has been previously found in several capuchin studies (Chapman and Fedigan 1990; Rose 1994; Fragaszy et al. 2004; McCabe and Fedigan 2007; Melin et al. 2014b). Our results also indicate that invertebrates are arguably a more important food source when considering the amount of time that females spent foraging on them and the number of items ingested, which is comparable to other capuchin species (*C. albifrons* and *C. apella*) as well as more insectivorous common squirrel monkeys (*Saimiri sciureus*) and emperor tamarins (*Saguinus imperator*) (Terborgh 1983). However, fruit contributes substantially more to the annual dry matter ingested than do invertebrates. Fruit also comprises a larger percentage of the total annual energy consumed by females than do invertebrates.

Foraging time and intake Differences between foraging time and energetic contribution highlight the importance of including nutritional analyses in assessments and comparisons of diet and behaviour, as well as considering the nutritional values of food items with respect to a food item's weight and moisture content rather than making comparisons of nutrients on a dry matter basis alone (Rothman et al. 2014). While this insight seems rather obvious, due to lack of available nutritional data, studies that do not include nutritional analysis must base data analysis on the assumption that the amount of food eaten and/or the energy ingested is proportional to the time spent eating or the number of items ingested. This point is particularly important when making comparisons across food types with the largest size/weight disparities (e.g. between fruit and invertebrates), since distribution, size, handling time, and food processing all may affect energy and macronutrient intake rates (Hladik 1977; Milton 1984; Schulke et al. 2006).

Nutritional composition by food type We analysed the variation in the nutritional composition among different food types (fruit, seed, flower, pith, non-caterpillar invertebrate, and caterpillar). Invertebrates were higher in protein and energy than fruit and seeds, although fruit and seeds contained higher levels of water-soluble carbohydrates. These values are expected given that invertebrates are primarily composed of protein, fat, and chitin (Rothman et al. 2014). They are also comparable to other studies that have analysed the nutritional composition of primate foods, including those specific to platyrrhines (reviewed by Norconk et al. 2009).

The energy and nutritional composition of invertebrates may be considerably offset by their significantly smaller weight when compared to fruits on a dry matter basis (Rothman et al. 2014). The more dispersed distribution of invertebrates in the environment relative to the generally more highly clumped distribution of fruits produced by angiosperms may also greatly affect intake rate and profitability in terms of nutritional gain per unit time foraging.

Future research that expands the nutritional dataset to include (1) liana fruits (which were not quantified in our phenological data collection or estimates of fruit abundance and energy availability), (2) fruits that are available but not consumed by capuchins, (3) invertebrate abundance, (4) mineral concentrations, and (5) plant defence compounds such as tannins and polyphenols would help to better determine nutritional factors important to food selection and how selection may vary in relation to overall nutrient and energy availability. Whether a resource is "important" in terms of energy, macronutrient, and mineral intake and how these factors affect ranging patterns and resource defence may not only relate to a food's nutritional composition but also to the availability of alternative resources (Janson 1988; Vogel and Janson 2006).

Nutritional intake by food type Females consumed fruit at a slightly higher rate than invertebrates on a per item basis; however, the energy intake rate (kJ/min) and sugar and protein intake rates (g/hr) were significantly higher from fruit versus invertebrate foraging. It is therefore important to emphasize that time spent foraging does not accurately reflect energy intake, at least for capuchins, due to the large degree of variation in weight and nutrient content across food items. Although there was no significant difference in fat content of invertebrates compared to other broad food-type categories, further investigation should be done to determine if capuchins might be consuming invertebrates to increase fat intake in addition to using them as a source of protein during the high-fruit season.

Evaluation of nutritional requirements and intake Like many other free-living animals, capuchins in Sector Santa Rosa face temporal variation in the abundance of food resources and exhibit variation in both foraging and intake patterns across study months. As expected, ripe fruit abundance significantly predicted both the overall proportion of time spent foraging and the proportion of energy intake from ripe fruit. As fruit abundance increased, the proportion of time spent by females foraging decreased, but the amount of energy consumed from fruit increased. With increased availability, they were likely able to find and consume this high-energy food type more quickly. When the range for 12-h mean energy intake of 637 kJ/day–2800 kJ/day for females is placed in the context of the minimum estimated requirement of 1000 kJ/day, and the requirement of *Cebus capucinus* females adjusted for the reproductive demands of 1400 kJ/day (inferred from studies of captive capuchins by Ausman and Hegsted 1980), there is the potential for female capuchins to far exceed or fall short of daily energy intake requirements at this field site. Although females appeared to meet estimated energy intake requirements during many months, and greatly exceed them in others when fruit abundance and intake rates were much higher than average (November 2009 and January 2011), it is important to note that they did not meet estimated requirements during the month with the lowest intake rate in June 2010. During this month, ripe fruit abundance was relatively very low, and consequently, females were focusing greater than 90% of their foraging time on invertebrates. Since invertebrates are low in mass and require additional time to search and capture (as indicated by intake rates), it is likely that females were simply unable to consume enough food during this period. Although less mass and energy were ingested in low-fruit months, the ingestion of protein-rich invertebrates led to an increase in protein intake during those months and potentially maintained fat intake; however, sugar and fibre intake did decrease during that time. Future analyses will assess variation in energy and macronutrient intake in more detail to advance our understanding of dietary variation when the abundance of fruit is low.

Analysis of plant toxins was out of the scope of this project, and therefore, we did not test the hypothesis that capuchin nutritional goals could be driven by the avoidance of secondary plant metabolites. Females did not seem to minimize energy expenditure as fruit abundance decreased, but instead increased the proportion of time spent foraging. When fruit abundance was high, females seemed to "overconsume" carbohydrates based on intake rates, but also seemingly exceeded protein requirements. In contrast, during periods of fruit scarcity, females increased the foraging of invertebrates and "overconsumed" protein to meet overall energy requirements. These patterns suggest that females may be attempting to maximize caloric intake in response to a decrease in fruit abundance. Further analyses, including those specific to assessing the intake geometry of various nutritional components, may help to elucidate patterns of capuchin foraging and the role of specific capuchin foods (Raubenheimer and Simpson 1993; Simpson and Raubenheimer 1993; Raubenheimer 2011), as has been successfully implemented in other primate studies (Felton et al. 2009b; Rothman et al. 2011; Johnson et al. 2013; Raubenheimer et al. 2014; Irwin et al. 2015).

In summary, female white-faced capuchins at Sector Santa Rosa, Costa Rica, focused foraging efforts on fruit and invertebrate food items. Fruit contributed the most to the overall energy gain despite the greater proportion of time devoted to searching for and consuming invertebrates. Although the nutritional composition of food types is variable, fruits were the most important source of sugar, whereas high proportions of protein intake came from invertebrates, particularly as fruit abundance decreased. Females were able to consume macronutrients at a much higher rate while foraging for fruit, likely due to the higher weight of food items compared to invertebrates. There was temporal variation in the types of foods consumed and in the ability of female capuchins to meet energy requirements, warranting a more detailed investigation into capuchin foraging patterns, variation in the utilization of other food types, and nutritional intake during this study.

In the context of the nutritional ecology of capuchin monkeys, this study adds to the documentation of behavioural responses to variation in fruit availability and the nutritional properties of food. Our understanding of behaviour and physiology in this context will continue to grow as we place our findings into a larger multi-year framework that is compared across field sites and species. Such a framework will help to determine the consistency of these responses, how they are linked to the fitness of individuals, and then how they were shaped by evolutionary pressures.

Acknowledgements Thank you to the Ministry of Environment, Energy and Technology (MINAET) of Costa Rica and the Animal Care Committee at the University of Calgary for permissions to conduct research. This project was also supported by Sr. R Blanco, MM Chavarria, and the staff of the Área de Conservación Guanacaste. Many field assistants and friends aided in the collection of valuable data for this project including M Myers, B Thankey, H Clouse, C Lees, B Kowalzik, Dr. C Turner Hogan, and S Chevez. R Espinoza, A Guadamuz, J Vargas, and R Franco provided valuable assistance with the identification of plant and invertebrate samples. We thank G Bridgett for logistical help with permits and Dr. J Addicott and J McQuade for assistance with data organization and database design. Thank you to Dr. E Vogel for contributing nutritional data and Drs. E Wikberg and F Campos for statistical guidance. The authors were supported by the following funding organizations: To MLB: The Leakey Foundation, Alberta Innovates Technology Futures, International Primatological Society, Animal Behaviour Society, Sigma Xi, and University of Calgary. To ADM: Natural Sciences and Engineering Research Council of Canada (NSERC) and Canada Research Chairs Program. To LMF: NSERC and Canada Research Chairs Program.

References

Altmann J (1974) Observational study of behavior: sampling methods. Behaviour 49:227–265

Altmann SA (1991) Diets of yearling female primates (*Papio cynocephalus*) predict lifetime fitness. Proc Natl Acad Sci U S A 88(2):420–423

Altmann SA (1998) Foraging for survival: yearling baboons in Africa. University of Chicago Press, Chicago

ANKOM Technology (2011) Method 6: neutral detergent fiber in feeds - filter bag technique (for A200 and A200I). https://www.ankom.com/sites/default/files/document-files/Method_6_NDF_A200.pdf. Accessed 31 Jan 2017

Antonow-Schlorke I, Schwab M, Cox LA, Li C, Stuchlik K, Witte OW, Nathanielsz PW, McDonald TJ (2011) Vulnerability of the fetal primate brain to moderate reduction in maternal global nutrient availability. Proc Natl Acad Sci U S A 108(7):3011–3016

AOAC International (2012a) AOAC official method 942.05 ash of animal feed. In: Latimer G (ed) Official methods of analysis of AOAC International, 19th edn. AOAC International, Gaithersburg

AOAC International (2012b) Official method 2003.05 crude fat in feeds, cereal grains, and forages. In: Latimer G (ed) Official methods of analysis of AOAC International, 19th edn. AOAC International, Gaithersburg

AOAC International (2012c) Official method 992.23 crude protein in cereal grains and oilseeds: generic combustion method. In: Latimer G (ed) Official methods of analysis of AOAC International, 19th edn. AOAC International, Gaithersburg

AOAC International (2012d) Official method 992.15 crude protein in meat and meat products including pet foods. In: Latimer G (ed) Official methods of analysis of AOAC International, 19th edn. AOAC International, Gaithersburg

AOAC International (2012e) Official method 930.15 loss on drying (moisture) for feeds (at 135 for 2 hours), dry matter on oven drying for feeds (at 135 for 2 hours). In: Latimer G (ed) Official methods of analysis of AOAC International, 19th edn. AOAC International, Gaithersburg

AOAC International (2012f) Official method 990.03 protein (crude) in animal feed: combustion method. In: Latimer G (ed) Official methods of analysis of AOAC International, 19th edn. AOAC International, Gaithersburg

Ausman LM, Hegsted DM (1980) Protein requirements of adult cebus monkeys (*Cebus albifrons*). Am J Clin Nutr 33(12):2551–2558

Bell RHV (1971) A grazing ecosystem in the Serengeti. Sci Am 225:86–93

Bentley GR (1999) Aping our ancestors: comparative aspects of reproductive ecology. Evol Anthr 7(5):175–185

Bergstrom ML, Fedigan LM (2010) Dominance among female white-faced capuchin monkeys (*Cebus capucinus*): hierarchical linearity, nepotism, strength and stability. Behaviour 147(7):899–931

Bergstrom ML (2015) Seasonal effects on the nutrition and energetic condition of female white-faced capuchin monkeys. Dissertation, University of Calgary

Bronson FH (1985) Mammalian reproduction: an ecological perspective. Biol Reprod 32(1):1–26

Campos FA, Bergstrom ML, Childers A, Hogan JD, Jack KM, Melin AD, Mosdossy KN, Myers MS, Parr NA, Sargeant E, Schoof VAM, Fedigan LM (2014) Drivers of home range characteristics across spatiotemporal scales in a Neotropical primate, *Cebus capucinus*. Anim Behav 91:93–109

Campos, FA (2018, this volume) A synthesis of long-term environmental change in Santa Rosa, Costa Rica. In: Kalbitzer U, Jack KM (eds) Primate life histories, sex roles, and adaptability - Essays in honour of Linda M. Fedigan. Developments in primatology: progress and prospects. Springer, New York, pp 331–354

Carnegie SD, Fedigan LM, Melin AD (2011a) Reproductive seasonality in female capuchins (*Cebus capucinus*) in Santa Rosa (Area de Conservación Guanacaste), Costa Rica. Int J Primatol 32(5):1076–1090

Carnegie SD, Fedigan LM, Ziegler TE (2011b) Social and environmental factors affecting fecal glucocorticoids in wild, female white-faced capuchins (*Cebus capucinus*). Am J Primatol 73(9):861–869

CEM Corportation (2014) Method notes: digestion of alfalfa. http://cem.com/en/digestion-of-alfalfa-80ml-quartz. Accessed 31 Jan 2017

CEM Corportation (2016) Methods notes: digestion of feed grain. http://cem.com/en/digestion-of-feed-grain. Accessed 31 Jan 2016

Chapman CA, Fedigan LM (1990) Dietary differences between neighboring *Cebus capucinus* groups: local traditions, food availability or responses to food profitability? Folia Primatol 54:177–186

Chapman CA, Chapman LJ, Rode KD, Hauck EM, McDowell LR (2003) Variation in the nutritional value of primate foods: among trees, time periods, and areas. Int J Primatol 24(2):317–333

Chivers DJ, Hladik CM (1980) Morphology of the gastrointestinal tract in primates: comparisons with other mammals in relation to diet. J Morphol 166:337–386

Clutton-Brock TH, Harvey PH (1977) Species differences in feeding and ranging behavior in primates. In: Clutton-Brock TH (ed) Primate ecology. Academic Press, London, pp 557–584

Clutton-Brock TH, Guinness FE, Albon SD (1982) Red deer: behavior and ecology of two sexes. Edinburgh University Press, Edinburgh

Conklin-Brittain NL, Knott CD, Wrangham RW (2006) Energy intake by wild chimpanzees and orangutans: methodological considerations and a preliminary comparison. In: Hohmann G, Robbins MM, Boesch C (eds) Feeding ecology in apes and other primates. Cambridge University Press, Cambridge, pp 445–471

Cummings JH, Jenkins DJ, Wiggins HS (1976) Measurement of the mean transit time of dietary residue through the human gut. Gut 17:210–218

Fedigan LM, Rose LM (1995) Interbirth interval variation in three sympatric species of neotropical monkey. Am J Primatol 37(1):9–24

Felton AM, Felton A, Lindenmayer DB, Foley WJ (2009a) Nutritional goals of wild primates. Funct Ecol 23(1):70–78

Felton AM, Felton A, Raubenheimer D, Simpson SJ, Foley WJ, Wood JT, Wallis IR, Lindenmayer DB (2009b) Protein content of diets dictates the daily energy intake of a free-ranging primate. Behav Ecol 20(4):685–690

Finke MD (2007) Estimate of chitin in raw whole insects. Zoo Biol 26(2):105–115

Ford SM, Davis LC (1992) Systematics and body size: implications for feeding adaptations in New World monkeys. Am J Phys Anthropol 88(4):415–468

Fragaszy DM (1990) Sex and age differences in the organization of behavior in wedge-capped capuchins, *Cebus olivaceus*. Behav Ecol 1(1):81–94

Fragaszy DM, Boinski S (1995) Patterns of individual diet choice and efficiency of foraging in wedge-capped capuchin monkeys (*Cebus olivaceus*). J Comp Psychol 109(4):339–348

Fragaszy D, Visalberghi E, Fedigan LM (2004) The complete capuchin: the biology of the genus *Cebus*. Cambridge University Press, Cambridge

Hall MB, Hoover WH, Jennings JP, Miller Webster TK (1999) A method for partitioning neutral detergent-soluble carbohydrates. J Sci Food Agr 79:2079–2086

Harborne JB (1984) Phytochemical methods: a guide to modern techniques of plant analysis. Chapman and Hall, London

Hinde K, Milligan LA (2011) Primate milk: proximate mechanisms and ultimate perspectives. Evol Anthropol 20:9–23

Hladik CM (1977) A comparative study of the feeding strategies of two sympatric species of leaf monkeys: *Presbytis senex* and *Presbytis entellus*. In: Clutton-Brock TH (ed) Primate ecology: studies of feeding and ranging behavior in lemurs, monkeys and apes. Academic Press, London, pp 324–353

Isbell LA (1991) Contest and scramble competition: patterns of aggression and ranging behavior among primates. Behav Ecol 2:143–155

Irwin MT, Raharison J, Raubenheimer DR, Chapman CA, Rothman J (2015) The nutritional geometry of resource scarcity: effects of lean seasons and habitat disturbance on nutrient intakes and balancing in wild sifakas. PLoS One 10(6):e0128046

Izar P, Verderane MP, Peternelli-dos-Santos L, Mendonça-Furtado O, Presotto A, Tokuda M, Visalberghi E, Fragaszy D (2012) Flexible and conservative features of social systems in tufted capuchin monkeys: comparing the socioecology of *Sapajus libidinosus* and *Sapajus nigritus*. Am J Primatol 74(4):315–331

Janson CH (1985) Aggressive competition and individual food consumption in wild brown capuchin monkeys (*Cebus apella*). Behav Ecol Sociobiol 18:125–138

Janson CH (1988) Experimental evidence for spatial memory in wild brown capuchin monkeys (*Cebus apella*). Anim Behav 55:1229–1243

Johnson CA, Raubenheimer D, Rothman JM, Clarke D, Swedell L (2013) 30 days in the life: daily nutrient balancing in a wild chacma baboon. PLoS One 8(7):e70383

Kay RF, Madden RH, Van Schaik C, Higdon D (1997) Primate species richness is determined by plant productivity: implications for conservation. P Natl Acad Sci USA 94(24):13023–13027

Key C, Ross C (1999) Sex differences in energy expenditure in non-human primates. P R Soc London B Bio 266(1437):2479–2485

Koenig A (2002) Competition for resources and its behavioral consequences among female primates. Int J Primatol 23:759–783

Lucas A (1998) Programming by early nutrition: an experimental approach. J Nutr 128(2):401S–406S

MacKinnon KC (2006) Food choice by juvenile capuchin monkeys (*Cebus capucinus*) in a tropical dry forest. In: Estrada A, Pavelka M, Garber P, Luecke L (eds) New perspectives in the study of Mesoamerican primates: distribution, ecology, behavior, and conservation. Springer Press, New York, pp 349–366

Matthews L (2009) Activity patterns, home range size, and intergroup encounters in *Cebus albifrons* support existing models of capuchin socioecology. Int J Primatol 30(5):709–728

McCabe GM (2005) Diet and nutrition in white-faced capuchins (*Cebus capucinus*): effects of group, sex and reproductive state. Thesis, University of Calgary

McCabe GM, Fedigan LM (2007) Effects of reproductive status on energy intake, ingestion rates, and dietary composition of female *Cebus capucinus* at Santa Rosa, Costa Rica. Int J Primatol 28:837–851

McCabe GM, Emery Thompson M (2013) Reproductive seasonality in wild Sanje mangabeys (*Cercocebus sanjei*), Tanzania: relationship between the capital breeding strategy and infant survival. Behaviour 150(12):1399–1429

McCabe G, Fernàndez D (2018, this volume) Patterns of infant mortality and seasonality in wild Sanje mangabeys, *Cercocebus sanjei*. In: Kalbitzer U, Jack KM (eds) Primate life histories, sex roles, and adaptability - Essays in honour of Linda M. Fedigan. Developments in primatology: progress and prospects. Springer, New York, pp 57–70

Melin, AD, Webb SE, Williamson RE, Chiou K (2018) Data Collection in Field Primatology: A Renewed Look at Measuring Foraging Behaviour. In: Kalbitzer U, Jack KM (eds) Primate life histories, sex roles, and adaptability - Essays in honour of Linda M. Fedigan. Developments in primatology: progress and prospects. Springer, New York, pp 161–187

Melin A, Fedigan L, Hiramatsu C, Hiwatashi T, Parr N, Kawamura S (2009) Fig foraging by dichromatic and *trichromatic Cebus capucinus* in a tropical dry forest. Int J Primatol 30(6):753–775

Melin AD, Fedigan LM, Young HC, Kawamura S (2010) Can color vision variation explain sex differences in invertebrate foraging by capuchin monkeys? Curr Zool 56(3):300–312

Melin AD, Hiramatsu C, Parr NA, Matsushita Y, Kawamura S, Fedigan LM (2014a) The behavioral ecology of color vision: considering fruit conspicuity, detection distance, and dietary importance. Int J Primatol 35(1):258–287

Melin AD, Young HC, Mosdossy KN, Fedigan LM (2014b) Seasonality, extractive foraging and the evolution of primate sensorimotor intelligence. J Hum Evol 71(0):77–86

Melin Meachem AD (2011) Polymorphic colour vision and foraging in white-faced capuchins: insights from field research and simulations of monkey vision. Dissertation, University of Calgary

Merrill AL, Watt BK (1955) Derivation of current calorie factors. In: USDA ARS (ed) Agricultural handbook no. 74: energy value of foods, basis and derivation. U.S. Government Printing Office, Washington D.C., p 24–43. https://naldc.nal.usda.gov/download/CAT40000007/. Accessed 31 Jan 2017

Miller KE, Bales KL, Ramos JH, Dietz JM (2006) Energy intake, energy expenditure, and reproductive costs of female wild golden lion tamarins (*Leontopithecus rosalia*). Am J Primatol 68(11):1037–1053

Milton K (1981) Food choice and digestive strategies of two sympatric primate species. The Am Nat 117(4):496–505

Milton K (1984) The role of food-processing factors in primate food choice. In: Rodman PS, Cant JGH (eds) Adaptations for foraging in nonhuman primates: contributions to an organismal biology of prosimians, monkeys and apes. Columbia University Press, New York, pp 249–279

Milton K (1987) Primate diets and gut morphology: implications for hominid evolution. In: Harris M, Ross EB (eds) Food and evolution: toward a theory of food habits. Temple University Press, Philadelphia, pp 93–115

Mosdossy KN, Melin AD, Fedigan LM (2015) Quantifying seasonal fallback on invertebrates, pith and bromeliad leaves by white-faced capuchin monkeys (*Cebus capucinus*) in a tropical dry forest. Am J Phys Anthropol 150:67–77

Mott GE, Jackson EM, McMahan CA, McGill HC Jr (1990) Cholesterol metabolism in adult baboons is influenced by infant diet. J Nutr 120(3):243–225

Mott GE, Lewis DS, McGill HC Jr (1991) Programming of cholesterol metabolism by breast or formula feeding. Ciba Found Symp 156:56–76

Moura ACdA (2004) The capuchin monkey and the Caatinga dry forest: a hard life in a harsh habitat. Dissertation, University of Cambridge

National Research Council (2003) Nutrient requirements of nonhuman primates. National Academy Press, Washington, DC

Norconk MA, Wright BW, Conklin-Brittain NL, Vinyard CJ (2009) Mechanical and nutritional properties of food as factors in platyrrhine dietary adaptations. In: Garber PA, Estrada A, Bicca-Marques JC, Heymann EW, Strier KB (eds) South American primates: comparative perspectives in the study of behavior, ecology, and conservation. Springer Science+Business Media, LLC, New York, pp 279–319

O'Malley R, Fedigan L (2005) Variability in food-processing behaviour among white-faced capuchins (*Cebus capucinus*) in Santa Rosa National Park, Costa Rica. Am J Phys Anthropol 128:63–73

Oftedal OT, Whiten A, Southgate DAT, Van Soest P (1991) The nutritional consequences of foraging in primates: the relationship of nutrient intakes to nutrient requirements. Philos T R Soc B 334(1270):161–170

Perry S (1996) Female-female social relationships in wild white-faced capuchin monkeys, *Cebus capucinus*. Am J Primatol 40(2):167–182

Peters RH, Cloutier S, Dubé D, Evans A, Hastings P, Kaiser H, Kohn D, Sarwer-Foner B (1988) The allometry of the weight of fruit trees and shrubs in Barbados. Oecologia 74(4):612–616

Raubenheimer D (2011) Toward a quantitative nutritional ecology: the right-angled mixture triangle. Ecol Monogr 81(3):407–427

Raubenheimer D, Simpson SJ (1993) The geometry of compensatory feeding in the locust. Anim Behav 45:953–964

Raubenheimer D, Rothman JM, Pontzer H, Simpson SJ (2014) Macronutrient contributions of insects to the diets of hunter–gatherers: a geometric analysis. J Hum Evol 71(0):70–76

Rose LM (1994) Sex differences in diet and foraging behavior in white-faced capuchins (*Cebus capucinus*). Int J Primatol 15(1):95–114

Rothman JM, Raubenheimer D, Chapman CA (2011) Nutritional geometry: gorillas prioritize non-protein energy while consuming surplus protein. Biol Lett 7:847–849

Rothman JM, Raubenheimer D, Bryer MAH, Takahashi M, Gilbert CC (2014) Nutritional contributions of insects to primate diets: implications for primate evolution. J Hum Evol 71:59–69

Schulke O, Chalise MK, Koenig A (2006) The importance of ingestion rates for estimating food quality and energy intake. Am J Phys Anthropol 68:951–965

Simpson SJ, Raubenheimer D (1993) A multi-level analysis of feeding behaviour: the geometry of nutritional decisions. Philos T R Soc B 342(1302):381–402

Smith RJ, Jungers WL (1997) Body mass in comparative primatology. J Hum Evol 32(6):523–559

Sterck EHM, Watts DP, van Schaik CP (1997) The evolution of female social relationships in nonhuman primates. Behav Ecol Sociobiol 41:291–309

Tardif SD, Jaquish CE (1997) Number of ovulations in the marmoset monkey (*Callithrix jacchus*): relation to body weight, age and repeatability. Am J Primatol 42(4):323–329

Terborgh J (1983) Five New World primates. Princeton University Press, Princeton

United States Department of Agriculture (2014) The PLANTS Database. http://plants.usda.gov. Accessed 7 Jul 2014

van Schaik CP, van Noordwijk MA (1989) The special role of male cebus monkeys in predation avoidance and its effect on group composition. Behav Ecol Sociobiol 24(5):265–276 van Schaik CP, Terborgh JW, Wright SJ (1993) The phenology of tropical forests: adaptive significance and consequences for primary consumers. Annu Rev Ecol Syst 24(1):353–377

Van Soest PJ, Robertson JB, Lewis BA (1991) Methods for dietary fiber, neutral deterent fiber, and nonstarch polysaccharides in relation to animal nutrition. J Dairy Sci 74(10):3583–3597

Vogel ER (2004) The ecological basis of aggression in white-faced capuchin monkeys, *Cebus capucinus*, in a Costa Rican dry forest. Dissertation, Stony Brook University

Vogel ER (2005) Rank differences in energy intake rates in white-faced capuchin monkeys, *Cebus capucinus*: the effects of contest competition. Behav Ecol Sociobiol 58:333–344

Vogel ER, Janson CH (2006) Predicting the frequency of food related agonism in white-faced capuchin monkeys (Cebus capucinus), using a novel focal tree method. Am J Primatol 69(5):533–550

Voland E (1998) Evolutionary ecology of human reproduction. Annu Rev Anthropol 27:347–374

Wrangham RW (1980) An ecological model of female-bonded primate groups. Behaviour 75:262–300

Chapter 12
Primate-Plant Mutualisms: Is There Evidence for Primate Fruit Syndromes?

Kim Valenta and Colin A. Chapman

Abstract Many researchers have posited the existence of fruit syndromes – sets of fruit traits such as colour, odour and size that match the behaviour, morphology and sensory adaptations of their key seed dispersal agents. Implicit in this hypothesis is the idea that dispersers have been the selective force behind fruit syndromes, based on their feeding preferences and behaviour. These hypotheses are contentious, as many argue that fruits, unlike flowers, are dispersed by a high diversity of animals and are thus unlikely to converge upon sets of fruit traits that are attractive to only a subset of the potential disperser population. Empirical evidence for the existence of fruit syndromes is mixed. For over 30 years, researchers have identified the traits of "primate fruits" – fruits that are exclusively or primarily dispersed by primates. Here, we review the primate fruit syndrome hypothesis in the history of primate seed dispersal studies. We additionally suggest that because of recent technological advances that allow for the quantification of fruit traits, coupled with the importance of primates as seed dispersers and the relative ease with which it is possible to quantify primate seed dispersal relative to other taxa, primates are excellent taxa with which to test the fruit syndrome hypothesis.

K. Valenta (✉)
Department of Anthropology and McGill School of Environment, McGill University, Montreal, QC, Canada

Makerere University Biological Field Station, Fort Portal, Uganda

Duke University, Department of Evolutionary Anthropology, Durham, NC, USA
e-mail: kim.valenta@duke.edu

C. A. Chapman
Department of Anthropology and McGill School of Environment, McGill University, Montreal, QC, Canada

Makerere University Biological Field Station, Fort Portal, Uganda

Wildlife Conservation Society, Bronx, NY, USA

Section of Social Systems Evolution, Primate Research Institute, Kyoto University, Kyoto, Japan

U. Kalbitzer, K. M. Jack (eds.), *Primate Life Histories, Sex Roles, and Adaptability*, Developments in Primatology: Progress and Prospects,
https://doi.org/10.1007/978-3-319-98285-4_12

Keywords Fruit syndromes · Colour · Odour · Seed dispersal · Colour vision

12.1 Introduction

Fruiting plants and animals are involved in complex and often mutualistic interactions, with animals relying on fruiting plants for food and fruiting plants relying on animals for seed dispersal (Howe and Smallwood 1982). Plant reliance on animals for reproduction is particularly important in tropical forests, where it is estimated that vertebrates disperse seeds from 70% to 94% of woody species (Jordano 2000). The interdependence of fruiting plants and frugivorous animals has led some researchers to characterize fruit-frugivore mutualisms as co-evolutionary relationships – interspecific interactions in which an evolutionary trait change in one population is a response to traits in the second population, followed by an evolutionary response by the second population to the changes in the first (Janzen 1980). A key proposition stemming from the proposed co-evolutionary nature of this plant-animal interaction is the fruit syndrome hypothesis. Fruit syndromes are suites of fruit traits, like colour, odour and size, that are well-matched to the behaviour, morphology, physiology and sensory adaptations of the animals that disperse their seeds (Janson 1983).

Because the definition of a fruit syndrome hinges on plant traits that have adapted to seed dispersing animals, fruit syndromes are necessarily localized phenomena – only those animals which provide a reproductive benefit to plants via seed dispersal are potentially contributing to the evolution of fruit traits aimed at attracting those animals. Thus, the question of the existence of primate fruit syndromes is necessarily a spatially and temporally discrete one. And because primates as an order are highly diverse morphologically, and have a great range of sensory adaptations (Jacobs 2008; Nevo and Heymann 2015), the evolution of primate fruit syndromes should be considered in light of the frugivorous primates in a given system, and not primates universally. Effectively, the question of whether or not primate fruit syndromes exist is a question of whether or not primates in a given system are exerting sufficient selective pressure on plants via seed dispersal as to alter the evolutionary trajectories of those plants.

Interestingly, while the first empirical work on seed dispersal syndromes centred on primate-dispersed fruits (Janson 1983; Gautier-Hion et al. 1985), and much of the current work addressing seed dispersal syndromes is focused on primate-dispersed fruits (Valenta et al. 2013; Nevo et al. 2015), in the decades between, primates were only rarely considered relative to other taxa (e.g. birds, bats) in the dispersal syndrome literature. In part, this may result from the fact that it was not until the 1990s that primates were thought to be important seed dispersers, which is reflected in the paucity of primate seed dispersal studies prior to that time (Fig. 12.1). Indeed, in van der Pijl's (1969) seminal work defining seed dispersal syndromes, he noted that primates "…are mostly destructive, eating everything edible… they may or may not be instrumental in dispersal" (p. 46). In the 1990s, so many researchers

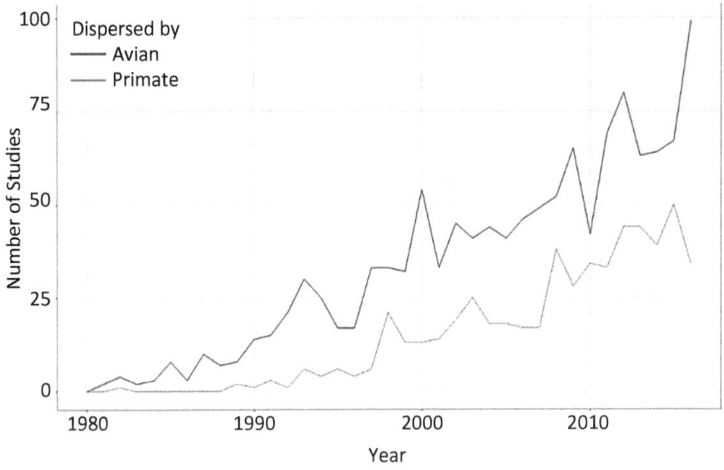

Fig. 12.1 Results of search in Google Scholar for "primate seed dispersal" ($N = 553$) and "avian seed dispersal" ($N = 1238$) by year between 1980 and 2016

turned to the subject of primate seed dispersal that the *American Journal of Primatology* presented a special issue on the topic in 1998 (Lambert and Garber 1998). While the special issue contained details of the ecological importance of primates in multiple systems, the editors of the special issue expressed doubt regarding the potential for primates to exert strong selective pressures on fruit traits, writing that "although it is clear that primates are likely to have an important ecological impact on seed fate and the movement of seeds away from the parent tree, a major point raised in several papers in this volume is that a co-evolutionary relationship between a given primate taxon and the fruit species it exploits is extremely unlikely" (Lambert and Garber 1998, p. 7).

Here, we first discuss the history of the fruit syndrome hypothesis in the primatological literature, considering the seminal research behind it, as well as the well-stated and appealing logic of the hypothesis. Secondly, we review the evidence that arose to challenge the hypothesis, which we propose has left the field in a state of theoretical limbo for the last 30 years. Finally, we propose ways forward that may break this deadlock by bringing new empirical tools to the question of primate-plant mutualisms and their ecological and evolutionary outcomes.

12.2 The Development of the Fruit Syndrome Hypothesis

One of the first treatments linking seed dispersal to fruit phenotypes is found in the book *The Dispersal of Plants Throughout the World* (Ridley 1930). This work presented exhaustive empirical descriptions of different forms of seed dispersal – dispersal by wind, water, adhesion to animal fur and ingestion and subsequent defecation by animals. Importantly, the author also detailed the morphological traits associated with each dispersal method (Ridley 1930). While Ridley did not explicitly infer co-evolutionary relationships between fruits and the animals that disperse their seeds, 40 years later, his observations of seed dispersal mechanisms and their associated morphological traits were used by van der Pijl (1969) to define water, wind and animal dispersal syndromes – clusters of co-occurring traits based on dispersal mode. Soon afterwards, following field research on primates in the tropical rainforests of Peru, Charles Janson further parsed the category of fruits dispersed by animal ingestion, noting that fruits dispersed by mammals – primarily primates – differed in several key aspects when compared to fruits dispersed by birds (Janson 1983). He found that bird-dispersed fruits tended to be small, unprotected by a husk and red, white, black or multicoloured, while mammal-dispersed fruits tended to be large, protected by a husk and orange, brown, yellow or green (Janson 1983).

At the same time as Charles Janson was working on this evolutionary question in Peru, Annie Gautier-Hion and her colleagues were investigating them in Gabon (Gautier-Hion 1984; Gautier-Hion et al. 1985). Gautier-Hion and colleagues suggested that suites of fruit characteristics evolved as a result of pressure from seed dispersers and seed predators, and they suggested that fruit traits are more diverse amongst mammal-dispersed plants than between mammal- and bird-dispersed plants, causing them to lump monkey- and bird-dispersed species together. Unlike Janson, this study identified monkey- and bird-dispersed fruits as brightly coloured and either arillate or containing succulent flesh. While the results of these two studies were quite different, these seminal publications together form the basis for what is today known as the fruit syndrome hypothesis – a hypothesis that, while intuitive, is also highly controversial (Herrera 1985; Lomascolo et al. 2010).

12.3 Arguments Against the Fruit Syndrome Hypothesis

The controversy surrounding the fruit syndrome hypothesis has to do with a number of evolutionary and ecological issues. First, plant and animal evolution can occur on very different time scales. Woody angiosperms are slower to reach reproductive age and persist much longer than most animals. Thus, multiple generations of an animal species may coexist with an individual tree, and co-evolution will be constrained by this temporal mismatch. Secondly, relative to animals, the gene flow of most tropical trees is very extensive, and some have argued that it is unlikely that animals are able to exert strong consistent selective pressure on fruit traits across a populations range (Lomascolo et al. 2010), though evidence has been found that selection by animals can lead to rapid phenotypic changes over short periods of time (Galetti et al. 2013; Brodie 2017). Third, in many tropical systems, there are many species of animals (e.g. bats, birds and monkeys) with a wide variety of sensory systems that disperse the seeds of the same fruiting species (Chapman 1995; Chapman and Chapman 1996; Lambert and Garber 1998). Some have argued that given the diversity of frugivores and their phenotypes, any selective pressure on plants to specialize on attracting any single disperser species, or any group of closely related species with similar phenotypes (e.g. all diurnal monkeys in a system), is weak, as this should be countered by selective pressure from other species (Herrera 1982).

The controversy surrounding the fruit syndrome hypothesis may also partially result from the difficulty of inferring selective pressure on plants given the long time scale of plant reproduction, particularly woody angiosperms. To date, the vast majority of research on primate seed dispersal has provided compelling evidence that monkeys, apes and lemurs remove large numbers of fruits from tropical trees and the seeds they defecate are viable (Lieberman et al. 1979; Lieberman and Lieberman 1986; Corlett and Lucas 1990; Wrangham et al. 1994; Estrada and Coates-Estrada 1996; Julliot 1996; Lambert 1997; Dew and Wright 1998). However, only rarely are data on fruit removal and seed handling linked with data on post-dispersal seed fate to fully understand the potential selective importance of primates. Ideally, studies of primate-dispersed fruits would include data on plant life stages from seed ingestion by primates, through germination, establishment and survival until the age of reproduction. However, given the slow maturation time of woody angiosperms – that can include long periods of dormancy and arrested development – this can take many decades (Connell and Green 2000; Zanne et al. 2005), which is well outside of the logistical capabilities of many research programs.

In addition to theoretical issues with the fruit syndrome hypothesis, several empirical studies seeking evidence of fruit syndromes, particularly those that did so using phylogenetic methods, failed to detect clear evidence for syndromes (Herrera 1987, 1992; Fischer and Chapman 1993; Jordano 1995; Valenta et al. 2016a). Some argue that perhaps the lack of a tight match between fruit traits and frugivores results from diffuse co-evolution (Fleming et al. 1987). Under diffuse co-evolution, rather than paired co-evolutionary trajectories, selective pressure from multiple mutualists is expected to lead to highly generalized suites of traits (Iwao and Rausher 1996). In the case of animal seed dispersal, diffuse co-evolution may describe any system wherein multiple animals with diverse sensory phenotypes and/or preferences disperse the seeds of a given tree species. In these cases, since many different species of birds, mammals, and invertebrates can disperse the seeds of the same plant species, there should be little selective pressure for a plant to evolve specialized fruit morphology to increase detection by just one disperser species or guild of disperser species. While diffuse co-evolution doesn't necessarily preclude the existence of fruit syndromes, the resulting prediction is that fruits will converge upon the production of signals that are salient to multiple taxa, rather than just one. This may be a successful strategy from the point of view of the plant, because when more than one species offers equivalent ecological services, then the loss of one of those species should be offset by the presence of the remaining species (Loiselle et al. 2007). Given the large differences in the evolutionary rates of tropical trees and the animals that are reported to disperse their seeds, this may be a strategy that plants widely adopt (Herrera 1985; Barraclough and Savolainen 2001). In these cases of so-called ecological redundancy, plants can be expected to exhibit adaptations and strategies favouring dispersal by multiple species (Walker 1992; Loiselle et al. 2007; Valenta et al. 2015a), and the relative contribution of a single taxa, or phenotypically similar taxa, becomes difficult to parse out.

12.4 Evidence for the Fruit Syndrome Hypothesis in Primatology

Despite theoretical problems and logistical constraints, there is some clear and compelling evidence that frugivorous primates may exert sufficient selective pressure to shape fruit phenotypes. Plants are sessile organisms that rely on the dispersal of their seeds for successful reproduction, and they have been found to evolve mechanisms to increase the likelihood of attracting effective dispersal vectors (Lomascolo et al. 2010). Frugivores affect plant fitness (Harms et al. 2001; Howe and Miriti 2004), and frugivorous primates have morphological and behavioural adaptations that enable them to successfully exploit fruit (Lambert 1997). Additionally, primate's preferences for certain suites of fruit traits have been documented, and fruit and seed morphology has been linked to specific guilds of frugivores (Valenta et al. 2016b). Compellingly, several parameters of fruit and seed morphology have been found to affect fruit choice by frugivorous primates, including exocarp hardness (Janson 1983), ripe fruit quantity (Curran and Leighton 2000), phenological stage (van Schaik et al. 1993; Chapman et al. 2005), chemical fruit traits (Worman and Chapman 2005), fruit odour (Valenta et al. 2013; Nevo et al. 2015; Nevo et al. 2016), pulp richness and seed size (Chapman et al. 1992; Zanne et al. 2005), fruit size (Janson 1983) and fruit colour (Stevens et al. 2009; Valenta et al. 2013; Melin et al. 2014; Valenta 2014; Valenta et al. 2015a). If these primate fruit preferences translate into increased reproductive success for plants, then primate seed dispersal should affect fruit traits.

While several studies have found that primates show clear preferences for certain fruit traits during foraging, linking these preferences with plant reproductive outcomes can be daunting, but not impossible. For example, a study on the fate of the seeds of *Monodora myristica* dispersed by primates illustrate the importance of understanding seed fate and its relationship to primary seed dispersal (Balcomb and Chapman 2003). The fruit of *Monodora myristica* is extremely large, with an average fruit diameter of 18.5 cm, multiple large seeds over 2 cm in length and a very thick and difficult-to-penetrate pericarp (Hamilton 1991; Lambert 1997). The fruit morphology of this species suggests dispersal by only the largest arboreal frugivores. Balcomb and Chapman (2003) documented fruit and seed removal of *M. myristica* during tree focal sampling and quantified the fate of dispersed and fallen seeds, seedling, sapling, and pole densities and survivorship of this species. These data were collected over 2 years at two sites to assess spatial and temporal variation. They found that large-bodied primates played a critical role in primary seed dispersal; chimpanzees (*Pan troglodytes*) and mangabeys (*Lophocebus albigena*) were the only frugivores that opened the hard-husked fruits, and they dispersed over 85% of the mature seeds that trees produced in a year. Another study of capuchin monkey-dispersed fruits in Costa Rica found that fruits of three species which were dominant in the diet of capuchin monkeys and had higher germination probabilities after ingestion by monkeys also had a higher chance of survival and germination at monkey deposition sites than seeds under parent tree canopies, up to 1 year after deposition (Valenta and Fedigan 2010).

Theoretically, if primate contributions to the dispersal and post-dispersal success of seeds are singular, or greater than those of non-primate mutualists, fruit traits should be under selection to maximize traits that are attractive to primates. However, understanding the ecological and evolutionary outcomes of primate frugivory vis-à-vis plants requires quantifying primate sensory phenotypes, behavioural preferences during fruit foraging and, ideally, fitness outcomes for primate-dispersed and non-primate-dispersed fruits of plant species.

12.5 Future Directions

The current deadlock between the intuitive appeal of the primate fruit syndrome hypothesis, the theoretical controversies discussed here and the logistical constraints of testing the hypothesis will not be easily overcome. However, we present two suggestions of how to advance testing this hypothesis. First, ideally data on the fitness consequences of each dispersal event affecting the population of trees of interest could be quantified. However, given the diversity of frugivores often visiting a tree, the difficulty inherent in quantifying post-dispersal survivorship, the slow life history of plants and the often large geographical range of tropical trees, this is very difficult if not logistically impossible. However, if one assumes that the critical life history stage for tropical tree recruitment is dispersal and early establishment, the logistics of tests becomes manageable. Making this assumption it becomes possible to quantitatively contrast the effectiveness of the different types of dispersers. If such studies are conducted in a few locations that represent the geographical spread of the tree species, then it is possible to characterize the relative selection pressure of the different disperser types and test the concordance between a disperser's contribution to the next generation of the tree species and the traits of the fruits.

The second promising advancement in studies of fruit syndromes is the quantification of traits. For example, in early studies, fruit colour was usually described according to a series of human-relevant categories like red, yellow and blue (Gautier-Hion et al. 1985). Similarly, fruit odour was often treated categorically (e.g. musky, sweet) or as a binary (e.g. odorous, non-odorous (Fischer and Chapman 1993; Tamboia et al. 1996; Nevo and Heymann 2015). Thus, descriptions of fruit traits were qualitatively rather than quantitatively assessed based on human's limited sensory capacities to receive them. Such subjective assessments of fruit traits can lead to both identifying fruit traits that are not salient to the intended receiver (e.g. classifying fruit colour as red in a system with colour-blind frugivores) and failing to identify fruit traits that are relevant to the intended receiver (e.g. reflectance in the ultraviolet range of the spectrum that mammals cannot detect).

Technological advances in the measurement of reflectance spectra mean that fruit colour can now be measured quantitatively and compared to the actual visual capabilities of frugivores (Stevens et al. 2007; Stevens et al. 2009). The few studies that have quantified fruit colouration in relation to foraging and feeding behaviour have found mixed evidence about the role of fruit colour in primate fruit choice and foraging efficiency (Riba-Hernández et al. 2005; Hiramatsu et al. 2008; Melin et al. 2009; Valenta et al. 2015b). However, the quantification of fruit colour and its analysis in light of primate colour vision phenotypes are an exciting technological advancement that has the potential to yield important new insights into the evolutionary and ecological outcomes of primate-plant mutualisms.

In addition to quantifying colour, research into primate frugivory has also adopted methods to quantify and qualify fruit odours via active sampling and analytical chemistry (Valenta et al. 2013). These techniques involve capturing volatile organic compounds (VOCs) in sorbent media, extracting VOCs using a gas chromatographer-mass spectrometer (GC-MS) and analysing the resulting spectrographs to identify the overall amount of odorants produced by a fruit, as well as the individual compounds that comprise those odorants (Valenta et al. 2013). Because of the high number of individual compounds that fruits produce, the process of identifying them can be daunting, but not impossible (Nevo et al. 2016). The next step in this research program will be linking the presence of odorant compounds to primate's ability to detect them, either molecularly or via behavioural choice experiments. Overall, the role of odour in primate foraging, and seed dispersal, has great potential to inform the ongoing debate about the merits of the fruit syndrome hypothesis.

The fruit syndrome hypothesis was proposed over a century ago, and it received wide acceptance despite poor data to support it. Given that primate foraging is relatively easy to quantify relative to other dispersers, that they often disperse large seeds that are suitable for germination and establishment experiments and that there are new technologies to quantify colour and odour, it is an exciting time for studies of primate fruit foraging and seed dispersal. We encourage researchers to adopt our suggested approach and methods and improve upon them to test the primate fruit syndrome hypothesis. This may produce a database that is sufficiently robust to allow a rigorous comparative test of this intriguing hypothesis.

Acknowledgements We dedicate this paper to Dr. Linda M. Fedigan who was influential in the careers of both authors despite the fact that they represent two generations of scientists. We are grateful to Dr. Eckhard Heymann and one anonymous reviewer for their helpful comments to improve the manuscript. Funding during the preparation of the manuscript was provided by the Canada Research Chairs Program, Natural Sciences and Engineering Research Council of Canada and the Fonds Québécois de la Recherché sur la Nature et les Technologies.

References

Balcomb SR, Chapman CA (2003) Bridging the gap: influence of seed deposition on seedling recruitment in a primate-tree interaction. Ecol Monogr 73(4):625–642

Barraclough TG, Savolainen V (2001) Evolutionary rates and species diversity in flowering plants. Evolution 55(4):677–683

Brodie JF (2017) Evolutionary cascades induced by large frugivores. Proc Natl Acad Sci 114(45):11998–12002

Chapman CA (1995) Primate seed dispersal: coevolution and conservation implications. Evol Anthropol 4:74–82
Chapman CA, Chapman LJ (1996) Frugivory and the fate of dispersed and non-dispersed seeds of six African tree species. J Trop Ecol 12:491–504
Chapman LJ, Chapman CA, Wrangham RW (1992) *Balanites-wilsoniana:* elephant dependent dispersal. J Trop Ecol 8:275–283
Chapman CA, Chapman LJ, Struhsaker TT, Zanne AE, Clark CJ, Poulsen JR (2005) A long-term evaluation of fruiting phenology: importance of climate change. J Trop Ecol 21:31–45
Connell JH, Green PT (2000) Seedling dynamics over thirty-two years in a tropical rain forest tree. Ecology 81(2):568–584
Corlett RT, Lucas PW (1990) Alternative seed-handling strategies in primates - seed spitting by long-tailed macaques (*Macaca fascicularis*). Oecologia 82:166–171
Curran LM, Leighton M (2000) Vertebrate responses to spatiotemporal variation in seed production of mast-fruiting dipterocarpaceae. Ecol Monogr 70(1):101–128
Dew JL, Wright P (1998) Frugivory and seed dispersal by four species of primates in Madagascar's eastern rain forest. Biotropica 30(3):425–437
Estrada A, Coates-Estrada R (1996) Frugivory by howling monkeys (*Alouatta palliata*) at los Tuxtlas, Mexico: dispersal and the fate of seeds. In: Estrada A, Fleming TH (eds) Frugivores and seed dispersal. Dordrecht Publishers, New York
Fischer K, Chapman C (1993) Frugivores and fruit syndromes: differences in patterns at the genus and species level. Oikos 66(3):472–482
Fleming T, Breitwisch R, Whitesides G (1987) Patterns of tropical vertebrate frugivore diversity. Annu Rev Ecol Syst:91–109
Galetti M, Guevara R, Côrtes MC, Fadini R, Von Matter S, Leite AB, Labecca F, Ribeiro T, Carvalho CS, Collevatti RG (2013) Functional extinction of birds drives rapid evolutionary changes in seed size. Science 340(6136):1086–1090
Gautier-Hion A (1984) La dissemination des graines par les cercopithecides forestier Africans. Terre et Vie 39:159–165
Gautier-Hion A, Duplantier JM, Quris R, Feer F, Sourd C, Decous JP, Doubost G, Emmons L, Ehrard C, Hecketswiler P, Moungazi A, Roussilhon C, Thiollay JM (1985) Fruit characters as a basis of fruit choice and seed dispersal in a tropical forest vertebrate community. Oecologia 65:324–337
Hamilton AC (1991) A field guide to Uganda forest trees. Makerere University Printery, Kampala
Harms KE, Condit R, Hubbell SP, Foster RB (2001) Habitat associations of trees and shrubs in a 50-ha neotropical forest plot. J Ecol 89:947–959
Herrera CM (1982) Seasonal variations in the quality of fruits and diffuse coevolution between plants and avian dispersers. Ecology 63:772–785
Herrera C (1985) Determinants of plant-animal coevolution: the case of mutualistic dispersal of seeds by vertebrates. Oikos 44:132–141
Herrera C (1987) Vertebrate-dispersed plants of the Iberian Peninsula: a study of fruit characteristics. Ecol Monogr 57(4):305–331
Herrera C (1992) Interspecific variation in fruit shape: allometry, phylogeny, and adaptation to dispersal agents. Ecology 73(5):1832–1841
Hiramatsu C, Melin AD, Aureli F, Schaffner CM, Vorobyev M, Matsumoto Y, Kawamura S (2008) Importance of achromatic contrast in short-range fruit foraging of primates. PLoS One 3:e3356
Howe HF, Miriti MN (2004) When seed dispersal matters. Bioscience 54(7):651–660
Howe HF, Smallwood J (1982) Ecology of seed dispersal. Annu Rev Ecol Syst 13:201–228
Iwao K, Rausher MD (1996) Evolution of plant resistance to multiple herbivores: quantifying diffuse coevolution. Am Nat 149:316–335
Jacobs GH (2008) Primate color vision: a comparative perspective. Vis Neurosci 25:619–633
Janson CH (1983) Adaptation of fruit morphology to dispersal agents in a neotropical rainforest. Science 219:187–189
Janzen DH (1980) When is it coevolution? Evolution 34:611–612

Jordano P (1995) Angiosperm fleshy fruits and seed dispersers: a comparative analysis of adaptation and constraints in plant-animal interactions. Am Nat 145(2):163–191
Jordano P (2000) Fruits and frugivory. In: Fenner M (ed) Seeds: the ecology of regeneration in plant communities. CAB International, Wallingford, pp 125–165
Julliot C (1996) Seed dispersal by red howling monkeys (Alouatta seniculus) in the tropical rain forest of French Guiana. Int J Primatol 17(2):239–258
Lambert JE (1997) Fruit processing and seed dispersal by chimpanzees (*Pan troglodytes schweinfurthii*) and redtail monkeys (*Cercopithecus ascanius schmidti*) in the Kibale National Park, Uganda. University of Illinois, Urbana
Lambert JE, Garber PA (1998) Evolutionary and ecological implications of primate seed dispersal. Am J Primatol 45:9–28
Lieberman M, Lieberman D (1986) An experimental study of seed ingestion and germination in a plant-animal assemblage in Ghana. J Trop Ecol 2(02):113–126
Lieberman D, Hall J, Swaine M, Lieberman M (1979) Seed dispersal by baboons in the Shai Hills, Ghana. Ecology 60(1):65–75
Loiselle BA, Blendinger PG, Blake JG, Ryder TB (2007) Ecological redundancy in seed dispersal systems: a comparison between Manakins (Aves: Pipridae) in two tropical forests. In: Dennis AJ, Schupp EW, Green R, Westcott DW (eds) Seed dispersal: theory and its application in a changing world. CABI Publishing, Oxfordshire, pp 178–196
Lomascolo SB, Levey D, Kimball RT, Bolker BM, Alborn HT (2010) Dispersers shape fruit diversity in *Ficus* (Moraceae). Proc Natl Acad Sci 107:14668–14672
Melin AD, Fedigan LM, Hiramatsu C, Hiwatashi T, Parr N, Kawamura S (2009) Fig foraging by dichromatic and trichromatic *Cebus capucinus* in a tropical dry forest. Int J Primatol 30(6):753–775
Melin AD, Hiramatsu C, Parr NA, Matsushita Y, Kawamura S, Fedigan LM (2014) The behavioral ecology of color vision: considering fruit conspicuity, detection distance and dietary importance. Int J Primatol 35(1):258–287
Nevo O, Heymann EW (2015) Led by the nose: olfaction in primate feeding ecology. Evolutionary Anthropology: Issues, News, and Reviews 24(4):137–148
Nevo O, Garri RO, Salazar LTH, Schulz S, Heymann E, Ayasse M, Laska M (2015) Chemical recognition of fruit ripeness in spider monkeys (*Ateles geoffroyi*). Sci Rep 5:14895
Nevo O, Heymann EW, Schulz S, Ayasse M (2016) Fruit odor as a ripeness signal for seed-dispersing primates? A case study on four Neotropical plant species. J Chem Ecol 42(4):323–328
Riba-Hernández P, Stoner KE, Lucas PW (2005) Sugar concentration of fruits and their detection via color in the Central American spider monkey (*Ateles geoffroyi*). Am J Primatol 67(4):411–423
Ridley HN (1930) The dispersal of plants throughout the world. Reeve, Ashford
Stevens M, Paraga CA, Cuthill IC, Partridge JC, Troscianko TOMS (2007) Using digital photography to study animal coloration. Biol J Linn Soc 90(2):211–237
Stevens M, Stoddard MC, Higham JP (2009) Studying primate color: towards visual system-dependent methods. Int J Primatol 30(6):893–917
Tamboia T, Cipollini ML, Levey DJ (1996) An evaluation of vertebrate seed dispersal syndromes in four species of black nightshade (Solanum sect. Solanum). Oecologia 107(4):522–532
Valenta K (2014) Endemic fruit signals in Madagascar drive variation in Eulemur fulvus foraging behaviour and efficiency. Toronto,
Valenta K, Fedigan LM (2010) Spatial patterns of seed dispersal by white-faced capuchins in Costa Rica: evaluating distant-dependent seed mortality. Biotropica 42(2):223–228
Valenta K, Burke RJ, Styler SA, Jackson DA, Melin AD, Lehman SM (2013) Colour and odour drive fruit selection and seed dispersal by mouse lemurs. Sci Rep 3:1–5
Valenta K, Brown KA, Melin AD, Monckton SK, Styler SA, Jackson DA, Chapman CA (2015a) It's not easy being blue: are there olfactory and visual trade-offs in plant signalling? PLoS One 10(6):e0131725

Valenta K, Brown KA, Rafaliarison RR, Styler SA, Jackson D, Lehman SM, Chapman CA, Melin AD (2015b) Sensory integration during foraging: the importance of fruit hardness, colour, and odour to brown lemurs. Behav Ecol Sociobiol 69(11):1855–1865
Valenta K, Miller CN, Monckton SK, Melin AD, Lehman SM, Styler SA, Jackson DA, Chapman CA, Lawes MJ (2016a) Fruit ripening signals and cues in a Madagascan dry forest: haptic indicators reliably indicate fruit ripeness to dichromatic lemurs. Evol Biol:1–12
Valenta K, Nevo O, Martel C, Chapman CA (2016b) Plant attractants: integrating insights from pollination and seed dispersal ecology. Evol Ecol: 1–19
Van der Pijl L (1969) Evolutionary action of tropical animals on the reproduction of plants. Biol J Linn Soc 1(1-2):85–96
van Schaik CP, Terborgh JW, Wright SJ (1993) The phenology of tropical forests: adaptive significance and consequences for primary consumers. Annu Rev Ecol Syst 24:353–377
Walker BH (1992) Biodiversity and ecological redundancy. Conserv Biol 6(1):18–23
Worman CO, Chapman CA (2005) Seasonal variation in the quality of a tropical ripe fruit and the response of three frugivores. J Trop Ecol 21:689–697
Wrangham RW, Chapman CA, Chapman LJ (1994) Seed dispersal by forest chimpanzees in Uganda. J Trop Ecol 10:355–368
Zanne AE, Chapman CA, Kitajima K (2005) Evolutionary and ecological correlates of early seedling morphology in east African trees and shrubs. Am J Bot 92(6):972–978

Chapter 13
Colour Vision Genetics Learned from New World Monkeys in Santa Rosa, Costa Rica

Shoji Kawamura

Abstract By the mid-1980s, it was known that colour vision was polymorphic in New World monkeys due to allelic variation of the L/M opsin gene. However, until the early 2000s, it was unknown whether this polymorphism existed within social groups of wild monkeys, other than mixed-species troops of tamarins. In 2003, I embarked on a collaborative project with Linda Fedigan and colleagues in Santa Rosa National Park. We collected faecal samples from white-faced capuchin monkeys (*Cebus capucinus*) and black-handed spider monkeys (*Ateles geoffroyi*) that were individually identified by researchers. The major findings of our genetic studies were (1) the confirmation of the allelic polymorphism of the L/M opsin within social groups of each of the two species, (2) the discovery of a novel spectral tuning mechanism in ateline L/M alleles, (3) population genetic evidence for balancing selection on the L/M opsin alleles in the two species, (4) unequal allele frequencies of L/M opsins and (5) the discovery of hybrid L/M opsins in sympatric howler monkeys. Of equal importance has been the ecological side of our colour vision study. In this chapter I summarize basic knowledge on colour vision and visual opsin genes in primates and then describe the contribution of our studies in Santa Rosa to our understanding of primate colour vision evolution.

Keywords Colour vision · Opsin · Capuchin monkeys · Spider monkeys · Polymorphism

S. Kawamura (✉)
Department of Integrated Biosciences, Graduate School of Frontier Sciences, The University of Tokyo, Kashiwa, Chiba, Japan
e-mail: kawamura@k.u-tokyo.ac.jp

U. Kalbitzer, K. M. Jack (eds.), *Primate Life Histories, Sex Roles, and Adaptability*, Developments in Primatology: Progress and Prospects,
https://doi.org/10.1007/978-3-319-98285-4_13

13.1 Summary of Basic Knowledge on Primate Colour Vision and Visual Opsin Genes

13.1.1 Specialization of Visual System in Primates

The visual system of primates is characterized by forward-facing eyes, which enables stereoscopic vision (Fleagle 2013). This is one of the primate pattern traits, together with grasping hands and feet with nails and an opposable thumb/toe, retention of the collar bone allowing for a flexible forelimb movement and the enlarged brain, which are essential for agile movement and jumping locomotion in predominantly arboreal and highly social life (Lambert 1987). The visual system of anthropoids and, in a less developed form, tarsiers is further characterized by the postorbital septum, fovea and increased representation of the visual centres in the brain cortex. The postorbital septum serves to improve visual acuity by preventing the chewing muscles from disrupting eye position (Heesy et al. 2007). The fovea is a retinal region where the inner retinal layers migrate laterally to create a pit, and the cone photoreceptor cells migrate centrally to increase the density for high visual acuity.

The unique evolution of trichromatic colour vision also characterizes primates, which arose from a dichromatic ancestor (Davies et al. 2012). Colour vision is based on the ability to distinguish light by wavelength (hue), and two or more spectral classes of cone photoreceptors are required in the retina for colour vision. The number of distinguishable colours increases as the number of spectrally distinct photoreceptors increases and as the spectral overlap among them is reduced (Vorobyev 2004).

13.1.2 L/M and S Opsins

The visual sensory cells of vertebrates are the rod and cone cells in the retina. The rods work for dim light vision and cones work for daylight and colour vision. Visual opsins are protein moiety of photosensory molecules (visual pigments) produced in the visual sensory cells. Visual opsins are classified into five phylogenetic types in vertebrates, RH1 (rhodopsin or rod opsin for dim light vision) and four cone opsins: RH2 (rhodopsin-like or green opsin), SWS1 (short-wavelength-sensitive type 1 or ultraviolet blue opsin), SWS2 (short-wavelength-sensitive type 2 or blue opsin) and M/LWS (middle- to long-wavelength-sensitive or red-green opsin) (Yokoyama 2000). These five types arose in the common ancestor of all vertebrates including jawless fish (Yokoyama 2000; Collin et al. 2003; Pisani et al. 2006; Collin et al. 2009; Davies et al. 2009a, b, 2012), which implies that early vertebrates already had four-dimensional colour vision via the four cone opsin types (tetrachromacy). Placental and marsupial mammals (therians) maintain only two types of cone visual opsins, SWS1 and M/LWS, in addition to the RH1 rod opsin, and are hence basically dichromatic in colour vision (Jacobs 1993). However, mono- and trichromatic exceptions are also known (Arrese et al. 2002; Ebeling et al. 2010; Jacobs 2013). Monotremes (platypuses and echidnas) also maintain two types of cone opsins, but theirs are SWS2 and M/LWS (Davies et al. 2007; Wakefield et al. 2008). Conventionally, in the case of therians, SWS1 opsin is called "S" or "blue" opsin, with the wavelength of maximal absorption (λ_{max}) at around 410–430 nm, and M/LWS opsins are collectively called "L/M" or "red-green" opsins, with λ_{max} at around 530–560 nm among primates (Kawamura et al. 2012). While the S opsin gene is autosomal, the L/M opsin genes are X chromosomal.

Primates are the exception among placental mammals in attaining trichromatic colour vision (Davies et al. 2012). This is achieved either by differentiating alleles in spectral sensitivity or juxtaposing spectrally distinct loci of the L/M opsin gene on the X chromosome (Fig. 13.1). Allelic differentiation is seen in several lemuriform primates (Tan and Li 1999; Heesy and Ross 2001; Jacobs et al. 2002; Jacobs and Deegan 2003; Veilleux and Bolnick 2009; Veilleux et al. 2016; Jacobs et al. 2017) and in the majority of New World monkeys (Jacobs 1984; Mollon et al. 1984; Jacobs 2007; Matsumoto et al. 2014). In this system, all males are dichromatic, but females are either dichromatic or trichromatic ("allelic" or "polymorphic" trichromacy) (Fig. 13.2). In catarrhine primates (Old World monkeys, apes and humans)

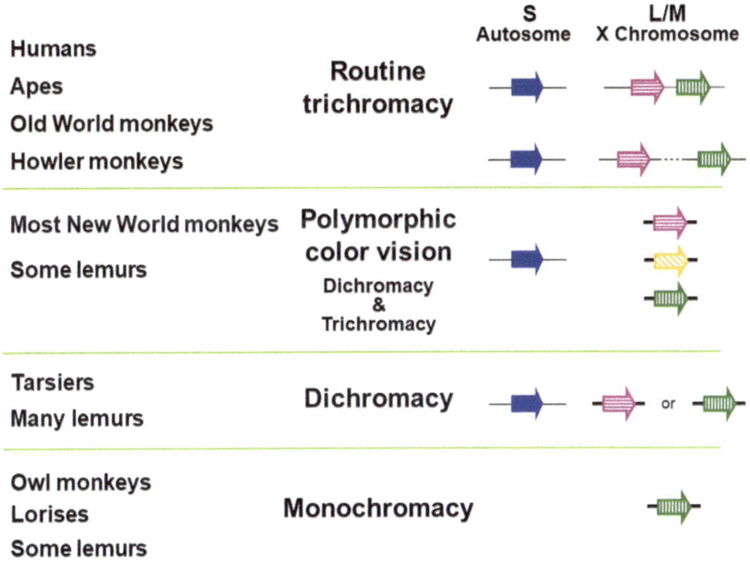

Fig. 13.1 Composition of the L/M and S opsin genes and colour vision types in primates. The S opsin gene is located on an autosome (the seventh chromosome in humans), and the L/M opsin gene is located on the X chromosome in therians. Routine trichromacy is attained independently in catarrhines and in howler monkeys by juxtaposition of the L and M opsin genes. Polymorphic colour vision (trichromacy and dichromacy) is attained by allelic polymorphism of the L/M opsin gene in the New World monkeys (with the exception of howler monkeys and owl monkeys) and some lemurs. The extant tarsiers and many lemur species are dichromatic due to a monomorphic L/M opsin gene. The owl monkeys, lorises and some lemurs are monochromatic due to a monomorphic L/M opsin gene and the loss of functional S opsin gene

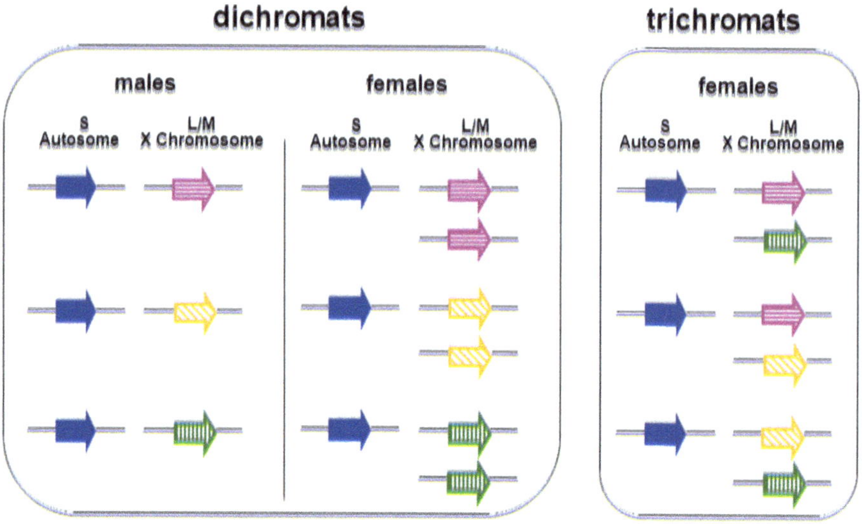

Fig. 13.2 Example of colour vision variation created by allelic polymorphism of L/M opsins. In this example, three alleles of the L/M opsin genes are assumed. All males are dichromatic due to the hemizygosity of X chromosome. However, there are three dichromatic phenotypes among males. Females with homozygous L/M opsins are also dichromatic like the males. Females with heterozygous L/M opsins are trichromatic. Three trichromatic phenotypes are possible among females. In total six colour vision types are possible because of the presence of three alleles of the L/M opsin gene

and the New World monkey genus *Alouatta* (howler monkeys), trichromacy was achieved through juxtaposition (duplication) of the spectrally differentiated L/M opsin genes on the same X chromosome (Jacobs et al. 1996a; Dulai et al. 1999; Surridge et al. 2003; Jacobs and Nathans 2009; Matsushita et al. 2014) (Fig. 13.1). In this system, both males and females are trichromatic ("routine" trichromacy).

13.1.3 Spectral Variation of Primate L/M Opsins

The majority of the spectral variation of primate L/M opsin subtypes is predicted by amino acid composition at the residues 180, 277 and 285 ("three-site" rule) (Yokoyama and Radlwimmer 1998, 1999, 2001; Hiramatsu et al. 2004; Yokoyama et al. 2008; Matsumoto et al. 2014). The λ_{max} of the L/M opsins with serine, tyrosine and threonine at residues 180, 277 and 285, respectively (denoted SYT), is expected to be approximately 560 nm (Yokoyama et al. 2008). The λ_{max} values of L/M opsins with other three-site combinations can be estimated by subtracting 5, 10 and 17 nm from 560 nm in the case of alanine, phenylalanine and threonine at residues 180, 277 and 285, respectively (Yokoyama et al. 2008). In addition, interactions between these mutations are estimated to be −2 nm for S180A/T285A, +1 nm for Y277F/T285A and +4 nm for S180A/Y277F/T285A (Yokoyama et al. 2008). All eight possible combinations of the three sites are reported for primate L/M opsins (Fig. 13.3).

Prosimians: Strepsirrhini (Lorisiformes, Lemuriformes); Tarsiiformes
Simians (Anthropoidea)
Haplorhini: Tarsiiformes; Platyrrhini (NWM): Cebidae (Callitrichinae, Aotinae, Cebinae), Pitheciidae, Atelidae (Atelinae, Alouattinae); Catarrhini (OWM & Apes, Humans)

180	277	285	Expected λ_{max} (nm)	Lorisiformes	Lemuriformes	Tarsiiformes	Callitrichinae	Aotinae	Cebinae	Pitheciidae	Atelinae	Alouattinae	OWM & Apes	Humans
S	Y	T	560		✔		✔		✔	✔	✔(553)	✔	✔	✔
A	Y	T	555		✔	✔	✔		✔	✔				✔
S	F	T	550		✔				✔	✔	✔(538)			
A	F	T	545						✔	✔	✔(537)	✔		
S	Y	A	543									✔		✔
A	Y	A	536	✔	✔	✔	✔	✔		✔				
S	F	A	534										✔	✔
A	F	A	532		✔				✔	✔	✔	✔	✔	✔

Fig. 13.3 L/M opsin subtypes in primates distinguished on the basis of the "three-site" composition. At each amino acid site, the longer-wave residue is indicated with black, and the shorter-wave residue is indicated with grey font type. The expected λ_{max} values are estimated for each subtype according to the "three-site" rule. The opsin subtypes reported to be found in each taxon are indicated with check marks. In Lemuriformes, two alleles have been identified in ten species (AYT and AYA, *Varecia variegata*, *V. rubra*, *Propithecus coquereli*, *P. coronatus*, *P. deckenii*, *P. tattersalli*, *P. verreauxi* and *Eulemur macaco flavifrons*; AYT and AFA, *P. diadema*, *P. edwardsi*), and at least three alleles (SYT, SFT, AFA) have been identified in *Indri indri*, while one of AYT, AYA or AFA has been identified in many others (Tan and Li 1999; Heesy and Ross 2001; Jacobs et al. 2002; Jacobs and Deegan 2003; Veilleux and Bolnick 2009; Valenta et al. 2016; Veilleux et al. 2016; Jacobs et al. 2017). In Tarsiiformes, extant species have either of the two subtypes (AYT or AYA), but their common ancestral species is suspected to have had both alleles (Tan and Li 1999; Melin et al. 2013b). In Pitheciidae of New World monkeys (NWM), six alleles have been found in bald uakaris (*Cacajao calvus*) (Corso et al. 2016). In Atelinae of NWM, λ_{max} values are indicated in parentheses for subtypes, which are shifted to shorter wavelength from the "three-site" rule expectation (Matsumoto et al. 2014). Here, two alleles are typically found in each species: SYT and SFT in *Ateles* (Hiramatsu et al. 2005) and *Brachyteles* (Talebi et al. 2006) and SYT and AFT in *Lagothrix lagotricha* (Matsumoto et al. 2014). AFA is additionally found in *Brachyteles hypoxanthus* (Talebi et al. 2006). In Alouattinae, AFT and SYA are recombinant variants recently reported (Matsushita et al. 2014). AYT and SFT are also reported for *Saimiri boliviensis* (Cropp et al. 2002) and for *Cebus* (*Sapajus*) *apella* (Soares et al. 2010), respectively, in Cebinae. In Old World monkeys (OWM) and apes, SFA is reported as a rare recombinant variant in a macaque species, *Macaca fascicularis* (Onishi et al. 1999, 2002), and in chimpanzees, *Pan troglodytes* (Terao et al. 2005). In humans, a variety of variants are reported (Deeb 2005, 2006; Hayashi et al. 2006)

13.1.4 S Opsin Loss and Monochromacy

Among primates, variation of S and L/M opsin genes results in various modes of colour vision (monochromacy, dichromacy, polymorphism with dichromacy and trichromacy, routine but not uniform trichromacy and routine and uniform trichromacy) at various taxonomic levels. Monochromacy (colour blindness) arose independently in a New World monkey genus, *Aotus* (owl monkeys), as well as in the common ancestor of Lorisiformes (lorises and galagos/bushbabies), and in various species of cheirogaleid lemuriformes [in genera *Phaner* (fork-marked lemurs), *Cheirogaleus* (dwarf lemurs) and *Allocebus* (hairy-eared mouse lemur)] due to loss of functional S opsin gene by deleterious mutations (Fig. 13.1) (Wikler and Rakic 1990; Jacobs et al. 1993; Deegan II and Jacobs 1996; Jacobs et al. 1996b; Kawamura and Kubotera 2004; Tan et al. 2005; Levenson et al. 2007; Veilleux et al. 2013). These species all share a nocturnal activity pattern.

13.1.5 Dichromacy and Nocturnality

Many other nocturnal primates retain the S opsin gene and therefore maintain dichromacy (Jacobs 2013). In fact, the functionality of the S opsin gene appears to be maintained by natural selection that removes deleterious mutations (i.e. purifying selection) in these species (Kawamura and Kubotera 2004; Tan et al. 2005; Moritz et al. 2017). The retention of dichromacy in nocturnal strepsirrhines and tarsiers was once suggested as evidence against the conventional view that ancestral primates were nocturnal (Tan et al. 2005). Tan et al. (2005) proposed that ancestral primates were diurnal or cathemeral and that nocturnality has evolved several times, first in the lorisiforms but much later in other lineages, reflecting different time periods of functional relaxation among lineages.

However, recent studies do not provide support for the diurnal/cathemeral origin hypothesis. A population genetic study in lemuriform nocturnal aye-ayes detected a signature of ongoing purifying selection maintaining the S opsin gene (*Daubentonia madagascariensis*) (Perry et al. 2007). Aye-ayes have a short-wave shifted S opsin (λ_{max} at 406 nm) (Carvalho et al. 2012). It has also been reported that twilight is enriched in short-wavelength (bluish) light with sufficient intensity for aye-ayes with the short-wave shifted S opsin to perform cone-mediated colour vision for their twilight activities (Melin et al. 2012). Another study regards the openness of forest canopy to the sky and the nocturnal activity under moonlight as the main factor influencing the retention of S opsin and colour vision in nocturnal lemuriform prosimians (Veilleux et al. 2013). Nocturnal light intensity, particularly short-wave light, is much greater in open canopy forests than in the understory of closed canopy forests (Veilleux and Cummings 2012). Veilleux et al. (2013) found that lemuriform nocturnal species under open canopy habitats generally experience strong purifying selection to maintain the S opsin gene, while, in contrast, those under closed canopy habitats experience weaker purifying selection or a relaxation of selection. These studies suggest that dichromatic colour vision can be compatible with the nocturnality of ancestral primates (Moritz et al. 2017).

13.1.6 Origin of Trichromacy Under Dim Light

Among strepsirrhines, occasional trichromacy due to the allelic polymorphism of the single-locus X-linked L/M opsin gene has been observed in 11 diurnal lemurid species (diurnal *Varecia*, *Propithecus* and *Indri* species and cathemeral *Eulemur* species) (see caption of Fig. 13.3). (Tan and Li 1999; Heesy and Ross 2001; Jacobs et al. 2002; Jacobs and Deegan 2003; Veilleux and Bolnick 2009; Veilleux et al. 2016; Jacobs et al. 2017). On the basis of the pattern of non-synonymous, synonymous and intron nucleotide substitutions in L/M opsin genes among Bornean (*Tarsius bancanus*), Philippine (*T. syrichta*) and Sulawesi (*T. tarsier*) tarsiers, we inferred that the L and M opsin alleles coexisted and thus that occasional trichromacy arose in the last common ancestor of crown tarsiers, which had hyper-enlarged eye orbits and is considered to be active in low light (Melin et al. 2013b). With the findings that full moonlight and twilight in tropical forest are sufficient for cone-mediated colour vision (Melin et al. 2012), we proposed that the origin of primate trichromacy was in activities under dim (mesopic) light conditions (Melin et al. 2013b). Although more data on the genetic variation of opsin genes and colour vision are necessary for diurnal, cathemeral and nocturnal prosimians, these recent studies challenge the traditional and simplistic view of the diurnal origin of primate trichromacy.

13.1.7 Visual Opsin Variation in New World Monkeys

Platyrrhines (New World monkeys) are known for their extensive inter- and intra-species variation in colour vision (Figs. 13.1, 13.2, and 13.3). As reviewed above, howler monkeys (*Alouatta*) attained sex-independent trichromacy through juxtaposition of the spectrally differentiated L/M opsin genes on the X chromosome (Jacobs et al. 1996a; Dulai et al. 1999). At the other extreme, owl monkeys (*Aotus*), the sole nocturnal anthropoid primates, exhibit cone monochromacy due to the loss of functional S opsin gene (Wikler and Rakic 1990; Jacobs et al. 1993, 1996b; Levenson et al. 2007). All the other genera encompassing three families of New World monkeys (Cebidae, Atelidae and Pitheciidae: Wildman et al. 2009) are reported to have allelic polymorphism of the single-locus X-linked L/M opsin gene and exhibit colour vision variation (Fig. 13.2) (Jacobs 2007; Matsumoto et al. 2014).

A wide variation of allelic composition occurs among New World monkeys (Fig. 13.3), ranging from two alleles, seen typically in *Ateles* (spider monkeys) and *Lagothrix* (woolly monkeys) (Jacobs and Deegan II 2001; Hiramatsu et al. 2005;

Hiwatashi et al. 2010; Matsumoto et al. 2014), up to six alleles in *Cacajao calvus* (bald uakari) (Corso et al. 2016). Similarly, in Pitheciidae, to which *Cacajao* belongs, many (five) alleles are suspected for *Callicebus moloch* (dusky titi monkeys), based on an electroretinogram study (Jacobs and Deegan II 2005). Three-allele composition is widely observed in Cebidae (SYT, AFT and AFA in Cebinae; SYT, AYT and AYA in Callitrichinae; Fig. 13.3) (Jacobs 2007; Matsumoto et al. 2014; de Lima et al. 2015).

13.1.8 Non-human Catarrhines in Contrast to Platyrrhines

The L and M opsin genes of catarrhines are highly similar in nucleotide sequence (~96% identity) and are closely juxtaposed (Nathans et al. 1986). Thus, they are intrinsically susceptible to recombination and gene conversion between them, which could cause hybrid L/M opsin genes, gene loss and gene multiplication (Drummond-Borg et al. 1989; Jorgensen et al. 1990; Balding et al. 1992; Ibbotson et al. 1992; Winderickx et al. 1992; Winderickx et al. 1993; Deeb et al. 1994; Dulai et al. 1994; Shyue et al. 1994; Zhou and Li 1996; Zhao et al. 1998; Verrelli and Tishkoff 2004). Furthermore, a recombination hotspot chi element is conserved in the exon 3 of the L/M opsin gene among primates (Winderickx et al. 1993). Nevertheless, in contrast to New World monkeys, the incidence of colour vision variation is remarkably low in non-human catarrhines (Fig. 13.3). We surveyed 152 gibbons from 3 genera (*Hylobates*, *Nomascus* and *Symphalangus*) and 8 species and found only normal L and M opsin genes in all individuals and no hybrid type L/M opsin genes (Hiwatashi et al. 2011). In 744 male long-tailed macaques (*Macaca fascicularis*) and 455 male monkeys from other macaque species, only three long-tailed macaques were found to carry single SFA-type (L-M) hybrid opsin gene and were consequently protanope dichromats (Onishi et al. 1999; Hanazawa et al. 2001). One chimpanzee was found to carry one SFA-type (L-M) hybrid opsin gene and one normal M opsin gene (protanomalous trichromat) in 58 male chimpanzees from breeding colonies (Saito et al. 2003; Terao et al. 2005). Other studies have reported the absence of colour vision variation in Old World monkeys and apes (Jacobs and Williams 2001; Verrelli et al. 2008).

Multiple copies of M opsin genes are likely to increase the frequency of unequal recombination events. Regarding non-human catarrhines, some studies report that multiple M copies are rare (Onishi et al. 1999, 2002; Terao et al. 2005; Verrelli et al. 2008), yet other studies report that they are common (Ibbotson et al. 1992; Dulai et al. 1994; Hiwatashi et al. 2011). Thus, among Old World monkeys and apes, there seems to be no clear trend to the copy number variation of the M opsin gene.

Our study of gibbon population samples showed that gene conversion has homogenized introns of L and M opsin genes (Hiwatashi et al. 2011). However, purifying selection against the homogenization has protected the nucleotide difference between L and M opsin genes in centrally located exons, particularly exons 3 and 5, which include the spectrally crucial "three sites" (Hiwatashi et al. 2011). This confirms that gene conversions (and perhaps other forms of recombination) do occur between L and M opsin genes in non-human catarrhines, but these genes are eliminated from the population by natural selection if gene conversions affect the gene region relevant to spectral difference between L and M opsins. In non-human catarrhines, even mildly anomalous trichromats have not been found, suggesting a severe selective disadvantage to colour vision variants.

13.1.9 Uniqueness of Humans in Colour Vision

Among catarrhines with routine and normal trichromacy, humans constitute a notable exception with high incidence (approximately 3–8% of males) of colour vision variation (often referred to as "defective") (Verhulst and Maes 1998; Verrelli and Tishkoff 2004; Bosten et al. 2005; Deeb 2006; Hood et al. 2006; Sharpe et al. 2006). The AYT-type (M-L) and SFA-type (L-M) hybrid L/M opsins (Fig. 13.3) are reported to be 38% and 6%, respectively, in males of European origin with normal colour vision (Winderickx et al. 1993). In humans, multiple M copies are found in 66% of males of European origin (Drummond-Borg et al. 1989) and 56% of males of Japanese origin (Hayashi et al. 2001). The most upstream gene is typically L and the others are M. Only the upper two genes are expressed, and when a hybrid gene occupies either position, it causes slightly or severely anomalous trichromacy (Hayashi et al. 1999). When there is only one L/M opsin gene on an X chromosome or when the two positions are occupied by identical genes in the only one X chromosome of male or in both X chromosomes in females, this causes dichromacy (red-green colour blindness: more specifically, protanope when L is lost and deuteranope when M is lost). These are more commonly found in men than in women as their two X chromosomes make them more likely to have a "normal" gene array in either one. There are rare cases of individuals, irrespective of sex, who lack functional blue cones (tritanopes, <1:10,000) due to mutations in S opsin gene on chromosome 7 (Sharpe et al. 1999).

13.2 Contribution of Studies in Santa Rosa to Our Understanding of Colour Vision Evolution in Primates

13.2.1 Demonstration of Colour Vision Polymorphism in Wild Populations of New World Monkeys

It was not known until the early 2000s whether polymorphism of the L/M opsin, and hence colour vision, existed within groups of wild New World monkeys, other than mixed-species troops of tamarins (Surridge et al. 2002; Smith et al. 2003a, b). In 2003 we started a colour vision study in Santa Rosa, Costa Rica. We collected faecal samples from individually identified white-faced capuchin monkeys (*Cebus capucinus*) and black-handed spider monkeys (*Ateles geoffroyi*) (Hiramatsu et al. 2005).

We conducted colour vision typing by analysing the faecal DNA. Based on the "three-site" rule, we found three (SYT, AFT and AFA) and two (SYT and SFT) L/M opsin alleles in *C. capucinus* and *A. geoffroyi*, respectively. We also demonstrated that these alleles were distinct in spectral sensitivity by reconstituting their photopigments in vitro (Hiramatsu et al. 2005, 2008). While our previous research on captive capuchin monkeys had found a correlation of L/M opsin genotype with colour vision phenotype (Saito et al. 2005a), our Santa Rosa study was the first to demonstrate colour vision polymorphism in single-species social groups of New World monkeys.

13.2.2 Discovery of a Novel Allele and a Novel Mechanism of Spectral Tuning for the L/M Opsin

One of the two alleles of the spider monkeys in Santa Rosa, SFT, was not previously reported for any primate at the time of Hiramatsu et al. (2005). Existence of SFT allele was subsequently confirmed in another ateline species, muriquis (*Brachyteles arachnoides*) (Talebi et al. 2006). By reconstituting the two alleles of spider monkeys (SYT and SFT), we found that their absorption spectra were deviated from the "three-site" expectation to shorter wavelength (Hiramatsu et al. 2008), making their spectral separation 1.5 times larger than expected. This was the first clear exception of the "three-site" rule in primates.

Motivated by this finding, we examined other ateline species, long-haired spider monkey (*Ateles belzebuth*) and common woolly monkey (*Lagothrix lagotricha*) (Matsumoto et al. 2014). We found that the woolly monkey had the AFT instead of the SFT allele (Fig. 13.3) and that its absorption spectrum was also deviated from the "three-site" rule. We inferred ancestral amino acid sequences of ateline L/M opsin alleles using the maximum-likelihood method and applied a series of site-directed mutagenesis to these opsins artificially synthesized. We showed that a mutation Y213D in the ancestral opsin of the two alleles disabled spectral effect of S180A and enabled N294K to have a large spectral effect, which occurred in one allele of the ateline ancestor and increased the spectral separation between the two alleles. We then demonstrated that chromatic discrimination of fruit from leaves was significantly enhanced by these mutations by modelling the chromaticity of dietary fruits and background leaves in a natural habitat of spider monkeys (Matsumoto et al. 2014).

13.2.3 Balancing Selection on the L/M Opsin Alleles

Because the importance of colour vision is generally not questioned, and because the polymorphism of colour vision appears evolutionarily stable among New World monkeys, the mode and strength of natural selection behind the polymorphism had not been rigorously tested. Trans-specific nature and, therefore, long persistence of L/M opsin polymorphism had been taken as the evidence of the maintenance of trichromacy via natural selection (balancing selection) (Boissinot et al. 1998; Surridge and Mundy 2002; Surridge et al. 2003). However, as summarized later, our behavioural observations of capuchin and spider monkeys in Santa Rosa questioned the trichromat advantage hypothesis. This motivated us to reevaluate the assumption that this polymorphism was being maintained by natural selection.

Assuming that the last common ancestor of all New World monkeys originated 26 million years ago (Schneider 2000), likewise one must assume that the opsin polymorphism has also persisted over this time period. In theory, the expected survival time (in number of generations) for a neutral X-linked allele is $3N_e$ in a stationary population (N_e stands for the effective population size) (Hartl and Clark 2007). If N_e of New World monkeys had been large enough (e.g. in the order of 10^6), then the polymorphism could have persisted for this length of time without natural selection. The estimated value of N_e depends on the accuracy of estimates of the mutation rate and generation time and is confounded by demographic effects such as the historical dynamism of population size, migration pattern and population structure. Thus, the argument for balancing selection based on allele age (and thus population size) has an inherent weakness.

Although demographic effects influence genetic variation of all genes in the genome alike, the pattern and the intensity of natural selection can vary among genomic regions depending on direct or indirect fitness effects of mutations in that region. We thus applied a method that compared the pattern of intraspecific genetic variation between a focal region (i.e. the L/M opsin gene) and other reference regions in the same genome using the same population samples to cancel out the effects of demographic factors that both regions share (Hiwatashi et al. 2010). We showed that the nucleotide sequence of the L/M opsin gene was significantly more polymorphic than the sequences of the neutral references in the same genome. In a coalescence simulation that took into account the observed nucleotide diversity of the neutral references, the Tajima's D value (Tajima 1989) of the L/M opsin gene deviated significantly in a positive direction from the expected range. These results are the first to statistically demonstrate balancing selection acting on the polymorphic L/M opsin gene of New World monkeys.

This study logically leads us to the next question: what is the nature of balancing selection? The simplest explanation is that trichromatic individuals experience a heterozygote advantage. However, other mechanisms underlying the balancing selection assume benefits not only for trichromats but also for dichromats, such as condition-dependent order of superiority among colour vision types including dichromacy, mutual benefit of association between different colour vision types, niche divergence among colour vision types and negative frequency-dependent selection. The studies summarized in the following three sections all address this question.

13.2.4 Unequal Allele Frequencies of L/M Opsins

Using allele frequency data, we can assess whether the frequency pattern is consistent with the simple heterozygote advantage model or with more complex superiority order of vision types. If trichromats have the best fitness, allele frequencies of the L/M opsin subtypes are expected to be equal to maximize the number of trichromats. However, in both capuchin and spider monkeys at Santa Rosa, the longest-wave allele is the most frequent, and the shortest-wave allele is the least frequent (Hiramatsu et al. 2005; Hiwatashi et al. 2010). The same trend is observed in studies of other populations of spider monkeys (Jacobs and Deegan II 2001) and muriquis (Talebi et al. 2006). Unequal allele frequencies are also observed in squirrel monkeys (Cropp et al. 2002) and callitrichines (marmosets and tamarins) (Surridge et al. 2005), although exhibiting different patterns than those observed in Santa Rosa monkeys. Deviation from equality could result from selective neutrality among alleles (Hartl and Clark 2007) or from a selection process that does not simply maximize the number of trichromats. Our population genetic study (Hiwatashi et al. 2010) rejected the neutrality hypothesis.

The skewed allele frequencies towards longer-wave alleles across the spider monkeys, muriquis and capuchins could imply that the selection consists of complex opposing processes. For trichromats, red-green colour discrimination would be greater in individuals having the longest- and the shortest-wave-sensitive L/M alleles than in individuals having an intermediate-wave-sensitive allele. On the other hand, for dichromats, the blue-yellow colour resolution would be worst in individuals having the shortest-wave-sensitive L/M allele and be best in those having the longest-wave-sensitive allele (Osorio et al. 2004). Thus, the longest-wave-sensitive allele would be favoured by both trichromats and dichromats, whereas the shortest one would be favoured only by trichromats and disfavoured by dichromats. The observed common skew towards longer-wave alleles could indicate that the trichromat benefit does not always exceed an opposing dichromat benefit and that different alleles could be maintained by different demands among vision types.

Using the chromaticity data of fruit, figs and leaves in Santa Rosa and the spectral sensitivity data of L/M opsin alleles of capuchins, we reported that shorter-wave alleles could be favoured by dichromats in a context where it allows them to distinguish bluish fruits from background leaves in their long-distance vision (Melin et al. 2014). Bluish fruits tend to be small, and their importance could be larger for small-bodied primates such as squirrel monkeys and callitrichines. This may explain why the longer-wave skewed pattern is not obvious in these species. Shorter-wave alleles are also found to be favourable over short distances in computer simulation studies of primate foraging tasks (Rowe and Jacobs 2004, 2007; Melin et al. 2013a), although increased utility of other senses, such as luminance vision and olfaction, could lessen their advantage during short-range foraging (Hiramatsu et al. 2008, 2009; Melin et al. 2009).

Complexity is also manifested in the evolutionary history of ateline L/M opsin alleles. We showed that in most atelines, the shortest-wave allele AFA was lost or rare (Fig. 13.3) (Matsumoto et al. 2014), which could imply that dichromat benefit surpasses the opposing trichromat benefit. On the other hand, the spectral separation between the remaining two alleles (SYT and SFT in *Ateles* and *Brachyteles*; SYT and AFT in *Lagothrix*) is enlarged by mutations that occurred in the ateline common ancestor, resulting in significant improvement in the ability to discriminate conspicuous dietary fruits from leaves in the natural habitat of spider monkeys under both bright and dim light conditions (Matsumoto et al. 2014). This would benefit trichromats in that they may tolerate the loss of the shortest-wave allele if the spectral separation of the longest- and intermediate-wave alleles is still sufficient in discriminating stimuli (Saito et al. 2005a) and in foraging performance as shown in capuchin monkeys (Melin et al. 2009). Conversely, in callitrichines, the spectral separation (~5 nm) between the longest and the intermediate alleles is comparable to deuteranomalous human trichromats severely impaired in red-green chromatic discrimination (Deeb 2006) (Fig. 13.3). This may also explain why the shortage of shortest-wave allele is not obvious in callitrichines.

13.2.5 Unexpected Hybrid L/M Opsins in Howler Monkeys

We extended our genetics study to another sympatric primate species in Santa Rosa, the mantled howler monkey (*Alouatta palliata*). Howler monkeys are in a special position in the study of primate colour vision because they are exceptional among New World monkeys by having L and M opsin genes on the same X chromosome, as catarrhines do, and being routine trichromats (Jacobs et al. 1996a). The L and M opsin genes in howler monkeys were originally reported to be the longest (λ_{max} at ~560 nm)- and the shortest (λ_{max} at ~532 nm)-wave subtypes of L/M opsins in primates, respectively (Jacobs et al. 1996a). This finding was taken as supporting evidence for the evolutionary advantage of trichromacy in primates, and polymorphic colour vision in most New World monkeys was seen as an intermediate stage of primate evolution from dichromacy to trichromacy with the spectrally most separated L/M opsin subtypes (Bowmaker et al. 1987; Jacobs et al. 1996a).

However, it was not known whether there was a variation in gene composition of L/M opsins in howlers as is in humans. We thus surveyed the L/M opsin genes of mantled howlers in Santa Rosa, as well as those in Nicaragua, and of Yucatan black howlers (*A. pigra*) in Belize (Matsushita et al. 2014). We found a hybrid L/M opsin gene in each species ("Apa_ML" and "Api_LM", corresponding to AFT and SYA in Fig. 13.3, respectively) with >10% of frequencies (Matsushita et al. 2014; Melin et al. 2017b). The λ_{max} values of the reconstituted hybrid photopigments Apa _ML and Api_LM are 547 nm and 546 nm, respectively (Matsushita et al. 2014; Melin et al. 2017b). These values are comparable to that of AFT allele seen in cebine and pitheciid New World monkeys (Fig. 13.3). The trichromatic capuchin monkeys having AFT allele are mildly anomalous trichromats who can discriminate stimuli using Ishihara pseudo-isochromatic plates (Saito et al. 2005a). Thus, even though howler monkeys once acquired the acutest trichromacy with SYT- and AFA-type L/M opsin genes in the same X chromosome, they created less acute trichromacy, replacing one with AFT-type hybrid L/M opsin gene on the gene array.

13.2.6 A Brief Summary of Our Behavioural Studies on Colour Vision of Santa Rosa Monkeys

To further address the question of trichromat advantage, we conducted field studies of feeding behaviour in Santa Rosa. Our study on spider monkeys showed that dichromats were not inferior to trichromats in frequency, accuracy and unit-time intake efficiency of detecting fruits (Hiramatsu et al. 2008). This result was surprising because spider monkeys are highly frugivorous, and the trichromat advantage was generally expected for foraging on fruit. Our explanation of this finding on the basis of colorimetric measurement of fruits and background leaves is that the luminance contrast of fruits compared to background leaves is the main determinant of fruit detection in both dichromats and trichromats (Hiramatsu et al. 2008). We further found that, irrespective of colour vision phenotypes, the spider monkeys sniff and reject visually cryptic fruits more often than visually conspicuous fruits (Hiramatsu et al. 2009). This indicates that colour vision is not the sole determinant of decisions to ingest or reject certain fruits.

Regarding capuchin monkeys, with extensive behavioural data accumulated over years in Santa Rosa, we observed that, for yellowish-to-reddish (conspicuous) fruits, trichromats had a higher intake rate than dichromats (Melin et al. 2017a). This holds when we analysed foraging data of juvenile monkeys. This also holds for mature monkeys, but only in the model in which social dominance rank was included. The juvenile result suggests that effects of colour vision type may be particularly strong during juvenescence, perhaps as a result of increased demands for resource acquisition in this critical window of development. In addition, as they age, capuchins may learn to effectively use nonvisual senses during foraging, lessening the importance of trichromacy among older monkeys (Melin et al. 2017a). Dichromat capuchins sniff more figs and exhibit longer foraging sequences than trichromats (Melin et al. 2009).

Furthermore, our field observation of capuchin monkeys revealed a clear dichromat advantage in foraging on camouflaged insects (Melin et al. 2007, 2010). In fact, dichromats are reported to be superior to trichromats at breaking camouflage caused by variegated backgrounds (Morgan et al. 1992; Saito et al. 2005b). Our study showed that this superiority really appeared in the wild.

These findings from observational studies in natural environments suggest that the superior ability of trichromats to see the red-green colour contrast may not translate into a selective advantage in the wild. Fedigan et al. (2014) tested whether colour vision phenotype is a significant predictor of female fitness in a population of wild capuchins, using 26 years of long-term survival and fertility data. They found no advantage to trichromats over dichromats for three fitness measures (fertility rates, offspring survival and maternal survival).

Here, I go back to the question: what is the nature of balancing selection? These studies reject the simplest explanation by heterozygote (i.e. trichromatic) advantage. I now direct more attention to other mechanisms requiring benefits not only for trichromats but also for dichromats. Further studies are needed for testing the alternative mechanisms of balancing selection that have been proposed to explain opsin polymorphism: condition-dependent order of superiority among colour vision types including dichromacy, mutual benefit of association, niche divergence and frequency dependence (Hiwatashi et al. 2010; Fedigan et al. 2014).

13.2.7 Advantage of Trichromacy Is Conditional

The strict conservation of normal trichromacy in non-human catarrhines is in sharp contrast to what is found in New World monkeys. The higher frequency of anomalous trichromacy in New World howler monkeys implies that the selective pressure to maintain "normal" trichromacy is lower in the Neotropics (Matsushita et al. 2014). It is an open question as to whether the difference between non-human catarrhines (uniform and normal trichromacy) and New World monkeys (polymorphic colour vision) is attributable to (1) *biogeographic differences* among continents, for example, the severity of seasonality or a prevalence of drably coloured fruits and asynchronous species, such as figs and palm fruits (Dominy et al. 2003); (2) *dietary variability*, for example, the degree of dependence on insects, leaves or colourful fruits and different food patch sizes (Melin et al. 2014); or (3) *variation in social colour signals* (Changizi et al. 2006; Fernandez and Morris 2007; Kamilar et al. 2013; Hiramatsu et al. 2017). Thus, the advantage of trichromacy is not always obvious and could be conditional on flora and fauna in the habitat of different primates, feeding habits, colorimetric and photic/scotopic properties of different environments and variation in the utility of colour signals for communication.

The persistence of colour vision variants in the human population may reflect advantage in achromatic visual tasks, such as detecting camouflaged preys and predators during hunting, as inferred from studies of New World monkeys. Effects of latitude on daily light availability are also discussed for potential advantages of colour vision variants in hunting at northern latitudes (Reimchen 1987; Pearce and Dunbar 2012). Benefits for females having L/M hybrid opsins and possibly enlarged colour space may also include a higher efficiency while gathering fruit resources (Jameson et al. 2001; Verrelli and Tishkoff 2004) (a question remains, however, as to why this is not selected in non-human catarrhines). In human evolutionary history, having different colour vision morphs in hunting-gathering lifeways may confer a selective advantage in a mutual benefit manner in highly cooperative human populations. By analysing L/M opsin nucleotide diversity collected from global ethnic groups, we should be able to infer at what point in human evolution colour vision polymorphism occurred and spread across human populations as seen today.

13.3 Ending Remarks

Our studies of New World monkeys at Santa Rosa, Costa Rica, on colour vision and L/M opsin genes have made significant contribution to promoting our understanding of colour vision evolution in primates by transdisciplinary approaches. Major findings among those explained above are the role of balancing selection in colour vision variation and the dichromat advantage in wild primates. Further studies are required for thus far unexplored subjects, such as long-distance detection of visual targets in foliage (Melin et al. 2014), foraging under dim light (Melin et al. 2013b), foraging performance in severe dry season and non-foraging tasks (predator detection, social recognition) in field studies. Future studies should also be directed towards deciphering the genetic diversity of other sensory genes, such as genes encoding for olfactory and taste receptors, and the interplay among different sensory modalities in primate behaviours. A deeper knowledge of primate colour vision and other senses will also facilitate our understanding of sensory evolution in non-human and human primates.

Acknowledgements I cannot express my gratitude with words to Professor Linda Marie Fedigan. I simply thank my luck that I encountered Linda during my search for a field primatologist collaborator in the early 2000s. Without her assistance, guidance and cooperation, my research would have taken a very different path and reached no better place. I also thank from the bottom of my heart the many friends and excellent collaborators who I got to know through Linda, especially Amanda Melin and Chihiro Hiramatsu, for making the research so productive. I also greatly appreciate excellent suggestions in revising the manuscript from anonymous reviewers and the editors, Dr. Urs Kalbitzer and Dr. Katharine Jack.

References

Arrese CA, Hart NS, Thomas N et al (2002) Trichromacy in Australian marsupials. Curr Biol 12:657–660

Balding DJ, Nichols RA, Hunt DM (1992) Detecting gene conversion: primate visual pigment genes. Proc R Soc B 249:275–280

Boissinot S, Tan Y, Shyue SK et al (1998) Origins and antiquity of X-linked triallelic color vision systems in New World monkeys. Proc Natl Acad Sci U S A 95:13749–13754

Bosten JM, Robinson JD, Jordan G et al (2005) Multidimensional scaling reveals a color dimension unique to 'color deficient' observers. Curr Biol 15:R950–R952

Bowmaker JK, Jacobs GH, Mollon JD (1987) Polymorphism of photopigments in the squirrel monkey: a sixth phenotype. Proc R Soc Lond B 231:383–390

Carvalho LS, Davies WL, Robinson PR et al (2012) Spectral tuning and evolution of primate short-wavelength-sensitive visual pigments. Proc R Soc B 279:387–393

Changizi MA, Zhang Q, Shimojo S (2006) Bare skin, blood and the evolution of primate colour vision. Biol Lett 2:217–221

Collin SP, Knight MA, Davies WL et al (2003) Ancient colour vision: multiple opsin genes in the ancestral vertebrates. Curr Biol 13:R864–R865

Collin SP, Davies WL, Hart NS et al (2009) The evolution of early vertebrate photoreceptors. Phil Trans R Soc B 364:2925–2940

Corso J, Bowler M, Heymann EW et al (2016) Highly polymorphic colour vision in a New World monkey with red facial skin, the bald uakari (*Cacajao calvus*). Proc R Soc B 283:20160067

Cropp S, Boinski S, Li W-H (2002) Allelic variation in the squirrel monkey X-linked color vision gene: biogeographical and behavioral correlates. J Mol Evol 54:734–745

Davies WL, Carvalho LS, Cowing JA et al (2007) Visual pigments of the platypus: a novel route to mammalian colour vision. Curr Biol 17:R161–R163

Davies WL, Carvalho LS, Tay BH et al (2009a) Into the blue: gene duplication and loss underlie color vision adaptations in a deep-sea chimaera, the elephant shark *Callorhinchus milii*. Genome Res 19:415–426

Davies WL, Collin SP, Hunt DM (2009b) Adaptive gene loss reflects differences in the visual ecology of basal vertebrates. Mol Biol Evol 26:1803–1809

Davies WI, Collin SP, Hunt DM (2012) Molecular ecology and adaptation of visual photopigments in craniates. Mol Ecol 21:3121–3158

de Lima EM, Pessoa DM, Sena L et al (2015) Polymorphic color vision in captive Uta Hick's cuxius, or bearded sakis (*Chiropotes utahickae*). Am J Primatol 77:66–75

Deeb SS (2005) The molecular basis of variation in human color vision. Clin Genet 67:369–377

Deeb SS (2006) Genetics of variation in human color vision and the retinal cone mosaic. Curr Opin Genet Dev 16:301–307

Deeb SS, Jorgensen AL, Battisti L et al (1994) Sequence divergence of the red and green visual pigments in great apes and humans. Proc Natl Acad Sci U S A 91:7262–7266

Deegan JF II, Jacobs GH (1996) Spectral sensitivity and photopigments of a nocturnal prosimian, the bushbaby (*Otolemur crassicaudatus*). Am J Primatol 40:55–66

Dominy NJ, Svenning JC, Li W-H (2003) Historical contingency in the evolution of primate color vision. J Hum Evol 44:25–45

Drummond-Borg M, Deeb SS, Motulsky AG (1989) Molecular patterns of X chromosome-linked color vision genes among 134 men of European ancestry. Proc Natl Acad Sci U S A 86:983–987

Dulai KS, Bowmaker JK, Mollon JD et al (1994) Sequence divergence, polymorphism and evolution of the middle-wave and long-wave visual pigment genes of great apes and Old World monkeys. Vis Res 34:2483–2491

Dulai KS, von Dornum M, Mollon JD et al (1999) The evolution of trichromatic color vision by opsin gene duplication in New World and Old World primates. Genome Res 9:629–638

Ebeling W, Natoli RC, Hemmi JM (2010) Diversity of color vision: not all Australian marsupials are trichromatic. PLoS One 5:e14231

Fedigan LM, Melin AD, Addicott JF et al (2014) The heterozygote superiority hypothesis for polymorphic color vision is not supported by long-term fitness data from wild neotropical monkeys. PLoS One 9:e84872

Fernandez AA, Morris MR (2007) Sexual selection and trichromatic color vision in primates: statistical support for the preexisting-bias hypothesis. Am Nat 170:10–20

Fleagle JG (2013) Primate adaptation and evolution, 3rd edn. Academic Press, San Diego

Hanazawa A, Mikami A, Sulistyo Angelika P et al (2001) Electroretinogram analysis of relative spectral sensitivity in genetically identified dichromatic macaques. Proc Natl Acad Sci U S A 98:8124–8127

Hartl DL, Clark AG (2007) Principles of population genetics, 4th edn. Sinauer Associates, Sunderland

Hayashi T, Motulsky AG, Deeb SS (1999) Position of a 'green-red' hybrid gene in the visual pigment array determines colour-vision phenotype. Nat Genet 22:90–93

Hayashi S, Ueyama H, Tanabe S et al (2001) Number and variations of the red and green visual pigment genes in Japanese men with normal color vision. Jpn J Ophthalmol 45:60–67

Hayashi T, Kubo A, Takeuchi T et al (2006) Novel form of a single X-linked visual pigment gene in a unique dichromatic color-vision defect. Vis Neurosci 23:411–417

Heesy CP, Ross CF (2001) Evolution of activity patterns and chromatic vision in primates: morphometrics, genetics and cladistics. J Hum Evol 40:111–149

Heesy CP, Ross CF, Demes B (2007) Oculomotor stability and the functions of the postorbital bar and septum. In: Ravosa MJ, Dagosto M (eds) Primate origins: adaptations and evolution. Springer, New York, pp 257–283
Hiramatsu C, Radlwimmer FB, Yokoyama S et al (2004) Mutagenesis and reconstitution of middle-to-long-wave-sensitive visual pigments of New World monkeys for testing the tuning effect of residues at sites 229 and 233. Vis Res 44:2225–2231
Hiramatsu C, Tsutsui T, Matsumoto Y et al (2005) Color-vision polymorphism in wild capuchins (*Cebus capucinus*) and spider monkeys (*Ateles geoffroyi*) in Costa Rica. Am J Primatol 67:447–461
Hiramatsu C, Melin AD, Aureli F et al (2008) Importance of achromatic contrast in short-range fruit foraging of primates. PLoS One 3:e3356
Hiramatsu C, Melin AD, Aureli F et al (2009) Interplay of olfaction and vision in fruit foraging of spider monkeys. Anim Behav 77:1421–1426
Hiramatsu C, Melin AD, Allen WL et al (2017) Experimental evidence that primate trichromacy is well suited for detecting primate social colour signals. Proc R Soc B 284:20162458
Hiwatashi T, Okabe Y, Tsutsui T et al (2010) An explicit signature of balancing selection for color-vision variation in New World monkeys. Mol Biol Evol 27:453–464
Hiwatashi T, Mikami A, Katsumura T et al (2011) Gene conversion and purifying selection shape nucleotide variation in gibbon L/M opsin genes. BMC Evol Biol 11:312
Hood SM, Mollon JD, Purves L et al (2006) Color discrimination in carriers of color deficiency. Vis Res 46:2894–2900
Ibbotson RE, Hunt DM, Bowmaker JK et al (1992) Sequence divergence and copy number of the middle- and long-wave photopigment genes in Old World monkeys. Proc R Soc Lond B 247:145–154
Jacobs GH (1984) Within-species variations in visual capacity among squirrel monkeys (*Saimiri sciureus*): color vision. Vis Res 24:1267–1277
Jacobs GH (1993) The distribution and nature of colour vision among the mammals. Biol Rev 68:413–471
Jacobs GH (2007) New World monkeys and color. Int J Primatol 28:729–759
Jacobs GH (2013) Losses of functional opsin genes, short-wavelength cone photopigments, and color vision-A significant trend in the evolution of mammalian vision. Vis Neurosci 30:39–53
Jacobs GH, Deegan JF 2nd (2003) Photopigment polymorphism in prosimians and the origins of primate trichromacy. In: Mollon JD, Pokorny J, Knoblanch K (eds) Normal and defective colour vision. Oxford University Press, Oxford, pp 14–20
Jacobs GH, Deegan JF II (2001) Photopigments and colour vision in New World monkeys from the family Atelidae. Proc R Soc Lond B 268:695–702
Jacobs GH, Deegan JF II (2005) Polymorphic New World monkeys with more than three M/L cone types. J Opt Soc Am A 22:2072–2080
Jacobs GH, Nathans J (2009) The evolution of primate color vision. Sci Am 300:56–63
Jacobs GH, Williams GA (2001) The prevalence of defective color vision in Old World monkeys and apes. Col Res Appl 26(Suppl):S123–S127
Jacobs GH, Deegan JF II, Neitz J et al (1993) Photopigments and color vision in the nocturnal monkey, *Aotus*. Vis Res 33:1773–1783
Jacobs GH, Neitz M, Deegan JF et al (1996a) Trichromatic colour vision in New World monkeys. Nature 382:156–158
Jacobs GH, Neitz M, Neitz J (1996b) Mutations in S-cone pigment genes and the absence of colour vision in two species of nocturnal primate. Proc R Soc Lond B 263:705–710
Jacobs GH, Deegan JF II, Tan Y et al (2002) Opsin gene and photopigment polymorphism in a prosimian primate. Vis Res 42:11–18
Jacobs RL, MacFie TS, Spriggs AN et al (2017) Novel opsin gene variation in large-bodied, diurnal lemurs. Biol Lett 13:20170050
Jameson KA, Highnote SM, Wasserman LM (2001) Richer color experience in observers with multiple photopigment opsin genes. Psychon Bull Rev 8:244–261

Jorgensen AL, Deeb SS, Motulsky AG (1990) Molecular genetics of X chromosome-linked color vision among populations of African and Japanese ancestry: high frequency of a shortened red pigment gene among Afro-Americans. Proc Natl Acad Sci U S A 87:6512–6516

Kamilar JM, Heesy CP, Bradley BJ (2013) Did trichromatic color vision and red hair color coevolve in primates? Am J Primatol 75:740–751

Kawamura S, Kubotera N (2004) Ancestral loss of short wave-sensitive cone visual pigment in lorisiform prosimians, contrasting with its strict conservation in other prosimians. J Mol Evol 58:314–321

Kawamura S, Hiramatsu C, Melin AD et al (2012) Polymorphic color vision in primates: evolutionary considerations. In: Hirai H, Imai H, Go Y (eds) Post-genome biology of primates. Springer, Tokyo, pp 93–120

Lambert D (1987) The Cambridge guide to prehistoric man. Cambridge University Press, Cambridge

Levenson DH, Fernandez-Duque E, Evans S et al (2007) Mutational changes in S-cone opsin genes common to both nocturnal and cathemeral *Aotus* monkeys. Am J Primatol 69:757–765

Matsumoto Y, Hiramatsu C, Matsushita Y et al (2014) Evolutionary renovation of L/M opsin polymorphism confers a fruit discrimination advantage to ateline New World monkeys. Mol Ecol 23:1799–1812

Matsushita Y, Oota H, Welker BJ et al (2014) Color vision variation as evidenced by hybrid L/M opsin genes in wild populations of trichromatic *Alouatta* New World monkeys. Int J Primatol 35:71–87

Melin AD, Fedigan LM, Hiramatsu C et al (2007) Effects of colour vision phenotype on insect capture by a free-ranging population of white-faced capuchins (*Cebus capucinus*). Anim Behav 73:205–214

Melin AD, Fedigan LM, Hiramatsu C et al (2009) Fig foraging by dichromatic and trichromatic *Cebus capucinus* in a tropical dry forest. Int J Primatol 30:753–775

Melin AD, Fedigan LM, Young HC et al (2010) Can color vision variation explain sex differences in invertebrate foraging by capuchin monkeys? Curr Zool 56:300–312

Melin AD, Moritz GL, Fosbury RAE et al (2012) Why aye-ayes see blue. Am J Primatol 74:185–192

Melin AD, Kline DW, Hickey C et al (2013a) Food search through the eyes of a monkey: a functional substitution approach for assessing the ecology of primate color vision. Vis Res 87:87–96

Melin AD, Matsushita Y, Moritz GL et al (2013b) Inferred L/M cone opsin polymorphism of ancestral tarsiers sheds dim light on the origin of anthropoid primates. Proc R Soc B 280:20130189

Melin AD, Hiramatsu C, Parr NA et al (2014) The behavioral ecology of color vision: considering fruit conspicuity, detection distance and dietary importance. Int J Primatol 35:258–287

Melin AD, Chiou KL, Walco ER et al (2017a) Trichromacy increases fruit intake rates of wild capuchins (*Cebus capucinus imitator*). Proc Natl Acad Sci U S A 114:10402–10407

Melin AD, Khetpal V, Matsushita Y et al (2017b) Howler monkey foraging ecology suggests convergent evolution of routine trichromacy as an adaptation for folivory. Ecol Evol 7:1421–1434

Mollon JD, Bowmaker JK, Jacobs GH (1984) Variations of colour vision in a New World primate can be explained by polymorphism of retinal photopigments. Proc R Soc Lond B 222:373–399

Morgan MJ, Adam A, Mollon JD (1992) Dichromats detect colour-camouflaged objects that are not detected by trichromats. Proc R Soc Lond B 248:291–295

Moritz GL, Ong PS, Perry GH et al (2017) Functional preservation and variation in the cone opsin genes of nocturnal tarsiers. Phil Trans R Soc B 372:20160075

Nathans J, Thomas D, Hogness DS (1986) Molecular genetics of human color vision: the genes encoding blue, green, and red pigments. Science 232:193–202

Onishi A, Koike S, Ida M et al (1999) Dichromatism in macaque monkeys. Nature 402:139–140

Onishi A, Koike S, Ida-Hosonuma M et al (2002) Variations in long- and middle-wavelength-sensitive opsin gene loci in crab-eating monkeys. Vis Res 42:281–292

Osorio D, Smith AC, Vorobyev M et al (2004) Detection of fruit and the selection of primate visual pigments for color vision. Am Nat 164:696–708

Pearce E, Dunbar R (2012) Latitudinal variation in light levels drives human visual system size. Biol Lett 8:90–93

Perry GH, Martin RD, Verrelli BC (2007) Signatures of functional constraint at aye-aye opsin genes: the potential of adaptive color vision in a nocturnal primate. Mol Biol Evol 24:1963–1970

Pisani D, Mohun SM, Harris SR et al (2006) Molecular evidence for dim-light vision in the last common ancestor of the vertebrates. Curr Biol 16:R318–R319

Reimchen TE (1987) Human color vision deficiencies and atmospheric twilight. Soc Biol 34:1–11

Rowe MP, Jacobs GH (2004) Cone pigment polymorphism in New World monkeys: are all pigments created equal? Vis Neurosci 21:217–222

Rowe MP, Jacobs GH (2007) Naturalistic color discriminations in polymorphic platyrrhine monkeys: effects of stimulus luminance and duration examined with functional substitution. Vis Neurosci 24:17–23

Saito A, Mikami A, Hasegawa T et al (2003) Behavioral evidence of color vision deficiency in a protanomalia chimpanzee (*Pan troglodytes*). Primates 44:171–176

Saito A, Kawamura S, Mikami A et al (2005a) Demonstration of a genotype-phenotype correlation in the polymorphic color vision of a non-callitrichine New World monkey, capuchin (*Cebus apella*). Am J Primatol 67:471–485

Saito A, Mikami A, Kawamura S et al (2005b) Advantage of dichromats over trichromats in discrimination of color-camouflaged stimuli in nonhuman primates. Am J Primatol 67:425–436

Schneider H (2000) The current status of the New World monkey phylogeny. An Acad Bras Ci 72:165–172

Sharpe LT, Stockman A, Jagle H et al (1999) Opsin genes, cone photopigments, color vision, and color blindness. In: Gegenfurtner KR, Sharpe LT (eds) Color vision: from genes to perception. Cambridge University Press, Cambridge, pp 3–51

Sharpe LT, de Luca E, Hansen T et al (2006) Advantages and disadvantages of human dichromacy. J Vis 6:213–223

Shyue SK, Li L, Chang BH et al (1994) Intronic gene conversion in the evolution of human X-linked color vision genes. Mol Biol Evol 11:548–551

Smith AC, Buchanan-Smith HM, Surridge AK et al (2003a) Leaders of progressions in wild mixed-species troops of saddleback (*Saguinus fuscicollis*) and mustached tamarins (*S. mystax*), with emphasis on color vision and sex. Am J Primatol 61:145–157

Smith AC, Buchanan-Smith HM, Surridge AK et al (2003b) The effect of colour vision status on the detection and selection of fruits by tamarins (*Saguinus* spp.). J Exp Biol 206:3159–3165

Soares JGM, Fiorani M, Araujo EA et al (2010) Cone photopigment variations in *Cebus apella* monkeys evidenced by electroretinogram measurements and genetic analysis. Vis Res 50:99–106

Surridge AK, Mundy NI (2002) Trans-specific evolution of opsin alleles and the maintenance of trichromatic colour vision in Callitrichine primates. Mol Ecol 11:2157–2169

Surridge AK, Smith AC, Buchanan-Smith HM et al (2002) Single-copy nuclear DNA sequences obtained from noninvasively collected primate feces. Am J Primatol 56:185–190

Surridge AK, Osorio D, Mundy NI (2003) Evolution and selection of trichromatic vision in primates. Trends Ecol Evol 18:198–205

Surridge AK, Suarez SS, Buchanan-Smith HM et al (2005) Color vision pigment frequencies in wild tamarins (*Saguinus* spp.). Am J Primatol 67:463–470

Tajima F (1989) Statistical method for testing the neutral mutation hypothesis by DNA polymorphism. Genetics 123:585–595

Talebi MG, Pope TR, Vogel ER et al (2006) Polymorphism of visual pigment genes in the muriqui (Primates, Atelidae). Mol Ecol 15:551–558

Tan Y, Li W-H (1999) Trichromatic vision in prosimians. Nature 402:36

Tan Y, Yoder AD, Yamashita N et al (2005) Evidence from opsin genes rejects nocturnality in ancestral primates. Proc Natl Acad Sci U S A 102:14712–14716

Terao K, Mikami A, Saito A et al (2005) Identification of a protanomalous chimpanzee by molecular genetic and electroretinogram analyses. Vis Res 45:1225–1235

Valenta K, Edwards M, Rafaliarison RR et al (2016) Visual ecology of true lemurs suggests a cathemeral origin for the primate cone opsin polymorphism. Funct Ecol 30:932–942
Veilleux CC, Bolnick DA (2009) Opsin gene polymorphism predicts trichromacy in a cathemeral lemur. Am J Primatol 71:86–90
Veilleux CC, Cummings ME (2012) Nocturnal light environments and species ecology: implications for nocturnal color vision in forests. J Exp Biol 215:4085–4096
Veilleux CC, Louis EE Jr, Bolnick DA (2013) Nocturnal light environments influence color vision and signatures of selection on the *OPN1SW* opsin gene in nocturnal lemurs. Mol Biol Evol 30:1420–1437
Veilleux CC, Scarry CJ, Di Fiore A et al (2016) Group benefit associated with polymorphic trichromacy in a Malagasy primate (*Propithecus verreauxi*). Sci Rep 6:38418
Verhulst S, Maes FW (1998) Scotopic vision in colour-blinds. Vis Res 38:3387–3390
Verrelli BC, Tishkoff SA (2004) Signatures of selection and gene conversion associated with human color vision variation. Am J Hum Genet 75:363–375
Verrelli BC, Lewis CM Jr, Stone AC et al (2008) Different selective pressures shape the molecular evolution of color vision in chimpanzee and human populations. Mol Biol Evol 25:2735–2743
Vorobyev M (2004) Ecology and evolution of primate colour vision. Clin Exp Optom 87:230–238
Wakefield MJ, Anderson M, Chang E et al (2008) Cone visual pigments of monotremes: filling the phylogenetic gap. Vis Neurosci 25:257–264
Wikler KC, Rakic P (1990) Distribution of photoreceptor subtypes in the retina of diurnal and nocturnal primates. J Neurosci 10:3390–3401
Wildman DE, Jameson NM, Opazo JC et al (2009) A fully resolved genus level phylogeny of neotropical primates (Platyrrhini). Mol Phylogenet Evol 53:694–702
Winderickx J, Lindsey DT, Sanocki E et al (1992) Polymorphism in red photopigment underlies variation in colour matching. Nature 356:431–433
Winderickx J, Battisti L, Hibiya Y et al (1993) Haplotype diversity in the human red and green opsin genes: evidence for frequent sequence exchange in exon 3. Hum Mol Genet 2:1413–1421
Yokoyama S (2000) Molecular evolution of vertebrate visual pigments. Prog Retin Eye Res 19:385–419
Yokoyama S, Radlwimmer FB (1998) The "five-sites" rule and the evolution of red and green color vision in mammals. Mol Biol Evol 15:560–567
Yokoyama S, Radlwimmer FB (1999) The molecular genetics of red and green color vision in mammals. Genetics 153:919–932
Yokoyama S, Radlwimmer FB (2001) The molecular genetics and evolution of red and green color vision in vertebrates. Genetics 158:1697–1710
Yokoyama S, Yang H, Starmer WT (2008) Molecular basis of spectral tuning in the red- and green-sensitive (M/LWS) pigments in vertebrates. Genetics 179:2037–2043
Zhao Z, Hewett-Emmett D, Li W-H (1998) Frequent gene conversion between human red and green opsin genes. J Mol Evol 46:494–496
Zhou YH, Li W-H (1996) Gene conversion and natural selection in the evolution of X-linked color vision genes in higher primates. Mol Biol Evol 13:780–783

Part IV
Primate Adaptations to Changing Environments

Peter Henzi

At the beginning of the 1970s, with the acceleration of interest in primate fieldwork, choice of a species to focus on was determined in large part by whether anyone else had studied – or was studying – them. In my case, to give a single example, I was warned off savanna baboons because, so much having been written about them (using data from perhaps five different sites), I would have had to come up with something 'really new' if I wanted to make a sensible contribution to the discipline. At the same time, it was also noted with amusement that PhD students invariably described their 9 or 12 month stint in the field as constituting the first 'long-term study' of species 'x'. The consequent dividend of this first wave of field studies, therefore, was a valuable and necessary snapshot of the order's behavioural ecology that delivered species-level breadth at the expense of any serious appreciation of variability, either spatial or temporal. To go back to all that work on baboons, a report that animals at Site A had been observed doing things not seen at Site B led to an acrimonious discussion at an international get-together back in 1978 as to who was right and who was wrong, rather than being seen as an opportunity to get at the underlying mechanisms that might account for such different outcomes at the two locations.

Seeing variability as an opportunity and not as a stumbling block comes more easily now. In the 40 years since that initial proliferation of field research, the discipline has matured and honed its focus. Funding and logistical constraints have meant that the choice of study species is more often driven by practical considerations than by the lure of the completely unknown. The result is that more work has become concentrated on fewer species at more sites, opening a window into spatial variation. And what began as circumscribed PhD studies have, as students became

P. Henzi (✉)
Department of Psychology, University of Lethbridge, Lethbridge, AB, Canada
e-mail: peter.henzi@uleth.ca

faculty members, turned into longer-term research programs that have made it increasingly difficult to avoid factoring temporal dynamics into the interrogation of the relationship between primate groups and their environments. Not only difficult to avoid but, with the less-than subtle effects of rapid anthropogenic climate change on the environment, compelling to witness and in urgent need of comprehension. Here, then, we are on new ground and, to paraphrase Socrates, all that we know is that we know nothing (certainly not nearly enough...). Still, this is ground that must be broken and to do so we will need appropriate conceptual and analytical tools. Fortunately, the four chapters in this part all provide timely and valuable pointers as to how we might set about the task of understanding not only how to address the consequences of changing environments for primate populations but also the ways in – and extent to – which they might respond.

Urs Kalbitzer and Colin Chapman (Chap. 14) begin with the proposition that behavioural flexibility – a hallmark trait at least of the simians – might make it possible for primates to address the problem of very rapid climate and environmental change. This then raises for them the obvious question of how one might go about doing so and they identify three intersecting possibilities: comparative studies, niche modelling, and a behavioural ecological approach in which individual behaviour is tied to environmental change data via local population dynamics. Although this last is central to their initial proposal, and clearly reflects their own sets of expertise, they make the undeniable point that any attempt to determine the extent to which shifts in conditions can be adaptively tracked by local behavioural adjustment necessitates – or is significantly improved by – the accumulation of long-term data at as many sites as possible.

Steig Johnson and Kerry Brown (Chap. 15) speak directly to Kalbitzer and Chapman's methodological and conceptual points with a case study that applies ecological niche modelling to the distribution of Meso-American primates. In addition to using the technique to describe the niche breadth of the two atelines (Geoffroy's spider monkey, *Ateles geoffroyi*; the mantled howler monkey, *Alouatta palliata*) and the Panamanian white-throated capuchin (*Cebus capucinus imitator*), they assess whether the relatively restricted geographical distribution of capuchins can be linked to a lack of suitable habitat. What ties this to Kalbitzer and Chapman's argument is that we might otherwise expect capuchins, being adaptive generalists (i.e. behaviourally and ecologically flexible) to have the widest regional distribution. What they show is that all that local flexibility counts for nought in the broader scheme of things if, like the capuchin, you are a species whose habitat requirements are set by a narrow climate niche. Johnson and Brown use this to make the case that capuchins might, as a consequence, be surprisingly vulnerable to the environmental consequences of climate change.

Which, if we wish to be usefully forewarned, makes it all the more important to be able to track climate and ecology, and deduce informative patterns. Which then is what, in a masterclass on data analysis and presentation, *Fernando Campos* (Chap. 16) does with 40 years of climate and phenological data from Linda Fedigan's capuchin site at Santa Rosa. His demonstration of the complex climate and vegetation patterns observed at the site, and the importance of appreciating the influence of the El Niño Southern Oscillation on these, provides a model of how to go about this, while making it abundantly clear that you can never have too much data in this line of work, especially if you want to go from what you know of the past to what you might need to know about the future.

The remaining chapter (Chap. 17) – by *Colin Chapman, Kim Valenta and Sarah Bortolamiol* – picks up on this, pointing out that where we do have long-term ecological and behavioural data (the relative paucity of such sites being roundly bemoaned by Campos), we finally come to understand Socrates' point: we now know that we know nothing. More precisely, the theoretical framework of primate behavioural ecology, derived from short-term studies and specified in terms of species averages, simply cannot countenance the sheer extent of variability: the diet of a single species can vary markedly both within and across populations, as food availability at a single site can change almost an order of magnitude across years. As the authors make clear, if we are to properly embrace and address variability, and then fold it into our theoretical models, we need, at the least, to identify the appropriate spatial and temporal scales at which to collect and analyse data.

The threads of Linda Fedigan's influence – whether subliminal or obvious – that bind these chapters beneath the surface are easier to identify. As a colleague or mentor, her good habits – of rigour, precision and stamina – and her appreciation of the value of commitment, of being in it for the long haul, have clearly been passed on. As an exemplar of how a field primatologist might conduct herself in a changing world, she has served as a model not only for the authors in this part but also for those of us at a greater remove.

Chapter 14
Primate Responses to Changing Environments in the Anthropocene

Urs Kalbitzer and Colin A. Chapman

Abstract Most primate habitats are undergoing intense and rapid changes due to anthropogenic influences resulting in many primate populations being threatened. Habitat loss and fragmentation are already extensive; thus dispersal to unoccupied habitats is an unlikely adaptive response to these changes. Furthermore, most primates have slow life histories and long generation times, and because environmental change is occurring at an unprecedented rate, gene-based adaptations are also unlikely to evolve fast enough to offer successful responses to these changes. However, long primate life histories are linked to well-developed brains, which may allow primates to respond to environmental change through behavioural flexibility. Here we ask: What are the most common challenges of changing environments for primates and what do we know about their behavioural abilities to respond to such changes? To answer this question, we first review the most common types of habitat/landscape alterations, the extent of human-primate interactions, and the impact of climate change. Next, we evaluate how primates respond to these changes via behavioural flexibility, and using different approaches and datasets, we discuss how

U. Kalbitzer (✉)
Department of Anthropology and Archaeology, University of Calgary, Calgary, AB, Canada

Department of Anthropology and McGill School of Environment, McGill University, Montreal, QC, Canada
e-mail: urs.kalbitzer@mail.mcgill.ca

C. A. Chapman
Department of Anthropology and McGill School of Environment, McGill University, Montreal, QC, Canada

Makerere University Biological Field Station, Fort Portal, Uganda

Wildlife Conservation Society, Bronx, NY, USA

Section of Social Systems Evolution, Primate Research Institute, Kyoto University, Kyoto, Japan

U. Kalbitzer, K. M. Jack (eds.), *Primate Life Histories, Sex Roles, and Adaptability*, Developments in Primatology: Progress and Prospects,
https://doi.org/10.1007/978-3-319-98285-4_14

to investigate if these responses are beneficial with regard to population persistence. Finally, we discuss how comparisons across species, space, and time can be used to draw generalizations about primate responses to environmental change while considering their behavioural flexibility and the data derived from case studies. We demonstrate how understanding behavioural flexibility as a response to environmental change will be crucial to optimize conservation efforts by constructing informed management plans.

Keywords Conservation · Comparative studies · Behavioural ecology · Behavioural flexibility · Habitat change

14.1 Introduction

14.1.1 The Anthropocene

Humans are dramatically changing the world. Our population has increased from 2.5 billion in 1950 to 7.4 billion in 2015 and is predicted to go up to 9.8 billion by 2050 (Fig. 14.1a; FAO 2017). This growth is associated with increasing resource use, including a dramatic increase in the number of livestock (Fig. 14.1b) and in the use of large areas for agriculture (Fig. 14.1c). Conversely, the proportion of areas with forest, including primary forest, is decreasing globally (Fig. 14.1d, e; FAO 2016).

Furthermore, the concentration of "greenhouse gases" has dramatically increased since the mid-nineteenth century, and these contribute significantly to increasing global temperature and associated changes in climate (IPCC 2014). It is well documented that over the last 130 years, the global climate has already warmed by approximately 0.85 °C, and the temperature increase is likely to exceed 2 °C by the end of the century (IPCC 2014; Raftery et al. 2017).

Finally, the current species extinction rate is much faster than the estimated evolutionary "background" extinction rate (Barnosky et al. 2011; Ceballos et al. 2015) and has been referred to as the "sixth mass extinction" (Ceballos et al. 2015). For example, conservative estimates suggest that the average rate of loss of vertebrate species over the last century is 100 times higher than the background rate (Ceballos et al. 2015). Beyond the extinction of entire species, populations of many species are in considerable decline (Ceballos et al. 2017).

Given the magnitude and range of anthropogenic impacts, Crutzen and Stoermer (2000) suggested that we label the current geological epoch as the "Anthropocene". While the starting point of this epoch, and whether the Anthropocene even qualifies as a geological epoch, is hotly debated (Carey 2016), the term is commonly used to summarize humanity's profound impact on the environment.

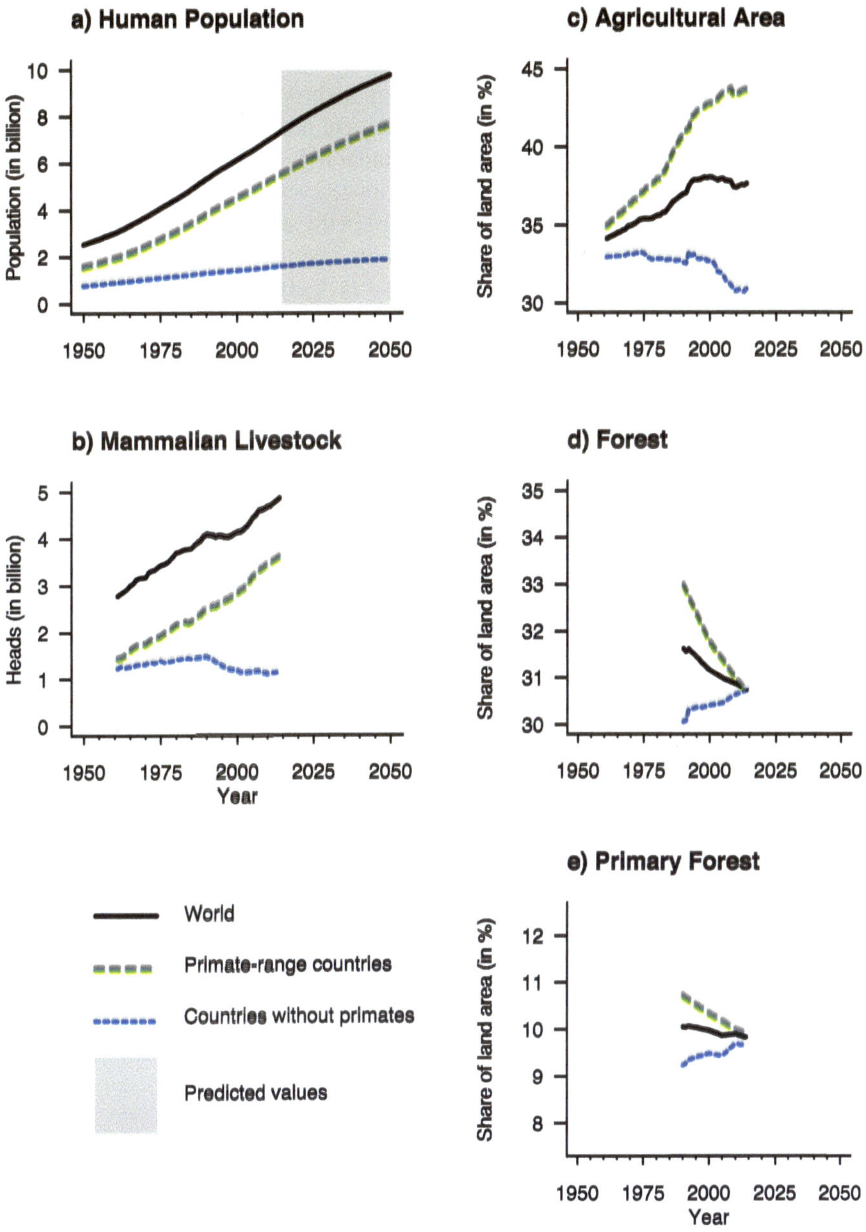

Fig. 14.1 Indicators of the human impact on the world, globally and summarized for countries with and without primates. (**a**) Human population size with predicted growth until 2050. (**b**) Number of livestock including the most important mammalian breeds (cattle, sheep, pigs, goats, and buffaloes). (**c**) Percentage of agricultural area per total land area. (**d**) Percentage of forest area per total land area. (**e**) Percentage of primary forest area per total land area. Note that the scale for (d) and (e) differs from the scale of (c). Data from the FAOSTAT Database (FAO 2017). A country was considered a primate-range country if its boundaries were spatially intersecting with the global distribution of primates, which was determined using the "Terrestrial Mammals" shapefile from the IUCN Red List of Threatened Species (IUCN 2016)

14.1.2 Primates in the Anthropocene

As a result of the increasing human population, anthropogenic landscape changes, and climate change, about 60% of the ~500 species of non-human primates (hereafter primates) are estimated to be threatened by extinction (Estrada et al. 2017). However, the impact of different human activities on primates varies across regions (Estrada et al. 2017; Almeida-Rocha et al. 2017), and within the same region different species may exhibit different population dynamics in response to the same changes (Chapman et al. 2000, 2010, 2018; Fedigan and Jack 2012).

The ability to make accurate and repeatedly reliable predictions about the population dynamics of species in response to current and anticipated environmental change will be central to creating effective conservation strategies. This will require a detailed understanding of environmental changes, their causes and cascading effects, and the biology of the species impacted by these changes. However, given the ongoing alterations to primate habitats and the bleak outlook for most primates (Estrada et al. 2017), there will not be enough time to evaluate comprehensively the ability of all primate species to respond. This will make extrapolations from existing datasets or those quickly gathered from lesser-known species and/or populations a necessity.

In this chapter, we review what we know about the ability of primates to respond to changing environments by addressing four questions: (1) What are the most common changes that occur in primate environments? (2) How do primates respond to environmental changes? (3) How can we best investigate the vulnerability of different primates to environmental changes considering that responses appear flexible? (4) How can long-term datasets be used to improve our understanding of primate vulnerability and, therefore, our ability to conserve primate populations? We then discuss the applicability and benefits of comparisons across species, space, and time and give an outlook for the integration of behavioural ecological studies into primate conservation.

14.2 What Are the Most Common Changes that Occur in Primate Environments?

14.2.1 Landscape Alterations in Primate Habitats

With very few exceptions, primates occur in tropical or subtropical regions (Fig. 14.2), and the human population is currently expanding faster in primate-range countries than in countries without primates (Fig. 14.1a). Associated with this greater human population increase in primate-range countries is the faster increase in the number of livestock (Fig. 14.1b) and expansion of agricultural land (Fig. 14.1c) compared to countries without primates. Furthermore, logging, mining, and construction of transportation networks, such as new roads, are rising in many tropical countries (Laurance et al. 2009, 2014; Weng et al. 2013). These landscape

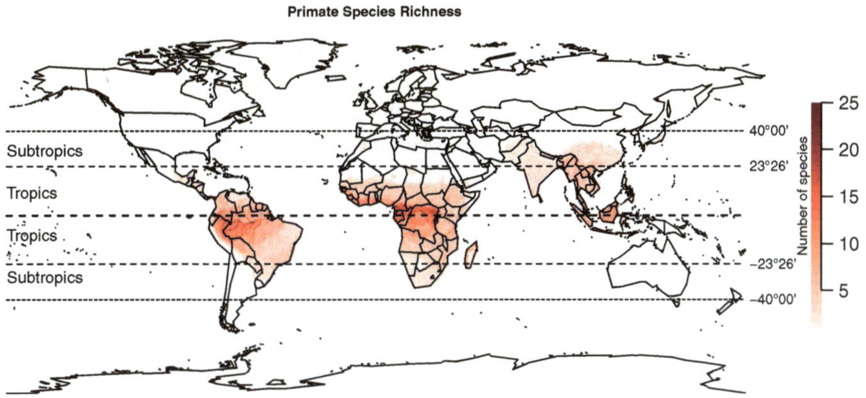

Fig. 14.2 Global distribution and species richness of primates. Distribution data are taken from the IUCN Red List of Threatened Species (IUCN 2016) and the map was created with the function lets.presabs from the letsR package (Vilela and Villalobos 2015) in R ver. 3.3.2 (R Core Team 2016)

alterations often lead to disturbances, degradations, fragmentations, or the entire loss of forest areas, which is the primary habitat of primates (Reed and Fleagle 1995; Chapman et al. 2006; Lovett and Marshall 2006). For example, cropland in tropical countries expanded by 48,000 km^2 per year between 1999 and 2008, largely at the expense of forest (Phalan et al. 2013). One estimate suggests that approximately 1 billion ha of additional agricultural land, primarily in developing countries, will be needed by 2050 to meet the demands of the growing human populations – an area larger than Canada (Laurance et al. 2014). As a result, forest cover is in considerable decline in primate-range countries, where, according to FAO data, 3.37% (or 72 of 2138 million ha) of forest cover has been lost between 2000 and 2014 (Fig. 14.1d). Primary forest, which is more important for many primates than disturbed forest (Gouveia et al. 2014; Chapman et al. 2018), only represented 32.33% of the total forest area in primate-range countries in 2014 and has decreased by 4.22% between 2010 and 2014 (Fig. 14.1e; see also Emrich et al. 2000; Wright and Muller-Landau 2006). In their most recent Global Forest Resources Assessment, the FAO (2016) estimated that the *rate* of global annual forest loss has decreased. However, in some important primate regions, such as Southeast Asia, the Congo basin, or the Brazilian Amazon, deforestation rates have increased in recent years (FAO 2017; Rigby and White 2017).

14.2.2 Primate Interactions with Humans

Increases in human population size are likely to lead to more frequent interactions between primates and people in both natural and anthropogenic habitats (McLennan et al. 2017). For example, monkey temples in Asia, which are important refuges for primates in human-dominated landscapes, are visited by millions of people annually and bring people and primates in close, and largely peaceful, contact (e.g. (Jones-Engel et al. 2005; Conly and Johnston 2008). However, greater human population density can also lead to an increase in the hunting of "bushmeat" for local and commercial uses (Fa and Brown 2009). Primates are especially vulnerable to bushmeat hunting because they commonly have slow life histories, and many primates have relatively large bodies, which makes them a preferred target for hunters (Linder and Oates 2011; Chapman and Gogarten 2012; Wilkie et al. 2016). The encroachment of human settlements to remote areas is also leading to increasing conflicts between humans and primates, including "crop raiding" (Hill 2000; Marchal and Hill 2009; see below).

Furthermore, more frequent encounters between primates and humans can lead to an increase in parasite and disease transmission from humans and domestic animals to primates and vice versa (Woodford et al. 2002; Chapman et al. 2005b). Documented cases include the transmission of human respiratory paramyxoviruses, *Streptococcus pneumoniae*, and the cold (human rhinovirus C) to chimpanzees (*Pan troglodytes*; Köndgen et al. 2008, 2017; Boesch 2008; Scully et al. 2018) or the transmission of human measles to rhesus macaques (*Macaca mulatta*; Jones-Engel et al. 2006b). In Kenya, an emergence of bovine tuberculosis in baboons (*Papio* spp.) was caused by infected meat consumed by the monkeys from the dump at a tourist lodge (Sapolsky and Share 2004).

There are also numerous examples for the transmission of diseases from primates to humans, such as the transmission of *Simian foamy virus* from macaques (*Macaca* spp.) or Ebola from gorillas (*Gorilla gorilla*) and chimpanzees to humans (Leroy et al. 2004; Jones-Engel et al. 2005, 2008; Bermejo et al. 2006). Perhaps, the most prominent case is the spread of Simian immunodeficiency viruses (SIV) from different African primates to humans, which gave rise to the global AIDS pandemic (Hahn et al. 2000).

An example of transmission in both directions is the virus causing yellow fever, which can be transmitted by mosquitos between humans and primates (Fernandes et al. 2017). The most recent outbreak of yellow fever in Brazil has resulted in the death of several hundred humans and thousands of monkeys (Bicca-Marques et al. 2017; Fernandes et al. 2017). Diseases that can be transmitted from primates to humans bear an additional risk for primates: during the recent outbreak of yellow fever outbreak in Brazil, some people afraid of yellow fever harassed and killed primates (Bicca-Marques et al. 2017).

14.2.3 Effect of Climate Change on Primate Habitats

In comparison to the extensive effects of logging, agricultural clearing, and hunting, the impact of climate change has been previously considered to be minor, but this view has been replaced by the recognition that current climate change is having significant impacts on tropical ecosystems, primates, and biota in general, and this impact is likely to increase in the near future (Parmesan 2006; Brook et al. 2008; Dunham et al. 2011; Corlett 2012; Pacifici et al. 2017). Generally, the increase of temperature in primate habitats is predicted to be higher than the average increase in global temperature, while the predicted change in rainfall patterns depends on the region (Graham et al. 2016). For example, East Africa is predicted to receive more rainfall, while rainfall is predicted to decline in Mesoamerica (Altmann et al. 2002; Chapman et al. 2005a; Graham et al. 2016). Even today, some long-term primate studies have already documented temperature changes of over 4 °C in the last 40–50 years and change in annual rainfall of as much as 300 mm (Altmann et al. 2002; Chapman et al. 2005a).

The frequency and intensity of extreme weather events, such as floods, heavy precipitation, hurricanes (also called typhoons or cyclones), heat waves, droughts, and fires, are expected to become more frequent and intense in many regions (IPCC 2013; Diffenbaugh et al. 2017). Furthermore, much of the interannual climatic variation in the tropics is driven by El Niño Southern Oscillations (ENSOs; Campos, Chap. 16, this volume; Dunham et al. 2011; Corlett 2012; Campos et al. 2015), and extreme "El Niños" (the warm phases of ENSOs) are expected to become more frequent, resulting in severe droughts in some areas, large amounts of rainfall in other areas, and more intense hurricanes in the Pacific (Cai et al. 2014, 2015).

Such changes in climate can have significant impacts on primate populations. For example, on Barro Colorado Island, Panama, fruiting, flowering, and leaf set were disrupted on six occasions between 1929 and 1994 when seasonal rains deviated from their typical pattern. In 1970, one such unusual rainfall event led to severe fruit crop failure and the mass mortality of howler monkeys (*Alouatta palliata*) and other animals (Foster 1982; Milton 1982; Wright et al. 1999; Wright and Calderón 2006). Additional extensive long-term monitoring of primate populations, climate, and phenology are needed to understand the future effect of climate change. For example, regression modelling of annual fruiting revealed solar irradiance and ENSO as the strongest predictors of fruiting in Kibale National Park, Uganda (Chapman et al. 2018). The projected changes in rainfall associated with climate change and coincident variation in cloud cover suggest that phenophase dynamics may be affected by climate change. As of yet, however, there is no clear signal as to how primate populations in Kibale will change, despite over 40 years of monitoring (Chapman et al. 2010, 2018).

14.2.4 Indirect, Synergistic, and Cascading Effects of Anthropogenic Changes on Primates

Perhaps the most significant changes in the future will be the result of various indirect, cascading, and synergistic effects on primate habitats that have not been anticipated. For example, climate change, habitat loss, and fragmentation can all affect plant phenology (Parmesan 2006; Morellato et al. 2016). This means that temporal and spatial variation in food abundance can change, potentially leading to the immigration of new competitors, predators, or pathogens into primate habitats. For example, Rothman et al. (2015) showed a general decline in the nutritional value of leaves in Kibale, Uganda, over as little as 15 years, and such changes have the potential to decrease habitat suitability for local populations of leaf-eating red colobus (*Procolobus rufomitratus*) and black-and-white colobus (*Colobus guereza*). Some environmental changes also have the potential to lead to other perturbations (i.e. cascading effects) and to amplify one another (i.e. synergistic effects; (Brook et al. 2008). For example, logging can affect primates by (1) decreasing food availability and (2) creating roads that fragment primate habitats and facilitate (3) bushmeat hunting and serve to (4) increase the opportunities for disease transmission (Chapman et al. 2005b; Goldberg et al. 2008; Remis and Jost Robinson 2012).

14.2.5 Protection and Restoration of Primate Habitats

Some disturbed primate habitats are changing because humans undertake efforts to protect and/or restore them, and such efforts can have significant benefits to primate populations (Fedigan and Jack 2001; Robbins et al. 2011; Strier and Ives 2012; Wheeler et al. 2016; Omeja et al. 2016). Ideally, these efforts will provide new protected areas or improve existing habitats and lead to viable populations of primates and other organisms. Since the 1990s, protected areas have increased in size globally, primate regions included (Butchart et al. 2010; Rands et al. 2010; Estrada et al. 2017). However, these positive developments need to be viewed realistically (Joppa et al. 2008; Andam et al. 2008; Joppa and Pfaff 2009, 2010). For example, many primates do not live inside protected areas (Meijaard et al. 2010; Estrada et al. 2017). Also, although protected areas are normally effective at protecting land from being cleared, they are less effective at eliminating logging, human-created fire, and bushmeat hunting (Oates 1996; Chapman and Peres 2001; Bruner et al. 2001; Hartter et al. 2011; Gaveau et al. 2016). Researching how to make the largest conservation gains for primates from existing and new conservation areas is a clear priority that will necessitate working closely with the local communities.

In primate-range countries, human population growth is much greater in urban than in rural areas as people move from the farms to the cities, and populations in rural areas are predicted to be on a general decline in the next few years (Fig. 14.3). With these trends, abandoned areas that were occupied by primates prior to being converted to human uses are increasing (Jacob et al. 2008). There are a variety of trajectories for these lands: they could be converted into huge agricultural monocultures, like palm oil plantations (Linder 2013), or to agroecosystems where some form of primate conservation is possible (Estrada et al. 2012), or to agricultural land with fragments and corridors (Pozo-Montuy et al. 2013; see also Meijaard et al. 2010). Alternatively, the land could be allowed to regenerate to natural forest, which offers greater potential for the persistence of primates (Chapman 2018). For example, Baya and Storch (2010) surveyed a village site in Korup National Park, Cameroon,

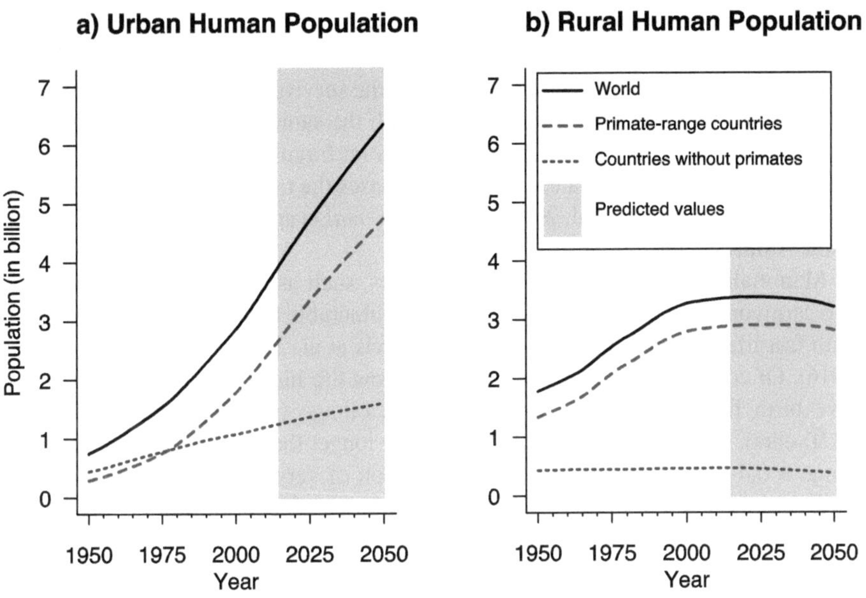

Fig. 14.3 Changes in human populations size with predicted growth until 2050 for (**a**) urban and for (**b**) rural areas. Data sources are the same as for Fig. 14.1

that was abandoned 7–8 years previously, and they found that all eight species of endemic primates had repopulated the area; in addition, sighting frequency was not significantly different from other sectors of the park surveyed in 2004–2005 (Linder 2008). In Kibale National Park, Uganda, 7 years after an area of grassland was replanted with trees as part of a carbon offset programme, all species of diurnal primates were present in high numbers, including the endangered red colobus and chimpanzee (Omeja et al. 2012, 2016; Chapman et al. 2018).

The outcome of such conservation efforts is, however, not always predictable because other factors, such as climate and the immigration of competitors (e.g. elephants; Omeja et al. 2014, 2016), can change, and different species of primates respond in different ways to habitat regeneration (Fedigan and Jack 2001, 2012; Chapman et al. 2018).

14.3 How Do Primates Respond to Environmental Changes?

14.3.1 General Ways to Respond to Changing Environments

Animals possess three key possible ways to responding to changing environments (Wong and Candolin 2015): (1) moving to other areas that fit their requirements, i.e. *dispersal*; (2) evolving adaptations to the new conditions, i.e. *evolutionary (or genetic) change*; or (3) exhibiting behavioural responses that are already in their repertoire (or reaction norms) to cope with new conditions, i.e. *phenotypic plasticity*.

Habitat loss is already a major concern for the survival of primates (Estrada et al. 2017), and, therefore, unoccupied habitat with the same characteristics as the current habitat is typically nonexistent. Furthermore, fragmentation results in physical barriers between suitable patches, which increases the risks for dispersing individuals (Arroyo-Rodríguez et al. 2013). Thus, *dispersal* is an unlikely suitable response to the Anthropocene.

Mammalian species with slow life histories, such as many even-toed ungulates and carnivores, are assumed to be even more vulnerable to extinction than mammals with fast life histories, like most rodents (Purvis et al. 2000; Gonzalez-Voyer et al. 2016). Of course, many primates have very slow life histories: on average, females give birth for the first time around 4 years of age (ranging from 256 days to 13.9 years), inter-birth intervals are commonly longer than a year (mean = 1.5 years, range = 0.4–5.5 years), and, with the exception of very few species, females give birth to only a single infant per birth (values based on data from the PanTHERIA database[1]; Jones et al. 2009). Most anthropogenic changes, however, occur very rapidly, and for primates and other long-lived animals, it is very unlikely that *evolutionary changes* are able to keep pace with these changes (Wong and Candolin 2015).

However, primates have relatively large brains, and it has been suggested that large brains are associated with higher behavioural flexibility, conferring advantages in dealing with both social and ecological challenges (Strier, Chap. 2, this volume; Reader and Laland 2002; Sol et al. 2008). Thus, given the constraints on dispersal and evolutionary change as a response to environmental change, behavioural adjustments appear to be the most likely possibility with which primates might respond to the Anthropocene. What kind of flexible, behavioural responses can primates exhibit to cope with the anthropogenic environmental changes thus becomes a critical question that must be answered to inform conservation strategies.

14.3.2 Primate Behavioural Responses to Landscape Alterations

In response to habitat fragmentations, disturbances, or degradations, primates can change their ranging patterns, activity budgets, and diet (Wong et al. 2006; Wong and Sicotte 2007; Pebsworth et al. 2012; Mekonnen et al. 2017; McLennan et al. 2017). Furthermore, the spatial and temporal distribution of food resources is considered to be one of the most important factors affecting social behaviour and mating patterns (i.e. primate socioecological theory; Wrangham 1980; Isbell 1991; Sterck et al. 1997; Snaith and Chapman 2007; Koenig et al. 2013). Thus, if human

[1] The calculation of age of first birth includes data from 102 primate species, and the calculation of inter-birth intervals included data from 108 primate species. These averages are not corrected for phylogenetic relatedness and might be biased depending on the inclusion of varying number of primates from different taxonomic groups.

impact on primate habitats affects the abundance and distribution of their food resources changes in group size and composition, reproductive patterns or social relationships may occur (langurs, Sterck 1999; tana river colobus, *Cercocebus galeritus*, Mbora et al. 2009; red colobus, Gogarten et al. 2015; muriquis, *Brachyteles hypoxanthus*, Strier and Mendes 2012).

Some of these behavioural responses allow primates to survive in human-modified landscapes (Schwitzer et al. 2011; Bonilla-Sánchez et al. 2012; Chapman et al. 2016; McLennan et al. 2017), such as tree plantations or suburban settings (Moore et al. 2010; Hoffman and O'Riain 2011). In some cases, primates even thrive in human-modified landscapes by supplementing their natural diet with human-cultivated resources via crop feeding (Hill 2000; Marchal and Hill 2009; Chapman et al. 2016); feeding on garbage, food items in houses, or fruit trees in gardens (Hoffman and O'Riain 2012); or being voluntarily provisioned with food from humans. For example, some primates obtain large parts of their diet from local people and tourists at monkey temples in Asia (Jones-Engel et al. 2005; Fuentes and Gamerl 2005).

14.3.3 Primate Behavioural Responses to Interactions with Humans

Where humans do not pose a threat to them, primates can easily habituate to humans. For example, the ursine colobus (*Colobus vellerosus*) at the Boabeng-Fiema Monkey Sanctuary in Ghana (Wong et al. 2006), different macaque species at monkey temples in Asia (Fuentes and Gamerl 2005; Jones-Engel et al. 2006a), and baboons close to tourist lodges and human settlements in Senegal or Botswana (UK, personal observation) are all examples where primates peacefully coexist with humans, and, in some cases, this coexistence can even result in economic benefits to local human populations (Fuentes and Gamerl 2005). However, if encounters are less peaceful, such as occurs in the context of bushmeat hunting, primates can become more cryptic, vigilant, try to avoid risky areas, and become more aggressive towards humans and dogs (Remis and Jost Robinson 2012; McLennan et al. 2017). For example, vervets (*Chlorocebus tantalus*) in areas of Cameroon where they were heavily hunted by humans with dogs suppress loud, conspicuous alarm calls directed at dogs, possibly to avoid detection (Kavanagh 1980).

14.3.4 Primate Behavioural Responses to Climate Change

In response to increasing temperatures, which can make metabolic costs of moving or foraging unsustainable during the hottest time of the day, primates can change their daily activity patterns. For example, they may rest or socialize in the shade during the middle of the day and spend more time foraging during cooler periods

(Hill 2005). Primates also show a high degree of variability in birth seasonality, ranging from strictly seasonal breeding species, such as macaques, to species in which females can give birth throughout the year, such as baboons (*Papio* spp.; Janson and Verdolin 2005). In such flexible species, females may adjust their breeding behaviour by only reproducing during favourable times. For example, yellow baboons in Kenya (*P. cynocephalus*) breed throughout the year, but females are less likely to cycle or to conceive after periods of extreme heat (Beehner et al. 2006). However, how quickly a population modifies their seasonal reproductive patterns remains to be determined.

While some primates rarely drink water, other species need year-round access to water sources (Hill 2005; McDougall et al. 2010; Fedigan and Jack 2012). If the habitat of these obligate drinkers receives less rainfall, water sources can become rare and widely dispersed, requiring changes in ranging patterns in order to maintain access to drinking water, as has been documented for vervet monkeys (*Chlorocebus aethiops*; McDougall et al. 2010). Such shifts could become permanent if previously favourable areas become unsuitable year-round. This may increase encounters between groups, leading to higher energy expenditures, increased stress, and even elevated mortality levels. In contrast, more rainfall and shorter dry periods may have no direct, but rather indirect, effects on primates, such as causing temporal shifts of food availability, or fruit crop failure, which has been documented to cause increased mortality (Milton 1982; Wright et al. 1999; Wright and Calderón 2006).

Hurricanes represent another extreme weather event with potentially severe impact on primates and their habitats that may increase in frequency and intensity with climate change (Erhart and Overdorff 2008; Dunham et al. 2011; Johnson et al. 2011; Schaffner et al. 2012). Primates respond in various ways to the impact of hurricanes. For example, black howlers (*Alouatta pigra*) and spider monkeys (*Ateles geoffroyi*) in Belize and toque macaques (*Macaca sinica*) in Sri Lanka responded to the habitat destruction following hurricanes by changes in activity budgets, diet, and social behaviour (Behie et al. in press; Dittus 1988; Behie and Pavelka 2005, 2013).

14.3.5 Primate Behavioural Responses to Indirect and Synergistic Effects of Changing Environments

Changes in plant phenology may represent one of the most important indirect effects of anthropogenic activities on primates because it can lead to a change in temporal and spatial distribution of food resources. As discussed above, primates can respond to such alterations by adjusting group size, social structure, and mating patterns. Furthermore, they can shift their diets and rely on fallback food when their preferred resources become scarce during some (or all) periods of the year (Hanya and Chapman 2012). Temporal variation in food abundance is also one of the main factors suspected to determine timing of reproduction (Janson and Verdolin 2005; Carnegie et al. 2011). For example, capuchin monkeys in Santa Rosa can give birth throughout the year but do so mostly during the period of highest fruit abundance (Carnegie et al. 2011). Thus, if temporal variation in food availability changes, some primates may shift their birth peak. Cascading and synergistic effects describe processes rather than specific types of environmental change; thus, primate responses will depend on the nature of the changes involved in such processes (e.g. habitat fragmentation, hunting), which, to a large extent, are unknown.

14.4 How Can We Best Investigate the Vulnerability of Different Primates to Environmental Changes Considering that Responses Appear Flexible?

Primates show highly flexible behaviour which they can adjust to various types of environmental change, but are these behavioural responses beneficial and sufficient to ensure their survival? And why are some species doing well in human-dominated landscapes, while other species already have gone extinct[2] or are threatened by extinction?

There are three general approaches to investigating the vulnerability of primates to anthropogenic change: (1) *comparative studies* of different populations and species, (2) *niche modelling studies*, and (3) *behavioural ecological studies* based on the behaviour of individuals.

14.4.1 Comparative Studies

Studies using the comparative method commonly aim to investigate the relationships among morphological, life history, behavioural, and ecological variables across species while controlling for the effects associated with phylogenetic relatedness (Harvey and Pagel 1991; Nunn and Barton 2001). With regard to species or population persistence, such analyses aspire to understand the relationships among biological variables (e.g. body weight, age at first reproduction) and variables reflecting the vulnerability of species to extinction. Using this approach, Purvis et al. (2000) showed that primates with small geographic distributions, large body mass, and low population density are at higher risk of extinction than are species with large distribution, small body mass, and high population density. Furthermore, mammalian species that are highly specialized with regard to diet or habitat are more threatened by habitat changes, while species with slow life histories, such as primates, are more threatened by hunting and other direct effects (González-Suárez et al. 2013).

[2] It seems almost a certainty that with the disappearance of Miss Waldron's red colobus (*Procolobus waldroni*), the first primate species has been driven to extinction in modern times (McGraw 2005; Oates et al. 2016).

The notion that behavioural flexibility may be beneficial for coping with environmental change has received support from a comparative study showing that within-species variability in life history (e.g. age at first reproduction) and population density appears to reduce the vulnerability of mammals to extinction (González-Suárez and Revilla 2013). Furthermore, mammals with relatively large brains (and therefore presumably greater behavioural flexibility) are more successful in novel environments than mammals with relatively small brains (Sol et al. 2008). However, a different study indicated that larger brains in primates and other mammals are associated with increased vulnerability to extinction (Gonzalez-Voyer et al. 2016). Thus, the behavioural flexibility resulting from large brains in primates may not outweigh the costs of slow life histories and the higher energy demands that large brains necessitate and, therefore, may even represent a disadvantage in this heavily human-dominated world.

Phylogenetic comparative studies require large datasets (e.g. PanTHERIA or AnAge; Jones et al. 2009; Tacutu et al. 2013), but there are often problems related to accuracy and comparability (Borries et al. 2016). Furthermore, animals are constantly making behavioural adjustments to environmental conditions, and, therefore, species parameters, such as group size, vary considerably depending on the study location, the provisioning of the group, or the study period (Strier 2009, 2017; Borries et al. 2016). To address this concern, some studies include the coefficient of variation (CV) of a variable to explicitly assess the effect of within-species variation (González-Suárez and Revilla 2013; Kamilar and Baden 2014). However, to observe a large proportion of the possible variation within a species, samples from long periods of time and different locations are necessary (Chapman et al., Chap. 17, this volume; Hogan and Melin, Chap. 10, this volume), and for most species this is simply not available. In such analyses it is also important to consider that variation assessed by CVs is positively related to the number of data points and the duration of the study (Strier, Chap. 2, this volume; González-Suárez and Revilla 2013; Strier et al. 2014). Finally, comparative studies are limited to the variables included in comparative databases, and these may not necessarily be the variables of interest with regard to the survival of a population.

Thus, while comparative studies can lead to important insights and generalizations as to how different species or populations of primates may or may not respond to anthropogenic changes, these limitations should be kept in mind, and it will remain important to investigate single species as this information is "essential to understand the species-specific aspects of vulnerability and potential for recovery" (Fedigan and Jack 2012, p. 181).

14.4.2 *Niche Modelling Studies*

One approach to assessing vulnerability at the species level is through niche modelling (e.g. Johnson and Brown, Chap. 15, this volume) which investigates broader factors (e.g. rainfall patterns, elevation, temperature) that affect the geographical distribution of a species. By extrapolating from the results of such studies, it is possible to predict which unoccupied habitats might be suitable for a specific species (Vidal-García and Serio-Silva 2011) and how this species might be able to cope with anticipated environmental change in their current habitats.

For example, Vidal-Garcia and Serio-Silva (2011) used a niche modelling approach to develop a distribution model for the primates of Southern Mexico. Using records of the presence of the 3 endemic primates and 19 potential environmental predictors of their distribution, they found strong relationships (e.g. *Alouatta palliata* was strongly associated with precipitation during the coldest quarter of the year). By using this modelling approach, these authors located areas with a high probability of the presence of the target primate, information that is now being used in conservation planning. However, such studies do not typically have the data required to investigate the proximate causes limiting population distributions, such as daily access to drinking water sources or required plant resources.

14.4.3 Behavioural Ecological Approach

Investigating the mechanisms underlying the persistence or extinction of populations on the individual level can be achieved by combining behavioural ecology with conservation biology. This behavioural ecological approach to conservation has also been labelled as "Conservation Behaviour" (Caro and Durant 1995; Blumstein and Fernández-Juricic 2010; Blumstein 2012), and it can contribute to many critical aspects of conservation, including predictions of population persistence, design of protected reserves, and management of populations (Caro and Durant 1995).

The idea behind conservation behaviour is that individual behavioural responses to environmental changes determine individual survival, reproduction, and migration, which ultimately determines population dynamics (Blumstein 2012; Wong and Candolin 2015). In other words, the focus on individual survival and reproduction of behavioural ecological studies is shifted to a focus on the broader survival of populations in conservation behaviour. Blumstein (2012) stressed the importance of social and mating behaviour in such investigations, as the link from environmental variation to demographic success often goes through social structure and breeding system. For example, environmental factors affect dispersal and mating patterns, which determine effective, and, therefore, minimal viable population sizes (Caro and Durant 1995). Knowing these parameters is important when designing protected areas of the appropriate size, shape, and connectivity. Anthropogenic changes can also affect the occurrence of sexually selected behaviours, such as male infanticide in primates, which can impede population growth (Jack and Fedigan, Chap. 6, this volume; Sterck 1999). Thus, environmental conditions such as the distribution and abundance of resources, predator density, cover from predators, and pathogens are linked to demographic structure through individual behaviour. Investigating this link can reveal insights into the adaptiveness of primate behavioural flexibility, uncover key factors that determine population persistence, and predict population responses to anthropogenic change, all of which can help improve conservation efforts.

14.5 How Can Long-Term Datasets Be Used to Improve Our Understanding of Primate Vulnerability and, Therefore, Our Ability to Conserve Primate Populations?

As a result of the slow life histories of primates, long-term datasets spanning several generations are often necessary to observe the relationship between environmental change and population dynamics (Strier 2009; Fedigan and Jack 2012; Campos et al. 2015; Chapman et al. 2018). Such datasets require a lot of effort and dedication to collect; thus only a few primate populations have now been continuously observed for the needed duration (e.g. the seven populations of different primate species that are part of the "Primate Life History Database"; see Strier et al. 2010; Campos et al. 2017). The collection of such long-term data also requires substantial continuous funding, a requirement that is getting harder to achieve despite a very apparent need.

Long-term data can be supplemented from other sources to generate more comprehensive datasets with which to investigate the impact of environmental change on population dynamics. To assess *environmental change*, questionnaires can be used to assess human activity and its impact on primates and their habitats (McKinney 2015). Aerial photographs and historic and contemporary satellite imagery can reveal information about changes in forest structure and coverage (Harper et al. 2007). Satellite imagery can also be used to estimate human-primate interactions by assessing human population density and the distance between human settlements and roads (e.g. logging routes) to primate habitats (Espinosa et al. 2014). Long-term data on past, current, and predicted temperature or rainfall are available from online resources, such as WorldClim (http://www.worldclim.org/), or from the Data Distribution Centre of the Intergovernmental Panel on Climate Change (IPCC DDC; http://www.ipcc-data.org/). Some long-term datasets also include plant phenology data (e.g. Santa Rosa, Costa Rica, Hogan and Melin, Chap. 10, this volume; Kibale, Uganda, Chapman et al. 2018; Cabang Panti Research Station, Borneo, Dillis et al. 2015), which, in combination with plant transects and food lists, can be used to assess variation in food availability over space and time.

Data on *individual behaviour* is often the central piece of long-term primate behavioural ecological studies and can be used to assess changes in both activity budgets and more specific patterns of social, feeding, and ranging behaviour in response to environmental change. Furthermore, long-term studies also put a great deal of effort into collecting data on life-history events such as emigration, death, or birth of individuals. These data are essential to assess the effect of individual behaviour on *survival and reproduction*, which is of central interest to behavioural ecologists. The same long-term life-history data can also be used to assess the effects of environmental change on survival and fertility rates (Campos et al. 2017). However, to assess changes in *demographic structure* of an entire population, regular area-wide censuses are necessary (e.g. Barro Colorado Island, Panama, Milton and Giacalone 2014; Beza Mahafaly, Sussman et al. 2012; Hacienda La Pacifica, Costa Rica, Clarke and Glander 2010; Hato Masaguaral, Venezuela, Rudran and

Fernandez-Duque 2003; Kibale, Uganda, Chapman et al. 2018; Kinkazan, Japan, Yamagiwa 2010; Santa Rosa, Costa Rica, Fedigan and Jack 2012) ideally including knowledge about sex and age composition of individuals in the encountered groups. Ultimately, changes in demographic structure can be used to make predictions about the *survival of populations*.

Such comprehensive datasets can be used to investigate how primates respond to changing environments, to make predictions about the viability of populations, and to implement conservation measures. For example, if females can breed throughout the year, how does this affect individual fitness and the dynamics within the population? Such questions can be addressed by using a comprehensive, longitudinal dataset to (1) assess fluctuation in food availability, temperature, and rainfall patterns; (2) determine the timing of life-history events with regard to environmental conditions for the entire populations; (3) investigate whether individuals have greater reproductive success, better health, and longer life expectancy if they adjust the timing of life-history events from year to year according to environmental conditions; and finally (4) assess how this individual flexibility affects the demographic structures of the entire population and improves the population viability in comparison to non-flexible species (see also Campos et al. 2017; Strier, Chap. 2, this volume). If flexibility is adaptive, these populations should be less affected by periods of unfavourable conditions, such as periods of resource scarcity. The knowledge acquired could be used to either provision susceptible populations with water or food during crucial periods (but this creates issues as well; e.g. Asquith 1989) or to specifically design protected areas that ensure sufficient resources and protection during these periods.

Building comprehensive and meaningful species models that are generalizable to a variety of habitats that link environmental change to population persistence through individual behaviour is challenging and currently only possible for very few primate species. Nevertheless, investigating the link between only some of the factors at a time, for example, the consequence of changes in temperature and rainfall on demography (Campos et al. 2015) or the fluctuation in resource availability or logging on population abundance (Chapman et al. 2018), can also be informative with regard to understanding primate adaptations to changing environments. Importantly, based on such studies, researchers could go back to their datasets and ask more specific questions about the behavioural mechanisms underlying the observed link (Fedigan and Jack 2001, 2012).

14.6 Comparisons Across Species, Space, and Time

While a single-species approach is necessary to identify the mechanism underlying primate population responses to environmental change, responding to major issues like climate change, deforestation, or bushmeat requires generalizations that are applicable to sets of species (phylogenetic and functional groups), times, and locations. Environmental factors, individual behaviour, and the demographic structure of populations can be compared along three different dimensions to disentangle flexibility and phylogenetic constraints in primate responses to changing environments (Chapman and Rothman 2009). First, long-term studies on a single population can be considered to be a comparative study over time (*temporal dimension*) because a population at one given time is compared with the same population at another time when conditions have changed. Second, populations of the same species can be compared across different habitats (*spatial dimension*). Third, different species can be compared within the same habitat (*phylogenetic dimension*). Furthermore, these three dimensions can be combined, for example, by comparing different species across different habitats (i.e. *spatial* and *phylogenetic dimension*). When investigating responses to environmental change, we think that the following three types of comparisons as especially useful: (1) *spatial comparisons*, (2) *phylogenetic and temporal comparisons*, and (3) *phylogenetic*, *spatial*, and *temporal comparisons.*

14.6.1 Spatial Comparisons

Comparisons within the same species across different habitats that have experienced different types and degrees of modification can advance our understanding of the potential of within-species flexibility (Struhsaker 1999; Chapman and Peres 2001; Chapman et al. 2010). For example, Meijaard et al. (2010) found that population densities of Bornean orangutans between conservation areas and pulp and paper plantations were similar and suggested that behavioural flexibility facilitated these apes surviving in modified landscapes. Such contrasts can be considered natural experiments that allow us to study animal adaptions (Schroeder et al. 2011; but see Caro and Sherman 2011 for limitations of this approach).

14.6.2 Phylogenetic and Temporal Comparisons

By comparing different species in the same habitat over time, it is possible to directly compare responses across species towards the same environmental changes. For example, howler and capuchin monkeys (*Cebus capucinus imitator*) in Santa Rosa, Costa Rica, showed differences in population growth in the same regenerating forest. In this case, both species faced the same changes in the environment, and differences in population dynamics can probably be attributed to differences in diet, life-history pace, dispersal patterns, and behavioural flexibility (Fedigan and Jack 2001, 2012). Additionally, studying several species within the same habitat enables the investigation of interactive effects at the community level, such as density compensation (Peres and Dolman 2000).

14.6.3 Phylogenetic, Spatial, and Temporal Comparisons

Comparisons of different species across different locations over time can be insightful if the locations are undergoing similar changes. For example, Campos et al. (2017) analysed long-term data to investigate the impact of climate variability on fertility and survival rates in seven species of primates and found out that highly seasonal species appear to be more vulnerable to climate change than non-seasonally breeding species. The challenge of such comparisons is to control for ecological and demographic differences between locations that potentially affect the observed response (Strier 2009), such as the number of receptive females when comparing male behaviour across species and habitats (e.g. Kalbitzer et al. 2015).

14.7 Integrating Behavioural Ecological Studies into Primate Conservation

The integration of conservation biology and behavioural ecology faces many challenges (Caro and Sherman 2013), yet a better incorporation of these two theoretical frameworks could help to improve efforts to protect primates. Primate behavioural ecologists can help to improve this integration by specifically considering (1) the effects of anthropogenic changes on behaviour and (2) the effects of specific behavioural responses, for example, the alteration of breeding seasonality, on population dynamics (McLennan et al. 2017). It is also important to make relevant data more easily available to other conservation biologists, managers, and the public. Similar to some conservation-oriented journals (e.g. *Biological Conservation*), behavioural ecological journals could include a dedicated space at the end of their published articles in which authors are asked to include information on the relevance, if any, of their findings to conservation. This could include information on predicted changes in population size as a result of environmental change or a description of the size and quality of habitat necessary to preserve the future populations. Furthermore, determining and communicating possible key factors that limit population growth or increase mortality in its current habitat, such as a lack of drinking water or protein-rich food (Milton and Giacalone 2014), may be critical to the implementation of measures, such as the artificial provisioning of these resources, or to guide regeneration projects that ensure the viability of these populations despite anticipated environmental change.

14.8 Conclusion

Our world is undergoing anthropogenic changes at an unprecedented pace and scale; thus it is crucial to understand how primates respond to these changes to prevent further extinctions. Behavioural flexibility will be vital for many primates to survive, and careful comparative investigations using data spanning generations of primates are required to determine whether this flexibility is sufficient to prevent population decline and to improve conservation efforts. Fortunately, some ongoing long-term studies have already collected data over a few decades, which can be supplemented with available data on environmental change to conduct analyses as to how primates are able to respond to the Anthropocene. However, based on the finding of the current long-term studies that are just now emerging, we highlight the need for more long-term studies that are explicitly designed to quantify change in behaviours and identify potential drivers of changes. For the many primate species without longitudinal data, extrapolating from other long-term studies offers the potential to obtain valuable conservation insights.

While it is important and often exciting to investigate questions about the ability of organisms to respond to environmental change, we should also make use of our own potential for flexibility and modify our own behaviour to preserve the exciting diversity of primates and biodiversity that we find in this world.

Acknowledgements Most importantly, we want to thank Linda Fedigan for her inspiring research and the great collaboration during the last 3+ years (UK) and for the last 40+ years (CAC) and for giving us a good reason to having a great symposium in honour of her academic career (Jack and Kalbitzer 2017). Our motivation for this chapter came from our work in different primate habitats across the globe and the observation that some species are very successful in dealing with human pressures, while others disappear. Furthermore, UK is grateful that he had the opportunity to work with the PACE database at the University of Calgary for the last few years, which was very inspiring regarding the discovery of the potential lying in such datasets. We also thank Katharine Jack, Peter Henzi, and one anonymous reviewer for providing us with constructive feedback on earlier drafts of this manuscript. UK was funded by an Eyes High Postdoctoral Fellowship from the University of Calgary while writing this manuscript.

References

Altmann J, Alberts SC, Altmann SA, Roy SB (2002) Dramatic change in local climate patterns in the Amboseli basin, Kenya. Afr J Ecol 40:248–251

Andam KS, Ferraro PJ, Pfaff A et al (2008) Measuring the effectiveness of protected area networks in reducing deforestation. Proc Natl Acad Sci 105:16089–16094

Arroyo-Rodríguez V, Moral EC, Mandujano S et al (2013) Assessing habitat fragmentation effects on primates: the importance of evaluating questions at the correct scale. In: Marsh LK, Chapman CA (eds) Primates in fragments. Springer, New York, pp 13–28

Asquith PJ (1989) Provisioning and the study of free-ranging primates: history, effects, and prospects. Am J Phys Anthropol 32:129–158

Barnosky AD, Matzke N, Tomiya S et al (2011) Has the Earth's sixth mass extinction already arrived? Nature 471:51–57

Baya L, Storch I (2010) Status of diurnal primate populations at the former settlement of a displaced village in Cameroon. Am J Primatol 72:645–652

Beehner JC, Onderdonk DA, Alberts SC, Altmann J (2006) The ecology of conception and pregnancy failure in wild baboons. Behav Ecol Behav Ecol 17:741–750

Behie AM, Pavelka MSM (2005) The short-term effects of a hurricane on the diet and activity of black howlers (*Alouatta pigra*) in Monkey River, Belize. Folia Primatol (Basel) 76:1–9

Behie AM, Pavelka MSM (2013) Interacting roles of diet, cortisol levels, and parasites in determining population density of Belizean howler monkeys in a hurricane damaged forest fragment. In: Marsh LK, Chapman CA (eds) Primates in fragments. Springer New York, New York, pp 447–456

Behie AM, Pavelka MSM, Hartwell K et al (in press) Alas the storm has come again! The impact of frequent natural disasters on primate conservation. In: Behie AM, Teichroeb JA, Malone N (eds) Primate Research and Conservation in the Anthropocene. Cambridge University Press, Cambridge/New York

Bermejo M, Rodríguez-Teijeiro JD, Illera G et al (2006) Ebola outbreak killed 5000 gorillas. Science 314:1564–1564

Bicca-Marques JC, Calegaro-Marques C Rylands AB, et al. (2017) Yellow fever threatens Atlantic Forest primates – eLetter in response to "Impending extinction crisis of the world's primates: why primates matter" (Estrada et al. 2017)

Blumstein DT (2012) Social behaviour. In: Candolin U, Wong BBM (eds) Behavioural responses to a changing world: mechanisms and consequences. Oxford University Press, Oxford

Blumstein DT, Fernández-Juricic E (2010) A primer of conservation behavior, 1st edn. Sinauer Associates, Sunderland

Boesch C (2008) Why do chimpanzees die in the forest? The challenges of understanding and controlling for wild ape health. Am J Primatol 70:722–726

Bonilla-Sánchez YM, Serio-Silva JC, Pozo-Montuy G, Chapman CA (2012) Howlers are able to survive in eucalyptus plantations where remnant and regenerating vegetation is available. Int J Primatol 33:233–245

Borries C, Sandel AA, Koenig A et al (2016) Transparency, usability, and reproducibility: guiding principles for improving comparative databases using primates as examples. Evol Anthropol 25:232–238

Brook BW, Sodhi NS, Bradshaw CJA (2008) Synergies among extinction drivers under global change. Trends Ecol Evol 23:453–460

Bruner AG, Gullison RE, Rice RE, da Fonseca GAB (2001) Effectiveness of parks in protecting tropical biodiversity. Science 291:125–128

Butchart SHM, Walpole M, Collen B et al (2010) Global biodiversity: indicators of recent declines. Science 328:1164–1168

Cai W, Borlace S, Lengaigne M et al (2014) Increasing frequency of extreme El Nino events due to greenhouse warming. Nat Clim Change 4:111–116

Cai W, Santoso A, Wang G et al (2015) ENSO and greenhouse warming. Nat Clim Change 5:849

Campos FA (2018) A synthesis of long-term environmental change in Santa Rosa, Costa Rica. In: Kalbitzer U, Jack KM (eds) Primate life histories, sex roles, and adaptability – essays in honour of Linda M. Fedigan. Springer, New York, pp 331–354

Campos FA, Jack KM, Fedigan LM (2015) Climate oscillations and conservation measures regulate white-faced capuchin population growth and demography in a regenerating tropical dry forest in Costa Rica. Biol Conserv 186:204–213

Campos FA, Morris WF, Alberts SC et al (2017) Does climate variability influence the demography of wild primates? Evidence from long-term life-history data in seven species. Glob Change Biol 23:4907–4921

Carey J (2016) Core concept: are we in the "Anthropocene"? Proc Natl Acad Sci 113:3908–3909

Carnegie SD, Fedigan LM, Melin AD (2011) Reproductive seasonality in female capuchins (*Cebus capucinus*) in Santa Rosa (Area de Conservaci6n Guanacaste), Costa Rica. Int J Primatol 32:1076–1090

Caro TM, Durant SM (1995) The importance of behavioral ecology for conservation biology: examples from Serengeti carnivores. In: Serengeti II: dynamics, management, and conservation of an ecosystem. Chicago University Press, Chicago, pp 451–472

Caro T, Sherman PW (2011) Behavioural ecology cannot profit from unstructured environmental change. Trends Ecol Evol 26:321–322

Caro T, Sherman PW (2013) Eighteen reasons animal behaviourists avoid involvement in conservation. Anim Behav 85:305–312

Ceballos G, Ehrlich PR, Barnosky AD et al (2015) Accelerated modern human–induced species losses: entering the sixth mass extinction. Sci Adv 1:e1400253
Ceballos G, Ehrlich PR, Dirzo R (2017) Biological annihilation via the ongoing sixth mass extinction signaled by vertebrate population losses and declines. Proc Natl Acad Sci 114:E6089–E6096
Chapman CA (2018) A road for a promising future for China's primates: The potential for restoration. Zoological Research 39:244–248
Chapman CA, Gogarten JF (2012) Primate conservation: is the cup half empty or half full? Nat Educ Knowl 4:7
Chapman CA, Peres CA (2001) Primate conservation in the new millennium: the role of scientists. Evol Anthropol Issues News Rev 10:16–33
Chapman CA, Rothman JM (2009) Within-species differences in primate social structure: evolution of plasticity and phylogenetic constraints. Primates 50:12–22
Chapman CA, Balcomb SR, Gillespie TR et al (2000) Long-term effects of logging on African primate communities: a 28-year comparison from Kibale National Park, Uganda. Conserv Biol 14:207–217
Chapman CA, Chapman LJ, Struhsaker TT et al (2005a) A long-term evaluation of fruiting phenology: importance of climate change. J Trop Ecol 21:31–45
Chapman CA, Gillespie TR, Goldberg TL (2005b) Primates and the ecology of their infectious diseases: how will anthropogenic change affect host-parasite interactions? Evol Anthropol 14:134–144
Chapman CA, Lawes MJ, Eeley HAC (2006) What hope for African primate diversity? Afr J Ecol 44:116–133
Chapman CA, Struhsaker TT, Skorupa JP et al (2010) Understanding long-term primate community dynamics: implications of forest change. Ecol Appl 20:179–191
Chapman CA, Twinomugisha D, Teichroeb JA et al (2016) How do primates survive among humans? Mechanisms employed by vervet monkeys at Lake Nabugabo, Uganda. In: Waller MT (ed) Ethnoprimatology – primate conservation in the 21st century. Springer, New York, pp 77–94
Chapman CA, Bortolamiol S, Matsuda I, Omeja PA, Paim FP, Reyna-Hurtado R, Sengupta R, Valenta K (2018) Primate population dynamics: variation in abundance over space and time. Biodivers Conserv 27:1221–1238
Chapman CA, Valenta K, Bonnell TR, Brown KA, Chapman LJ (2018). Solar radiation and ENSO predict fruiting phenology patterns in a 15-year record from Kibale National Park. Uganda. Biotropica 50:384–395
Chapman CA, Valenta K, Bortolamiol S (2018) How variable is a primate's world: spatial and temporal variation in potential ecological drivers of behaviour? In: Kalbitzer U, Jack KM (eds) Primate life histories, sex roles, and adaptability – essays in honour of Linda M. Fedigan. Springer, New York, pp 359–371
Clarke MR, Glander KE (2010) Secondary transfer of adult mantled howlers (*Alouatta palliata*) on Hacienda La Pacifica, Costa Rica: 1975–2009. Primates 51:241–249
Conly JM, Johnston BL (2008) The infectious diseases consequences of monkey business. Can J Infect Dis Med Microbiol 19:12–14
Corlett RT (2012) Climate change in the tropics: the end of the world as we know it? Biol Conserv 151:22–25
Crutzen PJ, Stoermer EF (2000) The "anthropocene", global change. IGBP Newsl 41:17–18
de Almeida-Rocha JM, Peres CA, Oliveira LC (2017) Primate responses to anthropogenic habitat disturbance: a pantropical meta-analysis. Biol Conserv 215:30–38
Diffenbaugh NS, Singh D, Mankin JS et al (2017, 114) Quantifying the influence of global warming on unprecedented extreme climate events. Proc Natl Acad Sci:4881–4886
Dillis C, Beaudrot L, Feilen KL et al (2015) Modeling the ecological and phenological predictors of fruit consumption by gibbons (*hylobates albibarbis*). Biotropica 47:85–93
Dittus WPJ (1988) Group fission among wild toque macaques as a consequence of female resource competition and environmental stress. Anim Behav 36:1626–1645

Dunham AE, Erhart EM, Wright PC (2011) Global climate cycles and cyclones: consequences for rainfall patterns and lemur reproduction in southeastern Madagascar. Glob Change Biol 17:219–227

Emrich A, Pokorny B, Sepp C (2000) The significance of secondary forest management for development policy. Deutsche Gesellschaft für Technische Zusammenarbeit (GTZ), Eschborn, Germany

Erhart EM, Overdorff DJ (2008) Population demography and social structure changes in *Eulemur fulvus rufus* from 1988 to 2003. Am J Phys Anthropol 136:183–193

Espinosa S, Branch LC, Cueva R (2014) Road development and the geography of hunting by an Amazonian indigenous group: consequences for wildlife conservation. PLOS ONE 9:e114916

Estrada A, Raboy BE, Oliveira LC (2012) Agroecosystems and primate conservation in the tropics: a review. Am J Primatol 74:696–711

Estrada A, Garber PA, Rylands AB et al (2017) Impending extinction crisis of the world's primates: why primates matter. Sci Adv 3:e1600946

Fa JE, Brown D (2009) Impacts of hunting on mammals in African tropical moist forests: a review and synthesis. Mammal Rev 39:231–264

FAO (2016) Global Forest Resources Assessment 2015: how are the world's forest changing? 2nd edn. Food and Agriculture Organization of the United Nations, Rome

FAO (2017) Food and Agriculture Organization of the United Nations – FAOSTAT Database. http://www.fao.org/faostat/en/#data. Accessed 27 Nov 2017

Fedigan LM, Jack K (2001) Neotropical primates in a regenerating Costa Rican dry forest: a comparison of howler and capuchin population patterns. Int J Primatol 22:689–713

Fedigan LM, Jack KM (2012) Tracking neotropical monkeys in Santa Rosa: lessons from a regenerating Costa Rican dry forest. In: Kappeler PM, Watts DP (eds) Long-term field studies of primates. Springer, Berlin/Heidelberg, pp 165–184

Fernandes NCC d A, Cunha MS, Guerra JM et al (2017) Outbreak of yellow fever among nonhuman primates, Espirito Santo, Brazil, 2017. Emerg Infect Dis 23:2038–2041

Foster RB (1982) The seasonal rhythm of fruitfall on Barro Colorado Island. In: Leigh EG Jr, Rand AS, Windsor DM (eds) The ecology of a tropical forest: seasonal rhythms and long-term changes. Smithsonian Institution Press, Washington, DC, pp 151–172

Fuentes A, Gamerl S (2005) Disproportionate participation by age/sex classes in aggressive interactions between long-tailed macaques (*Macaca fascicularis*) and human tourists at Padangtegal Monkey Forest, Bali, Indonesia. Am J Primatol 66:197–204

Gaveau DLA, Wich SA, Marshall AJ (2016) Are protected areas conserving primate habitat in Indonesia? In: Wich SA, Marshall AJ (eds) An introduction to primate conservation. Oxford University Press, Oxford, pp 193–204

Gogarten JF, Jacob AL, Ghai RR et al (2015) Group size dynamics over 15+ years in an African forest primate community. Biotropica 47:101–112

Goldberg TL, Gillespie TR, Rwego IB et al (2008) Forest fragmentation as cause of bacterial transmission among nonhuman primates, humans, and livestock, Uganda. Emerg Infect Dis 14:1375–1382

González-Suárez M, Revilla E (2013) Variability in life-history and ecological traits is a buffer against extinction in mammals. Ecol Lett 16:242–251

González-Suárez M, Gómez A, Revilla E (2013) Which intrinsic traits predict vulnerability to extinction depends on the actual threatening processes. Ecosphere 4:1–16

Gonzalez-Voyer A, González-Suárez M, Vilà C, Revilla E (2016) Larger brain size indirectly increases vulnerability to extinction in mammals. Evolution 70:1364–1375

Gouveia SF, Villalobos F, Dobrovolski R et al (2014) Forest structure drives global diversity of primates. J Anim Ecol 83:1523–1530

Graham TL, Matthews HD, Turner SE (2016) A global-scale evaluation of primate exposure and vulnerability to climate change. Int J Primatol 37:158–174

Hahn BH, Shaw GM, De KM et al (2000) AIDS as a zoonosis: scientific and public health implications. Science 287:607–614

Hanya G, Chapman CA (2012) Linking feeding ecology and population abundance: a review of food resource limitation on primates. Ecol Res 28:183–190

Harper GJ, Steininger MK, Tucker CJ et al (2007) Fifty years of deforestation and forest fragmentation in Madagascar. Environ Conserv 34:325–333

Hartter J, Ryan SJ, Southworth J, Chapman CA (2011) Landscapes as continuous entities: forest disturbance and recovery in the Albertine Rift landscape. Landsc Ecol 26:877

Harvey PH, Pagel MD (1991) The comparative method in evolutionary biology. Oxford University Press, Oxford

Hill CM (2000) Conflict of interest between people and baboons: crop raiding in Uganda. Int J Primatol 21:299–315

Hill R (2005) Day length seasonality and the thermal environment. In: Brockman DK, van Schaik CP (eds) Seasonality in primates – studies of living and extinct human and non-human primates. Cambridge University Press, Cambridge/New York, pp 197–213

Hoffman TS, O'Riain MJ (2011) The spatial ecology of chacma baboons (*Papio ursinus*) in a human-modified environment. Int J Primatol 32:308–328

Hoffman TS, O'Riain MJ (2012) Landscape requirements of a primate population in a human-dominated environment. Front Zool 9:1

Hogan JD, Melin AD (2018) Intra- and interannual variation in the fruit diet of wild capuchins: impact of plant phenology. In: Kalbitzer U, Jack KM (eds) Primate life histories, sex roles, and adaptability – essays in honour of Linda M. Fedigan. Springer, New York, pp 193–210

IPCC (2013) Summary for policymaker. In: Stocker TF, Qin D, Plattner GK et al (eds) Climate change 2013: the physical science basis. Contribution of working group I to the fifth assessment report of the Intergovernmental Panel on Climate Change. Cambridge University Press, Cambridge/New York, pp 1–30

IPCC (2014) Climate change 2014: synthesis report. Contribution of working groups I, II and III to the fifth assessment report of the Intergovernmental Panel on Climate Change. IPCC, Geneva, Switzerland

Isbell LA (1991) Contest and scramble competition: patterns of female aggression and ranging behavior among primates. Behav Ecol 2:143–155

IUCN (2016) IUCN Red List of Threatened Species. Version 2016–1

Jack KM, Fedigan LM (2018) Alpha male capuchins (*Cebus capucinus imitator*) as keystone individuals. In: Kalbitzer U, Jack KM (eds) Primate life histories, sex roles, and adaptability – essays in honour of Linda M. Fedigan. Springer, New York, pp 91–109

Jack KM, Kalbitzer U (2017) How to cultivate a tree: celebrating the career of Linda Marie Fedigan. Evol Anthropol Issues News Rev 26:139–142

Jacob AL, Vaccaro I, Sengupta R et al (2008) Integrating landscapes that have experienced rural depopulation and ecological homogenization into tropical conservation planning. Trop Conserv Sci 1:307–320

Janson C, Verdolin J (2005) Seasonality of primate births in relation to climate. In: Brockman DK, van Schaik CP (eds) Seasonality in primates – studies of living and extinct human and non-human primates. Cambridge University Press, Cambridge/New York, pp 307–350

Johnson SE, Brown KA (2018) The specialist capuchin? Using ecological niche models to compare niche breadth in Mesoamerican primates. In: Kalbitzer U, Jack KM (eds) Primate life histories, sex roles, and adaptability – essays in honour of Linda M. Fedigan. Springer, New York, pp 311–326

Johnson SE, Ingraldi C, Ralainasolo FB et al (2011) Gray-headed lemur (*Eulemur cinereiceps*) abundance and forest structure dynamics at Manombo, Madagascar. Biotropica 43:371–379

Jones KE, Bielby J, Cardillo M et al (2009) PanTHERIA: a species-level database of life history, ecology, and geography of extant and recently extinct mammals: ecological archives E090-184. Ecology 90:2648–2648

Jones-Engel L, Engel GA, Schillaci MA et al (2005) Primate-to-human retroviral transmission in Asia. Emerg Infect Dis 11:1028

Jones-Engel L, Engel GA, Heidrich J et al (2006a) Temple monkeys and health implications of commensalism, Kathmandu, Nepal. Emerg Infect Dis 12:900–906

Jones-Engel L, Engel GA, Schillaci MA et al (2006b) Considering human–primate transmission of measles virus through the prism of risk analysis. Am J Primatol 68:868–879

Jones-Engel L, May CC, Engel GA et al (2008) Diverse contexts of zoonotic transmission of simian foamy viruses in Asia. Emerg Infect Dis 14:1200–1208

Joppa LN, Pfaff A (2009) High and far: biases in the location of protected areas. PLOS ONE 4:e8273

Joppa L, Pfaff A (2010) Reassessing the forest impacts of protection. Ann N Y Acad Sci 1185:135–149

Joppa LN, Loarie SR, Pimm SL (2008) On the protection of "protected areas". Proc Natl Acad Sci 105:6673–6678

Kalbitzer U, Heistermann M, Cheney D et al (2015) Social behavior and patterns of testosterone and glucocorticoid levels differ between male chacma and Guinea baboons. Horm Behav 75:100–110

Kamilar JM, Baden AL (2014) What drives flexibility in primate social organization? Behav Ecol Sociobiol 68:1677–1692

Kavanagh M (1980) Invasion of the forest by an African savannah monkey: behavioural adaptations. Behaviour 73:238–260

Koenig A, Scarry CJ, Wheeler BC, Borries C (2013) Variation in grouping patterns, mating systems and social structure: what socio-ecological models attempt to explain. Philos Trans R Soc B Biol Sci 368:20120348–20120348

Köndgen S, Kühl H, N'Goran PK et al (2008) Pandemic human viruses cause decline of endangered great apes. Curr Biol 18:260–264

Köndgen S, Calvignac-Spencer S, Grützmacher K et al (2017) Evidence for human Streptococcus pneumoniae in wild and captive chimpanzees: a potential threat to wild populations. Sci Rep 7:14581

Laurance WF, Goosem M, Laurance SGW (2009) Impacts of roads and linear clearings on tropical forests. Trends Ecol Evol 24:659–669

Laurance WF, Sayer J, Cassman KG (2014) Agricultural expansion and its impacts on tropical nature. Trends Ecol Evol 29:107–116

Leroy EM, Rouquet P, Formenty P et al (2004) Multiple Ebola virus transmission events and rapid decline of central African wildlife. Science 303:387–390

Linder JM (2008) The impact of hunting on primates in Korup National Park, Cameroon: implications for primate conservation. City University of New York, New York

Linder JM (2013) African primate diversity threatened by "new wave" of industrial oil palm expansion. Afr Primates 8:25–38

Linder JM, Oates JF (2011) Differential impact of bushmeat hunting on monkey species and implications for primate conservation in Korup National Park, Cameroon. Biol Conserv 144:738–745

Lovett JC, Marshall AR (2006) Why should we conserve primates? Afr J Ecol 44:113–115

Marchal V, Hill C (2009) Primate crop-raiding: a study of local perceptions in four villages in north Sumatra, Indonesia. Primate Conserv 24:107–116

Mbora DNM, Wieczkowski J, Munene E (2009) Links between habitat degradation, and social group size, ranging, fecundity, and parasite prevalence in the Tana River mangabey (*Cercocebus galeritus*). Am J Phys Anthropol 140:562–571

McDougall P, Forshaw N, Barrett L, Henzi SP (2010) Leaving home: responses to water depletion by vervet monkeys. J Arid Environ 74:924–927

McGraw WS (2005) Update on the search for Miss Waldron's red colobus monkey. Int J Primatol 26:605–619

McKinney T (2015) A classification system for describing anthropogenic influence on nonhuman primate populations. Am J Primatol 77:715–726

McLennan MR, Spagnoletti N, Hockings KJ (2017) The implications of primate behavioral flexibility for sustainable human–primate coexistence in anthropogenic habitats. Int J Primatol 38:105–121

Meijaard E, Albar G, Nardiyono et al (2010) Unexpected ecological resilience in Bornean orangutans and implications for pulp and paper plantation management. PLOS ONE e12813:5
Mekonnen A, Fashing PJ, Bekele A et al (2017) Impacts of habitat loss and fragmentation on the activity budget, ranging ecology and habitat use of Bale monkeys (*Chlorocebus djamdjamensis*) in the southern Ethiopian Highlands. Am J Primatol 79:e22644
Milton K (1982) Dietary quality and demographic regulation in a howler monkey population. In: Leigh EG Jr, Rand AS, Windsor DM (eds) The ecology of a tropical forest: seasonal rhythms and long-term changes. Smithsonian Institution Press, Washington, DC, pp 273–289
Milton K, Giacalone J (2014) Differential effects of unusual climatic stress on capuchin (*Cebus capucinus*) and howler monkey (*Alouatta palliata*) populations on Barro Colorado Island, Panama. Am J Primatol 76:249–261
Moore RS, Nekaris KAI, Eschmann C (2010) Habitat use by western purple-faced langurs *Trachypithecus vetulus nestor* (Colobinae) in a fragmented suburban landscape. Endanger Species Res 12:227–234
Morellato LPC, Alberton B, Alvarado ST et al (2016) Linking plant phenology to conservation biology. Biol Conserv 195:60–72
Nunn CL, Barton RA (2001) Comparative methods for studying primate adaptation and allometry. Evol Anthropol Issues News Rev 10:81–98
Oates JF (1996) Habitat alteration, hunting and the conservation of folivorous primates in African forests. Aust J Ecol 21:1–9
Oates JF, Struhsaker T, McGraw S (2016) *Piliocolobus waldronae*. The IUCN red list of threatened species 2016: e.T18248A92649220. International Union for Conservation of Nature and Natural Resources (IUCN)
Omeja PA, Obua J, Rwetsiba A, Chapman CA (2012) Biomass accumulation in tropical lands with different disturbance histories: contrasts within one landscape and across regions. For Ecol Manag 269:293–300
Omeja PA, Jacob AL, Lawes MJ et al (2014) Changes in elephant abundance affect forest composition or regeneration. Biotropica 46:704–711
Omeja PA, Lawes MJ, Corriveau A et al (2016) Recovery of tree and mammal communities during large-scale forest regeneration in Kibale National Park, Uganda. Biotropica 48:770–779
Pacifici M, Visconti P, Butchart SHM et al (2017) Species' traits influenced their response to recent climate change. Nat Clim Change 7:205
Parmesan C (2006) Ecological and evolutionary responses to recent climate change. Annu Rev Ecol Evol Syst 37:637–669
Pebsworth PA, MacIntosh AJJ, Morgan HR, Huffman MA (2012) Factors influencing the ranging behavior of chacma baboons (*Papio hamadryas ursinus*) living in a human-modified habitat. Int J Primatol 33:872–887
Peres CA, Dolman PM (2000) Density compensation in neotropical primate communities: evidence from 56 hunted and nonhunted amazonian forests of varying productivity. Oecologia 122:175–189
Phalan B, Bertzky M, Butchart SHM et al (2013) Crop expansion and conservation priorities in tropical countries. PLOS ONE 8:e51759
Pozo-Montuy G, Serio-Silva JC, Chapman CA, Bonilla-Sánchez YM (2013) Resource use in a landscape matrix by an arboreal primate: evidence of supplementation in black howlers (*Alouatta pigra*). Int J Primatol 34:714–731
Purvis A, Gittleman JL, Cowlishaw G, Mace GM (2000) Predicting extinction risk in declining species. Proc R Soc Lond B Biol Sci 267:1947–1952
R Core Team (2016) R: a language and environment for statistical computing. R Foundation for Statistical Computing, Vienna, Austria
Raftery AE, Zimmer A, Frierson DMW et al (2017) Less than 2 °C warming by 2100 unlikely. Nat Clim Change 7:637–641
Rands MRW, Adams WM, Bennun L et al (2010) Biodiversity conservation: challenges beyond 2010. Science 329:1298–1303

Reader SM, Laland KN (2002) Social intelligence, innovation, and enhanced brain size in primates. Proc Natl Acad Sci 99:4436–4441

Reed KE, Fleagle JG (1995) Geographic and climatic control of primate diversity. Proc Natl Acad Sci 92:7874–7876

Remis MJ, Jost Robinson CA (2012) Reductions in primate abundance and diversity in a multiuse protected area: synergistic impacts of hunting and logging in a congo basin forest. Am J Primatol 74:602–612

Rigby HT White CJ (2017) Amazon deforestation, once tamed, comes roaring back. N. Y. Times

Robbins MM, Gray M, Fawcett KA et al (2011) Extreme conservation leads to recovery of the Virunga mountain gorillas. PLOS ONE 6:e19788

Rothman JM, Chapman CA, Struhsaker TT et al (2015) Long-term declines in nutritional quality of tropical leaves. Ecology 96:873–878

Rudran R, Fernandez-Duque E (2003) Demographic changes over thirty years in a red howler population in Venezuela. Int J Primatol 24:925–947

Sapolsky RM, Share LJ (2004) A pacific culture among wild baboons: its emergence and transmission. PLoS Biol 2:534–541

Schaffner CM, Rebecchini L, Ramos-Fernandez G et al (2012) Spider monkeys (*Ateles geoffroyi yucatenensis*) cope with the negative consequences of hurricanes through changes in diet, activity budget, and fission–fusion dynamics. Int J Primatol 33:922–936

Schroeder J, Nakagawa S, Hinsch M (2011) Behavioural ecology is not an endangered discipline. Trends Ecol Evol 26:320–321. https://doi.org/10.1016/j.tree.2011.03.013

Schwitzer C, Glatt L, Nekaris K, Ganzhorn J (2011) Responses of animals to habitat alteration: an overview focussing on primates. Endanger Species Res 14:31–38

Scully EJ, Basnet S, Wrangham RW et al (2018) Lethal respiratory disease associated with human rhinovirus C in wild chimpanzees, Uganda, 2013. Emerg Infect Dis J 24:2

Snaith TV, Chapman CA (2007) Primate group size and interpreting socioecological models: do folivores really play by different rules? Evol Anthropol Issues News Rev 16:94–106

Sol D, Bacher S, Reader SM, Lefebvre L (2008) Brain size predicts the success of mammal species introduced into novel environments. Am Nat 172:S63–S71

Sterck EHM (1999) Variation in langur social organization in relation to the socioecological model, human habitat alteration, and phylogenetic constraints. Primates 40:199–213

Sterck EHM, Watts DP, van Schaik CP (1997) The evolution of female social relationships in nonhuman primates. Behav Ecol Sociobiol 41:291–309

Strier KB (2009) Seeing the forest through the seeds: mechanisms of primate behavioral diversity from individuals to populations and beyond. Curr Anthropol 50:213–228

Strier KB (2017) What does variation in primate behavior mean? Am J Phys Anthropol 162:4–14

Strier KB (2018) The temporal scale of behavioral and demographic flexibility: implications for comparative analyses and conservation. In: Kalbitzer U, Jack KM (eds) Primate life histories, sex roles, and adaptability – essays in honour of Linda M. Fedigan. Springer, New York, pp 19–33

Strier KB, Ives AR (2012) Unexpected demography in the recovery of an endangered primate population. PLOS ONE 7:e44407

Strier KB, Mendes SL (2012) The northern muriqui (*brachyteles hypoxanthus*): lessons on behavioral plasticity and population dynamics from a critically endangered species. In: Kappeler PM, Watts DP (eds) Long-term field studies of primates. Springer, Berlin/Heidelberg, pp 125–140

Strier KB, Altmann J, Brockman DK et al (2010) The primate life history database: a unique shared ecological data resource. Methods Ecol Evol 1:199–211

Strier KB, Lee PC, Ives AR (2014) Behavioral flexibility and the evolution of primate social states. PLoS ONE 9:e114099

Struhsaker TT 1999 Ecology of an African rain forest: logging in Kibale and the conflict between conservation and exploitation. University Press of Florida, Gainesville, Fla

Sussman RW, Richard AF, Ratsirarson J et al (2012) Beza Mahafaly Special Reserve: long-term research on lemurs in southwestern Madagascar. In: Long-Term Field Studies of Primates. Springer, Berlin, Heidelberg, pp 45–66

Tacutu R, Craig T, Budovsky A et al (2013) Human ageing genomic resources: integrated databases and tools for the biology and genetics of ageing. Nucleic Acids Res 41:D1027–D1033
Vidal-García F, Serio-Silva JC (2011) Potential distribution of Mexican primates: modeling the ecological niche with the maximum entropy algorithm. Primates 52:261
Vilela B, Villalobos F (2015) letsR: a new R package for data handling and analysis in macroecology. Methods Ecol Evol 6:1229–1234
Weng L, Boedhihartono AK, Dirks PHGM et al (2013) Mineral industries, growth corridors and agricultural development in Africa. Glob Food Secur 2:195–202
Wheeler CE, Omeja PA, Chapman CA et al (2016) Carbon sequestration and biodiversity following 18 years of active tropical forest restoration. For Ecol Manag 373:44–55
Wilkie DS, Wieland M, Boulet H et al (2016) Eating and conserving bushmeat in Africa. Afr J Ecol 54:402–414
Wong BBM, Candolin U (2015) Behavioral responses to changing environments. Behav Ecol 26:665–673
Wong SNP, Sicotte P (2007) Activity budget and ranging patterns of *Colobus vellerosus* in forest fragments in central Ghana. Folia Primatol (Basel) 78:245–254
Wong SNP, Saj TL, Sicotte P (2006) Comparison of habitat quality and diet of *Colobus vellerosus* in forest fragments in Ghana. Primates 47:365–373
Woodford MH, Butynski TM, Karesh WB (2002) Habituating the great apes: the disease risks. Oryx 36(2)
Wrangham RW (1980) An ecological model of female-bonded primate groups. Behaviour 75:262–300
Wright SJ, Calderón O (2006) Seasonal, El Niño and longer term changes in flower and seed production in a moist tropical forest. Ecol Lett 9:35–44
Wright SJ, Muller-Landau HC (2006) The future of tropical forest species. Biotropica 38:287–301
Wright SJ, Carrasco C, Calderón O, Paton S (1999) The El Niño Southern Oscillation, variable fruit production, and famine in a tropical forest. Ecology 80:1632–1647
Yamagiwa J (2010) Research history of Japanese macaques in Japan. In: The Japanese macaques. Springer, Tokyo, pp 3–25

Chapter 15
The Specialist Capuchin? Using Ecological Niche Models to Compare Niche Breadth in Mesoamerican Primates

Steig E. Johnson and Kerry A. Brown

Abstract The high degree of dietary breadth and flexibility in capuchins (*Cebus* and *Sapajus*), coupled with their advanced cognitive abilities, is well documented. Owing to these characteristics, capuchins may be considered highly adaptable generalists, capable of occupying a wide range of habitats. The Panamanian white-throated capuchin (*Cebus* (*capucinus*) *imitator*) coexists with several other primate genera in Mesoamerica. Some taxa, such as the mantled howler monkey (*Alouatta palliata*) and Geoffroy's spider monkey (*Ateles geoffroyi*), have broader geographic ranges than *C. imitator*. This may be due to historical biogeography (e.g. earlier colonization of the region by *Ateles* and *Alouatta*) or, alternatively, because some habitats may be less suitable for capuchins. We investigated the latter hypothesis using ecological niche models (ENMs) to predict range extent based on climate and elevation, as well as niche breadth. Our results suggest a narrow climate niche relative to the more wide-ranging *Ateles geoffroyi*, as well as *Alouatta* spp. Precipitation in the coldest quarter and temperature seasonality were the two most important climate variables for determining *C. imitator* habitat suitability. These findings may have implications for capuchins' ability to persist through environmental changes (climate change and forest degradation, and loss) and warrant consideration in assessing their extinction risk.

Keywords Niche models · Mesoamerica · Capuchin · Howler monkey · Spider monkey

---

Electronic supplementary material: The online version of this chapter (https://doi.org/10.1007/978-3-319-98285-4_15) contains supplementary material, which is available to authorized users.

S. E. Johnson (✉)
Department of Anthropology and Archaeology, University of Calgary, Calgary, AB, Canada
e-mail: steig.johnson@ucalgary.ca

K. A. Brown
Department of Geography and Geology, Kingston University, Kingston Upon Thames, UK

U. Kalbitzer, K. M. Jack (eds.), *Primate Life Histories, Sex Roles, and Adaptability*, Developments in Primatology: Progress and Prospects,
https://doi.org/10.1007/978-3-319-98285-4_15

15.1 Introduction

Capuchins (*Cebus* and *Sapajus*) are found throughout the Neotropics and are often described as remarkably adaptable: 'flexibility, opportunism, and adaptability are the hallmarks of capuchin success' (Fragaszy et al. 2004: 29). This may be attributed primarily to their ability to exploit many kinds of foods, including difficult to access resources through extractive foraging and advanced cognition (e.g. Chapman 1987, 1988; Chapman and Fedigan 1990; Fragaszy et al. 1990; Fragaszy et al. 2013; Janson 1998).

The flexibility of capuchin monkeys apparently extends to habitat selection, as they are found in a variety of forest types, from humid to dry, and across a wide elevation range, from 0 to over 2000 m above sea level (Fragaszy et al. 2004). Their geographic distribution is also extensive: from northern Argentina to Honduras, including throughout the Amazon basin (Lynch Alfaro et al. 2012; Nascimento et al. 2015). However, while overlapping in much of South America, the 'gracile' forms (*Cebus* spp.) have a smaller and more westerly distribution than the 'robust' capuchins (*Sapajus* spp.), reflecting distinct points of origin and a rapid expansion of the latter genus westward from the Atlantic Forest (Lynch Alfaro et al. 2012; but see Nascimento et al. 2015). The robust capuchins occupy a wider latitudinal range (Lynch Alfaro et al. 2012) and may outcompete gracile capuchins at local scales (Defler 1985), indicating potential differences in the degree of adaptability or flexibility in the two genera.

Nonetheless, only the gracile capuchins are found in Mesoamerica, the northwestern extent for nonhuman primates in the Neotropics (Rylands et al. 2006). Indeed, only a single species is found from western Panama up through Honduras: the Panamanian white-throated capuchin (*Cebus* (*capucinus*) *imitator*) (Boubli et al. 2012; see Methods). However, they share this region with up to five other primate genera: *Aotus*, *Saguinus*, and *Saimiri* (all of which maintain very small ranges that overlap narrowly or perhaps not at all with *C. imitator*), as well as the more widespread *Alouatta* and *Ateles* (Rylands et al. 2006). The black-handed spider monkey (*Ateles geoffroyi*) is the sole species in this genus through much of the region, while *Alouatta* maintains two parapatric species: the brown howler monkey (*A. palliata*) and the black howler monkey (*A. pigra*). Both *Alouatta palliata* and *Ateles geoffroyi* are widely sympatric with *C. imitator* in this region but are also found substantially further north (Ford 2006; Rylands et al. 2006).

The differences in the extent of these three wide-ranging genera may reflect historical biogeography. *Alouatta* and perhaps *Ateles* may have colonized Mesoamerica first (e.g. Ford 2006; Fragaszy et al. 2004; Lynch Alfaro et al. 2012). Under Ford's (2006) model, the ancestors to *Alouatta pigra* were among the first primates to arrive in the region across the emergent Isthmus of Panama from northwestern South America, approximately 3.5–3 mya. Subsequent periods of rising sea levels would have again isolated Mesoamerica until approximately 2 mya, after which source populations for *Alouatta palliata*, *Ateles geoffroyi*, and *Cebus imitator* would have arrived. Other analyses support this later time frame for *C. imitator* (ca. 1.9

mya; Lynch Alfaro et al. 2012), but there are various scenarios placing the colonization of the ateline species as either substantially earlier or roughly contemporaneous with *C. imitator* (3–2 mya; Collins and Dubach 2000; Cortés-Ortiz et al. 2003). Therefore, it remains uncertain that the more extensive range of *Ateles* and *Alouatta* in Mesoamerica relative to *C. imitator* is simply due to an earlier arrival and thus having more time to successfully disperse further north.

It is also plausible that a physical barrier may have limited the northern extent of *Cebus imitator*. For example, Fragaszy et al. (2004) suggest potential river barriers to the northern expansion of *C. imitator* into Guatemala or Belize. Furthermore, there are mountain ranges in northwestern Nicaragua and southwestern El Salvador that limit *Cebus*, along with the atelines (Ford 2006). Ford (2006) also notes that the northern limit for *Cebus* appears to align with a depression that would have been below sea level during the Pleistocene. However, none of these barriers were sufficient to ultimately impede the expansion of the ateline primates, especially *Ateles geoffroyi*, into more northern regions – leaving the question of a physical barrier for *C. imitator* a matter of conjecture at present (Fragaszy et al. 2004).

In this chapter, we investigate an alternative, ecological explanation for why *Cebus imitator* maintains a more limited distribution than the ateline genera in Mesoamerica. It is possible that *C. imitator* cannot successfully colonize or persist in certain habitat types inhabited by *Alouatta* spp. and especially *Ateles geoffroyi* – effectively restricting their expansion into the more northern regions. This suggestion seems paradoxical given capuchins' famous flexibility in diet, incorporating primarily fruit but also seasonally large proportions of insect or other animal prey (e.g. Chapman 1987, 1988; Mosdossy et al. 2015; Rose 1994), and broad habitat selection (e.g. Chapman et al. 1989). However, the primarily frugivorous *Ateles geoffroyi* and folivore-frugivore *Alouatta* spp. also maintain a high degree of ecological flexibility, if perhaps not quite the dietary breadth of capuchins (Amato and Garber 2014; Behie and Pavelka 2005; Bonilla-Sánchez et al. 2012; Chapman 1987, 1988; Milton 1981). Furthermore, while *C. imitator* may be a habitat generalist at local scales, it is possible that this versatility does not extend to broader spatial scales – which is more apparent in the atelines (Ford 2006).

We used ecological niche models (ENMs) to test the hypothesis that *Cebus imitator* has a narrower ecological niche than *Ateles geoffroyi* and *Alouatta* spp., which may limit their distribution in Mesoamerica. ENMs typically use occurrence data and associated climatic and landscape variables to model species' distributions (Elith et al. 2006; Phillips et al. 2006). The output of these models may be interpreted as the spatial or climate niche of the species in question and include a niche breadth metric, Levins' B (Levins 1968; Warren et al. 2010). Such methods have been increasingly applied in studies of primate biogeography to understand both current and future species distributions (e.g. Blair et al. 2013; Brown and Yoder 2015; Holzmann et al. 2015; Johnson et al. 2016; Junker et al. 2012; Kamilar et al. 2016; Kamilar and Tecot 2016), while complementary approaches have been used to investigate how dietary niches, including niche overlap, may influence broad-scale biogeographic patterns across primate species (e.g. Kamilar and Ledogar 2011). We specifically predicted that *C. imitator* would have a narrower niche breadth using ENM outputs than *Ateles geoffroyi* and *Alouatta palliata* (the most wide-ranging species); we had no a priori prediction for the niche breadth of *C. imitator* relative to *Alouatta pigra*, which maintains a relatively small geographic range in southern Mexico and Belize (Rylands et al. 2006). In addition, we examined which climate variables had the strongest influence on species distribution models to highlight potentially important ecological constraints for each of these primates.

15.2 Methods

15.2.1 Study Species and Region

The study species included the brown howler monkey (*Alouatta palliata*), the black howler monkey (*Alouatta pigra*), the black-handed spider monkey (*Ateles geoffroyi*), and the Panamanian white-throated capuchin (*Cebus imitator*). We apply the taxonomy currently adopted by the IUCN (2017) for *Alouatta* spp. and *Ateles geoffroyi*. However, due to recent revisions in the genus, *Cebus* is awaiting reassessment by the IUCN (A. Rylands, pers. comm.); therefore, we use the taxonomy proposed by Boubli et al. (2012). Importantly, this revision contrasts with the nomenclature commonly applied to the Mesoamerican capuchin species (i.e. *C. capucinus* or *C. c. imitator*), including in this volume. In particular, Boubli et al. (2012) suggested elevating *C. imitator* to full species, distinct from its sister taxon *C. capucinus* which is distributed primarily in South America. This proposed taxonomy will be adopted by the IUCN (A. Rylands, pers. comm.). We note that choice of taxonomic scheme will affect distributions and therefore species' potential climate niches. Boubli et al. (2012) indicate well-supported monophyly in *C. imitator* (unlike in *C. capucinus*). Recent molecular analyses have also upheld the monotypic status of *Ateles geoffroyi* (despite the need for subspecific revision; Morales-Jimenez et al. 2015), *Alouatta palliata*, and *Alouatta pigra* (Cortés-Ortiz et al. 2003). Thus, we contend that the taxa included are appropriate for the analysis. Several other Mesoamerican nonhuman primates with more restricted ranges were excluded due to low sample size: *Aotus zonalis*, *Saguinus geoffroyi*, and *Saimiri oerstedii*; the former two species also show little or no overlap with focal species within the study region (Rylands et al. 2006).

The study region included most of Mesoamerica, from western Panama through southern Mexico (Fig. 15.1). Specifically, we bounded the study area to the southeastern limit for *Cebus imitator* and *Ateles geoffroyi* up to the northern limit of *A. geoffroyi* (Boubli et al. 2012; Rylands et al. 2006). The area incorporates the entirety of the distributions of three study species; however, we excluded the portion of the range of *Alouatta palliata* that extends from eastern Panama into South America as we were primarily interested in comparing distributions and niche breadth within

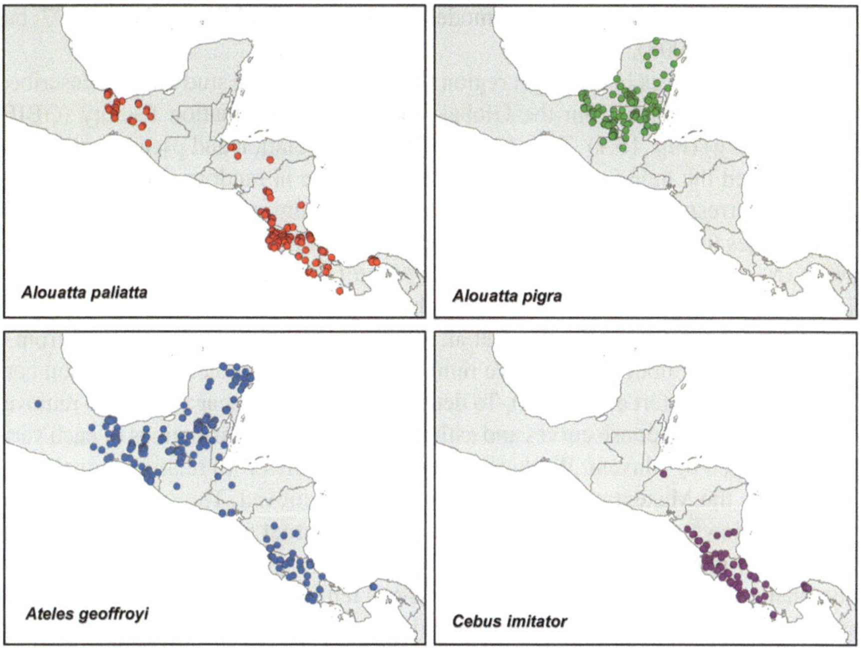

Fig. 15.1 Presence locations for the study species obtained from GBIF (www.gbif.org) (*Alouatta palliata*, $N = 162$; *Alouatta pigra*, $N = 123$; *Ateles geoffroyi*, $N = 237$; *Cebus imitator*, $N = 86$). The study region was truncated west of the Panama Canal (southeastern extent of *Cebus imitator* and *Ateles geoffroyi*) and at the most northerly latitude for *Ateles geoffroyi*

the region potentially occupied by *C. imitator*. The total study region spanned 1,415,125 km^2.

15.2.2 Ecological Niche Models

We developed ecological niche models (ENMs) using Maxent v.3.3.3 k. These models incorporate presence-only locality records to estimate habitat suitability (probability of presence) for a given species (Phillips et al. 2006). Maxent determines the probability distribution that is closest to uniform (i.e. maximum entropy) and constrains the distribution via environmental variables corresponding to species locations (Phillips et al. 2006). Maxent has been demonstrated to perform well in comparison to other methods (e.g. Elith et al. 2006; Pittman and Brown 2011) and with relatively few presence locations (Pearson et al. 2007). We used default settings for all model parameters. We used fivefold cross validation to assess model performance, based on the held-out (test) folds (Elith et al. 2011). We constructed receiver operating characteristic (ROC) curves for each fold and used the area under the curve (AUC) as the measure of model performance (Fielding and Bell 1997; but see Lobo et al. 2008).

Species presences in the target region (i.e. bounded by the study region described above) were obtained from the Global Biodiversity Information Facility (GBIF: www.gbif.org) (Fig. 15.1). To reduce spatial autocorrelation and pseudoreplication, we truncated the number of records by selecting one at random within each 1 km grid cell, corresponding to the resolution of the environmental variables. Subsequent sample sizes for each species were as follows: *Alouatta palliata* (*N* = 162), *Alouatta pigra* (*N* = 123), *Ateles geoffroyi* (*N* = 237), and *Cebus imitator* (*N* = 86). To build ENMs for the target region, we initially selected 19 climatic variables obtained from the WorldClim database (Hijmans et al. 2005), as well as elevation derived from a digital elevation model (DEM). We removed collinear variables with Pearson correlation >0.90 (Syfert et al. 2013). To determine which collinear variable to remove, we assessed the response curves and estimates of relative contributions of each variable generated by Maxent. Weak predictors and variables with relatively small contribution to the Maxent model were preferentially removed. The variables retained were as follows: annual mean temperature, mean diurnal temperature range (mean of (monthly maximum temperature – monthly minimum temperature)), temperature seasonality (standard deviation of annual mean temperature *100), maximum temperature in the warmest month, minimum temperature in the coldest month, annual temperature range (maximum temperature in the warmest month – minimum temperature in the coldest month), annual precipitation, precipitation in the driest month, precipitation seasonality (coefficient of variation of monthly precipitation), precipitation in the warmest quarter, precipitation in the coldest quarter (Hijmans et al. 2005; see also www.worldclim.org), and elevation (see Results, Table 15.1). Outputs from ENMs indicated the proportion each climatic and elevation variable contributed to habitat suitability for each species in the target region. However, the contributions of each variable to ENMs must be interpreted with caution, as different percentage values may be obtained via alternate paths to best solutions in different model runs using the same data sets; this is exacerbated when predictor variables are collinear (Phillips et al. 2006), which we reduced per the above procedures.

15.2.3 Climate Niche Breadth

Estimates of climate niche breadth were derived from Maxent ENM outputs. We quantified niche breadth with Levins' inverse concentration metric (Levins' B; Levins 1968), which is scaled from 0 (minimum breadth) to 1 (maximum breadth) over the modelled landscape (Nakazato et al. 2010). We also generated 95% confidence intervals via bootstrapping with 100 iterations in Maxent. Measurements were implemented in ENMTools (Warren et al. 2010).

Table 15.1 Percentage contribution of climate and elevation variables to habitat suitability according to ENMs for the study species

Variable	*Alouatta palliata*	*Alouatta pigra*	*Ateles geoffroyi*	*Cebus imitator*
Annual mean temperature	2.0	8.4	1.1	0.7
Mean diurnal temperature range	9.7	0.9	4.4	2.3
Temperature seasonality	**23.6**	**25.9**	**12.6**	**28.6**
Maximum temperature in the warmest month	3.3	0.3	**10.9**	0
Minimum temperature in the coldest month	3.7	4.0	2.9	1.3
Annual temperature range	1.3	0.5	**39.7**	1.8
Annual precipitation	**13.6**	**11.1**	9.1	0.4
Precipitation in the driest month	3.1	8.7	2.1	3.6
Precipitation seasonality	2.8	**32.7**	6.7	0.2
Precipitation in the warmest quarter	6.8	3.2	5.1	2.8
Precipitation in the coldest quarter	**28.1**	1.9	3.0	**57.0**
Elevation	2.0	2.4	2.3	1.2

Contributions >10% for each species are indicated in bold. Climate variables obtained from WorldClim (Hijmans et al. 2005)

15.3 Results

15.3.1 Ecological Niche Models

ENMs indicating the distributions of the study species predicted from environmental variables are presented in Fig. 15.2. Highly suitable portions of these distributions are generally consistent with observed ranges (e.g. see Fig. 15.1; Rylands et al. 2006). Models showed strong discrimination on held-out folds, with a mean cross validated area under the curve (AUC) of 0.812 for *Ateles geoffroyi*, 0.910 for *Alouatta palliata*, 0.922 for *Alouatta pigra*, and 0.948 for *Cebus imitator*. The total areas of suitable habitat (probability of presences >0.25) within the study region for each species were the following: *Alouatta palliata* (167,401 km^2), *Alouatta pigra* (218,712 km^2), *Ateles geoffroyi* (525, 132 km^2), and *C. imitator* (111,913 km^2).

15.3.2 Climate Niche Breadth

Niche breadth varied across taxa, with a high of 0.533 in *Ateles geoffroyi* and a low of 0.141 in *Cebus imitator* (Fig. 15.3). *Alouatta palliata* and *A. pigra* had similar, intermediate values (0.250 and 0.203, respectively; Fig. 15.3). Confidence intervals did not overlap for any of the species.

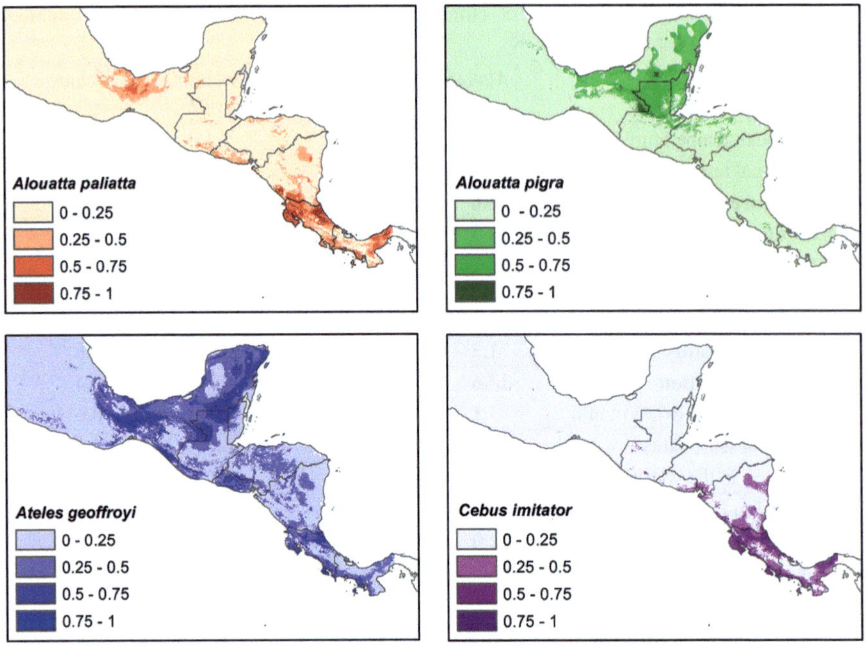

Fig. 15.2 Ecological niche models (ENMs) for the study species. Legends indicate habitat suitability (probability of species presence) within the study region

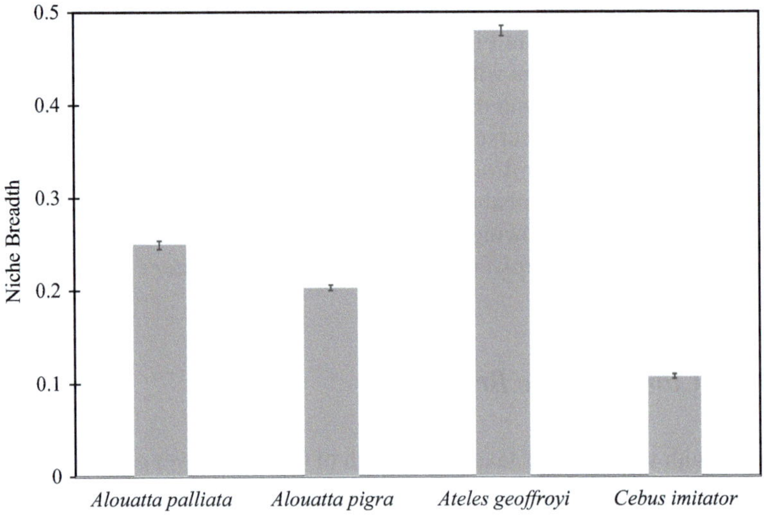

Fig. 15.3 Niche breadth (Levins' B) ± 95% confidence intervals for the study species. Confidence intervals were generated via bootstrapping with 100 iterations in Maxent and calculated in ENMTools

15.3.3 Variables Predicting Habitat Suitability

Each environmental predictor's percentage contribution to the habitat suitability in the ENM for each species is included in Table 15.1. Figures indicating the relationship between each predictor and habitat suitability for each species are also shown in supplemental materials (Figs. S15.1, S15.2, S15.3 and S15.4). Temperature seasonality had the strongest effect in the final model. This variable was a strong predictor in each of the four species' ENMs, ranging from 12.6% to 28.6% contribution to habitat suitability (Table 15.1). Habitats with higher temperature seasonality were generally more suitable for *Alouatta palliata* (Fig. 15.4a). Meanwhile, the relationship in the other three species was broadly similar: nonlinear, with a marked peak in habitat suitability at intermediate values (Fig. 15.4b–d). This peak was located at relatively low levels of seasonality in *Cebus imitator*, indicating a narrow optimal range for shifts in temperature (Fig. 15.4d).

Alouatta palliata distribution was also predicted by precipitation in the coldest quarter (28.1%) and annual precipitation (13.6%; Fig. S15.1), while precipitation seasonality (32.7%) and annual precipitation (11.1%) were also important predictors for *A. pigra* distribution (Fig. S15.2). *Ateles geoffroyi* presences were most strongly predicted by annual temperature range (39.7%), along with maximum temperature in the warmest month (10.9%; Fig. S15.3). Precipitation in the coldest quarter (57%) was the strongest predictor of *Cebus imitator* presences, showing a logistic curve reaching an asymptote at intermediate rainfall values (Fig. 15.5).

15.4 Discussion

15.4.1 Testing the 'Specialist' Capuchin Hypothesis

We constructed ENMs through Maxent for the four widest-ranging primate species in Mesoamerica to compare the ecological niche breadth of these taxa in terms of climate and elevation (i.e. solely in terms of ENM outputs). We found that *Cebus imitator* had the narrowest niche of any of the species examined, which may be interpreted as this species having the most specialized requirements in terms of habitat suitability. This finding may result from the fact that *Ateles geoffroyi* and *Alouatta* (as a genus) occupy larger geographic distributions, which may contain more climatic and elevation gradients. However, it is noteworthy that *Alouatta pigra*, which has an observed range similar in area to *C. imitator* (though the species are likely allopatric; Rylands et al. 2006), still had a substantially broader niche – and similar to its more expansive congener, *Alouatta palliata.* Thus, the correlation between extent of the range and niche breadth is not perfect (also reflected in comparing results for the two *Alouatta* spp.); niche breadth may instead reflect variable adaptation to or preference for particular environmental types within the overall range.

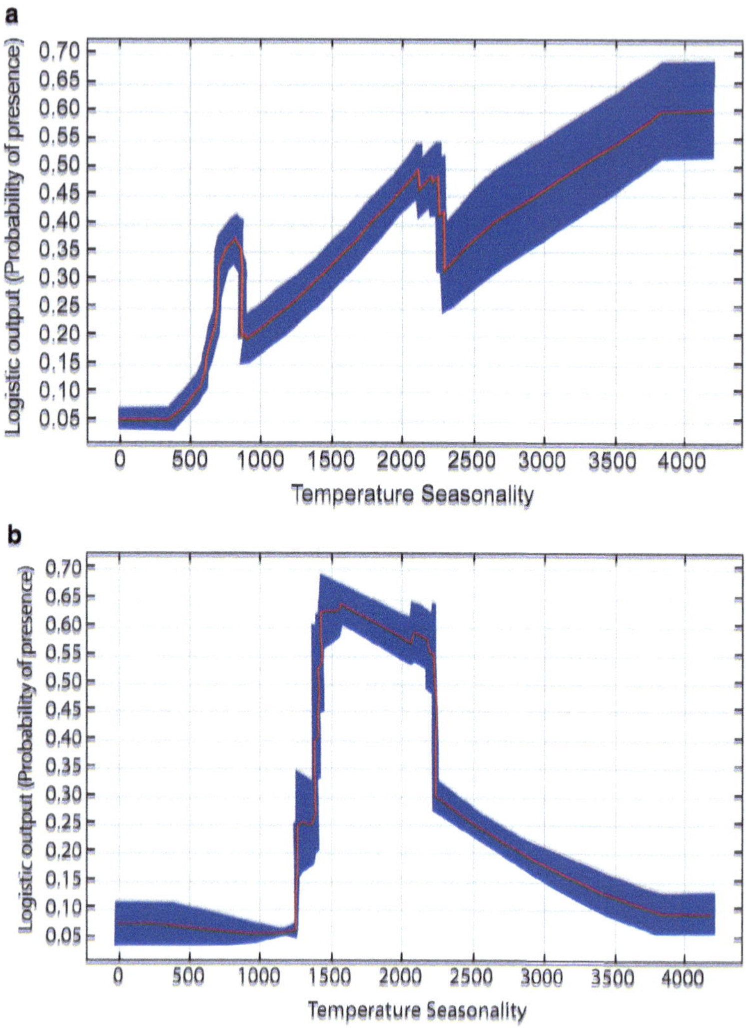

Fig. 15.4 Effects of temperature seasonality (standard deviation of annual mean × 100) on habitat suitability (probability of presence) in ENMs for (**a**) *Alouatta palliata*, (**b**) *Alouatta pigra*, (**c**) *Ateles geoffroyi*, and (**d**) *Cebus imitator*. Curves indicate how habitat suitability changes according to temperature seasonality, maintaining other variables at their average sample value. The red lines indicate the mean response of the five replicate Maxent runs, and the blue bands represent +/− one standard deviation

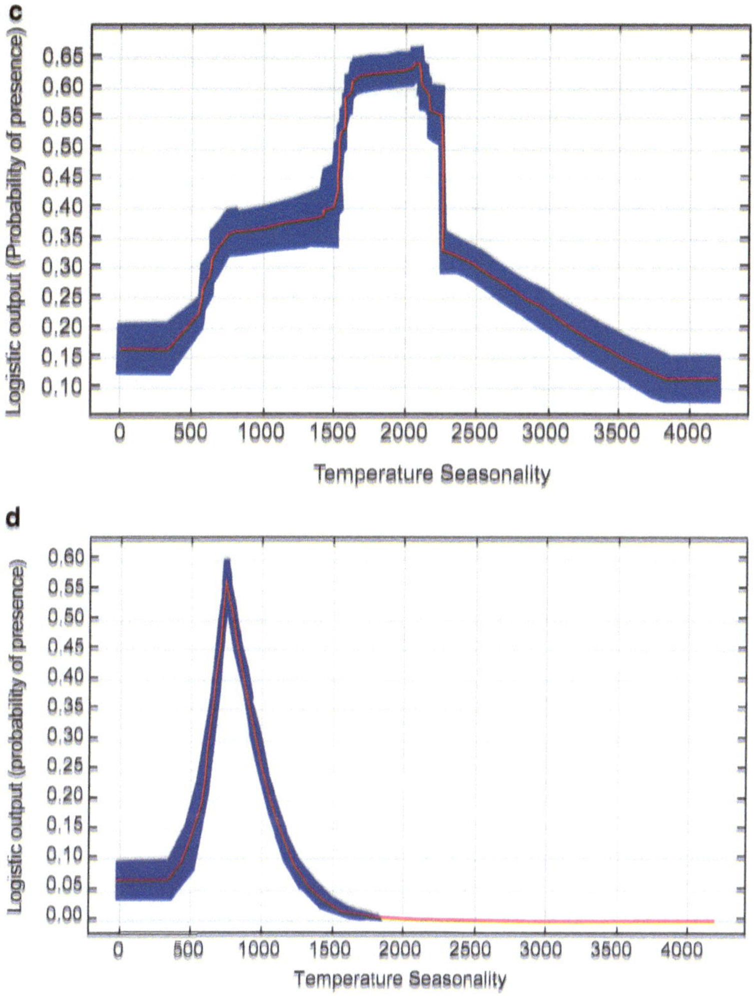

Fig. 15.4 (continued)

Our results therefore support the hypothesis that *Cebus imitator* has a smaller range than species such as *Ateles geoffroyi* and *Alouatta palliata* due to its narrower climate niche limits; this, in turn, may limit the expansion of capuchins into certain habitats occupied by the latter species. As noted, *C. imitator* may also be more specialized in terms of habitat requirements than even *Alouatta pigra*, which is confined to the Yucatán Peninsula (Baumgarten and Williamson 2007; Cortés-Ortiz et al. 2003; Ford 2006; Rylands et al. 2006; see also Fig. 15.1). These findings stand in contrast to characterizations of *C. imitator* as highly flexible and adaptable based on diet and local-scale habitat use (e.g. Chapman et al. 1989).

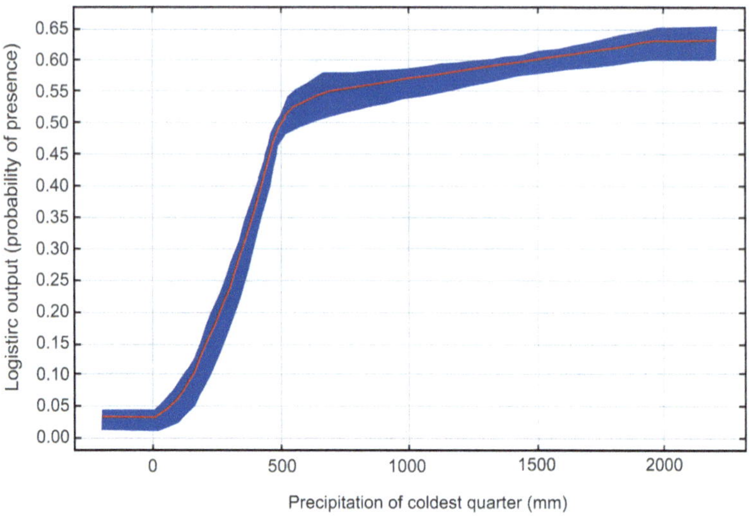

Fig. 15.5 Effect of precipitation in the coldest quarter on *Cebus imitator* habitat suitability (probability of presence) in ENMs. Curve indicates how habitat suitability changes according to precipitation, maintaining other variables at their average sample value. The red line indicates the mean response of the five replicate Maxent runs, and the blue band represents +/− one standard deviation

It is important to note that our analysis does not test other potential hypotheses regarding differences in range extents in Mesoamerican primates, and such hypotheses are not necessarily mutually exclusive. Foremost, we have no data to investigate the timing of when *Alouatta*, *Ateles*, or *Cebus* first colonized Mesoamerica from South America, nor can we speculate on variable rates of dispersal across the region. As mentioned, the potentially earlier arrival of the atelines (Lynch Alfaro et al. 2012) could explain their more northerly extent, even in the absence of more successful colonization of a greater variety of environmental types found across Mesoamerica. Yet the evidence for earlier arrival is somewhat equivocal, with some biogeographic analyses placing the capuchin migration into the region as roughly contemporaneous with those of *Ateles geoffroyi* and *Alouatta palliata* (e.g. Ford 2006). This is particularly suggestive that there might be mechanisms beyond simply the timing of arrival, since the latter two species have substantially larger and more northerly distributions (which do not entirely overlap themselves). Furthermore, even if the ancestors to *C. imitator* did colonize Mesoamerica following the ateline lineages, there may have been sufficient time – 1.7–2 mya (Boubli et al. 2012; Ford 2006; Lynch Alfaro et al. 2012) – for this lineage to reach available habitats in the region.

Further plausible mechanisms for the discordant distributions of the Mesoamerican primates could involve competitive exclusion (Gause 1934; Hardin 1960) or the atelines limiting the expansion of capuchins by outcompeting them in potentially shared environments. The latter has been suggested, for instance, within the *Alouatta* spp.: *A. palliata* may have displaced *A. pigra* to their current, narrower range in the Yucatán Peninsula (Cortés-Ortiz et al. 2003; Ford 2006; but see Baumgarten and Williamson 2007). Such competition could reinforce the limitations imposed on the distribution of *C. imitator* already considered above. For instance, the earlier arrival and expansion of the atelines (Ford 2006; Lynch Alfaro et al. 2012) would have excluded less-competitive capuchins from the north – with or without a physical barrier, such as rising sea levels, that might have impeded the northern expansion of *C. imitator* in the past (Ford 2006). Alternatively, for the hypothesis tested here, the narrower climatic niche in *C. imitator* could make them less competitive with the more generalist atelines, leading to the capuchins' exclusion north of Honduras. However, *C. imitator* is sympatric with both *Ateles geoffroyi* and *Alouatta pigra* throughout much of its distribution (Rylands et al. 2006) and, at least at small spatial scales within this range, is found in *more* – not fewer – habitat types (Chapman et al. 1989). Dietary differences between capuchins and the atelines (Chapman 1987, 1988) would likely also reduce exploitative competition. Therefore, direct competition between capuchins and other Mesoamerican primates does not appear to be a primary mechanism in determining individual species' distributions; nevertheless, further research into competitive interactions via niche modelling (e.g. Peers et al. 2013) may be warranted for evaluating such processes at large spatial scales.

15.4.2 Environmental Factors Contributing to Mesoamerican Primate ENMs

Given the substantially narrower climate niche for *Cebus imitator* indicated in our models, it may be instructive to compare the environmental variables that contributed to each species' distributions. However, as indicated above, the contributions of each variable should be considered primarily for heuristic purposes, as values can vary across model solutions (Phillips et al. 2006). Habitat suitability for all four study species' ranges was relatively strongly predicted by temperature seasonality (Table 15.1). Suitable habitat for *C. imitator* appeared to peak within the narrowest range of values for this variable, and this peak was at the lowest value of any of the four species (Fig. 15.4). These findings suggest that thermoregulatory stress may have important implications for the ability of *C. imitator* to occupy various environments. At seasonally hot and dry Santa Rosa National Park in northwestern Costa Rica, Campos et al. (2014) found that increasing temperatures reduced capuchin home range size, particularly in high-use areas; this is consistent with observations of reduced activity associated with higher temperatures (Campos and Fedigan 2009). However, our models pinpointed (low) temperature seasonality – versus (low or high) temperature – as a particularly strong contributor to the climate niche of capuchins.

We also found that precipitation in particular seasons was a strong contributor to distribution models for *Cebus imitator* (57%; Table 15.1). Habitat suitability increased across the lower range of values for rainfall (0–500 m) during the coldest quarter (Fig. 15.5), indicating that perhaps a threshold for precipitation must be met during this season to ensure suitable habitat. Although the relationships can be highly variable, rainfall may be an important predictor of plant phenology and, consequently, food availability for primates (e.g. van Schaik et al. 1993; van Schaik and Pfannes 2005). *Cebus imitator*, whose reproduction is linked to annual peaks in fruit abundance (Carnegie et al. 2011), may be especially sensitive to such fluctuations. Alternatively, the impact of seasonal precipitation levels for capuchin presences could again be associated with thermoregulatory stresses. In particular, *C. imitator* range use may be strongly influenced by access to water sources during hot, dry conditions (Campos and Fedigan 2009; Campos et al. 2014; Fedigan et al. 1996). On the surface, our findings of the importance of precipitation levels in colder months may not appear relevant to such conditions; however, it is plausible that adequate rainfall must precede the drier months in order to replenish potential drinking water resources during the more critical season.

15.4.3 Climate Niche Specialization, Climate Change, and Capuchin Conservation

The results of this analysis add to our understanding of capuchin biogeography and ecological flexibility at larger spatial scales. They also speak to the theme of this section – primate adaptations to changing environments. Deforestation continues to transform capuchin habitats in Mesoamerica (de Albuquerque et al. 2015; DeClerck et al. 2010), and the persistence of *Cebus imitator* depends on its inherent abilities to occupy a sufficient area within these dynamic environments to support viable populations in the long term. Forest loss across the range will afford ever fewer options for a species which appears relatively limited in its climate niche and thus its ability to occupy remaining environments. Moreover, indicators of anthropogenic disturbance may be used as predictor variables to further refine ecological niche models for capuchins and the other primates of Mesoamerica, potentially adding to our understanding as to why these species have distinct distributions (Kamilar and Tecot 2016). Fortunately, capuchins have demonstrated strong resilience, rebounding rapidly at the population level when anthropogenic deforestation and forest degradation are controlled (Campos et al. 2015; Fedigan and Jack 2001).

Global climate change also threatens to alter environmental conditions, leading to range shifts and reductions (Lawler et al. 2006) – a threat to primate populations that is almost certainly combined with continued deforestation (e.g. Dunham et al. 2008). For example, El Niño Southern Oscillation (ENSO) events, which have strong and variable impacts on local climate conditions globally, are anticipated to increase in severity with greenhouse warming (Cai et al. 2014; Ingram and Dawson

2005). These events have been associated with significant declines in capuchin and other primate populations or reproduction (Campos et al. 2015; Dunham et al. 2008, 2011; Milton and Giacalone 2014; Wiederholt and Post 2010, 2011). Both increased (Milton and Giacalone 2014) and decreased (Campos et al. 2015) rainfalls linked to distinct ENSO phases have substantially impacted capuchin populations. In the former case at Barro Colorado Island, Panama, there was near-total population collapse (>70%) in *Cebus capucinus* following heavy rains, while sympatric *Alouatta palliata* remained stable (Milton and Giacalone 2014). In the latter instance, infant:adult female ratios of *C. imitator* declined significantly following ENSO-related droughts at Santa Rosa National Park, at least during periods of overall population instability (Campos et al. 2015). This case is particularly noteworthy given that our analysis suggested that *C. imitator* has a low habitat suitability where rainfall is low seasonally; an increase in frequency of ENSO-related droughts could therefore have significant, species-wide impacts.

In the IUCN Red List of Threatened Species, *C. imitator* has only recently been elevated from a subspecies of *C. capucinus* to full species (based on Boubli et al. 2012; Ruiz-Garcia et al. 2012), and it presently awaits publication of its conservation status reassessment (A. Rylands, pers. comm.). However, the determination of the IUCN is that *C. imitator* will be classified as Vulnerable (A. Rylands, pers. comm.). This places *C. imitator* at similar or lower risk of extinction compared to *Ateles geoffroyi* (Endangered), *Alouatta pigra* (Endangered), and some Mesoamerican subspecies of *Alouatta palliata* (*A. p. mexicana*, Critically Endangered; *A. p. coibensis*, Vulnerable) (IUCN 2017). Especially in the context of environmental change described above, the evidence that *C. imitator* maintains a relatively narrow climatic niche may be grounds for reconsideration of its conservation status. One of the criteria in IUCN assessments is: 'Geographic range in the form of either extent of occurrence OR area of occupancy OR both… [showing] continuing decline, observed, inferred or projected' (IUCN 2017). In this respect, the *C. imitator* may be at particularly elevated risk of extinction, at least compared to Mesoamerican atelines currently assessed as more endangered. More robust evidence for this assertion could be obtained through similar ecological niche modelling approaches that incorporate future climate change scenarios (Elith and Leathwick 2009).

In conclusion, we found that *Cebus imitator* may be considered a climate niche specialist, with a more restricted range, relative to other wide-ranging primates in Mesoamerica. This stands in stark contrast to their demonstrated flexibility in behavioural ecology within the environments that they occupy and highlights the importance of spatial scale when examining species' ecology, distributions, and extinction risk.

Acknowledgements We would like to thank Tracy Wyman for assistance with analyses and figures. In addition, we wish to express our appreciation to the editors and two anonymous reviewers for helpful improvements to the manuscript. We are also very grateful to the organizers of the Festschrift conference, held in Banff, Alberta, on December 2–3, 2016, for the opportunity to participate in this wonderful celebration of Dr. Fedigan's distinguished career.

This paper evolved from discussions following various committee meetings, exams, and defences of several of Dr. Fedigan's graduate students over the years and happily brought a lemurologist momentarily back to his capuchin roots. This is just one example of the great privilege and pleasure I (SEJ) have had to work with Linda over the past dozen years, and her intellectual inspiration long preceded that; I am grateful for her mentorship, leadership, and friendship.

References

Amato KR, Garber PA (2014) Nutrition and foraging strategies of the black howler monkey (*Alouatta pigra*) in Palenque National Park, Mexico. Am J Primatol 76(8):774–787. https://doi.org/10.1002/ajp.22268

Baumgarten A, Williamson GB (2007) The distributions of howling monkeys (*Alouatta pigra* and *A. palliata*) in southeastern Mexico and Central America. Primates 48(4):310–315. https://doi.org/10.1007/s10329-007-0049-y

Behie AM, Pavelka MSM (2005) The short-term effects of a hurricane on the diet and activity of black howlers (*Alouatta pigra*) in Monkey River, Belize. Folia Primatol 76:1–9

Blair ME, Sterling EJ, Dusch M, Raxworthy CJ, Pearson RG (2013) Ecological divergence and speciation between lemur (*Eulemur*) sister species in Madagascar. J Evol Biol 26:1790–1801. https://doi.org/10.1111/jeb.12179

Bonilla-Sánchez YM, Serio-Silva JC, Pozo-Montuy G, Chapman CA (2012) Howlers are able to survive in *Eucalyptus* plantations where remnant and regenerating vegetation is available. Int J Primatol 33(1):233–245. https://doi.org/10.1007/s10764-011-9569-9

Boubli JP, Rylands AB, Farias IP, Alfaro ME, Alfaro JL (2012) *Cebus* phylogenetic relationships: a preliminary reassessment of the diversity of the untufted capuchin monkeys. Am J Primatol 74(4):381–393. https://doi.org/10.1002/ajp.21998

Brown JL, Yoder AD (2015) Shifting ranges and conservation challenges for lemurs in the face of climate change. Ecol Evol 5(6):1131–1142. https://doi.org/10.1002/ece3.1418

Cai W, Borlace S, Lengaigne M, van Rensch P, Collins M, Vecchi G, Timmermann A, Santoso A, McPhaden MJ, Wu L, England MH, Wang G, Guilyardi E, Jin F-F (2014) Increasing frequency of extreme El Niño events due to greenhouse warming. Nat Clim Chang 4(2):111–116. https://doi.org/10.1038/nclimate2100

Campos FA, Fedigan LM (2009) Behavioral adaptations to heat stress and water scarcity in white-faced capuchins (*Cebus capucinus*) in Santa Rosa National Park, Costa Rica. Am J Phys Anthropol 138(1):101–111. https://doi.org/10.1002/ajpa.20908

Campos FA, Bergstrom ML, Childers A, Hogan JD, Jack KM, Melin AD, Mosdossy KN, Myers MS, Parr NA, Sargeant E, Schoof VAM, Fedigan LM (2014) Drivers of home range characteristics across spatiotemporal scales in a Neotropical primate, *Cebus capucinus*. Anim Behav 91:93–109. https://doi.org/10.1016/j.anbehav.2014.03.007

Campos FA, Jack KM, Fedigan LM (2015) Climate oscillations and conservation measures regulate white-faced capuchin population growth and demography in a regenerating tropical dry forest in Costa Rica. Biol Conserv 186:204–213. https://doi.org/10.1016/j.biocon.2015.03.017

Carnegie SD, Fedigan LM, Melin AD (2011) Reproductive seasonality in female capuchins (*Cebus capucinus*) in Santa Rosa (Area de Conservación Guanacaste), Costa Rica. Int J Primatol 32(5):1076. https://doi.org/10.1007/s10764-011-9523-x

Chapman CA (1987) Flexibility in diets of three species of Costa Rican primates. Folia Primatol 49(2):90–105

Chapman CA (1988) Patterns of foraging and range use by three species of neotropical primates. Primates 29(2):177–194

Chapman CA, Fedigan LM (1990) Dietary differences between neighboring *Cebus capucinus* groups: local traditions, food availability or responses to food profitability? Folia Primatol 54:177–186

Chapman CA, Chapman L, Glander KE (1989) Primate populations in northwestern Costa Rica: potential for recovery. Primate Conserv 10:37–44

Collins AC, Dubach JM (2000) Biogeographic and ecological forces responsible for speciation in *Ateles*. Int J Primatol 21(3):421–444. https://doi.org/10.1023/a:1005487802312

Cortés-Ortiz L, Bermingham E, Rico C, Rodríguez-Luna E, Sampaio I, Ruiz-García M (2003) Molecular systematics and biogeography of the Neotropical monkey genus, *Alouatta*. Mol Phylogenet Evol 26(1):64–81. https://doi.org/10.1016/S1055-7903(02)00308-1

de Albuquerque FS, Benito B, Beier P, Assunção-Albuquerque MJ, Cayuela L (2015) Supporting underrepresented forests in Mesoamerica. Nat Conservacao 13(2):152–158. https://doi.org/10.1016/j.ncon.2015.02.001

DeClerck FAJ, Chazdon R, Holl KD, Milder JC, Finegan B, Martinez-Salinas A, Imbach P, Canet L, Ramos Z (2010) Biodiversity conservation in human-modified landscapes of Mesoamerica: past, present and future. Biol Conserv 143(10):2301–2313. https://doi.org/10.1016/j.biocon.2010.03.026

Defler TR (1985) Contiguous distribution of two species of *Cebus* monkeys in El Tuparro National Park, Colombia. Am J Primatol 8:101–112

Dunham AE, Erhart EM, Overdorff DJ, Wright PC (2008) Evaluating effects of deforestation, hunting, and El Nino events on a threatened lemur. Biol Conserv 141(1):287–297. https://doi.org/10.1016/j.biocon.2007.10.006

Dunham AE, Erhart EM, Wright PC (2011) Global climate cycles and cyclones: consequences for rainfall patterns and lemur reproduction in southeastern Madagascar. Glob Chang Biol 17(1):219–227. https://doi.org/10.1111/j.1365-2486.2010.02205.x

Elith J, Leathwick JR (2009) Species distribution models: ecological explanation and prediction across space and time. Annu Rev Ecol Evol Syst 40:677–697. https://doi.org/10.1146/annurev.ecolsys.110308.120159

Elith J, Graham CH, Anderson RP, Dudik M, Ferrier S, Guisan A, Hijmans RJ, Huettmann F, Leathwick JR, Lehmann A, Li J, Lohmann LG, Loiselle BA, Manion G, Moritz C, Nakamura M, Nakazawa Y, Overton JM, Peterson AT, Phillips SJ, Richardson K, Scachetti-Pereira R, Schapire RE, Soberon J, Williams S, Wisz MS, Zimmermann NE (2006) Novel methods improve prediction of species' distributions from occurrence data. Ecography 29:129–151

Elith J, Phillips SJ, Hastie T, Dudik M, Chee YE, Yates CJ (2011) A statistical explanation of MaxEnt for ecologists. Divers Distrib 17:43–57. https://doi.org/10.1111/J.1472-4642.2010.00725.X

Fedigan LM, Jack K (2001) Neotropical primates in a regenerating Costa Rican dry forest: a comparison of howler and capuchin population patterns. Int J Primatol 22(5):689–713. https://doi.org/10.1023/a:1012092115012

Fedigan LM, Rose LM, Avila RM (1996) See how they grow: tracking capuchin monkey (*Cebus capucinus*) populations in a regenerating Costa Rican dry forest. In: Norconk MA, Rosenberger AL, Garber PA (eds) Adaptive radiations of Neotropical Primates. Springer, Boston, pp 289–307. https://doi.org/10.1007/978-1-4419-8770-9_17

Fielding AH, Bell JF (1997) A review of methods for the assessment of prediction errors in conservation presence/absence models. Environ Conserv 24(1):38–49

Ford SM (2006) The biogeographic history of Mesoamerican primates. In: Estrada A, Garber PA, Pavelka MSM, Luecke L (eds) New perspectives in the study of Mesoamerican Primates: distribution, ecology, behavior, and conservation. Springer, New York, pp 81–114

Fragaszy DM, Visalberghi E, Robinson JG (1990) Variability and adaptability in the genus *Cebus*. Folia Primatol 54:114–118

Fragaszy DM, Visalberghi E, Fedigan LM (2004) The complete capuchin: the biology of the genus *Cebus*. Cambridge University Press, Cambridge

Fragaszy DM, Liu Q, Wright BW, Allen A, Brown CW, Visalberghi E (2013) Wild bearded capuchin monkeys (*Sapajus libidinosus*) strategically place nuts in a stable position during nut-cracking. PLoS One 8(2):e56182. https://doi.org/10.1371/journal.pone.0056182

Gause GF (1934) The struggle for existence. Williams and Wilkins, Baltimore

Hardin G (1960) The competitive exclusion principle. Science 131(3409):1292–1297. https://doi.org/10.1126/science.131.3409.1292

Hijmans RJ, Cameron SE, Parra JL, Jones P, Jarvis A (2005) Very high resolution interpolated climate surfaces for global land areas. Int J Climatol 25:1965–1978

Holzmann I, Agostini I, DeMatteo K, Areta JI, Merino ML, Di Bitetti MS (2015) Using species distribution modeling to assess factors that determine the distribution of two parapatric howlers (*Alouatta* spp.) in South America. Int J Primatol 36(1):18–32. https://doi.org/10.1007/s10764-014-9805-1

Ingram JC, Dawson TP (2005) Climate change impacts and vegetation response on the island of Madagascar. Phil Trans R Soc A 363:55–59

IUCN (2017) The IUCN red list of threatened species. http://www.iucnredlist.org/. Accessed 14 July 2017

Janson CH (1998) Experimental evidence for spatial memory in foraging in wild capuchin monkeys (*Cebus apella*). Anim Behav 55:1229–1243

Johnson SE, Delmore KE, Brown KA, Wyman TM, Louis EE (2016) Niche divergence in a brown lemur (*Eulemur* spp.) hybrid zone: using ecological niche models to test models of stability. Int J Primatol 37(1):69–88. https://doi.org/10.1007/s10764-015-9872-y

Junker J, Blake S, Boesch C, Campbell G, Ld T, Duvall C, Ekobo A, Etoga G, Galat-Luong A, Gamys J, Ganas-Swaray J, Gatti S, Ghiurghi A, Granier N, Hart J, Head J, Herbinger I, Hicks TC, Huijbregts B, Imong IS, Kuempel N, Lahm S, Lindsell J, Maisels F, McLennan M, Martinez L, Morgan B, Morgan D, Mulindahabi F, Mundry R, N'Goran KP, Normand E, Ntongho A, Okon DT, Petre C-A, Plumptre A, Rainey H, Regnaut S, Sanz C, Stokes E, Tondossama A, Tranquilli S, Sunderland-Groves J, Walsh P, Warren Y, Williamson EA, Kuehl HS (2012) Recent decline in suitable environmental conditions for African great apes. Divers Distrib 18(11):1077–1091. https://doi.org/10.1111/ddi.12005

Kamilar JM, Ledogar JA (2011) Species co-occurrence patterns and dietary resource competition in primates. Am J Phys Anthropol 144(1):131–139. https://doi.org/10.1002/ajpa.21380

Kamilar JM, Tecot SR (2016) Anthropogenic and climatic effects on the distribution of *Eulemur* species: an ecological niche modeling approach. Int J Primatol 37(1):47–68. https://doi.org/10.1007/s10764-015-9875-8

Kamilar JM, Blanco MB, Muldoon KM (2016) Ecological niche modeling of mouse lemurs (*Microcebus* spp.) and its implications for their species diversity and biogeography. In: Lehman SM, Radespiel U, Zimmermann E (eds) The dwarf and mouse lemurs of Madagascar: biology, behavior and conservation biogeography of the Cheirogaleidae. Cambridge University Press, Cambridge, pp 449–461

Lawler JJ, White D, Neilson RP, Blaustein AR (2006) Predicting climate-induced range shifts: model differences and model reliability. Glob Chang Biol 12(8):1568–1584. https://doi.org/10.1111/j.1365-2486.2006.01191.x

Levins R (1968) Evolution in changing environments: some theoretical explorations. Monographs in population biology. Princeton University Press, Princeton

Lobo JM, Jiménez-Valverde A, Real R (2008) AUC: a misleading measure of the performance of predictive distribution models. Glob Ecol Biogeogr 17(2):145–151. https://doi.org/10.1111/j.1466-8238.2007.00358.x

Lynch Alfaro JW, Boubli JP, Olson LE, Di Fiore A, Wilson B, Gutiérrez-Espeleta GA, Chiou KL, Schulte M, Neitzel S, Ross V, Schwochow D, Nguyen MTT, Farias I, Janson CH, Alfaro ME (2012) Explosive Pleistocene range expansion leads to widespread Amazonian sympatry between robust and gracile capuchin monkeys. J Biogeogr 39(2):272–288. https://doi.org/10.1111/j.1365-2699.2011.02609.x

Milton K (1981) Food choice and digestive strategies by two sympatric primate species. Am Nat 117:496–505

Milton K, Giacalone J (2014) Differential effects of unusual climatic stress on capuchin (*Cebus capucinus*) and howler monkey (*Alouatta palliata*) populations on Barro Colorado Island, Panama. Am J Primatol 76(3):249–261. https://doi.org/10.1002/ajp.22229

Morales-Jimenez AL, Cortés-Ortiz L, Di Fiore A (2015) Phylogenetic relationships of Mesoamerican spider monkeys (*Ateles geoffroyi*): molecular evidence suggests the need for a revised taxonomy. Mol Phylogenet Evol 82(Part B):484–494. https://doi.org/10.1016/j.ympev.2014.08.025

Mosdossy KN, Melin AD, Fedigan LM (2015) Quantifying seasonal fallback on invertebrates, pith, and bromeliad leaves by white-faced capuchin monkeys (*Cebus capucinus*) in a tropical dry forest. Am J Phys Anthropol 158(1):67–77. https://doi.org/10.1002/ajpa.22767

Nakazato T, Warren DL, Moyle LC (2010) Ecological and geographic modes of species divergence in wild tomatoes. Am J Bot 97(4):680–693

Nascimento FF, Lazar A, Seuanez HN, Bonvicino CR (2015) Reanalysis of the biogeographical hypothesis of range expansion between robust and gracile capuchin monkeys. J Biogeogr 42:1349–1363

Pearson RG, Raxworthy CJ, Nakamura M, Peterson AT (2007) Predicting species distributions from small numbers of occurrence records: a test case using cryptic geckos in Madagascar. J Biogeogr 34:102–117

Peers MJL, Thornton DH, Murray DL (2013) Evidence for large-scale effects of competition: niche displacement in Canada lynx and bobcat. P Roy Soc B-Biol Sci 280(1773):1–10. https://doi.org/10.1098/rspb.2013.2495

Phillips SJ, Anderson RP, Schapire RE (2006) Maximum entropy modeling of species geographic distributions. Ecol Model 190:231–259

Pittman SJ, Brown KA (2011) Multi-scale approach for predicting fish species distributions across coral reef seascapes. PLoS One 6(5):e20583. https://doi.org/10.1371/journal.pone.0020583

Rose LM (1994) Sex differences in diet and foraging behavior in white-faced capuchins (*Cebus capucinus*). Int J Primatol 15(1):95–114. https://doi.org/10.1007/bf02735236

Ruiz-Garcia M, Castillo MI, Ledezma A, Leguizamon N, Sánchez R, Chinchilla M, Gutierrez-Espeleta GA (2012) Molecular systematics and phylogeography of Cebus capucinus (Cebidae, Primates) in Colombia and Costa Rica by means of the mitochondrial COII gene. Am J Primatol 74(4):366–380. https://doi.org/10.1002/ajp.20940

Rylands AB, Groves CP, Mittermeier RA, Cortés-Ortiz L, Hines JJH (2006) Taxonomy and distributions of Mesoamerican primates. In: Estrada A, Garber PA, Pavelka MSM, Luecke L (eds) New perspectives in the study of Mesoamerican primates: distribution, ecology, behavior, and conservation. Springer US, Boston, pp 29–79. https://doi.org/10.1007/0-387-25872-8_3

van Schaik CP, Pfannes KR (2005) Tropical climates and phenology: a primate perspective. In: Brockman DK, van Schaik CP (eds) Seasonality in primates: studies of living and extinct human and non-human primates. Cambridge University Press, New York, pp 23–54

Syfert MM, Smith MJ, Coomes DA (2013) The effects of sampling bias and model complexity on the predictive performance of MaxEnt species distribution models. PLoS One 8(2):e55158. https://doi.org/10.1371/journal.pone.0055158

van Schaik CP, Terborgh JW, Wright SJ (1993) The phenology of tropical forests: adaptive significance and consequences for primary consumers. Annu Rev Ecol Syst 24:353–377

Warren DL, Glor RE, Turelli M (2010) ENMTools: a toolbox for comparative studies of environmental niche models. Ecography 33(3):607–611. https://doi.org/10.1111/J.1600-0587.2009.06142.X

Wiederholt R, Post E (2010) Tropical warming and the dynamics of endangered primates. Biol Lett 6(2):257–260. https://doi.org/10.1098/rsbl.2009.0710

Wiederholt R, Post E (2011) Birth seasonality and offspring production in threatened neotropical primates related to climate. Glob Chang Biol 17(10):3035–3045. https://doi.org/10.1111/j.1365-2486.2011.02427.x

Chapter 16
A Synthesis of Long-Term Environmental Change in Santa Rosa, Costa Rica

Fernando A. Campos

Abstract Long-term monitoring is an essential component of primate conservation, and much of this research is explicitly concerned with how primates respond to and cope with diverse forms of environmental change. Here, I synthesize over four decades of data on environmental change in the Santa Rosa sector the Costa Rica's Área de Conservación Guanacaste, to stimulate new research on the impacts of environmental change beyond seasonality on Santa Rosa's primates. Focusing on climate variables and landscape-scale vegetation phenology, I describe and quantify typical seasonal patterns, interannual variability, and long-term trends. Santa Rosa's highly seasonal rainfall patterns show marked interannual variability that is largely driven by the El Niño Southern Oscillation (ENSO). The wettest and driest periods on record have occurred in association with powerful cold ENSO episodes (La Niña) and warm ENSO episodes (El Niño), respectively. Start dates for the wet season can vary by 40 days, but no long-term linear trend was evident in the wet season start dates or in total annual rainfall. Temperature anomalies in Santa Rosa are also strongly associated with ENSO conditions over a backdrop of long-term warming. The annual cycle of plant phenology is dominated by large-scale leaf shedding during the long dry season. The timing and degree of seasonal phenological peaks show complex relationships with rainfall. Long-term data, in combination with the site's natural environmental variability, provide uniquely quantitative context for understanding primate adaptations to changing environments – a framework that can be extended to ecological forecasting under future environmental change.

Keywords Climate change · Weather · Seasonality · Landscape · Habitat

---

F. A. Campos (✉)
Department of Anthropology, The University of Texas at San Antonio, San Antonio, TX, USA

Department of Biology, Duke University, Durham, NC, USA

U. Kalbitzer, K. M. Jack (eds.), *Primate Life Histories, Sex Roles, and Adaptability*, Developments in Primatology: Progress and Prospects,
https://doi.org/10.1007/978-3-319-98285-4_16

16.1 Introduction

The study of how primates adapt to changing environments has been a cornerstone of Linda Fedigan's primate research at the Santa Rosa research site, in northwestern Costa Rica, since its beginning in 1983. Linda had selected Santa Rosa to be the location for her long-term study of primate life histories and behavioural ecology, and at the urging of the Costa Rican National Park Service, she agreed to expand her research program to include periodic surveys of the park's entire capuchin and howler monkey populations to keep track of how these monkeys were faring in the nascent park (Fedigan and Jack 2012). Implicitly, this agreement marked the origin of her study of changing environments, a theme that continued to be a major organizing principle in Linda's research over the ensuing decades. The question of how primates respond to changing environments, with its clear links to conservation, had not yet acquired widespread attention in primatology (as a discipline, primatology had entered a phase in which sociobiology was the dominant theme). Linda's early interest in this topic proved prescient, as primatology has refocused around conservation-oriented concerns such as the roles of anthropogenic pressures and environmental change in driving the biology and behaviour of wild primates (see Chap. 14 by Kalbitzer and Chapman in this volume).

"Changing environments" can encompass an extremely broad range of processes in multiple interrelated domains – physical, ecological, and anthropogenic – over a vast range of time scales. Clearly, the definition must be narrowed to be useful in the context of this paper. The "changing environment" that interested the visionary Costa Rican park administrators who set the course for this research was most probably referring what I will call *long-term habitat variability*, which refers to variation over several years or decades in Santa Rosa's ecological environment – its forest structure, landscape configuration and dynamics, species composition, species interactions, fire ecology, etc. These are processes that land managers can directly influence by implementing conservation and restoration measures (Janzen 1988, 2000; Laurance et al. 2012). While some of these aspects of environmental change, such as species interactions, are complex and difficult to measure, others are relatively straightforward to quantify with the use of proxies, especially by taking advantage of remote-sensing technologies. For example, various biophysical properties of vegetation – including leaf area index (reviewed by Boyd and Danson 2005), forest age (Kalacska et al. 2004; Arroyo-Mora et al. 2005), above-ground biomass (Kalacska et al. 2007), and vertical canopy structure (Drake et al. 2002) – can be estimated quantitatively and tracked over time with high spatial precision and accuracy over large, continuous areas using sensor technologies borne on satellites or aircraft (Linke et al. 2007).

Research linking long-term habitat *temporal* variability with Santa Rosa's primates has generally focused on the effects of forest regeneration and associated habitat characteristics on primate population growth using long-term census data (Fedigan et al. 1985; Fedigan 1986; Fedigan et al. 1996, 1998; Fedigan and Jack 2001, 2012). A smaller set of studies used survey data to examine the effects of habitat *spatial* variability on primate density (Sorensen and Fedigan 2000; DeGama-Blanchet and Fedigan 2006), and some recent studies consider both temporal and spatial habitat variabilities on individual behaviour over relatively short monthly to seasonal time scales (Campos and Fedigan 2014; Campos et al. 2014). Relatively little research has been carried out in Santa Rosa on primate responses to *long-term* habitat variability using individual-based data (but see Chapman 1988, 1990; Campos et al. 2014); this is an area that is ripe for study.

It is also likely that the original conception of primate-focused environmental change research in Santa Rosa included *long-term climate variability*, a term that refers to variation in a location's climate statistics over several years or decades. While global climate change had not yet acquired its current urgency, Santa Rosa's highly variable rainfall patterns have always been abundantly clear. The research record that relates Santa Rosa's climate variability to its primates has focused almost entirely on the effects of seasonality by exploring different facets of how the primates' behavioural ecology varies between wet and dry seasons. This focus reflects a general pattern in primatology to emphasize research on the effects of short-term climate variability that has been coarsely characterized in dichotomous seasons. As in most other primate field studies, the effects of Santa Rosa's interannual climate variability, and of its long-term climate trends, have been relatively understudied (but see Fedigan et al. 2008; Campos et al. 2015, 2017). Here, I give particular attention to the El Niño Southern Oscillation (ENSO), which is the most globally influential driver of interannual climate variability throughout the tropics, including in Central America (Aceituno 1988). The ENSO is a coupled oscillation of ocean surface temperature and air surface pressure in the central Pacific Ocean that is associated with correlated weather patterns around the world. There are two distinct phases on ENSO, a warm phase known as "El Niño" and a cool phase known as "La Niña," that produce approximately opposite effects on local weather, which can differ regionally (e.g. some places get drier, while others get wetter during an El Niño phase) (Ropelewski and Halpert 1987). Weather patterns in Central America also depend to some extent on the Pacific Decadal Oscillation (PDO), which is associated with a pattern of climate variability very similar to that produced by the ENSO but that differs primarily in time scale: whereas ENSO events last on the order of 1 year, PDO phases persist for several decades (Mantua and Hare 2002; Chavez et al. 2003). Although ENSO is the dominant climatic driver in Central America, it is modulated by the PDO such that its effects are magnified when the two oscillations are in phase and reduced when the two oscillations are out of phase (Wang et al. 2014).

In this paper, I summarize approximately four decades of environmental change at Santa Rosa, with a particular focus on climate variables and large-scale vegetation characteristics. My aims are (1) to present a compendium of data on diverse forms of environmental change in multiple domains (temperature, rainfall, and vegetation), (2) to illustrate new ways of approaching and quantifying long-term environmental change beyond seasonality to stimulate research in these areas, and (3) to discuss future directions for climate change-integrated conservation planning for Santa Rosa's primates.

16.2 Environmental Changes in Santa Rosa

16.2.1 Rainfall

16.2.1.1 Data Sources and Preparation

Local rainfall measurements at Santa Rosa began in 1979 and continue to the present, and the rainfall record comprises data from several different rain gauges. Measurements of a single rain event are likely to vary from location to location due to the patchy nature of rainfall, which in Santa Rosa typically takes the form of short, intense showers in scattered locations. But unlike temperature measurements, rainfall measurements at a single site are unlikely to show systematic biases that depend on the placement of the instrument. In other words, when there are multiple rain gauges in the same general vicinity, none should read *consistently* higher than the others due to microhabitat differences (assuming that there is no vegetation overhanging any rain gauge). For this reason, I use the complete record of locally measured rainfall totals to examine long-term patterns of change in Santa Rosa.

In addition to locally measured rainfall, I examine long-term patterns in water balance using a drought index, the Standardized Precipitation-Evapotranspiration Index (SPEI) (Vicente-Serrano et al. 2009). The SPEI quantifies deficits and surpluses in water balance, including movement of water from the air to the earth's surface (precipitation) and movement of water to the air from sources such as soil and waterbodies (evaporation) as well as plants (transpiration). The SPEI combines information on local rainfall, temperature, and solar radiation, and it is interpolated over a 0.5 degree global grid (approximately 55 km × 55 km over Santa Rosa). There are different versions of the SPEI that correspond to different time scales in recognition of the fact that the relevant time scale over which water deficits or surpluses accumulate may differ among ecosystems in important ways. In the analysis of long-term wet and dry cycles described below, I have chosen to use the 48-month SPEI to examine cumulative water deficits or surpluses over multiple years. I obtained the SPEI data from the SPEI Global Drought Monitor (Beguería et al. 2017).

16.2.1.2 Typical Seasonal Patterns

The typical seasonal pattern of rainfall in Santa Rosa has been described in detail elsewhere (Janzen and Hallwachs 2016). In brief, the annual rainfall cycle can be neatly divided into two halves: a long dry season from mid-November to mid-May characterized by strong northeasterly trade winds and a long wet season from mid-May to mid-November. The long wet season has a bimodal character, with the smaller rainfall peak in May/June and the larger rainfall peak in October (Fig. 16.1); in some years, the relative sizes of these peaks can be reversed. There is sometimes a short dry season or *veranillo* in late July or August that is caused by a brief intensification of trade wind activity (Taylor and Alfaro 2005).

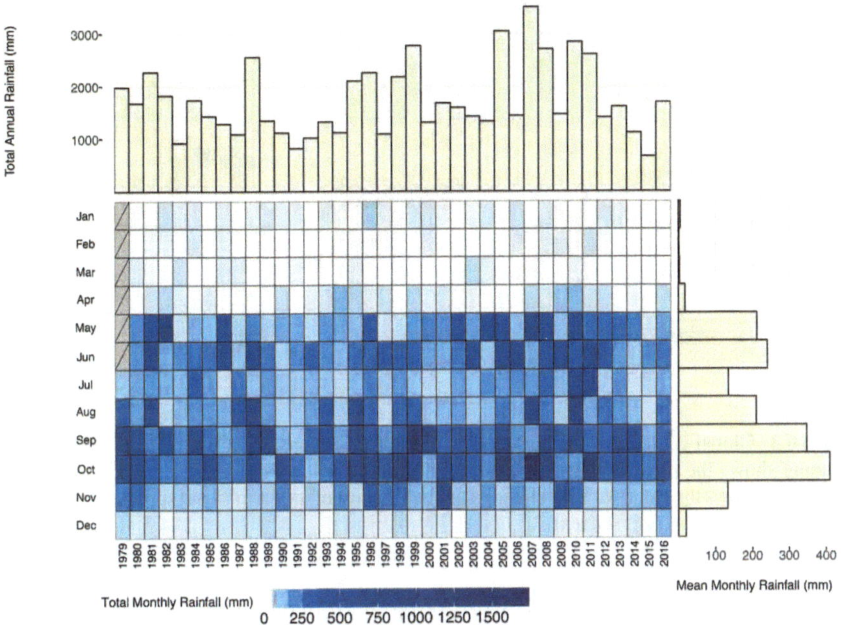

Fig. 16.1 Monthly and total annual rainfall in Santa Rosa since 1979. The coloured cells show rainfall totals for individual months, and the marginal bar plots show total annual rainfall (top) and mean monthly rainfall (right)

The transition from dry to wet seasons is usually abrupt and involves several days of heavy deluges. False starts can occur in March or April in which sporadic showers take place, only to be followed by weeks of further dry weather. The transition from wet to dry seasons after the heavy rains cease in late November is more prolonged and involves a gradual drying out of the landscape over a period of several weeks or longer. It should be noted that these are broad tendencies; any given year can deviate substantially from this pattern, and as detailed below, these deviations are principally driven by ENSO.

As in other tropical dry forests, the strongly seasonal pattern of rainfall is the dominant driver of many ecological processes in Santa Rosa. Rainfall seasonality poses unique challenges to organisms in tropical dry forests: they must tolerate both long dry periods and long wet periods, and they must cope with the variable timing of transitions between wet and dry seasons. There is growing evidence that dry forest-adapted trees will be sensitive to changes in future rainfall regimes, with probable ratcheting effects on other organisms in these ecosystems (Uriarte et al. 2016; Allen et al. 2017). From a global perspective, Santa Rosa, with its intermediate levels of highly clumped precipitation, falls in one of the world's rainfall seasonality "hotspots." Using the high-resolution gridded climate surfaces from WorldClim.org (Fick and Hijmans 2017), precipitation seasonality as measured by the Feng et al. (2013) index is greater in Santa Rosa than in 98.9% of the earth's land surface

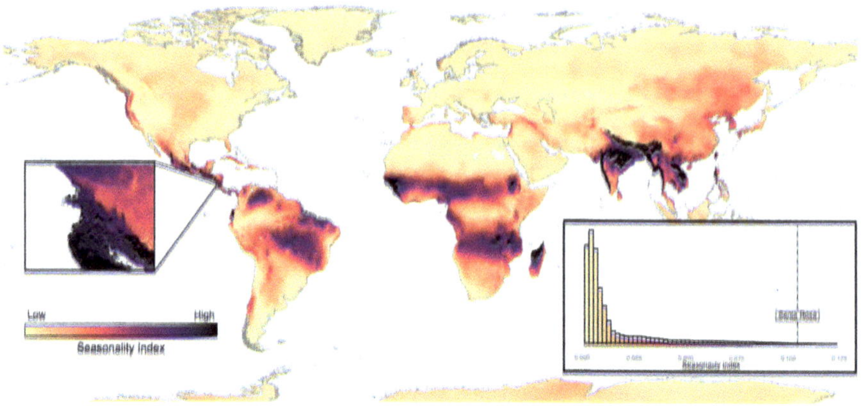

Fig. 16.2 Global context for the high degree of rainfall seasonality in Santa Rosa. The coloured shading shows the precipitation seasonality index described in Feng et al. (2013), and the inset histogram shows the value for Santa Rosa relative to the rest of the earth's land surface. Precipitation data from Worldclim.org (see text)

(Fig. 16.2). This index combines the amount of rainfall with its annual distribution; low seasonality index values can result from either low annual rainfall or evenly distributed rainfall (see Feng et al. 2013 for the formulation of this index).

16.2.1.3 Interannual Variability

Total annual rainfall in Santa Rosa is highly variable. The wettest years may receive as much as 5 times more rainfall than the driest years (3498 mm in 2007 vs. 661 mm in 2015, Fig. 16.1). Monthly rainfall anomalies – the difference between actual rainfall in a particular month and the long-term average for that month – have fluctuated erratically over time, showing wet and dry cycles that are irregular and sometimes prolonged. ENSO is a major driver of the interannual variability in rainfall. Monthly rainfall anomalies are strongly correlated with ENSO conditions in some months but not in others (Fig. 16.3). In the middle to late wet season (August to October), the link between ENSO and rainfall is especially strong. For example, the Oceanic Niño Index (ONI), the primary index used by the National Oceanic and Atmospheric Administration to monitor ENSO activity and to define phases of ENSO, explains 33% of the variance in August rainfall anomalies. The ONI is a 3-month running mean of sea-surface temperature anomalies in a region of the tropical Pacific Ocean known as Niño 3.4, and El Niño and La Niña conditions are present when the anomalies exceed +0.5 °C or −0.5 °C, respectively. The direction of the relationship between the ONI and rainfall in months for which the linear regression is significant is consistently negative, with El Niño phases leading to drier-than-usual conditions and La Niña phases leading to wetter-than-usual conditions (Fig. 16.3). The wettest year on record, 2007, occurred in conjunction with

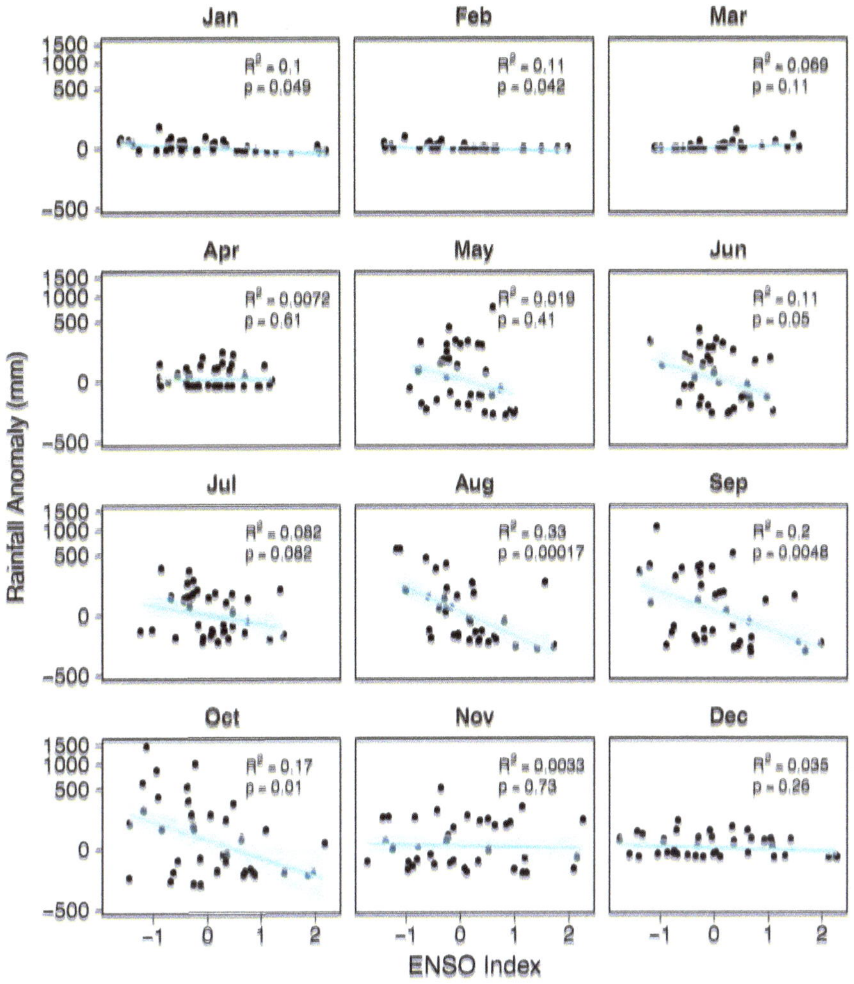

Fig. 16.3 Relationship between monthly rainfall anomalies in Santa Rosa and the Oceanic Niño Index, an index of ENSO activity. The negative relationships from June to October indicate that El Niño phases are associated with abnormally dry conditions in Santa Rosa during the wet season

a strong La Niña event, and the driest year on record, 2015, occurred in conjunction with a strong El Niño event.

The arrival of the persistent heavy rains that mark the beginning of the wet season typically occurs in May, but this event varies from year to year both in its timing and its attributes. The wet season can begin with a torrent of May rainfall, as in 1982, 1986, 2007, and 2010; more commonly it ramps up somewhat gradually (Fig. 16.4). The period immediately before the start of the wet season is critical for many organisms in Santa Rosa because temperature is at its annual peak, while access to shade and the availability of surface water are at their low points. These conditions maximize the risks of heat stress and dehydration, and a significant delay in the

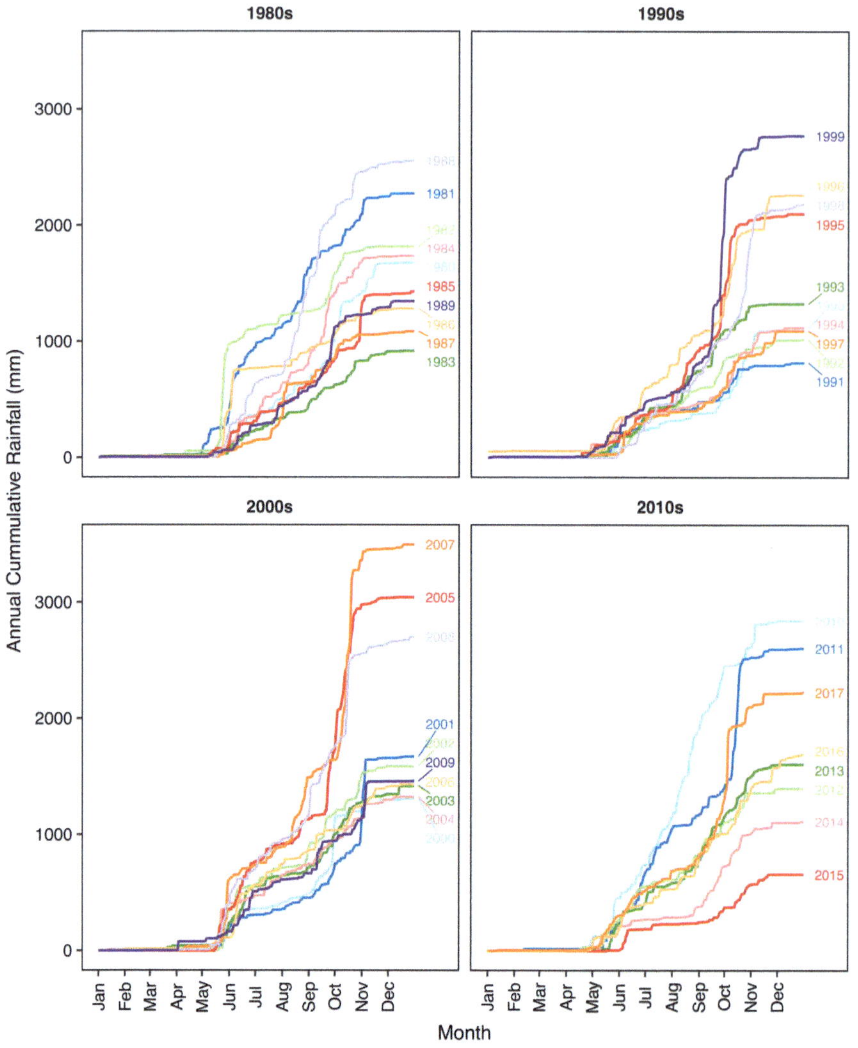

Fig. 16.4 Annual cumulative rainfall in Santa Rosa since 1980, the first complete year in which rainfall was measured locally

arrival of rains prolongs this challenging period. I identified the start of the wet season in each year by finding the first occurrence of a week in which median daily rainfall exceeded 1 mm (i.e. the wet season begins at the start of the first 7-day period in which there were more rainy days than dry days, with rainy days defined as those having ≥1 mm of rain). The resulting wet season start dates appear reasonable on visual inspection: most false starts are ignored, and all start dates fall between late April and early June. Based on this definition, the beginning of the wet season can vary by as much as 40 days, ranging from April 22 to June 1 (Fig. 16.5).

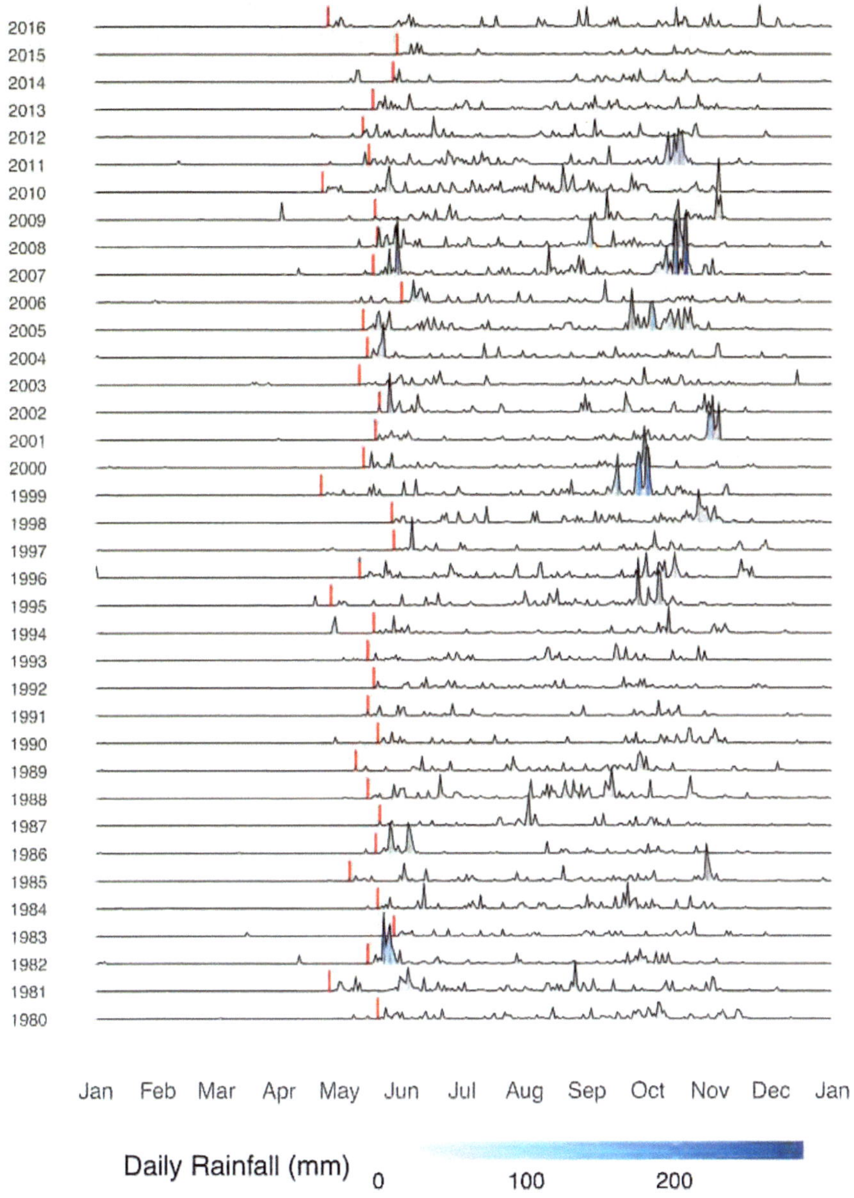

Fig. 16.5 Wet season start dates (red vertical lines) for each year shown over the daily record of rainfall. The wet season start date is defined for each year as the first occurrence of a 7-day period in which the median of daily rainfall over the 7 days exceeds 1 mm

As with other weather variables, the wet season start date is linked (somewhat weakly) to ENSO conditions, with El Niño phases leading to later-than-usual arrival of the first persistent heavy rains (linear regression, F = 8.76, $R^2 = 0.20$, p = 0.0055, n = 37).

16.2.1.4 Long-Term Trends

Since the rainfall records began in 1979, there have been extended wet and dry periods at Santa Rosa. The most notable long dry periods have occurred from approximately 1985 to 1994, 2000 to 2004, and 2012 to 2015; the most notable long wet periods occurred from 1979 to 1982, 1995 to 1999 and 2005 to 2011 (Fig. 16.6a). The long-term drought index, the 48-month SPEI, supports these observations. The dry period from 1985 to 1994 was the longest lasting, but the dry period from 2012 to 2015 was the most severe in terms of overall ecosystem water deficit (Fig. 16.6a). There was no long-term linear trend in total annual rainfall over the last four decades. This lack of a linear trend is in contrast to previous indications of

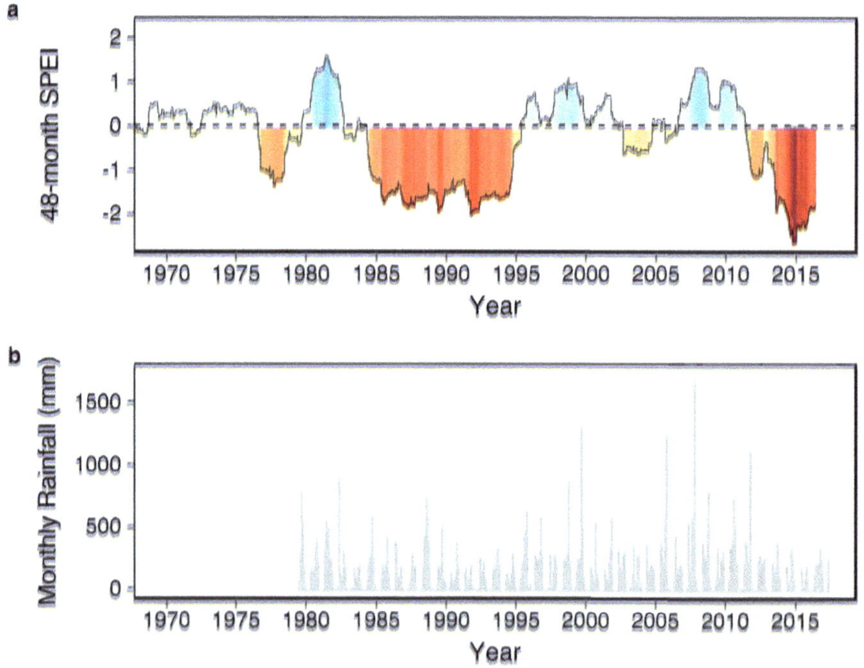

Fig. 16.6 (**a**) Fluctuations in ecosystem water balance at Santa Rosa balance as measured by the 48-month Standardized Precipitation-Evapotranspiration Index (SPEI), an index used to monitor drought. The coloured shading simply reiterates the SPEI value for each month and is intended to aid visual interpretation. (**b**) Monthly rainfall totals over the same time period for context. SPEI data obtained from the SPEI Global Drought Monitor (see text)

long-term drying in this region (Borchert 1998; Enquist and Enquist 2011). No long-term trends were evident in the wet season start dates (Fig. 16.5).

16.2.2 Temperature

16.2.2.1 Data Sources and Preparation

As in the case of rainfall, measurements of daily maximum and minimum temperature have been collected on site, sporadically at times, by different instruments since July 1979. Temperature data from different sources must be interpreted carefully because the measured values can depend on microclimate differences among the places where the thermometers have been located. For example, ambient temperature can be strongly influenced by exposure to sunlight and shade, as well as by proximity to heat-absorbing surfaces. Close examination of the temperature data collected at Santa Rosa reveals systematic variation among the different instruments (Fig. 16.7), which are all located within about 200 m of each other in Santa Rosa's administrative area. For example, the data sources labelled "Source 2" and "Source 4" in Fig. 16.7 have lower precision than the other data sources, and the data source labelled "Source 5" shows noticeably higher variance in T_{max} as well as reduced separation between daily T_{max} and T_{min}. These instrument-related differences raise difficult questions about how to interpret patterns of variability that are present in the data. Did Santa Rosa actually experience wildly

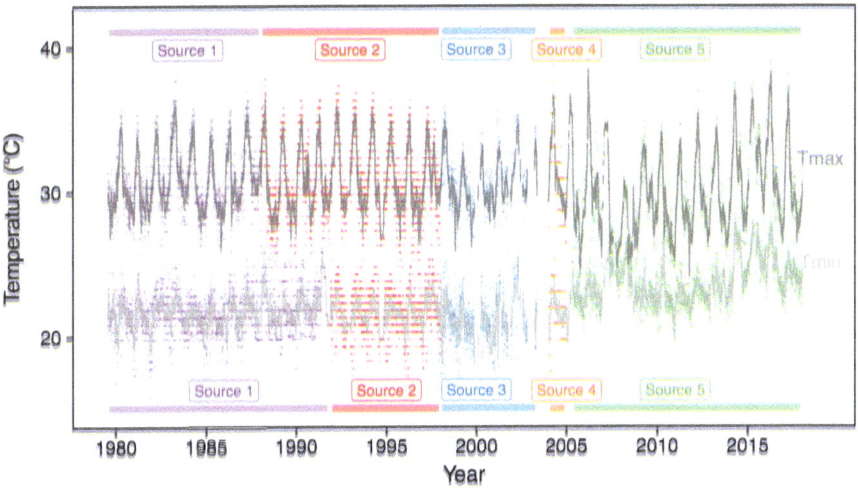

Fig. 16.7 Locally measured temperature data collected since 1979 by different thermometers in Santa Rosa. The small points show daily temperature measurements from the different data sources labelled at the top and bottom. The grey lines show 15-day rolling means that are intended to clarify the general aspect of the time series. Clear differences are evident among the data sources in their precision, central tendency, and variability

fluctuating temperatures from 2005 to 2008, including exceptionally cold T_{max} and exceptionally warm T_{min} in 2007 and 2008, or are these patterns just instrument artefacts? This is particularly problematic for analyses that involve interannual temperature patterns and long-term changes in temperature.

One way of sidestepping these issues is to use bias-corrected weather data from nearby weather stations. Several climate modelling groups have used such data to compile land surface temperature data sets with daily or monthly temporal resolution and that are interpolated over a global grid; these data products include GISTEMP produced by the NASA Goddard Institute for Space Studies (Hansen et al. 2010), HadCRUT4 produced by the Climatic Research Unit (University of East Anglia) in conjunction with the Hadley Centre (UK Met Office) (Morice et al. 2012), MLOST produced by the National Oceanic and Atmospheric Administration (Vose et al. 2012), and BEST produced by Berkeley Earth (Rohde et al. 2013). The results of the various climate modelling groups mirror each other despite their use of independent bias correction methods (Rohde et al. 2013); this high degree of correspondence increases confidence in the accuracy of the data. The data take the form of a grid indexed by latitude, longitude, and time. This results in a time series of temperatures for each grid cell. Although such data sets largely rectify the problems associated with sensor biases, they have other drawbacks. In particular, the accuracy of interpolated values depends to some extent on the locations of nearby weather stations: the climate conditions of locations nearer to weather stations and with higher density of weather stations will be more accurately represented in the grid. Additionally, the spatial resolution can be quite coarse; for example, the surface temperature data set produced by Berkeley Earth (Rohde et al. 2013) has 1×1 degree grid cells, which corresponds 111 km × 111 km at the equator. There is obviously a great deal of climate variation that can occur in a grid cell of that size, for example, due to elevation differences. But while temperature estimates derived from a bias-corrected global surface temperature grid may not be spot on for any particular location, they are indispensable for examining long-term climatic changes in places where locally measured conditions are either absent or problematic. Because the temperature data collected locally at Santa Rosa show instrument-related biases, I use the global land surface temperature data compiled by Berkeley Earth (BEST) to quantify interannual variability and long-term trends in temperature at this site.

16.2.2.2 Typical Seasonal Patterns

Temperatures in Santa Rosa typically reach their minimum in the late wet season, October to November, and then climb steadily throughout the dry season until reaching a peak in April or early May (Fig. 16.8). Although temperature shows a consistent seasonal pattern, this intra-annual variation is small in a global context. Using the high-resolution gridded climate surfaces from WorldClim.org (Fick and Hijmans 2017), only 5.0% of the earth's land surface experiences smaller temperature seasonality than Santa Rosa, with seasonality defined as the standard deviation of

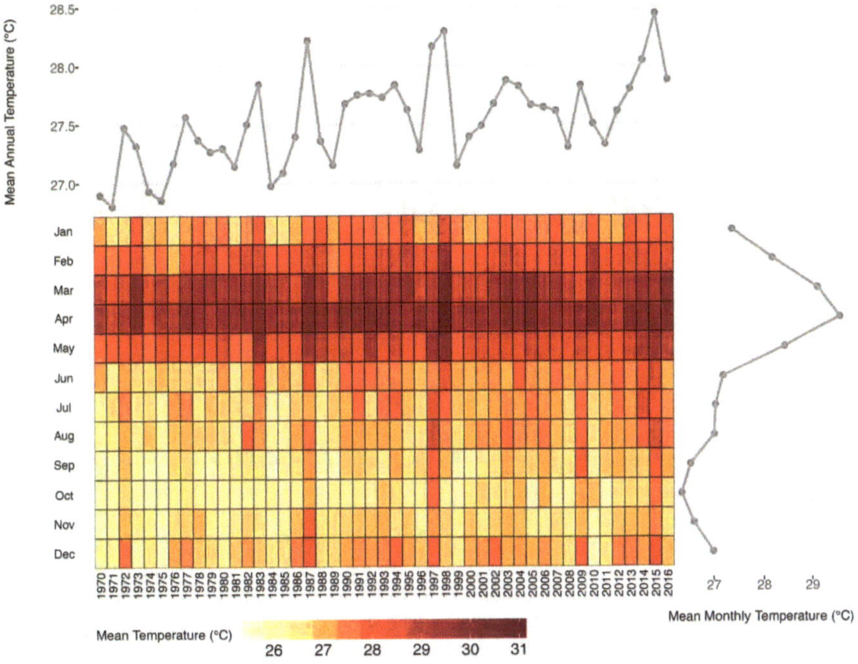

Fig. 16.8 Mean monthly and mean annual temperature in Santa Rosa since 1970. The coloured cells show temperature values for individual months, and the marginal plots show long-term patterns in mean annual temperature (top) and mean monthly temperature (right). Temperature data from Berkeley Earth (see text)

the 12 long-term means of monthly temperature for each 30 arc-second grid cell of the earth's surface over the period 1970 to 2000.

16.2.2.3 Interannual Variability

There is a strong correlation between indices of ENSO and temperature anomalies in Santa Rosa, defined as deviations relative to long-term mean of the period lasting from 1951 to 1980. The Oceanic Niño Index explains much of the total variance in temperature anomalies throughout the year (18% of variance in T_{min}, 41% of variance in T_{avg}, and 33% of variance in T_{max}) (Fig. 16.9a). In some months, this relationship is particularly strong. For example, ENSO explains 68% of the variance in March T_{avg} anomalies. Rare extreme phases of ENSO have occurred in conjunction within the hottest and coldest periods on record. The hottest temperatures recorded in this region occurred in 1983, 1997–1998, and 2016 and these periods coincide with historically powerful El Niño events (Fig. 16.9b).

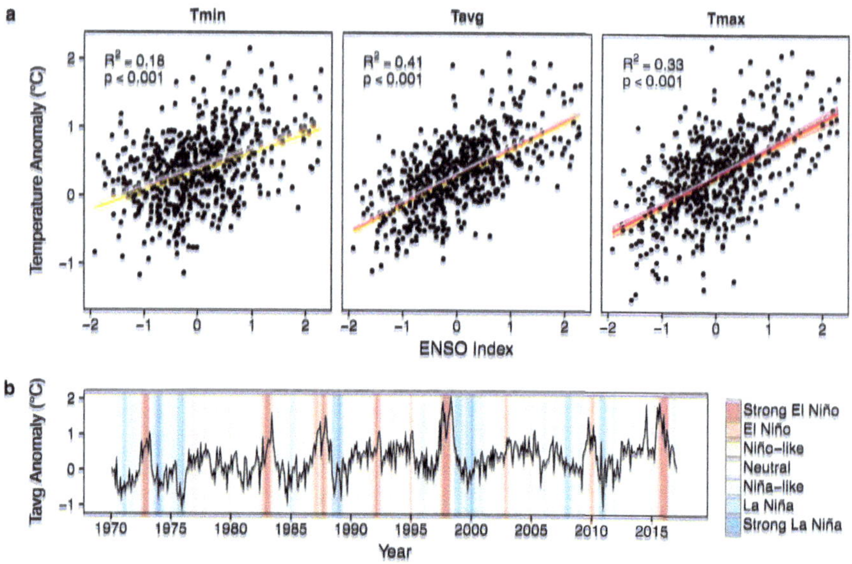

Fig. 16.9 (**a**) Linear relationship between monthly temperature anomalies in Santa Rosa and the Oceanic Niño Index, an index of ENSO activity. The strong positive relationships indicate that El Niño conditions are associated with abnormally warm conditions in Santa Rosa. (**b**) Anomalies in monthly mean of daily average temperature (Tavg) relative to phases of ENSO. Temperature data from Berkeley Earth (see text)

16.2.2.4 Long-Term Trends

Daily average temperature in Santa Rosa has increased by approximately 0.77 ° C since 1970, based on a 5-year running mean (Fig. 16.10a). Daily maximum temperature has increased slightly more (1.07 ° C), and daily minimum temperature has increased slightly less (0.70 ° C) (Fig. 16.10a). This is in keeping with global estimates of climate warming over the last four decades. The greatest relative increases have occurred during the wet season months of July and August, and the smallest relative increases have occurred in March and November (Fig. 16.10b).

16.2.3 Large-Scale Vegetation Phenology

16.2.3.1 Data Sources and Preparation

Like other tropical dry forests, Santa Rosa experiences profound changes in vegetation phenology over a typical annual cycle (Fig. 16.11). Remote sensing enables continuous monitoring of vegetation characteristics, such as the presence and abundance of green leaves on deciduous trees, over large-scale space. I used two satellite-based vegetation indices to make inferences about vegetation phenology,

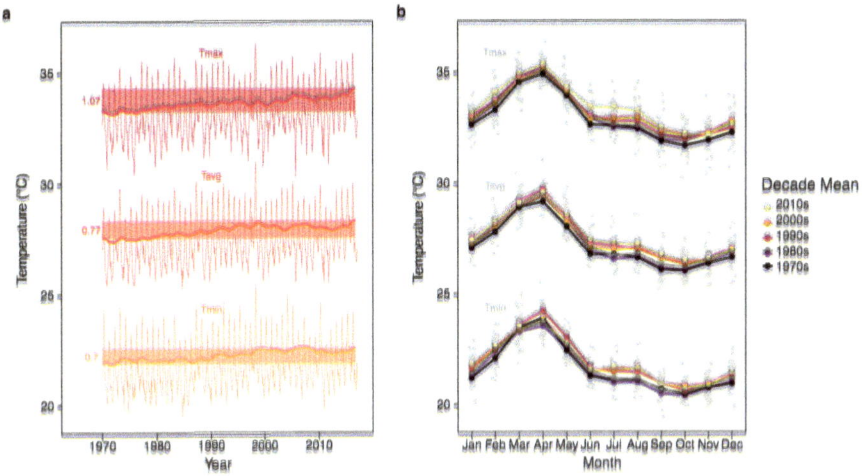

Fig. 16.10 (**a**) Long-term change in monthly means of daily maximum (T_{max}), average (T_{avg}), and minimum (T_{min}) temperature in Santa Rosa. The thin lines show the monthly values, and the thicker lines show a 5-year running mean to highlight the long-term trend. The shaded areas show the magnitude of warming since 1970. (**b**) Seasonal and decadal trends in monthly means of maximum, average, and minimum daily temperatures in Santa Rosa since 1970. The light grey points show all monthly values with a slight horizontal jitter applied to improve readability. The coloured points and lines show decade mean values. Temperature data from Berkeley Earth (see text)

the Normalized Difference Vegetation Index (NDVI) and the Enhanced Vegetation Index (EVI). Vegetation indices describe the "greenness" of each pixel of the scene, integrating information about the density, health, and phenophase of vegetation located in the pixel. Although correlated with each other (Fig. 16.13c), the NDVI and EVI are not entirely redundant. In particular, the EVI corrects for distortions caused by atmospheric conditions, and unlike NDVI it does not saturate when green biomass is high, but the NDVI has been more commonly used historically, and some older satellite sensors did not include the blue band that is required for the EVI calculation.

The satellite data used in this analysis comprise a time series of version 5 MODIS Terra and Aqua vegetation indices products (MOD13QA3 and MYD13Q1, respectively) from the NASA Land Processes Distributed Active Archive Center (LP DACC) and the USGS/Earth Resources Observation and Science Center (Didan 2015a, b). The MODIS data have low spatial resolution (250 × 250 m pixels), making them unsuitable for analysing fine spatial patterns, but they have continuous coverage since February 2000 with 16-day granularity. The 16-day compositing periods of the two different MODIS satellites are offset so that 1 begins every 8 days. Thus, by combining the Terra and Aqua data products, I obtained a single time series of vegetation indices with 8-day temporal granularity spanning 2000 to 2016. The continuity and broad coverage make the data valuable for assessing how the vegetation index values are affected by local conditions such as recent rainfall. Using the R package MODISTools (Tuck and Phillips, 2013), I obtained vegetation a

b

Fig. 16.11 High-resolution satellite images of Santa Rosa's administration area (**a**) in May 2013 at the end of the dry season and (**b**) in November 2002 at the end of the wet season. Images from Google Earth

indices for a square centred at the point 10.843° latitude and −85.613° longitude, which is the administration area of Santa Rosa, and that measures 2.25 km on each side; that is, the square extends for 1 km in each cardinal direction away from the central 250 m pixel. I excluded data for which the pixel reliability assessment indicated reduced quality, which was usually due to persistently cloudy conditions over the compositing period. I use these data to analyse patterns over time in large-scale vegetation phenology, focusing on the timing and magnitude of annual peaks. The vegetation indices are somewhat "noisy" and contain frequent gaps due to cloud cover, especially during the wet season. These characteristics mean that finding an accurate annual maximum and minimum is more complicated than it may appear. I imputed missing values in each phenology time series using the Kalman smoothing procedure provided in the function R package imputeTS (Moritz 2017). The steps used to find annual peaks are detailed in Fig. 16.12a–f.

16.2.3.2 Typical Seasonal Patterns

The annual lowest point of green vegetation as measured by the EVI and NDVI typically occurs in Santa Rosa during the first or second week of April (Fig. 16.13a,b). This point represents the peak of vegetation dormancy, when deciduous trees have lost most or all of their leaves. Although the peak of vegetation dormancy is variable, the timing in early to middle April precedes the average transition date from dry to wet seasons by approximately 1 month. This suggests that some deciduous plants anticipate the start of the wet season by putting out new leaves before the rains begin in earnest, perhaps by responding to changes in humidity, cloud cover, or the aforementioned wet season "false starts" that occur in some years. Soon after the

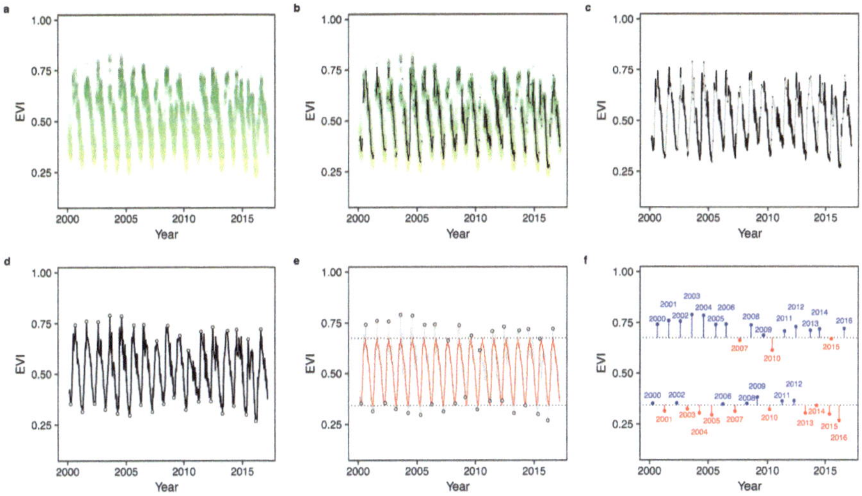

Fig. 16.12 Steps in the procedure used to find annual peaks in vegetation indices, here, for the Enhanced Vegetation Index (EVI). (**a**) Complete set of EVI measurements (coloured points) for the target area in Santa Rosa, consisting of 81 pixels for each compositing period that occurred at 8-day intervals from 2000 to 2016. (**b**) The thin black line connects the average of all EVI values on each collection date. The numerous gaps result from the removal of unreliable data due to cloud cover. (**c**) Imputation of missing values using Kalman Smoothing (grey line). (**d**) Identification of annual maximum and minimum values (small circles) on the imputed time series. (**e**) Comparison between imputed vegetation index time series (grey) and the variability in the EVI that can be attributed purely to seasonality (red). The seasonal component was obtained by seasonal decomposition using loess. (**f**) Deviation of each year's observed peak values (dots above and below the horizontal dotted line) from the expected peak values attributable to the purely seasonal component (dashed lines)

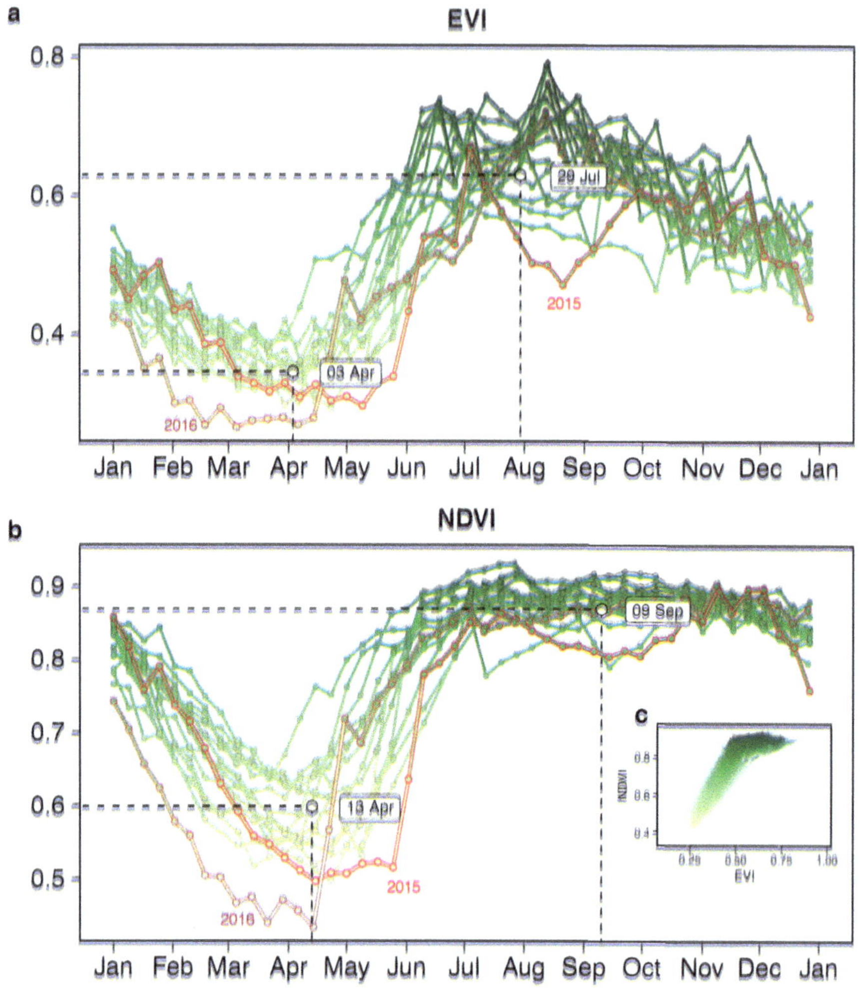

Fig. 16.13 Seasonal and interannual variation in vegetation indices. (**a**) EVI, Enhanced Vegetation Index, and (**b**) NDVI, Normalized Difference Vegetation Index. The points show landscape-scale averages of the vegetation index values measured every 8 days from 2000, when monitoring began, to 2016. The lines connect the values in each separate year. A year of exceptional drought, 2015, and the subsequent dry season of 2016 are highlighted in red and brown, respectively. Average dates for the annual minimum and maximum are labelled at the intersection of the dashed lines. The inset plot (**c**) shows the relationship between the NDVI and EVI values measured on the same date

rains arrive, there is a rapid greenup period during which most trees put out new leaves. The EVI tends to reach a peak in late July or mid-August, which is the middle of the wet season, and thereafter declines slowly through the remainder of the wet season (Fig. 16.13a). This slow decline may reflect the end of the primary growing season and the beginning of leaf senescence in some species. The NDVI increases quickly with the start of the wet season but reaches a plateau by early July;

thereafter, the NDVI remains roughly constant until it begins to decrease in mid to late November in association with the onset of leaf shedding (Fig. 16.13b). This plateauing phenomenon is typical of the NDVI, which is sensitive to chlorophyll and saturates when plant biomass is high.

16.2.3.3 Interannual Variability and Long-Term Trends

The annual maximum of the EVI varies somewhat from year to year both in its timing and magnitude (Fig. 16.14, top left panel). There are some surprising and somewhat unintuitive patterns revealed by the EVI maxima. For example, the lowest EVI maxima occurred not during drought years but rather during the exceptionally wet years of 2010 and 2007. Some other very wet years, such as 2005, did not show a dampened maximum in the EVI. It may be that the timing of the EVI peak coincides with the *veranillo*, and I hypothesize that the peak's magnitude depends on the inconsistent formation of this short dry season. The clear, sunny weather during a well-formed *veranillo*, when most trees are already flush with leaves, could further promote plant growth and lead to greater peaks in the EVI. There was high

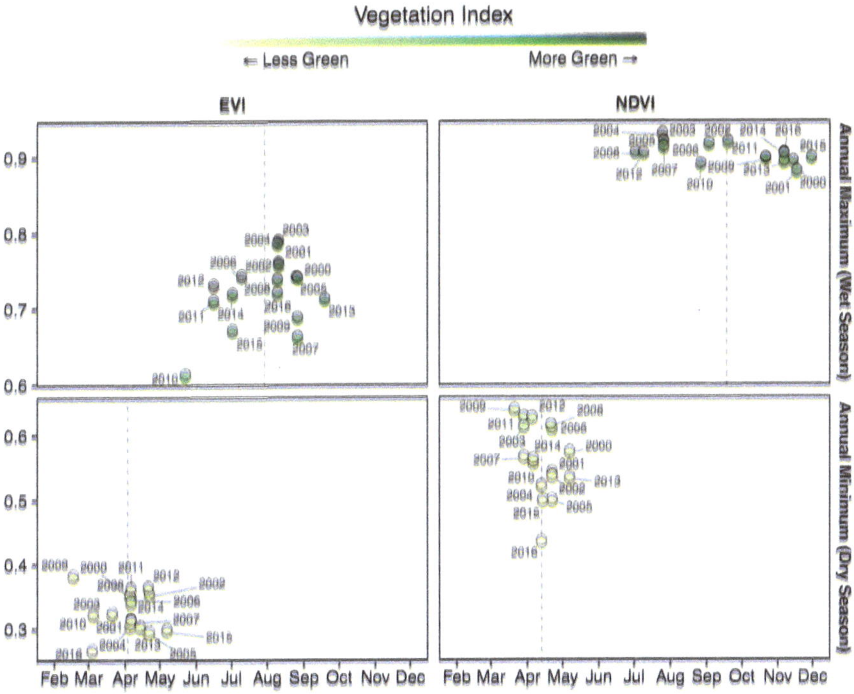

Fig. 16.14 Interannual variability in the timing (x-axis) and magnitude (y-axis) of the annual maximum values (top panels) and minimum (bottom panels) of the two vegetation indices, EVI (left panels) and NDVI (right panels). The dashed lines show the average dates for each peak

July–August rainfall during 2010 and 2007, with absent or ill-formed *veranillos* (see Figs. 16.1 and 16.4). Thus, the highest EVI peaks appear in years such as 2003 and 2004 that showed a rainfall lull during July and August and therefore a salient *veranillo*.

Distinct patterns in the EVI maxima can be observed during drought years. During recent severe drought in 2015, the EVI spiked in early July following low but unexceptional June rainfall; it then formed a second sharp low point in August as a powerful El Niño event led to extraordinarily low rainfall throughout the remainder of the wet season (Fig. 16.13a; see also Fig. 16.1). This agrees with ground observations by researchers who noted that many trees that would normally be full of leaves were bare and that there was high tree mortality during that time. Unlike the EVI, the annual maxima of the NDVI show relatively little variation in their magnitudes from year to year due to the aforementioned saturation phenomenon (Fig. 16.14, top right panel). Consequently, differences in the timing of the NDVI maxima are likely to be spurious and uninformative. However, during recent severe drought in 2015, the NDVI did not plateau but rather oscillated during the normal wet season (Fig. 16.13b).

The annual minima of the EVI and NDVI are variable both in their magnitude and timing (Fig. 16.14, bottom panels). The magnitude of each year's minimum value can be viewed as a measure of the "severity" of the dry season. There is a very strong positive linear relationship between these minimum values and total rainfall in the previous year (EVI: $F = 33.6$, $R^2 = 0.69$, $p < 0.001$, $n = 17$; NDVI: $F = 31.4$, $R^2 = 0.68$, $p < 0.001$, $n = 17$). In other words, the key variable that determines dry season severity, in terms of lack of green vegetation, is the amount of rain received in the previous wet season. As an example, the most barren dry season on record occurred in 2016 following the severe drought of 2015 (Fig. 16.13a,b).

To analyse what determines the *timing* of the annual minima in the vegetation indices, I considered two possible influences: the wet season start date in the current year and total rainfall in the previous year. The expected effect of the wet season start date is straightforward: earlier arrival of the wet season should end the current dry season sooner, leading to an earlier date for the minimum in the vegetation index. The expected effect of rainfall in the previous year is not so straightforward. On the one hand, more rainfall during the previous wet season could essentially delay the gradual drying-out process by which the dry season progresses, leading to a later date for the minimum EVI/NDVI. On the other hand, more rainfall could provide deciduous trees with a critical store of water that enables them to flush early in anticipation of the arrival of the wet season, leading to an earlier date for the minimum EVI or NDVI. To test these hypotheses, I created four candidate models of the annual minimum date for each vegetation index: a model with wet season start date only, a model with rainfall in the previous year only, a model that included additive effects of both variables, and a null model that included no predictor variables (Table 16.1). No model of NDVI minima dates received support over the null model. The best model of EVI minima dates, based on comparison of AIC values and model R^2 values, included both rainfall in the previous year and wet season start date ($\Delta AIC = -2$ compared to null model). As expected, a delay in the

Table 16.1 Model selection table for models of the date of annual minima in the vegetation indices NDVI and EVI

Vegetation index	Fixed effects	AIC	ΔAIC	R^2
EVI	RainPreviousYear + WetSeasonStartDate	144.9	0	0.299
EVI	WetSeasonStartDate	145.3	0.448	0.190
EVI	None (null model)	146.9	2.038	0.000
EVI	RainPreviousYear	148.4	3.508	0.031
NDVI	None (null model)	140.2	0	0.000
NDVI	WetSeasonStartDate	142.0	1.756	0.014
NDVI	RainPreviousYear	142.2	1.998	0.000
NDVI	RainPreviousYear + WetSeasonStartDate	143.9	3.711	0.017

arrival of the wet season had a nearly 1:1 effect in bringing about a delay in the annual minimum EVI, controlling for rainfall in the previous year (estimate: 0.90, SE = 0.39). The coefficient estimate for rainfall in the previous year was small but still improved the model fit (estimate: -0.0071, SE = 0.0048). The interpretation for this coefficient estimate would be that every $1/0.0071 \approx 140$ mm of additional rainfall received in the previous year hastens the timing of leaf flushing by 1 day, controlling for the arrival of the wet season.

Forest succession processes, in which regenerating forest stands acquire increasing proportions of evergreen vs. deciduous species over time, could produce long-term trends in the vegetation indices. However, no linear, long-term "greening" trend was evident (e.g. Fig. 16.11f), and the relatively brief 17-year time span of the data set suggests that species turnover is unlikely to contribute significantly to the interannual variation in vegetation indices described here.

16.3 Future Directions

16.3.1 Primate Responses to Environmental Change Beyond Seasonality

I have conveyed in this synthesis and review that there are many aspects of environmental change beyond rainfall and temperature seasonality and that, in general, the effects of such changes on primates have been understudied despite their resounding urgency. There are several factors that have contributed to this situation. First, and perhaps most importantly, it is invariably difficult to meet the data requirements for carrying out long-term environmental change research on primates. The extraordinarily slow life histories and low reproductive rates of wild primates mean that decades of study are typically necessary to achieve adequate sample sizes (see, e.g., Campos et al. 2017). In some animal taxa, it may be possible to use "replicate" populations that exist along an environmental gradient to investigate these questions using a space-for-time substitution (Fukami and Wardle

2005; Blois et al. 2013), but this is rarely true for primates. Second, the roots of North American primatology in the field of anthropology rather than ecology or biology have probably contributed to the tendency to overlook collecting long-term, detailed ecological data at many primate field sites (e.g. on food abundance, small-scale climate variability, and vegetation community structure). This historical tendency has been changing rapidly over recent decades as conservation and ecology have become ascendant topics in primatology. The pioneering research of Linda Fedigan and her students and collaborators on the effects of environmental change on Santa Rosa's primates helped to pilot this shift in primatology, along with research carried out at a few other long-term primate research sites including Kibale (Chapman et al. 2010), Amboseli (Bronikowski and Altmann 1996; Alberts et al. 2005; Alberts and Altmann 2012), and Caratinga (Strier and Boubli 2006). Third, scientific advances in fieldwork-based disciplines such as primatology often come incrementally from graduate-student projects that typically involve 1–2 years of fieldwork. Such time scales are far too short to address questions about the effects of interannual environmental variability and long-term environmental trends from any single student's primary data. Therefore, long-term databases that compile and synthesize many years of data can greatly facilitate research on primate responses to long-term environmental change. The Santa Rosa primate project has stayed at the forefront in this domain with the creation of the "PACE" database by Dr. John Addicott (Fedigan and Jack 2012); the "Babase" database from Amboseli is another notable example (Pinc et al. 2009).

16.3.2 Integrating Environmental Change into Conservation

Research on organisms' adaptations to long-term environmental variability aims not only to uncover important biological processes but ultimately to inform conservation planning. Primates are in the midst of a global extinction crisis that is driven by anthropogenic pressures (Estrada et al. 2017; see Chap. 14 by Kalbitzer and Chapman in this volume). There are numerous worrisome indications that primates will be particularly poor at adapting to changing climates and landscapes by moving to suitable locations or via rapid genetic changes because of their biological idiosyncrasies, including low dispersal capacities (Schloss et al. 2012) and low reproductive rates (Charnov and Berrigan 1993; Jones 2011). On the other hand, some primates and many other animals show remarkable abilities to cope with environmental variability in situ by altering their behaviour and/or life histories to cope with changing conditions (Brockman and van Schaik 2005; Campos et al. 2017; Beever et al. 2017). Rapid environmental changes can cause a wide variety of deleterious effects on natural populations, including changes in population size and/or density, changes in demographic rates, shifts in distributions, disruption of coevolved mutualisms, and extirpation or extinction (reviewed in Mawdsley et al. 2009). Increasingly, conservation biologists, land managers, and governing bodies must integrate these dauntingly complex contingencies in their conservation decision-making. To be successful, this task must be informed by sound research on responses to long-term environmental change (Urban et al. 2016).

To give one pressing example, climate change is a widely recognized threat to the earth's biodiversity (Thomas et al. 2004; Maclean and Wilson 2011; Urban 2015), including its primates (Graham et al. 2016). Rising temperatures are the aspect of global climate change that has most captured the public's attention. Whereas most forms of global environmental change show idiosyncratic and spatially heterogeneous patterns, temperature increases have already been and will continue to be an essentially universal consequence of climate change, experienced nearly everywhere on earth to greater or lesser degrees. The magnitude of future temperature increases will depend on complex sociopolitical factors related to greenhouse gas emissions. Nonetheless, further temperature increases are virtually certain for decades into the future. While the temperature increases observed in Santa Rosa are not exceptional in the context of global climate warming, it is a dangerous misjudgement to dismiss the ecological risks of small temperature increases in tropical ecosystems. Many tropical species are adapted to occupy narrow, stable thermal ranges. Species in tropical regions tend to live closer to their upper temperature limits, and even small increases in ambient temperatures may put them outside of their tolerance zone (Tewksbury et al. 2008). This means that species living in warm climates are likely to suffer disproportionately from small increases in temperature. Across mammals, the risks associated with climate warming increase from polar regions towards tropical regions, even though the magnitude of warming projected for temperate and polar regions exceeds that of the tropics (Khaliq et al. 2014).

Climate change-integrated conservation planning involves decision-making that is informed by climate change contingencies, such as those described above, and their possible effects on species and ecosystems (Hannah et al. 2002; Heller and Zavaleta 2009; Mawdsley et al. 2009; Gillson et al. 2013). The larger Área de Conservación Guanacaste (ACG) – a large, ecologically diverse "mega-park" that surrounds and includes the tropical dry forest of Santa Rosa as well as several adjacent biomes – exemplifies several important concepts in climate change-integrated conservation planning, and thus it may play an important role in evaluating possible adaptation pathways. For example, Mawdsley (2009) highlights 16 climate change adaptation strategies that conservation planners should adopt to mitigate negative effects of climate change, and several of these strategies are long-standing pillars of the ACG's *raison d'être*. These principles include increasing the extent of protected areas, positioning protected areas along elevational and climatic gradients to maximize resilience, focusing restoration efforts on ecosystems rather than on particular species, protecting movement corridors and refugia, establishing long-term monitoring programs, reducing other pressures that could act synergistically with climate change such as fire and hunting, and integrating conservation goals with the development needs of nearby human settlements (Janzen 1987, 2000; Janzen and Hallwachs 2016). Under Linda's directorship, the Santa Rosa primate project has been a steady contributor in this endeavour, and it is poised to continue in this role as its primates face new environmental challenges in the coming decades.

Acknowledgements This chapter grew out of many discussions with Linda Fedigan and her students and collaborators, past and present, about the important role that climate plays in the behaviour, health, and survival of many organisms, including primates. I was motivated to compile this long-term record of environmental change in Santa Rosa by a growing recognition of our planet's changing climate and a sense of unease about how these changes will affect its biological systems. I thank Linda for introducing me to this glorious corner of northwestern Costa Rica and for her wonderful mentorship and support.

Jeff Klemens and Maria Marta helped to compile the weather data prior to 2006 that were used in this study. Many students and field assistants contributed to the daily weather measurements after 2006. I thank Roger Blanco Segura and the Costa Rican Park Service for their ongoing support and permission to carry out research in the ACG. Richard Corlett, Urs Kalbitzer, and two anonymous reviewers provided very helpful feedback that improved the manuscript. Finally, I thank the organizers of the Fedigan Festschrift for their hard work in editing and assembling this volume and for bringing us together for a terrific conference in Banff to honour Linda's distinguished career.

References

Aceituno P (1988) On the functioning of the Southern Oscillation in the South American sector. Part I: surface climate. Mon Weather Rev 116:505–524. https://doi.org/10.1175/1520-0493(1988)116<0505:OTFOTS>2.0.CO;2

Alberts SC, Altmann J (2012) The Amboseli Baboon Research Project: 40 years of continuity and change. In: Kappeler PM, Watts DP (eds) Long-term field studies of primates. Springer, Berlin, pp 261–287

Alberts SC, Hollister-Smith JA, Mututua RS et al (2005) Seasonality and long-term change in a savanna environment. In: Brockman DK, van Schaik CP (eds) Seasonality in Primates: studies of living and extinct human and non-human Primates. Cambridge University Press, New York, pp 157–195

Allen K, Dupuy JM, Gei MG et al (2017) Will seasonally dry tropical forests be sensitive or resistant to future changes in rainfall regimes? Environ Res Lett 12:023001. https://doi.org/10.1088/1748-9326/aa5968

Arroyo-Mora JP, Sanchez-Azofeifa GA, Kalacska MER et al (2005) Secondary forest detection in a neotropical dry forest landscape using Landsat 7 ETM+ and IKONOS imagery. Biotropica 37:497–507. https://doi.org/10.1111/j.1744-7429.2005.00068.x

Beever EA, Hall LE, Varner J et al (2017) Behavioral flexibility as a mechanism for coping with climate change. Front Ecol Environ. https://doi.org/10.1002/fee.1502

Beguería S, Latorre B, Reig F, Vicente-Serrano SM (2017) SPEI Global Drought Monitor. http://spei.csic.es/map/maps.html. Accessed 20 Nov 2017

Blois JL, Williams JW, Fitzpatrick MC et al (2013) Space can substitute for time in predicting climate-change effects on biodiversity. Proc Natl Acad Sci 110:9374–9379. https://doi.org/10.1073/pnas.1220228110

Borchert R (1998) Responses of tropical trees to rainfall seasonality and its long-term changes. Clim Chang 39:381–393. https://doi.org/10.1023/A:1005383020063

Boyd DS, Danson FM (2005) Satellite remote sensing of forest resources: three decades of research development. Prog Phys Geogr 29:1–26. https://doi.org/10.1191/0309133305pp432ra

Brockman DK, van Schaik CP (eds) (2005) Seasonality in Primates: studies of living and extinct human and non-human Primates. Cambridge University Press, New York

Bronikowski AM, Altmann J (1996) Foraging in a variable environment: weather patterns and the behavioral ecology of baboons. Behav Ecol Sociobiol 39:11–25. https://doi.org/10.1007/s002650050262

Campos FA, Fedigan LM (2014) Spatial ecology of perceived predation risk and vigilance behavior in white-faced capuchins. Behav Ecol 25:477–486. https://doi.org/10.1093/beheco/aru005

Campos FA, Bergstrom ML, Childers A et al (2014) Drivers of home range characteristics across spatiotemporal scales in a Neotropical primate, *Cebus capucinus*. Anim Behav 91:93–109. https://doi.org/10.1016/j.anbehav.2014.03.007

Campos FA, Jack KM, Fedigan LM (2015) Climate oscillations and conservation measures regulate white-faced capuchin population growth and demography in a regenerating tropical dry forest in Costa Rica. Biol Conserv 186:204–213. https://doi.org/10.1016/j.biocon.2015.03.017

Campos FA, Morris WF, Alberts SC et al (2017) Does climate variability influence the demography of wild primates? Evidence from long-term life-history data in seven species. Glob Change Biol 23:4907. https://doi.org/10.1111/gcb.13754

Chapman CA (1988) Patterns of foraging and range use by three species of neotropical primates. Primates 29:177–194. https://doi.org/10.1007/BF02381121

Chapman CA (1990) Association patterns of spider monkeys: the influence of ecology and sex on social organization. Behav Ecol Sociobiol 26:409–414. https://doi.org/10.1007/BF00170898

Chapman CA, Struhsaker TT, Skorupa JP et al (2010) Understanding long-term primate community dynamics: implications of forest change. Ecol Appl 20:179–191. https://doi.org/10.1890/09-0128.1

Charnov EL, Berrigan D (1993) Why do female primates have such long lifespans and so few babies? or life in the slow lane. Evol Anthropol Issues News Rev 1:191–194. https://doi.org/10.1002/evan.1360010604

Chavez FP, Ryan J, Lluch-Cota SE, Niquen CM (2003) From anchovies to sardines and back: multidecadal change in the Pacific Ocean. Science 299:217–221. https://doi.org/10.1126/science.1075880

DeGama-Blanchet HN, Fedigan LM (2006) The effects of forest fragment age, isolation, size, habitat type, and water availability on monkey density in a tropical dry forest. In: Estrada A, Garber PA, Pavelka MSM, Luecke L (eds) Adaptive radiations of Neotropical Primates. Springer, New York, pp 165–188

Didan K (2015a) MOD13Q1 MODIS/Terra Vegetation Indices 16-Day L3 Global 250m SIN Grid V006

Didan K (2015b) MYD13Q1 MODIS/Aqua Vegetation Indices 16-Day L3 Global 250m SIN Grid V006

Drake JB, Dubayah RO, Clark DB et al (2002) Estimation of tropical forest structural characteristics using large-footprint lidar. Remote Sens Environ 79:305–319

Enquist BJ, Enquist CAF (2011) Long-term change within a Neotropical forest: assessing differential functional and floristic responses to disturbance and drought. Glob Change Biol 17:1408–1424. https://doi.org/10.1111/j.1365-2486.2010.02326.x

Estrada A, Garber PA, Rylands AB et al (2017) Impending extinction crisis of the world's primates: why primates matter. Sci Adv 3:e1600946. https://doi.org/10.1126/sciadv.1600946

Fedigan LM (1986) Demographic trends in the *Alouatta palliata* and *Cebus capucinus* populations in Santa Rosa Park, Costa Rica. In: Else J, Lee P (eds) Primate ecology and conservation. Cambridge University Press, Cambridge, pp 285–293

Fedigan LM, Jack KM (2001) Neotropical primates in a regenerating Costa Rican dry forest: a comparison of howler and capuchin population patterns. Int J Primatol 22:689–713. https://doi.org/10.1023/A:1012092115012

Fedigan LM, Jack KM (2012) Tracking Neotropical monkeys in Santa Rosa: lessons from a regenerating tropical dry forest. In: Kappeler P, Watts D (eds) Long-term field studies of Primates. Springer, Heidelberg, pp 165–184

Fedigan LM, Fedigan L, Chapman CA (1985) A census of *Alouatta palliata* and *Cebus capucinus* monkeys in Santa Rosa National Park. Costa Rica Brenesia 23:309–322

Fedigan LM, Rose LM, Avila RM (1996) See how they grow: tracking capuchin monkey (*Cebus capucinus*) populations in a regenerating Costa Rican dry forest. In: Norconk M, Rosenberger

A, Garber PA (eds) Adaptive radiations of Neotropical Primates. Plenum Press, New York, pp 289–307

Fedigan LM, Rose LM, Avila RM (1998) Growth of mantled howler groups in a regenerating Costa Rican dry forest. Int J Primatol 19:405–432. https://doi.org/10.1023/A:1020304304558

Fedigan LM, Carnegie SD, Jack KM (2008) Predictors of reproductive success in female white-faced capuchins (*Cebus capucinus*). Am J Phys Anthropol 137:82–90. https://doi.org/10.1002/ajpa.20848

Feng X, Porporato A, Rodriguez-Iturbe I (2013) Changes in rainfall seasonality in the tropics. Nat Clim Chang 3:811–815. https://doi.org/10.1038/nclimate1907

Fick SE, Hijmans RJ (2017) WorldClim 2: new 1-km spatial resolution climate surfaces for global land areas. Int J Climatol 37(12):4302–4315. https://doi.org/10.1002/joc.5086

Fukami T, Wardle DA (2005) Long-term ecological dynamics: reciprocal insights from natural and anthropogenic gradients. Proc R Soc Lond B Biol Sci 272:2105–2115. https://doi.org/10.1098/rspb.2005.3277

Gillson L, Dawson TP, Jack S, McGeoch MA (2013) Accommodating climate change contingencies in conservation strategy. Trends Ecol Evol 28:135–142. https://doi.org/10.1016/j.tree.2012.10.008

Graham TL, Matthews HD, Turner SE (2016) A global-scale evaluation of primate exposure and vulnerability to climate change. Int J Primatol 37:158–174. https://doi.org/10.1007/s10764-016-9890-4

Hannah L, Midgley GF, Millar D (2002) Climate change-integrated conservation strategies. Glob Ecol Biogeogr 11:485–495. https://doi.org/10.1046/j.1466-822X.2002.00306.x

Hansen J, Ruedy R, Sato M, Lo K (2010) Global surface temperature change. Rev Geophys 48:RG4004. https://doi.org/10.1029/2010RG000345

Heller NE, Zavaleta ES (2009) Biodiversity management in the face of climate change: a review of 22 years of recommendations. Biol Conserv 142:14–32. https://doi.org/10.1016/j.biocon.2008.10.006

Janzen DH (1987) How to grow a tropical national park: basic philosophy for Guanacaste National Park, northwestern Costa Rica. Experientia 43:1037–1038. https://doi.org/10.1007/BF01952233

Janzen DH (1988) Management of habitat fragments in a tropical dry forest: growth. Ann Mo Bot Gard 75:105–116. https://doi.org/10.2307/2399468

Janzen DH (2000) Costa Rica's Área de Conservación Guanacaste: a long march to survival through non-damaging biodevelopment. Biodiversity 1:7–20. https://doi.org/10.1080/14888386.2000.9712501

Janzen DH, Hallwachs W (2016) Biodiversity conservation history and future in Costa Rica: the case of Área de Conservación Guanacaste (ACG). In: Kappelle M (ed) Costa Rican Ecosystems. University of Chicago Press, Chicago, pp 290–341

Jones JH (2011) Primates and the evolution of long, slow life histories. Curr Biol 21:R708–R717. https://doi.org/10.1016/j.cub.2011.08.025

Kalacska M, Sanchez-Azofeifa GA, Calvo-Alvarado JC et al (2004) Species composition, similarity and diversity in three successional stages of a seasonally dry tropical forest. For Ecol Manag 200:227–247. https://doi.org/10.1016/j.foreco.2004.07.001

Kalacska M, Sanchez-Azofeifa GA, Rivard B et al (2007) Ecological fingerprinting of ecosystem succession: estimating secondary tropical dry forest structure and diversity using imaging spectroscopy. Remote Sens Environ 108:82–96. https://doi.org/10.1016/j.rse.2006.11.007

Kalbitzer U, Chapman CA (2018) Primate responses to changing environments in the Anthropocene. In: Kalbitzer U, Jack KM (eds) Primate life histories, sex roles, and adaptability – essays in honour of Linda M. Fedigan. Developments in primatology: progress and prospects. Springer, New York, pp 283–302

Khaliq I, Hof C, Prinzinger R et al (2014) Global variation in thermal tolerances and vulnerability of endotherms to climate change. Proc R Soc Lond B Biol Sci 281:20141097. https://doi.org/10.1098/rspb.2014.1097

Laurance WF, Useche DC, Rendeiro J et al (2012) Averting biodiversity collapse in tropical forest protected areas. Nature 489:290–294. https://doi.org/10.1038/nature11318

Linke J, Betts MG, Lavigne MB, Franklin SE (2007) Introduction: structure, function, and change of forest landscapes. In: Wulder MA, Franklin SE (eds) Understanding forest disturbance and spatial pattern: remote sensing and GIS approaches. CRC Press (Taylor & Francis Group), Boca Raton, Florida, pp 1–29

Maclean IMD, Wilson RJ (2011) Recent ecological responses to climate change support predictions of high extinction risk. Proc Natl Acad Sci 108:12337–12342. https://doi.org/10.1073/pnas.1017352108

Mantua NJ, Hare SR (2002) The Pacific Decadal Oscillation. J Oceanogr 58:35–44. https://doi.org/10.1023/A:1015820616384

Mawdsley JR, O'malley R, Ojima DS (2009) A review of climate-change adaptation strategies for wildlife management and biodiversity conservation. Conserv Biol 23:1080–1089. https://doi.org/10.1111/j.1523-1739.2009.01264.x

Morice CP, Kennedy JJ, Rayner NA, Jones PD (2012) Quantifying uncertainties in global and regional temperature change using an ensemble of observational estimates: the HadCRUT4 data set. J Geophys Res Atmospheres 117:D08101. https://doi.org/10.1029/2011JD017187

Moritz S, Bartz-Beielstein T (2017). "imputeTS: Time Series Missing Value Imputation in R." The R Journal 9(1):207–218. https://journal.r-project.org/archive/2017/RJ-2017-009/index.html

Pinc KO, Altmann J, Alberts SC (2009). BABASE: Technical specifications for the Amboseli Baboon Project Data Management System. http://papio.biology.duke.edu/babase_system.html

Rohde R, Muller RA, Jacobsen R et al (2013) A new estimate of the average earth surface land temperature spanning 1753 to 2011. Geoinformatics Geostat Overv 01:1–7. https://doi.org/10.4172/2327-4581.1000101

Ropelewski CF, Halpert MS (1987) Global and regional scale precipitation patterns associated with the El Niño/Southern Oscillation. Mon Weather Rev 115:1606–1626. https://doi.org/10.1175/1520-0493(1987)115<1606:GARSPP>2.0.CO;2

Schloss CA, Nuñez TA, Lawler JJ (2012) Dispersal will limit ability of mammals to track climate change in the Western Hemisphere. Proc Natl Acad Sci 109:8606–8611. https://doi.org/10.1073/pnas.1116791109

Sorensen TC, Fedigan LM (2000) Distribution of three monkey species along a gradient of regenerating tropical dry forest. Biol Conserv 92:227–240. https://doi.org/10.1016/S0006-3207(99)00068-3

Strier KB, Boubli JP (2006) A history of long-term research and conservation of northern muriquis (*Brachyteles hypoxanthus*) at the Estação Biológica de Caratinga/RPPN-FMA. Primate Conserv 20:53–63

Taylor MA, Alfaro EJ (2005) Central America and the Caribbean, Climate of. In: Oliver JE (ed) Encyclopedia of world climatology. Springer, Dordrecht, pp 183–189

Tewksbury JJ, Huey RB, Deutsch CA (2008) Putting the heat on tropical animals. Science 320:1296–1297. https://doi.org/10.1126/science.1159328

Thomas CD, Cameron A, Green RE et al (2004) Extinction risk from climate change. Nature 427:145–148. https://doi.org/10.1038/nature02121

Tuck, S., & Phillips, H. (2013). MODISTools: MODIS Subsetting Tools. R package version 0.93.5. Retrieved from http://CRAN.R-project.org/package=MODISTools

Urban MC (2015) Accelerating extinction risk from climate change. Science 348:571–573. https://doi.org/10.1126/science.aaa4984

Urban MC, Bocedi G, Hendry AP et al (2016) Improving the forecast for biodiversity under climate change. Science 353:aad8466. https://doi.org/10.1126/science.aad8466

Uriarte M, Schwartz N, Powers JS et al (2016) Impacts of climate variability on tree demography in second growth tropical forests: the importance of regional context for predicting successional trajectories. Biotropica 48:780–797. https://doi.org/10.1111/btp.12380

Vicente-Serrano SM, Beguería S, López-Moreno JI (2009) A multiscalar drought index sensitive to global warming: the standardized precipitation evapotranspiration index. J Clim 23:1696–1718. https://doi.org/10.1175/2009JCLI2909.1
Vose RS, Arndt D, Banzon VF et al (2012) NOAA's merged Land–Ocean surface temperature analysis. Bull Am Meteorol Soc 93:1677–1685. https://doi.org/10.1175/BAMS-D-11-00241.1
Wang S, Huang J, He Y, Guan Y (2014) Combined effects of the Pacific Decadal Oscillation and El Niño-Southern Oscillation on global land dry–wet changes. Sci Rep 4:srep06651. https://doi.org/10.1038/srep06651

Chapter 17
How Variable Is a Primate's World: Spatial and Temporal Variation in Potential Ecological Drivers of Behaviour?

Colin A. Chapman, Kim Valenta, and Sarah Bortolamiol

Abstract The field of primatology has reached the stage where there are sufficient long-term studies and many shorter investigations on the same species at many different locations, in which we are able to appreciate how variable the behaviour of primates can be and how predictable their environment is over space and time. For example, redtail monkeys (*Cercopithecus ascanius*) exhibit extreme flexibility in diet; i.e. within the same national park, the amount of time they spend eating fruit varies from 36 to 60% of their foraging time, and among populations, time spent eating fruit ranges from 13 to 61%. Similarly, long-term phenological data from the same area encompassing over two decades illustrate that fruit availability can vary among years by as much as eightfold. While data have steadily accumulated on how variable primate behaviour and proposed environmental predictors of behaviour can be, this information has not been used to effectively re-evaluate theory. For exam-

C. A. Chapman (✉)
Department of Anthropology and McGill School of Environment, McGill University, Montreal, QC, Canada

Wildlife Conservation Society, Bronx, NY, USA

Makerere University Biological Field Station, Fort Portal, Uganda

Section of Social Systems Evolution, Primate Research Institute, Kyoto University, Kyoto, Japan
e-mail: Colin.Chapman@McGill.ca

K. Valenta
Department of Anthropology and McGill School of Environment, McGill University, Montreal, QC, Canada

Makerere University Biological Field Station, Fort Portal, Uganda

Duke University, Department of Evolutionary Anthropology, Durham, NC, USA

S. Bortolamiol
Department of Geography, McGill University, Montreal, QC, Canada

UMR 7533 Laboratoire Dynamiques Sociales et Recomposition des Espaces, Paris Diderot University, Paris, France

UMR 7206 Eco-Anthropologie et Ethnobiologie (MNHN/CNRS/Paris Diderot), Paris, France

U. Kalbitzer, K. M. Jack (eds.), *Primate Life Histories, Sex Roles, and Adaptability*, Developments in Primatology: Progress and Prospects,
https://doi.org/10.1007/978-3-319-98285-4_17

ple, current primate socioecological theory has derived general frameworks using the average behavioural traits of species or genera, but these new data suggest it is inappropriate to use such averages. Similarly, environments have often been characterized by single studies of 2 years or less, which does not sufficiently account for environmental variation. Here, we present examples of behavioural and ecological variation and consider ways that our field could advance in the future by considering this variation.

Keywords Diet · Social organization · Ecological determinants · Group size · Standardized methods

17.1 Introduction

Socioecological research has traditionally been a central theme in behavioural studies of primates (Eisenberg et al. 1972; Clutton-Brock and Harvey 1977; Wrangham 1980; Terborgh 1983; Janson and van Schaik 1988; Sterck et al. 1997). In the broadest sense, this research has involved investigations into how primate behaviour, social structure, and organization are influenced by a species' or population's environment. Such research has also been used to develop theories concerning group size (Chapman et al. 1995; Gogarten et al. 2015), feeding strategies and competition (Janson 2000; Clutton-Brock and Janson 2012), and dominance relationships (Wright 1999), just to name a few. Thus, this is one area of research where primate studies have contributed most to the development of general theoretical models and gathering empirical data to test these models has stimulated extensive primate research and involved the quantification of their environment (Clutton-Brock and Janson 2012).

Perhaps one area of such research that deserves special mention is the development of what has become known as the "Primate Socioecological Model" (Clutton-Brock and Harvey 1977; Isbell 1991; Koenig and Borries 2006; Snaith and Chapman 2007). Since the early1980s, researchers have built theories concerning the consequences of feeding competition on group size and social relationships (Wrangham 1979, 1980; van Schaik 1983; van Schaik 1989). In brief, it became widely accepted that if food resources were uniformly distributed in small patches of even quality that could not be defended and monopolized, or in very large patches where the whole group could simultaneously feed, then food competition would not occur through outright contest. Rather, animals would scramble to acquire the food items first. Under these ecological conditions, it is suggested that female fitness would be affected primarily by group size, and individuals would not be able to increase food acquisition through aggression. In contrast, when food resources occurred in well-defined, defensible patches that varied in quality and were not large enough for all group members to feed simultaneously, then aggression over food would occur among group members. This logic leads to a series of predictions regarding female dispersal and social relationships under various competitive regimes (Janson and van Schaik 1988; Isbell 1991; Sterck et al. 1997; Koenig 2002; Snaith and Chapman 2007, 2008).

A second area of research on how the environment influences primate groups that deserves special mention deals with the expression of atypical patterns in social and sexual behaviour and morphology in the lemurs of Madagascar. It is suggested that Madagascar has an unpredictable climate, causing unreliable fruiting patterns, which results in these atypical patterns because female lemurs have an elevated energy need associated with reproduction compared to female primates from other geographical areas. This explanation has been called the energy frugality model (EFH; Wright 1999). To support this hypothesis, it is necessary to demonstrate that Madagascar, where female dominance is common, has a more unpredictable climate and fruiting patterns than other tropical regions where male dominance is the norm. To fully assess resource availability and predictability for testing the EFH hypothesis, the data must result from long-term phenology studies that span multiple cycles of climatic variation; but how many cycles? So, a question that needs to be asked is: How many years of monitoring are needed to assess interannual variability in fruiting?

To answer such questions, description of changing patterns of resource availability must be carefully considered. For example, a system could be highly variable (wet one season, dry the next), but highly predictable (the seasons occur at same time every year, and the amount of rain is similar from year to year), or it could exhibit the same variability but be unpredictable (rains come at any time of the year and with varying amounts). It is when resource conditions are predictable that we most likely expect to observe strategies develop for resource acquisition. Of course, the timeframe over which the data are available influences this characterization. For example, a system could have wet and dry seasons occurring at the same time every year, but a 20-year study could document the same annual rainfall for 4 years that is followed by a period of slowly decreasing rainfall for 8 years, 5 years of stable rainfall, and 3 years of slowly increasing rainfall. Examining any pair of adjacent years (many ecological studies are 2 years or less) would suggest that the rainfall was predictable, but over the two decades, it is evident that the rainfall is unpredictable. Colwell (1974) presents a detailed consideration of these terms with respect to plant phenology.

Testing any general theories such as these points to the importance of understanding how well the proposed ecological determinants can be assessed. Any such assessment depends on how well the variables of interest are represented by the sampling, both from temporal and spatial perspectives, and on the accuracy with which the ecological variable proposed as a selective force is measured. The field of primatology has reached the stage where there are sufficient long-term studies and many shorter investigations on the same species at many different locations, in which we are able to appreciate how variable the behaviour of primates can be and how predictable their environment is over space and time. Over the last three decades, there has been an impressive accumulation of data on the diets and behaviour of wild primates, enabling more informed intraspecific comparisons (Butynski 1990; Davies et al. 1999; Chapman and Rothman 2009). Similarly, there are now a growing number of long-term studies available to examine year-to-year variation in ecological variables used in theoretical arguments (Chapman et al. 2005b, Alberts and Altmann 2012; Fedigan and Jack 2012; Kappeler and Watts 2012; Campos et al. 2017; Chapman et al. 2017; Hogan and Melin 2018, this Volume).

Based on the perspective that the variation and predictability of ecological factors are important for understanding of the evolution of a trait, our objective is to address three issues:

1. At what temporal scale must monitoring be conducted to adequately represent the variability and predictability that systems typically experience? How often are any 1 or 2 years of sampling, as is typically done when developing theory, representative of reality?
2. At what spatial scale must monitoring be conducted to adequately represent the variability and predictability that the system experiences? How often is the sampling of one or two sites, as is typically done when developing theory, representative of reality?
3. Do the indices used to measure the ecological variables of interest represent reality?

This evaluation is done within a relatively simple evolutionary framework. We believe it may be helpful to state our framework because it is the basis with which ecological factors can influence behaviour and because the framework is very frequently not considered. Within the ecological timeframe that is used to test ideas, this framework assumes that if a trait within a population at one time or location is being favoured by a specific selective force, it cannot be countered by an opposing selective pressure at another point in time or at another location, if directional selection is to occur. Stated with respect to the situation under consideration, if one population at a specific location or time (ecological timeframe) is experiencing selection for a behaviour, and interbreeds with another population experiencing a very different or opposing selection of similar magnitude, no change in behaviour will occur. When studies are made only at one location or at one point in time and evolutionary conclusions are drawn, we are assuming that selection is operating in the same way for all members of the population which may be spread across hundreds or thousands of square kilometres and it is occurring over a meaningful duration. Is this a reasonable assumption? A meaningful duration from an evolutionary perspective will involve multiple generations, and given that chimpanzee generation time is 24.6 years (Langergraber et al. 2012), we are typically looking at consistent selection pressures over decades or centuries.

17.1.1 At What Temporal Scale Must Monitoring Be Conducted to Adequately Represent the Variability that Systems Typically Experience?

The availability of food resources is frequently the basis of hypotheses used to explain variation in primate behavioural and even anatomical characteristics. Numerous hypotheses have proposed that ecological variables, such as overall habitat wide food availability, seasonal shortages, and month-to-month variability in food, are important selective forces (Oates et al. 1990; Janson and Chapman 1999; Lambert 2009; Marshall et al. 2009). Testing such hypotheses involves evaluating food availability over time and thus involves the quantification of phenological patterns of plants. Proposed drivers of phenological patterns of tropical rain forests include rainfall, day length, irradiance, and temperature (Opler et al. 1976; Ashton et al. 1988; van Schaik et al. 1993; Newbery et al. 1998; but see Polansky and Boesch 2013; Chapman et al. In press), mode of seed dispersal (Smythe 1970; Wheelwright 1985), activity of pollinators or seed dispersers (Snow 1965; Frankie et al. 1974; Rathke and Lacey 1985), variation in germination conditions (Janzen 1967; Frankie et al. 1974), the potential to swamp the ability of seed predators to destroy seeds (Visser et al. 2011), life history traits (Davies and Ashton 1999; Pau et al. 2011), and relative abundance of the trees themselves (van Schaik et al. 1993). Each of these interacting drivers has the potential to be highly variable; thus it is not surprising that phenological parameters, like fruiting, have been documented to vary between weeks, months, years, and even longer periods (e.g. El Nino cycles, Wright and Calderón 2006; Hogan and Melin 2018, this Volume).

To investigate at what temporal scale primatologists typically need to consider, we will evaluate one theoretical argument in detail and subsequently briefly reflect on a number of other examples. A number of theories have been put forward to explain why lemurs characteristically differ from anthropoids with respect to features like female dominance, lack of sexual dimorphism regardless of mating system, sperm competition coupled with male-male aggression, high infant mortality, and strict seasonal breeding (Kappeler and Fichtel 2015). The most widely accepted hypothesis to explain such differences is the energy frugality hypothesis (EFH), and the article naming the hypothesis has been cited 422 times as of March 5, 2018 (Wright 1999). This hypothesis claims that these lemur traits are adaptations to conserve energy or maximize use of scarce resources that were selected to efficiently deal with the severe and unpredictable environment of Madagascar and the particularly high costs to females of food scarcity (Wright 1999; Pochron et al. 2003; Dewar and Richard 2007; Dunham 2008).

An assumption of this hypothesis is that food resource availability is less predictable in Madagascar as compared to areas where these "lemur traits" are not found, like mainland Africa. Support for this hypothesis is largely derived from the study of Overdorff and colleagues (Overdorff 1996a, b) that monitored for 24 months 104 trees belonging to 26 species that contributed to the diet of brown lemurs (*Eulemur fulvus* and *E. rubriventer*) in Ranomafana National Park. Over those 2 years, there were 5 months when fruit resources were considered to be scarce. Based on these observations, Overdorff (1996a, b) concluded that fruiting phenology is unpredictable, and resources are unreliable for primates in Malagasy forests, supporting the EFH.

These wet forest studies have been influential in subsequent research in Madagascar, despite their limited duration, small sample sizes, and, importantly, the lack of direct comparisons to tropical wet forests elsewhere. To test this difference in resource predictability between Madagascar and mainland Africa, Federman et al. (2017) analysed phenology datasets and their environmental correlates from two tropical wet forests, the Réserve Naturelle Intégrale Betampona, Madagascar, and Kibale National Park, Uganda. Temperature was a significant environmental predictor of fruit availability at both sites. However, contrary to what would be predicted by the EFH, long-term temperature data from both sites (15 years from Kibale and 14 from Betampona) indicated that temperature was more predictable at Betampona than at Kibale. Most importantly, however, they found no evidence of a significant difference in the predictability of fruit between the two sites, which would be essential as support for the EFH. Since this theory is largely based on the 2-year study of Overdorff, Federman et al. (2017) also analysed the Kibale data at 2-year intervals. Over a 2-year period, Betampona experienced no months when fruiting was lower than 10% of the maximum observed amount of available fruit, while over a 15-year timeframe, Kibale experienced 7 such months. When scarcity is defined as less than 25% of the maximum observed amount of fruit, Betampona experienced 3 consecutive months of scarcity (12.5% of the total period), while in the 15-year Kibale study, there were 71 months of at least 25% scarcity (39.4% of the total period); also there were nine times where periods of scarcity constituted 3 or more consecutive months. This amounts to a period of scarcity at Kibale roughly every 1.5 years. Thus, when comparing 2-year intervals at Kibale to the monitoring period at Betampona, fruiting predictability at Betampona falls within the range of 2-year intervals from Kibale. Further, during the 2 years examined at each site, Betampona showed no long periods of fruit scarcity analogous to those observed by Overdorff (Overdorff 1996a, b) at Ranomafana. Thus, these results conflict with the prevailing energy frugality hypothesis, which proposes that the selective force behind many of lemurs' unique features is the unpredictable resource availability or temperature in Madagascar.

Other examples of ideas being refuted when the data is sampled at different temporal scales are not difficult to find. For example, figs (*Ficus* spp.) have frequently been presented as examples of keystone plant resources in tropical forests (Terborgh 1986; Power et al. 1996), and textbooks have presented figs as a clear case of the keystone species concept (Bush 2000). Yet, to our knowledge, whenever the concept has been tested using data at longer temporal scale than the original study by Terborgh (1986), figs were not found to be keystone resources (Gautier-Hion and Michaloud 1989; Peres 2000; Chapman et al. 2005a). If figs served as a keystone food resource, they must be available during months when few trees were fruiting, which long-term data indicates does not occur (Chapman et al. 2005a).

The answer to our original question – at what temporal scale must monitoring be conducted to adequately represent the variability that systems typically experience? – is we do not know. The answer will depend on the question asked, the ecological variable of interest, and the primate species. With respect to questions involving the assessment of fruit availability, we have data from one location that provides some indication. Fruiting data from six species collected over the last 47 years from Kibale suggest that only after a decade does the assessment include times close (10%) of the overall maximum and minimum and include patterns of abundant and scarce years (Chapman unpublished data). Of course, researchers cannot wait a decade to publish!

As for solutions, we encourage updates to be presented, and with online publishing, this has become possible. Given the growing difficulty of maintaining funding over a decade, updates should not be required to obtain the original intensity of sampling, but casual observations should be allowed to be posted (e.g. long-term researchers who have worked in the area for 30 years say they have never seen a time with such fruit scarcity). Similarly, additions to online data sets should be encouraged. We strongly encourage that whenever possible researchers re-evaluate hypotheses that have become generally accepted but were developed from limited data sets. We must critically evaluate what our academic community wants to consider a valuable contribution. For example, many journals do not publish case studies, but long-term data often involve case studies (Chapman and Peres 2001).

17.1.2 At What Spatial Scale Must Monitoring Be Conducted to Adequately Represent the Variability that the System Experiences?

Many theoretical arguments have used a dietary classification to assess interspecific differences in diet to predict evolution of traits, such as day range (Clutton-Brock and Harvey 1977), brain size and intelligence (Dunbar 2003), morphology and physiology (Teaford and Walker 1984; McNab 2002), and life history (Leigh 1994). Such interspecific comparisons have been a foundation on which many primate theories have been advanced; however, there has been an accumulation of data on primate diets over the last four decades that has led to a growing appreciation of the magnitude of dietary variation within species (Butynski 1990; Chapman and Fedigan 1990; Davies et al. 1999; Chapman et al. 2002b). For example, redtail monkey groups (*Cercopithecus ascanius*) within the same national park varied in the amount of time they spent foraging on leaves (13–35%), fruit (36–60%), and insects (15–31%; Chapman et al. 2002a). On a larger spatial scale, contrasting the diets of 11 groups of redtail monkeys across a number of countries documented that the proportion of time spent eating different plant parts varies dramatically (leaves 7–74%, fruit 13–61%, and insects 1–16%; Chapman et al. 2002a). Even greater levels of variation were found in blue monkeys (*Cercopithecus mitis*). Butynski (1990) studied four groups of blue monkeys at the same site and found that the amount of time feeding on fruit varied from 22 to 35%. Fruit intake varied among populations from 26% to 91%, and leaf intake varied from 3% to 47% (Chapman et al. 2002a). One population of blue monkeys has even been described to rely on bamboo (*Arundinaria alpina*) for 60% of its foraging time (Twinomugisha and Chapman 2008). Similar differences can be found when folivores are considered (Davies et al. 1999; Chapman et al. 2002b). When the scientific community only has one study, of one population, at one location, there is no way to evaluate if values used in interspecific comparisons are good representatives of the species or extreme outliers.

Yet again we cannot answer our question what is the needed spatial scale to sample to represent the variability of a system, and again it will be dependent on the question, ecological variable of interest, and species. Of course, increasing the sample size will likely make it more representative, and having sampled several distinct sites should result in insights if more variation is to be expected. Also, for many questions, other mammals will be influenced in the same way as primates, and therefore considering non-primate models will be useful (Clutton-Brock and Janson 2012).

17.1.3 Do the Indices Used to Measure the Ecological Variables of Interest Represent Reality?

Lastly, we feel that it is important to emphasize that if the scientific community is to rely on indices to represent the ecological variables hypothesized to be selective pressures on primate behaviour, these indices must accurately represent the ecological variable of interest. Unfortunately, very little research exists to evaluate the methods used to quantify many of the variables of interest. Much of the theory that is central to primatological thinking was conceived and published prior to any significant quantification of the proposed driving ecological variables (Clutton-Brock and Janson 2012). At this time, researchers used natural history observations to state, for instance, the time of the year when food was scarce. For example, it was often assumed that the dry season was the period of low food availability; however, recent phenological studies have shown that this is often not the case (Wright and Calderón 2006; Melin et al. 2014; Mosdossy et al. 2015; Chapman et al. In press). These natural history observations were used to develop theory, but have they been subsequently substantiated with rigorous methods? Here we address the reliability of methods to quantify of food abundance.

For many questions, the first question that must be considered when quantifying food availability is “what is food?” Tropical rain forests are extraordinarily species-rich, often with over 1000 woody species in an area of a few square kilometres (Croat 1978). Only a fraction of this diversity is consumed by any one primate species, but knowing exactly which plants are potential foods is extremely difficult. Typical lists of plant foods for a single primate species vary from 60 to 200 species, and often these lists grow even after several years of study. Fig. 17.1 illustrates the number of food items (species and part, e.g. *Prunus africana* bark) that we identi-

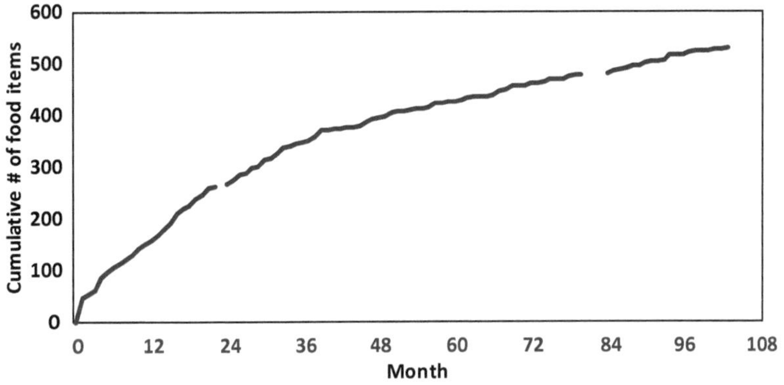

Fig. 17.1 The number of food items (species and part) that we identified as being eaten by a long-term red colobus study group over 9 years in Kibale National Park, Uganda

fied as being eaten by a long-term red colobus study group over 9 years in Kibale. The expectation would be that as the duration of the study increased the rate of accumulation of new foods would decline and reach an asymptote. Researcher may debate when the asymptote should be reached, but most field researchers would view that this should occur well before 9 years. But it does not (Fig. 17.1).

For most theoretical questions, not *all foods* are important for the question being asked, as they are rarely eaten. Furthermore, it is a logistic impossibility to measure temporal changes in the availability of *all foods*. As a result, researchers usually employ an operational definition of food, but there is no consistent operational definition of what is food, which is likely one of the most basic questions. At one extreme, some studies of frugivores assume that any fleshy fruit is a potential food, even when the study species has not been observed to eat that fruit (Terborgh 1983). This assumption is partly justified by the fact that there is relatively little differentiation in the fruit parts of the diets of coexisting primate species (Terborgh 1983) and the fact that primates eat many species of fruit that are evolved to be dispersed by other animals (Janson 1983). At the other extreme, some studies consider only those plant species the animals were seen to eat as food. Thus, possibly the top X number of food items eaten in a particular month is used as that month's index of food abundance. However, the operational definition of the X number can dramatically alter the estimate of the ecological parameter of interest. For example, during the first 5 years of research on spider monkey (*Ateles geoffroyi*) in Costa Rica (Chapman 1990), the animals were rarely seen to eat the fruits of *Bursera simaruba* (< 0.1 of their feeding time); however in the sixth year, the community fed on this fruit extensively (22.8%) during a month when little else was available. *Bursera simaruba* is a very common tree, and including it in the diet of spider monkeys for the month that it was eaten increases the food density estimates by 20.3 food trees/ha. If the observations of feeding on *B. simaruba* fruit had not been made in the sixth year, it would not have been considered a food source. Similarly, using 9 years of red colobus foraging data provides a second more robust example. The species to be included

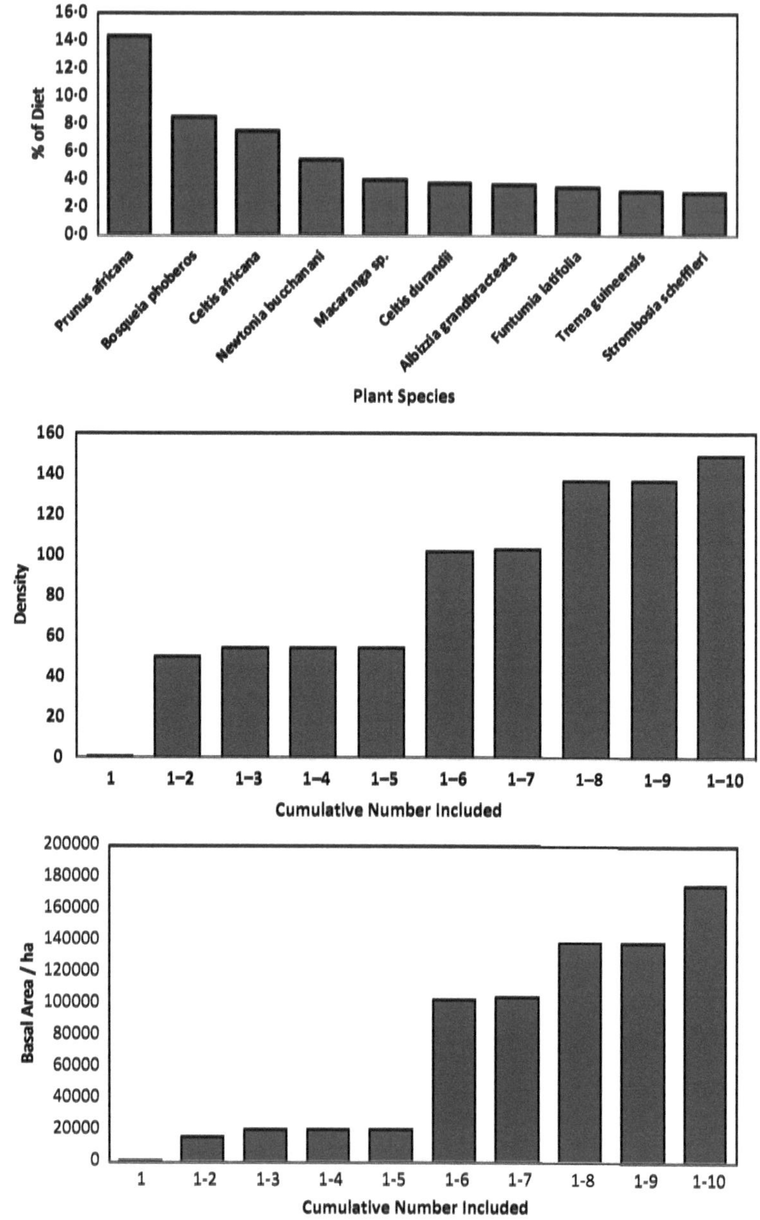

Fig. 17.2 Foraging behaviour of red colobus monkeys in Kibale National Park, Uganda. (**a**) The ten most frequently eaten species in their diet as an index of food abundance; (**b**) the cumulative density of the red colobus food trees in Kibale National Park, Uganda, including the top 1 species, top 1 and 2 species, top 1, top 2, top 3… top 10 as an index of food abundance; (**c**) the cumulating density of the red colobus food trees species weighting density by DBH as an index of food abundance

in a species diet can be indexed as including various indices of the "top species," for example, only the top species eaten for the most time (species 1), the top two most eaten species (species 1 + 2), top three eaten species (species 1 + 2 + 3), or top X eaten species (species 1 + 2 + 3 + 4...X). Figure 17.2a depicts the ten most frequently eaten species by red colobus, while Fig. 17.2b depicts the density of food trees including the top 1, top 2, and so on up to the top 10 species, and Fig. 17.2c depicts the same species included in an index of food abundance but weighting density by diameter of breast height (DBH). It is evident that an increase in the number of species used in the index can cause a small or large increase in the density or basal area of what are considered resources. This is because some species frequently consumed by the red colobus are very rare, while other species are common. Furthermore, the pattern of increase is different if the density is weighted by DBH or not, which is simply a result that some commonly eaten species are large as adults, while others are not.

Unfortunately, once food has been defined, determining its abundance and determining its value are both very difficult. Ideally a researcher would be able know the number of food items and their value to the primate in question in terms of how they regulate their population. Even for food items that occur in discrete patches such as trees, estimating abundance has proven extremely difficult. A number of indices have been proposed, but each varies in its accuracy and ease of measure (Chapman et al. 1995; Anderson et al. 2005; Bortolamiol et al. 2014). Estimates include a number of indices of the abundance of food item on a tree (e.g. DBH, basal area, modified basal area, ranked abundance) with DBH being the most common. With respect to evaluating the accuracy of DBH in predicting fruit biomass, Chapman et al. (1992) evaluated this relationship for four species of primate consumed trees in Kibale and found that *within a species* DBH explained an average of 47.2% of the variance in fruit biomass. However, generalizing this study to all primate foods should be done with extreme caution as it is limited to just four small understory species that one of the authors (CAC) could climb. Yet the limitations of this study have not been well considered, and it has been cited 369 times, typically as a suitable way to measure food availability (as of September 21, 2017, Google Scholar). Such estimates of per tree fruit abundance are then combined with estimates of the density of food trees calculated from vegetation plots (e.g. sum of the DBHs of all suitable food trees) and then weighted by the temporal variation in the abundance of the items (e.g. phenology trails, plots, fruit traps). In statistical terms, we only estimate the variable of interest once (e.g. fruit abundance in a particular month). If the index used only explains approximately 50% of the variance, as is the case with DBH and fruit biomass, and there are say three variables used to predict fruit abundance (e.g. DBH, density, and phenology) and each has even a small random error associated with measurement, then the accuracy of the estimate of fruit availability is low. Considering this, we are very surprised that primatologists have found statistically significant relationships as often as they have, which may imply that the relationships being examined are actually robust.

Lastly there is the question of the nutritional value of foods, which is very difficult to measure (Rothman et al. 2012) yet extremely important. The animal may be attempting to maximize energy (Rothman et al. 2011), protein to fibre balance (Chapman et al. 2004), or nutrients and minerals (Rode et al. 2003), and without a comprehensive study, it is not possible to know which. Ideally a researcher would like to know the nutrient that is limiting the population growth and consider it in detail; however, this requires a detailed nutritional analysis over a sufficiently long period where that nutrient's availability and the population abundance changed. Such data are rarely available.

Given such error and difficulties of defining food, it is not surprising that some of these measures do not even correlate. For example, estimates of fruit abundance derived from cumulative DBH of monthly fruiting trees do not correlate with estimates derived from fruit traps (Chapman et al. 1994). What this mean for advancing theory needs much more careful attention.

Of the three questions we posed at the start, it pleases us to be able to state that we can clearly answer this question. To the question "Do the indices used to measure the ecological variables of interest represent reality?" the answer is clearly NO. This means there are clear opportunities to advance the field by evaluating and comparing different methods. However, conducting the needed studies will take time. In the meantime, it only seems logical: IF the goal is to advance knowledge about primates and NOT to advance personal CVs, researcher should be encouraged to use the mostly commonly agreed upon method and provide the results of each of these methods on in supplementary material, until the time that the "best" method is agreed upon. A classic and unfortunate example involves how to estimate the number of primate species. It is our humble impression that a year does not go by without one or two papers are published.

Many of which are statistically elegant but biologically naïve.

17.2 Conclusions

Our intent is not to be critical of the theories or studies we have mentioned, rather it is to use this example to illustrate that science is a process where ideas grow and are modified with the inclusion of new investigations, the use of new methods, and the maturation of the particular field. Early primate socioecological studies arrived at some very interesting concepts that have proven to be useful to advance our understanding of these intriguing animals. With the gradual accumulation of research on the same species at many different locations and of a few long-term studies, it has become apparent that there is a great deal of variability in both ecological variables that potentially influence primates and in primate behaviour. Thus, there is a need to assess this variability, evaluate the flexibility of primates over time, and constantly re-evaluate our past theoretical advancements. Associated with this is a clear need to continue to improve the methods that we use to assess the ecology of primates.

Acknowledgements This paper is dedicated to Dr. Linda Fedigan who was so very influential to Colin Chapman starting his research career when he was an undergrad, supervised him through his graduate degrees, and then has provided critical mentorship and letters of recommendation to this day. Funding for the research in Kibale National Park was provided by the Canada Research Chairs

Program, Natural Science and Engineering Research Council of Canada, Fonds Québécois de la Recherché sur la Nature et les Technologies, and the National Geographic Society. Tom Struhsaker has been particularly helpful over the years with sharing long-term data and opinions, and Karen Strier provided stimulating ideas and discussion on variability and flexibility. Finally, we would like to thank Urs Kalbitzer, Kathy Jack, and the anonymous reviewers for helpful comments on the manuscript.

References

Alberts SC, Altmann J (2012) The Amboseli Baboon Research Project: 40 years of continuity and change. In: Kappeler P, Watts DP (eds) Long-term field studies of primates. Springer, Berlin, pp 261–288

Anderson DP, Nordheim E, Moermond TC, Bi G, Zoro B, Boesch C (2005) Factors Influencing Tree Phenology in Taï National Park, Côte d'Ivoire1. Biotropica 37:631–640

Ashton PS, Givinish T, Appanah S (1988) Staggered flowering in the Dipterocarpaceae: new insights into floral induction and the evolution of mast fruiting in the aseasonal tropics. Am Nat 132:44–66

Bortolamiol S, Cohen M, Potts K, Rwaburindore P, Kasenene J, Seguya A, Pennec F, Viguaud Q, S. K (2014) Suitable habitats for endangered frugivorous mammals: small-scale comparison, regeneration forest and chimpanzee density in Kibale National Park, Uganda. PLoS One 9:e102177

Bush MB (2000) Ecology of a changing planet. Prentice Hall, Upper Saddle River

Butynski TM (1990) Comparative ecology of blue monkeys (*Cercopithecus mitis*) in high- and low-density sub-populations. Ecol Monogr 60:1–26

Campos FA, Morris WF, Alberts SC, Altmann J, Brockman DK, Cords M, Pusey A, Stoinski TS, Strier KB, Fedigan LM (2017) Does climate variability influence the demography of wild primates? Evidence from long-term life-history data in seven species. Glob Chang Biol

Chapman CA (1990) Ecological constraints on group size in three species of neotropical primates. Folia Primatol 55:1–9

Chapman CA, Fedigan LM (1990) Dietary differences between neighboring cebus monkey groups: local traditions or responses to food availability? Folia Primatol 54:177–186

Chapman CA, Peres CA (2001) Primate conservation in the new millennium: the role of scientists. Evol Anthropol 10:16–33

Chapman CA, Rothman JM (2009) Within-species differences in primate social structure: evolution of plasticity and phylogenetic constraints. Primates 50:12–22

Chapman CA, Chapman LJ, Wrangham RW, Hunt K, Gebo D, Gardner L (1992) Estimators of fruit abundance of tropical trees. Biotropica 24:527–531

Chapman CA, Wrangham RW, Chapman LJ (1994) Indices of habitat-wide fruit abundance in tropical forest. Biotropica 26:160–171

Chapman CA, Wrangham RW, Chapman LJ (1995) Ecological constraints on group size: an analysis of spider monkey and chimpanzee subgroups. Behavioural Ecology and Sociobiology 36:59–70

Chapman CA, Chapman LJ, Cords M, Gauthua M, Gautier-Hion A, Lambert JE, Rode KD, Tutin CEG, White LJT (2002a) Variation in the diets of Cercopithecus species: differences within forests, among forests, and across species. In: Glenn M, Cords M (eds) The guenons: diversity and adaptation in African monkeys. Plenum, New York, pp 319–344

Chapman CA, Chapman LJ, Gillespie TR (2002b) Scale issues in the study of primate foraging: red colobus of Kibale National Park. Am J Phys Anthropol 117:349–363

Chapman CA, Chapman LJ, Naughton-Treves L, Lawes MJ, McDowell LR (2004) Predicting folivorous primate abundance: validation of a nutritional model. Am J Primatol 62:55–69

Chapman CA, Chapman LJ, Zanne AE, Poulsen JR, Clark CJ (2005a) A 12-year phenological record of fruiting: implications for frugivore populations and indicators of climate change. In: Dew JL, Boubli JP (eds) Tropical fruits and frugivores. Springer, The Netherlands, pp 75–92
Chapman CA, Struhsaker TT, Lambert JE (2005b) Thirty years of research in Kibale National Park, Uganda, reveals a complex picture for conservation. Int J Primatol 26:539–555
Chapman CA, Corriveau A, Schoof VAM, Twinomugisha D, Valenta K (2017) Long-term primate research and its significance for developing theory and conservation plans. J Mammal 98:652–660
Chapman CA, Valenta K, Bonnell TR, Brown KA, Chapman LJ (in press) Solar radiation drives fruit phenology: Evaluating a 16 year record from Kibale National Park, Uganda. Proceedings of the Royal Society of London B: Biological Sciences
Clutton-Brock TH, Harvey PH (1977) Primate ecology and social organization. Journal of Zoological Society of London 183:1–39
Clutton-Brock T, Janson C (2012) Primate socioecology at the crossroads: past, present, and future. Evolutionary Anthropology: Issues, News, and Reviews 21:136–150
Colwell RK (1974) Predictability, constancy, and contingency of periodic phenomena. Ecology 55:1148–1153
Croat TB (1978) Flora of Barro Colorado Island. Stanford University Press, Stanford
Davies SJ, Ashton PS (1999) Phenology and fecundity in 11 sympatric pioneer species of *Macaranga* (Euphorbiaceae) in Borneo. Am J Bot 86:1786–1795
Davies AG, Oates JF, Dasilva GL (1999) Patterns of frugivory in three West African colobine monkeys. Int J Primatol 20:327–357
Dewar RE, Richard AF (2007) Evolution in the hypervariable environment of Madagascar. Proc Natl Acad Sci 104:13723–13727
Dunbar RIM (2003) The social brain: mind, language, and society in evolutionary perspective. Annu Rev Anthropol 32:163–181
Dunham AE (2008) Battle of the sexes: cost asymmetry explains female dominance in lemurs. Anim Behav 76:1435–1439
Eisenberg JF, Muckenhirn NA, Rudran R (1972) The relation between ecology and social structure in primates. Science 176:863–874
Federman S, Sinnott-Armstrong M, Baden AL, Chapman CA, Daly DC, Richard AR, Valenta K, Donoghue MJ (2017) The paucity of frugivores in Madagascar may not be due to unpredictable temperatures or fruit resources. PLoS One 12:e0168943
Fedigan LM, Jack KM (2012) Tracking neotropical monkeys in Santa Rosa: lessons from a regenerating Costa Rican dry forest. In: Kappeler P, Watts D (eds) Long-term field studies of primates. Springer, Heidelberg, pp 165–184
Frankie GW, Baker HG, Opler PA (1974) Comparative phenological studies of trees in tropical wet and dry forests in the lowlands of Costa Rica. J Ecol 62:881–919
Gautier-Hion A, Michaloud G (1989) Are figs always keystone resources for tropical frugivorous vertebrates? A test in Gabon. Ecology 70:1826–1833
Gogarten JF, Jacob AL, Ghai RR, Rothman JM, Twinomugisha D, Wasserman MD, Chapman CA (2015) Group size dynamics over 15+ years in an African forest primate community. Biotropica 47:101–112
Hogan J, Melin A (2018, this volume) Intra- and interannual variation in the fruit diet of wild capuchins: impact of plant phenology. In: Kalbitzer U, Jack KM (eds) Primate life histories, sex roles, and adaptability – essays in Honour of Linda M Fedigan developments in primatology: progress and prospects. Springer, New York, pp 193–210
Isbell LA (1991) Contest and scramble competition: patterns of female aggression and ranging behaviour among primates. Behav Ecol 2:143–155
Janson CH (1983) Adaptation of fruit morphology to dispersal agents in a neotropical rainforest. Science 219:187–189
Janson CH (2000) Primate socio-ecology: the end of a golden age. Evol Anthropol 9:73–86

Janson CH, Chapman CA (1999) Resources and the determination of primate community structure. In: Fleagle JG, Janson CH, Reed KE (eds) Primate Communities. Cambridge University Press, Cambridge, pp 237–267
Janson CH, van Schaik CP (1988) Recognizing the many faces of primate food competition: methods. Behaviour 105:165–186
Janzen DH (1967) Synchronization of sexual reproduction of trees within a dry season in Central America. Evolution 21:620–637
Kappeler PM, Fichtel C (2015) Eco-evo-devo of the lemur syndrome: did adaptive behavioral plasticity get canalized in a large primate radiation? Front Zool 12:S15
Kappeler PM, Watts DP (2012) Long-term field studies of primates. Springer, Berlin
Koenig A (2002) Competition for resources and its behavioral consequences among female primates. Int J Primatol 23:759–783
Koenig A, Borries C (2006) The predictive power of socioecological models: a reconsideration of resource characteristics, agonism and dominance hierarchies. In: Hohmann G, Robbins MM, Boesch C (eds) Feeding ecology in apes and other primates. Cambridge University Press, Cambridge, pp 263–284
Lambert JE (2009) Primate fallback strategies as adaptive phenotypic plasticity: scale, process, and pattern. Am J Phys Anthropol 140:759–766
Langergraber KE, Prüfer K, Rowney C, Boesch C, Crockford C, Fawcett K, Inoue E, Inoue-Muruyama M, Mitani JC, Muller MN (2012) Generation times in wild chimpanzees and gorillas suggest earlier divergence times in great ape and human evolution. Proc Natl Acad Sci 109:15716–15721
Leigh SR (1994) Ontoaenetic correlates of diet in anthropoid primates. Am J Phys Anthropol 94:499–522
Marshall AJ, Boyko CM, Feilen KL, Boyko RH, Leighton M (2009) Defining fallback foods and assessing their importance in primate ecology and evolution. Am J Phys Anthropol, in press
McNab BK (2002) The physiological ecology of vertebrates: a view from energetics. Cornell University Press, Cornell
Melin AD, Young HC, Mosdossy KN, Fedigan LM (2014) Seasonality, extractive foraging and the evolution of primate sensorimotor intelligence. J Hum Evol 71:77–86
Mosdossy KN, Melin AD, Fedigan LM (2015) Quantifying seasonal fallback on invertebrates, pith, and bromeliad leaves by white-faced capuchin monkeys (Cebus capucinus) in a tropical dry forest. Am J Phys Anthropol 158:67–77
Newbery DM, Songwe NC, Chuyong GB (1998) Phenology and dynamics of an African rainforest at Korup, Cameroon. In: Newbery DM, Pins HHT, Brown ND (eds) Dynamics of tropical communities. Blackwell, Oxford
Oates JF, Whitesides GH, Davies AG, Waterman PG, S.M. G, DaSilva GL, Mole S (1990) Determinants of variation in tropical forest primate biomass: new evidence from West Africa. Ecology 71:328–343
Opler PA, Frankie GW, Baker HG (1976) Rainfall as a factor in the release, timing, and synchronization of anthesis by tropical trees and shrubs. J Biogeogr 3:231–236
Overdorff DJ (1996a) Ecological correlates to activity and habitat use of two prosimian primates: Eulemur rubriventer and Eulemur fulvus rufus in Madagascar. Am J Primatol 40:327–342
Overdorff DJ (1996b) Ecological correlates to social structure in two lemur species in Madagascar. Am J Phys Anthropol 100:487–506
Pau S, Wolkovich EM, Cook BI, Davies TJ, Kraft NJB, Bolmgren K, Betancourt JL, Cleland EE (2011) Predicting phenology by integrating ecology, evolution, and climate science. Glob Chang Biol 17:3633–3646
Peres CA (2000) Identifying keystone plant resources in tropical forests: the case of gums from *Parkia* pods. J Trop Ecol 16:287–317
Pochron ST, Fitzgerald J, Gilbert CC, Lawrence D, Grgas M, Rakotonirina G, Ratsimbazafy R, Rakotosoa R, Wright PC (2003) Patterns of female dominance in Propithecus diadema edwardsi of Ranomafana National Park, Madagascar. Am J Primatol 61:173–185

Polansky L, Boesch C (2013) Long-term changes in fruit phenology in a West African lowland tropical rain forest are not explained by rainfall. Biotropica 45:434–440
Power ME, Tilman D, Estes JA, Menge BA, Bond WJ, Mills LS, Daily G, Castilla JC, Lubchenco J, Paine RT (1996) Challenges in the quest for keystones. Bioscience 46:609–620
Rathke B, Lacey EP (1985) Phenological patterns of terrestrial plants. Annu Rev Ecol Syst 16:179–214
Rode KD, Chapman CA, Chapman LJ, McDowell LR (2003) Mineral resource availability and consumption by colobus in Kibale National Park, Uganda. Int J Primatol 24:541–573
Rothman JM, Raubenheimer D, Chapman CA (2011) Nutritional geometry: gorillas prioritize non-protein energy while consuming surplus protein. Biol Lett:rsbl20110321
Rothman JM, Chapman CA, van Soest PJ (2012) Methods in primate nutritional ecology: a user's guide. Int J Primatol 33:542–566
Smythe N (1970) Relationships between fruiting seasons and seed dispersal methods in a neotropical forest. Am Nat 104:25–35
Snaith TV, Chapman CA (2007) Primate group size and socioecological models: do folivores really play by different rules? Evol Anthropol 16:94–106
Snaith TV, Chapman CA (2008) Red colobus monkeys display alternative behavioural responses to the costs of scramble competition. Behav Ecol 19:1289–1296
Snow DW (1965) A possible selective factor in the evolution of fruiting seasons in tropical forests. Oikos 15:274–281
Sterck EHM, Watts DP, van Schaik CP (1997) The evolution of female social relationships in nonhuman primates. Behav Ecol Sociobiol 41:291–309
Teaford MF, Walker A (1984) Quantitative differences in dental microwear between primate species with different diets and a comment on the presumed diet of Sivapithecus. Am J Phys Anthropol 64:191–200
Terborgh J (1983) Five new world primates. Princeton University Press, Princeton
Terborgh J (1986) Keystone plant resources in the tropical forest. In: Soule ME (ed) Conservation biology: the science of scarcity and diversity. Sinauer, Sunderland, pp 330–344
Twinomugisha D, Chapman CA (2008) Golden monkey ranging in relation to spatial and temporal variation in food availability. Afr J Ecol 46:585–593
van Schaik CP (1983) Why are diurnal primates living in groups? Behaviour 87:120–144
van Schaik CP (1989) The ecology of social relationships amongst female primates. In: Standen V, Foley RA (eds) Comparative socioecology: the behavioural ecology of humans and other mammals. Blackwell Scientific Publications, Boston, pp 195–218
van Schaik CP, Terborgh JW, Wright SJ (1993) The phenology of tropical forests: adaptive significance and consequences for primary consumers. Annu Rev Ecol Syst 24:353–377
Visser MD, Jongejans E, van Breugelm N, Zuidema PA, Chen Y-Y, Kassim AR, de Kroon H (2011) Strict mast fruiting for a tropical dipterocarp tree: a demographic cost-benefit analysis of delayed reproduction and seed predation. J Ecol 99:1033–1044
Wheelwright NT (1985) Fruit size, gape width, and the diets of fruit-eating birds. Ecology 66:808–818
Wrangham RW (1979) On the evolution of ape social systems. Soc Sci Inf 18:335–368
Wrangham RW (1980) An ecological model of female-bonded primate groups. Behaviour 75:262–300
Wright PC (1999) Lemur traits and Madagascar ecology: coping with an island environment. Yearb Phys Anthropol 43:31–72
Wright S, Calderón O (2006) Seasonal, El Nino and longer term changes in flower and seed production in a moist tropical forest. Ecol Lett 9:35–44

Index

U. Kalbitzer, K. M. Jack (eds.), *Primate Life Histories, Sex Roles, and Adaptability*, Developments in Primatology: Progress and Prospects,
https://doi.org/10.1007/978-3-319-98285-4

MIX
Papier aus verantwortungsvollen Quellen
Paper from responsible sources
FSC® C105338

If you have any concerns about our products,
you can contact us on
ProductSafety@springernature.com

In case Publisher is established outside the EU,
the EU authorized representative is:
Springer Nature Customer Service Center GmbH
Europaplatz 3, 69115 Heidelberg, Germany

Printed by Libri Plureos GmbH
in Hamburg, Germany